GOBI   2|18   24 $\underline{18}$

D1481701

GALEN

HYGIENE

I

LCL 535

GALEN

HYGIENE

BOOKS 1–4

EDITED AND TRANSLATED BY
IAN JOHNSTON

HARVARD UNIVERSITY PRESS
CAMBRIDGE, MASSACHUSETTS
LONDON, ENGLAND
2018

Library of Congress Control Number 2017940160
CIP data available from the Library of Congress

ISBN 978-0-674-99712-7

Composed in ZephGreek and ZephText by
Technologies 'N Typography, Merrimac, Massachusetts.
Printed on acid-free paper and bound by
Maple Press, York, Pennsylvania

CONTENTS

To Susie Collis and Iain More

ACKNOWLEDGMENTS

Several people have made significant contributions toward the completion of this and the previous volume on Galen. It is my pleasure to acknowledge their help and thank them for their contributions.

Bob Milns read through the translations of Books 1 to 3 of the *Hygiene* against the Greek and made a number of very helpful corrections and suggestions. Greg Horsley and Niki Papavramidou looked at passages I was having difficulty with in the translations and offered their thoughts, which were invariably clarifying. Two doctors read through the translations with a critical medical eye—Steve Wilkinson, *A Method of Medicine to Glaucon,* and my daughter, Justine Johnston, the *Hygiene*—with the aim of helping to make some of Galen's more convoluted passages intelligible from the medical viewpoint.

I am especially grateful to my partner, Susie Collis, who read through the introductions and translations of both works at various stages in their evolution and was responsible for numerous improvements. She was also instrumental in obtaining various important materials. This work is dedicated to her and to my longtime friend Iain More, who has demonstrated so well the value of attention

to general hygienic measures as he grapples with a long and difficult illness. He presents an inspiring example of the importance of mental state to physical well-being, which bears on Galen's point about the relation of hygiene to the strength of the soul.

GENERAL INTRODUCTION

Galen (AD 129–ca. 216) is rightly recognized as one of the greatest figures in the history of Western medicine. Moreover, he and his revered predecessor, Hippocrates (ca. 440–370 BC), are the only ancient doctors to have a substantial body of surviving writing.[1] Almost all the many other medical men who practiced and wrote prior to AD 300 are known only through fragments and testimonia,[2] many of which are to be found in Galen's writings. So not only are Galen's copious writings on the subject of ancient medicine of very considerable interest in themselves, they are also the repository for much of the material on writers whose own works no longer exist. Despite this, Galen has been roundly criticized on a number of grounds: arrogance and self-aggrandizement, intemperate attacks on predecessors and contemporaries, prolixity, tedious rep-

[1] The only other sizable surviving medical work from that period is Celsus' *De medicina*. Celsus (fl. AD 15–35) was not a practicing doctor. There is a complete English translation of this work, which is an excellent source of information on medicine as it was at the turn of the millennium, by R. G. Spencer, *Celsus, De medicina*, 3 vols. LCL 292, 304, and 336.

[2] Two notable exceptions for whom there is some surviving material are Soranus of Ephesus (fl. AD 98–138) and Aretaeus of Cappadocia (ca. AD 150–200).

etition, and convoluted sentence construction in his writings. He has also been criticized for having a stultifying effect on the development of medicine. The first group of criticisms may have some justification; the second charge, however, cannot be laid at Galen's feet. This is more a testament to the scope and quality of his work and, to a degree, the inadequacy of his successors. Certainly, Galen's influence was profound and enduring, lasting a century or more after William Harvey's discovery of the circulation of the blood, which necessitated a major revision of existing physiological concepts.[3] Details of Galen's life, the medical milieu of his time, and the magnitude of his contribution have been well recounted in the excellent recent studies specifically on Galen by Veronique Boudon-Millot and Susan Mattern, and in general studies such as that by Vivian Nutton.[4] His great importance is also tacitly acknowledged by the very substantial number of translations of his works: into Syriac and Arabic during the first millennium AD, into Latin during the Middle Ages and the Renaissance, and into modern Western languages over the last century or so.[5]

[3] William Harvey's *On the Motion of the Heart and Blood in Animals (De motu cordis)* was published in Frankfurt in 1628.

[4] On Galen specifically, see Boudon-Millot, *Galien de Pergame,* and Mattern, *Galen and the Rhetoric of Healing* and *The Prince of Medicine.* On ancient medicine generally, see Nutton, *Ancient Medicine.*

[5] For full lists of Galen's extant works together with post-Kühn translations, see J. A. López-Férez, ed., *Galeno: Obra, Pensamiento e Influencia* (1991), 309–29, and Hankinson, *The Cambridge Companion to Galen,* 391–403.

Galen's *Hygiene*, often known by the Latin title *De sanitate tuenda* (*On the Preservation of Health*), is one of his most important works in terms of providing a comprehensive account of the practice of medicine. It is his substantial statement on one of the two major components of the art, which may be divided broadly into the preservation of health when it is present (hygiene) and the restoration of health when in it is vitiated by disease (therapeutics)—a division Galen himself explicitly makes. The *Hygiene* was written during one of Galen's most prolific decades (170–180); the date given by Bardong is 175.[6] It is not mentioned in Galen's *On My Own Books,* but it is referred to in *On the Order of My Own Books* (in a damaged passage) and in the list of works given at the end of *The Art of Medicine.*[7] In both these instances it is listed with the therapeutic works. The importance of the *Hygiene* is reflected in its inclusion in the Summaria Alexandaria,[8] the multiple editions of Linacre's Latin translation between 1517 (this was his first translation of a Galenic treatise) and 1559, and its mention in eighteenth-century writings. Indeed, it still has a practical relevance today in ways most of Galen's other works do not, although of course they still remain of considerable historical interest. Galen wrote two other short works on aspects of hygiene

[6] See Boudon, *Galien,* 392n4.

[7] In his translation of *De ordine librorum propriorum,* Singer, *Galen: Selected Works,* speaks of Galen's *Hygiene* as "this enormously influential work" (403–4). There are several references in *Ars M.,* in particular, I.404K.

[8] See Johnston and Horsley, *Galen: Method of Medicine,* 1.lii–liv.

(*Thrasybulus* and *On Exercise with a Small Ball*), which are considered further below, as are a number of related works that he refers to in the *Hygiene*.

HYGIENE BEFORE GALEN

There was clearly a significant body of writing on hygiene and related matters prior to Galen. It is also clear that a substantial amount of this was available to Galen, but almost all of it is now lost. What remains are the origins in the Hippocratic corpus and fragments only of the others, predominantly preserved in Galen's own writings. The concepts of what constitutes health and the methods of maintaining it appear to have been largely similar throughout the ancient period, although of course there was a theoretical divide on the grounds of continuum versus atomist theories. Basically, the idea was that the maintenance of health depended on diligent attention to diet and activity and the judicious use of exercise, massage, and bathing. The measures could be supplemented with medicinal preparations aimed at preserving a balanced state by ensuring the body's proper evacuation of superfluities and attention to environmental factors. A passage from Galen's *Thrasybulus* that considers some historically important contributors is given below, followed by brief comments on relevant individuals.

Therefore, dismissing these people,[9] for it is not base arts but arts we come to consider, let us call on

[9] Galen is referring to gymnastic trainers who train athletes in abnormal ways.

those who were truly knowledgeable in the gymnastic art—Hippocrates, Diocles, Praxagoras, Philotimus, Erasistratus and Herophilus, and those others who thoroughly learned the whole art concerning the body. Of course, we heard just now that Plato says there is no specific name. Accordingly, you should not seek one name relating to the whole art concerning the body, for you will not find it. But if at some time you are appointed to speak about this, let it suffice for you, imitating Plato, to recount [what he said]: "I say there is one treatment of the body which has two parts—the gymnastic and the medical."[10] Obviously the gymnastic is for those who are healthy, while the medical is for those who are sick. But what is more worthy of inquiry is that Plato did not distinguish on logical grounds the hygienic art from the medical, as all those men previously mentioned did, of whom, if you wish, let us call to mind one, since his writings are available to all.

Accordingly, Erasistratus, in the first book of his *Hygiene*, says: "It is impossible to find a doctor who has dedicated himself to the matter of hygiene." And subsequently, "The *apepsias* occurring along with some affection, and the treatment of these, fall to medicine and not to the matter of hygiene." Then, going on still further: "If there is some ill health involving the body, due to which those things administered will always be destroyed, and in this way will come to the same *kakochymia* as those previously existing, it is for a doctor to resolve

---

[10] Plato, *Gorgias* 464b.

such a condition and not a hygienist." Then, subsequently: "It is for the doctors to speak about these [conditions] and to resolve them; this should not fall to hygienists."

It appears that Erasistratus not only named something the art of hygiene similar to all the others, but also named the practitioner of this art a hygienist, just as, I think, he named the practitioner of the medical art a doctor, so that, of the therapeutic art concerning the body, for which there was no specific name among the Greeks, there is a division into two primary parts. And just as we name these same arts medicine and hygiene, so too we name their practitioners hygienist and doctor. Many other doctors similarly used these names. (V.879–81K)

Hippocrates (440–370 BC): As in his other works, Galen's main authority is Hippocrates, and his *Hygiene* contains a number of quotes (some recurring) from the Hippocratic corpus. The relevant works are *Regimen in Health,* the three books entitled *Regimen* I, II, and III, and *Airs, Waters, Places.* Reference is also made to *Aphorisms, Epidemics,* and *On Breaths.* The topics considered in the short *Regimen in Health,* a work which bears an uncertain relationship to *Nature of Man* (a work of major importance in Galen's theorizing generally),[11] are as follows: (1) seasonal variations in food and drink (particularly wines), seen in terms of *krasis;* (2) types of physique and ages; (3) regimen determined on the grounds of age, season, ethos, place, and physique; (4) hygienic measures

---

[11] See E. M. Craik, *The 'Hippocratic' Corpus* (2015), 207–13.

such as walking, working, and bathing; (5) regimen for those who are fat or thin; (6) emetics and purges (clysters); (7) care of infants and women; (8) regimen for athletes; (9) diseases arising from the brain. The conclusion stresses the importance of health: "It is necessary for a man who is wise to consider health the most worthwhile thing for people, and to learn from his own thought to be benefitted in disease."[12]

Diocles of Carystus (4th c. BC): A Dogmatic by persuasion, Diocles is credited with at least twenty works, all of which are now lost. Manetti (EANS, 256–57) describes his *Matters of Health, to Pleistarchus* as his "most influential work"—a work which, like his work on foods, wines, and herbs (*On Rootcutting*) and that on olive oil, to Archidamus, was aimed, at least in part, at a nonspecialist audience. Six of the eleven fragments on hygiene collected by Van Der Eijk are from Galen. One long fragment from Oribasius is of particular interest, detailing the appropriate activities for the preservation of health over one day, from waking in the morning to retiring at night.[13]

Praxagoras of Cos (ca. 325–275 BC) is regarded as a Dogmatic. He was particularly noted for his development of the concept of humors, subdividing the basic four, on grounds of color, taste, and other aspects, to give a total of ten. Alteration of these humors was seen as the major cause of disease. Among the works credited to him, none of which are now extant, there is a work on dietetics, but

[12] See Hippocrates, *Regimen in Health*, trans. W. H. S. Jones, *Hippocrates* IV, LCL 150, 44–59.
[13] See P. J. Van der Eijk, *Diocles of Carystus*, 2 vols. (2000), 1.292–331.

none on regimen more generally or on hygiene *per se*. In the work on dietetics, he is said to have considered the powers or faculties of different foods, and also that people differ in nature and constitution, and therefore require different diets. Galen describes him as being concerned with the issue of preservation of health and as being a master of gymnastic prescription. Philotimus, his pupil, is said to have written a work entitled *Art of Cooking* and perhaps to have contributed to Praxagoras' work on dietetics.[14]

Herophilus of Chalcedon (fl. 280–260 BC): Von Staden, in his collection of fragments of Herophilus, includes four only under hygiene. Two of these are from Galen's *Thrasybulus,* where he is linked with others in regard to hygiene (as in the quote given above). One of the other two is from Sextus Empiricus and is given by Von Staden as follows: "Herophilus says in his *Regimen* that, in the absence of health, wisdom cannot be displayed, science is non-evident, strength not exerted in contest, wealth useless, and rational speech powerless."[15]

Erasistratus (fl. 260–240 BC): Like Herophilus, Erasistratus is included by Galen in *Thrasybulus,* in his two lists of those interested in hygiene. In one of these, given above, there is reference to a work on hygiene by Erasistratus. In Garofalo's collection of fragments, there are fifteen under hygiene, as follows: Galen, 7; Caelius Aurelianus, 5; and one each from Oribasius, Athenasus Dip-

[14] F. Steckerl, *The Fragments of Praxagoras of Cos and His School* (Leiden: E. J. Brill, 1958).

[15] Von Staden, *Herophilus,* 406–7.

nosophus, and Celsus.[16] One reference in Galen's own *Hygiene* is as follows: "Anyway, the movements in cradles, cots and their own bent arms were discovered by the nursemaids of children. And we might also come in some way to the other issue of what is most essential for the maintenance of health, although Asclepiades opposed and gave the clearest condemnation of exercises, while Erasistratus spoke less vehemently but displayed the same opinion as Asclepiades" (VI.37K).

Asclepiades of Bithynia (fl. 120–90 BC): He developed a theory of basic structure in which fragile corpuscles (*anarmoi onkoi*) were thought to move through nonobservable channels (*poroi*) in the body; ill-health arose when this movement was reduced or increased. Although he is not credited with any work on hygiene, his treatment methods for diseases arising from variations in this process consisted largely of methods associated with hygiene—massage, diet, wine, mild exercise, and moderation in personal habits. Galen mentions Asclepiades several times in his *Hygiene*.[17]

Two noted Methodics, Themison of Laodicea (fl. 90–40 BC) and Thessalus of Tralles (1st c. AD) are credited with works on hygiene, although Galen doesn't mention such writings.

Theon of Alexandria (fl. AD 130–160) is described by Keyser (EANS, 795) as an "autodidact ex-athlete who

[16] Garofalo, *Erasistrati Fragmenta,* 115–23.

[17] For a general account of Asclepiades' theory and therapeutic methods, see J. Vallance, *The Lost Theory of Asclepiades of Bithynia* (1990).

wrote a work on exercise, and a longer work *Gymnastrion,* both only known from Galen." Galen gives detailed consideration to Theon's ideas on massage in his *Hygiene* and also mentions him in relation to baths. Although Galen is critical of Theon, his criticism is tempered to some degree with admiration.

GALEN'S SYSTEM OF HYGIENE

Although Galen's exposition of his system of hygiene is not altogether systematic due to his penchant for launching into substantial digressions from his stated intentions—something he explicitly acknowledges in the work itself—his conceptual approach is certainly systematic.

1. There are the foundations consisting of various definitions of health and its relation to the two other possible states, i.e., disease and a state intermediate between health and disease. The key components of these definitions are the structure and function of the body and its parts. Thus, there are bodies described as faultless in terms of structure and function that are entirely in accord with nature; there are bodies impaired in structure or function, which are contrary to nature; and there are those in between, which are deficient in structure or function to some degree, but not to the extent that there is actual disease.

2. There are the principles of hygiene that are the theoretical points and modes of the application of hygienic measures to the body.

3. There are the practical methods of hygiene.

Each of these components will be briefly examined.

Foundations

Concepts and Definitions

Galen offers several clear but different definitions of health, usually in association with disease. As of course a clear concept of what constitutes health is essential to a discipline devoted to its preservation, several of these definitions are given below.

On the Differentiae of Diseases: "Here one must accept the agreed principle that all men, when they have the functions of the parts of the body faultlessly directed to serving the actions of life, persuade themselves they are healthy. On the other hand, when they are damaged in any one of these [parts], they consider themselves diseased in that part. This being so, health is to be sought in these two things: (i) the functions which accord with nature; (ii) the constitutions of the organs by which we function. So disease is either damage of function or constitution" (VI.836K).

The Method of Medicine: "In respect of each of the things signified, one thing is the same for all. It is like this too with being diseased and being healthy. What is in all those who are diseased and all those who are healthy is one and the same in either case. Just as by the term 'man' one thing is signified, so too is it the case for 'health'" (X.130K).

Hygiene: "Since health is a kind of balance, and every balance is brought about and expressed in a twofold manner, at one time coming to the highest point and being truly a balance and at another wanting slightly in perfec-

tion, hygiene too is a twofold balance. On the one hand, it is exact, optimal, complete and perfect; on the other hand, it is lacking this perfection, although not yet to such a degree that the organism is distressed" (VI.13K).

"We call health that state in which we neither feel pain nor are impeded in the functions pertaining to life" (VI.18K).

Medical Definitions:[18] "Health is an *eukrasia* of the primary humors in us in accord with nature or functions of the physical capacities free from interference. Health is an *eukrasia* of the four primary elements from which the body is constituted—hot, cold, wet and dry. Other definitions include: a harmony of the heat and cold, wetness and dryness making up the person. What goes along with health? Three things: κάλλος (beauty), εὐεξία (a good bodily state) and ἀρτιότης (soundness, completeness) . . ." (XIX.382–83K). The remainder of this definition consists of an elaboration of these three terms.

A further quote from *Medical Definitions* concerns the terms κατὰ φύσιν and παρὰ φύσιν, which I have consistently translated "in accord with nature" and "contrary to nature," respectively:

> Health is that which is in accord with nature. Disease is that which is contrary to nature. What is "natural" (φύσει) but neither "in accord with nature" nor already "contrary to nature" is like someone very thin, or dry, or thick-set, or fat, or sharp-nosed, or grey, or snub-nosed, or grey-eyed. Those

[18] *Definitiones medicae* is regarded as spurious. Nonetheless, the definitions are relevant.

who are thus are not in a condition "in accord with nature" for they have gone beyond "balance," but neither are they "contrary to nature" for they are not hindered with respect to functions. Such a thing that is "non-natural" (οὐ φύσει) is neither "contrary to nature," nor "in accord with nature," nor "natural." Examples are those having *leuke*, leprous warts, warts and the like. For these are not "in accord with nature" as they are outside what accords with nature, but they are not "contrary to nature" for they do not hinder the functions that accord with nature. They are not, however, natural in that they do not occur from the beginning, nor are they from the initial genesis. They remain, therefore, "non-natural." What is "non-natural" by definition is close to what accords with nature and what is contrary to nature. (XIX.384–85K)

So health is something that exists; it is not simply the absence of disease. The key terms in the several definitions are constitution, as a stable state, function, accord with nature, and contrary to nature. In the terms of Galen's *On the Constitution of the Art of Medicine*, hygiene is one of the productive arts that either maintains something that already exists (i.e., health) or restores it if it is impaired, short of being actual disease. The issue of degrees or range of health is an important one and is inherent in the latitude of the component terms of his definitions.[19]

[19] On this matter see particularly, Galen's *Ars M.*, Books 1–2.

Structure

Galen, in his various works, considers the structure of the body at several levels. First, he held firm views on the basic structure of the matter of which animal bodies (and other things) are composed. Of the two theories current in his time (continuum theory and atomist theory), he was an unequivocal supporter of the former. This originated with Empedocles and was espoused by Galen's most respected medical and philosophical predecessors, Hippocrates, Plato, and Aristotle. Simply stated, according to the continuum theory, the substance of bodies is a mixture of the four elemental "substances," fire, air, water, and earth, and the four elemental qualities associated with them— hot, cold, wet, and dry. This is not, of course, an observable structure and so cannot be directly assessed, although inferences can be drawn from perceptible qualities. Second, related at a macroscopic level to the four elemental qualities are the four humors—blood (hot and wet), yellow bile (hot and dry), black bile (cold and dry), and phlegm (cold and wet). Under certain circumstance, these humors can be observed, but in part their presence and movement must be inferred from perceptible phenomena. Third, Galen made a threefold division of observable macroscopic structures: (1) *homoiomeres,* which are bodily tissues of uniform structure—for example, bone, muscle, cartilage (see the section on terminology below); (2) organic parts, which are composite structures subserving a particular function or functions—for example, heart, brain, lungs; (3) the whole body. In the ideal healthy body, *homoiomerous* structures, both those existing separately and those contributing to the structure of organic parts,

have a proper balance of the four elemental qualities (*krasis*), while organic parts are normal in terms of size, conformation, number, and position.

Function

Four aspects of Galen's ideas on function are considered briefly below.

1. Capacity, function, and action (*dunamis, energeia, ergon*): As seen in the definitions above, function is a key term in the definition of health. It is best understood in relation to capacity and action. In simple terms, *dunamis* (as "capacity" or "faculty") is the potential to carry out a function, the function is what is carried out, and the action is the observed application of the function. For example, the legs have the capacity or faculty of walking; their function is to walk; their action is the actual process of walking. These three terms have caused some confusion. The three quotes given below are added in an attempt to provide some clarity on the matter of function in the general sense.

Phillips says of Galen's use of *dunamis* in *On the Natural Faculties* that "the notion of δύναμις in this book is very pervasive and mostly verbal, being a development in medicine, not of δύναμις as known in *Ancient Medicine,* but of the Aristotelian δύναμις as potentiality as contrasted with ἐνέργεια, activity or actuality, also Aristotelian."[20]

The Soul's Dependence on the Body: "Many of the wise are openly in confusion on this matter, having an incorrect understanding of 'capacity' (δύναμις). They seem to me

[20] See E. D. Phillips, *Aspects of Greek Medicine* (1987), 176.

to wrongly conceive of capacity as something which dwells in substances, as we do in houses, not being aware that the effective cause of each thing that comes about is conceived of in relation to something else, and there is some name of this cause as of such a thing which is separate and *per se*. But in it, in relation to what is brought about from it, the capacity is of what is brought about, and because of this we say that substance has as many capacities as it has functions (ἐνέργειαι)" (IV.769K).

Hygiene: "One must not, therefore, determine those who are healthy and those who are diseased simply on the basis of the strength or weakness of the functions; one must apply the term 'in accord with nature' [κατὰ φύσιν] to those who are healthy and the term 'contrary to nature' [παρὰ φύσιν] to those who are diseased, as health is a condition in accord with nature productive of functions and disease a condition contrary to nature injurious to functions" (VI.21K).

2. Major functional systems: Galen identifies the following four major functional systems operating under the control of their principles:[21]

> Brain, spinal cord, and the nerves, both cranial and spinal, responsible for sensory and motor functions.
>
> Heart and arteries, responsible for the vital force, innate heat, and distribution of *pneuma.*
>
> Stomach, liver, and veins, responsible for the reception, concoction, and distribution of nutriment to all parts of the body.
>
> Testes and spermatic ducts, responsible for reproductive functions.

[21] See his *Ars M.* 5, I.318–19K for his discussion of this.

3. Innate heat: The concept of innate heat is in significant part derived from Aristotle. In Galen's formulation it is somehow provided at the time of the initial formation. This view, also held by Hippocrates, was by no means universal. Others, such as Erasistratus, Praxagoras, and Asclepiades, believed that the heat of the body was acquired from external sources. Galen's view is clearly expressed in the following statement from his *On Tremor, Palpitation, Convulsion, and Rigor*.

> We do not posit masses and pores as elements of the body, nor do we declare that heat comes from motion or friction or some other cause; rather we suppose the whole body breathing and flowing together, the heat not acquired nor subsequent to the generation of the animal, but itself first, original and innate. This is nothing other than the nature and soul of life, so you would not be wrong thinking heat to be a self-moving and constantly moving substance. (VII.616K)[22]

This innate heat, according to Galen, has its seat in the heart and is distributed to the rest of the body by the arteries, which were, at the time, also thought to carry *pneuma*.

4. *Pneuma*. This plays an important role in Galen's concepts of structure and function, and the relation of these

[22] This work has been translated by D. Sider and M. McVaugh, "Galen on Tremor, Palpitation, Spasm and Rigor," *Transactions and Studies of the College of Physicians, Philadelphia* (1979), 1.183–220, p. 199. See also May, *Galen On the Usefulness,* 1.50–52; and Siegel, *Galen's System of Physiology and Medicine*, 164–68.

to health and disease. *Pneuma,* in Galen's view, is derived from the external air, drawn in either through the nose and cribriform plate, to be changed in the *rete mirabile* (a structure not actually present in humans) to *psychic pneuma,* or into the lungs for subsequent distribution via the arteries. In his *Method of Medicine,* he writes:

> That syncope is an acute collapse of capacity has been stated by my predecessors. Since, however, the substance of the capacities controlling us lies in the *pneuma* and in the *krasis* of the solid bodies, what we must do is preserve these when they are present and restore them when they are weakened. How we must preserve them in a time of health has been shown in my work *Hygiene.* How we must preserve them in the diseases has already been stated in my earlier works and will be stated again now. Up to this point, however, the whole overview has not yet arrived at an appropriate general statement, nor does it have the kind of method that pertains to the other things considered earlier. Therefore, it is time to add what is lacking. What we must do is preserve the substance of the *pneuma* along with the solid bodies in diseases so that, in terms of both quality and quantity, they are in accord with nature as far as possible. If it is possible to make provision so that nothing of the substance of these is evacuated or changed in any way at all, this would be best.
>
> Since, however, in my *Hygiene* it was shown to be impossible that such a thing should ever exist in the mortal body, what we must attempt is to correct

the outflow of the substance by an addition and restore what is being changed to *eukrasia* through an opposite change. If both the evacuation and the change occur gradually, then the correction of both will be gradual, and this is the work of the art of hygiene, as was shown. If, however, not only the evacuations, but also the additions occur all at once and in large amounts, this would constitute a disease and we would require a therapeutic method for its cure. Just as the correction of a small deviation to a contrariety to nature is the task of the art of hygiene, so too, the correction of a large deviation is the task of the art of therapeutics. Certainly, I showed clearly that the brain is a fount, as it were, of the psychic *pneuma,* which is refreshed and nourished by the inspiration of air and from what the net-like plexus arrangement (*rete mirabile*) provides. My exposition of the vital *pneuma* was not, however, similarly clear. It is certainly not implausible that it seems to be contained in the heart and arteries, this too being nourished particularly from the respiration but now from the blood as well. If there is also a physical *pneuma,* it too would be contained in the liver and the veins. There was a very full discussion about the substance of capacities in my work *On the Opinions of Hippocrates and Plato.* [23]

The substance of the solid bodies needs, of course, to be of a certain quantity and, because of this, the

[23] See *Plac. Hippocr. Plat.,* V.181–805K—in particular, Book 6 (De Lacy, 1978, 2.360–427).

nutriments preserve the class of living things. No less does it need the *eukrasia* of the constituent elements themselves that compose it, and it has been stated often already by what means we must preserve this in an *eukratic* state. But the collapse of the capacity, about which I now propose to speak, frequently occurs after the destruction of substance of solid [bodies] in the course of the most chronic diseases, when the organism is thinned by atrophy, and in the colliquative (*syntectic*) fevers that are acute. It arises after a change in *krasis* when [the solid bodies] have been disproportionately heated, cooled, made moist, made dry or are affected by some conjunction of these. The change of the *pneuma* occurs due to humors in a bad state and the bad quality of the ambient air when it has been brought to this on occasion for one reason or another, and further, due to the noxious potencies or poisons of venomous animals. (X.837–40K)

There is also a nonphysiological *pneuma* as described in *On the Causes of Symptoms,* which is generated in the stomach from phlegmatous humors and foods. This is termed "flatulent *pneuma*" and is responsible for eructations and *borborygmi* (VII.239–42K).

Principles

In considering Galen's principles of hygiene, several aspects need to be taken into account, as follows:

1. where, in Galen's system of structure and function, hygienic measures are to be applied

2. what kind of people are the proper subjects of hygiene
3. what the overall aims of hygiene are

In the first case, the points of applications are, (1) the *krasis* of the *homoiomerous* structures, either alone, or where they exist as components of compound (organic) structures; (2) in organic structures as such, size and conformation of the four categories of variation in these (the other two being number and position); (3) the three major physiological systems (omitting the reproductive), in that the state of the *pneuma* is a potential point of application in relation to the nervous and cardiovascular systems, the innate heat in the case of the cardiovascular system, and the intake of food and drink, their processing, and the elimination of residual material in the case of the "digestive" system.

In the second case, Galen, in *Thrasybulus,* lists three groups of people to whom hygienic measures may properly be applied: (1) those who are in a good state already (*euektic*); (2) those who are recovering from illness (analeptic); (3) those who are in a state intermediate between health and disease, which includes those in the three states he gives particular attention to in his *Hygiene*—fatigues, old age, and "healthy" *dyskrasias.* To these, he adds a fourth group, which is of a different category than the first three—the prophylactic. On this, he makes the following statement in his *On the Constitution of the Art of Medicine:* ". . . there is a preventative part of the art, which they call specifically prophylactic . . . For whenever either some abundance or badness of humors, or a blockage, or some destructive power supervenes in the body, there is a

danger to a degree not hitherto present that the person will become diseased, and sometimes the danger will be extreme. Such causes are difficult to diagnose because the person doesn't yet suffer pain."[24]

In the third case, following on from the above, a basic twofold aim for hygiene may be identified: (1) there is the preservation of health in those who are already clearly healthy; (2) there is the restoration of health in those in whom there is a deviation from perfection that does not reach the stage of disease. In summary, Galen recognizes two groups of things that can adversely affect the structure and function of the body, so to preserve or restore health, hygienic measures must be able to prevent, ameliorate, or correct them. These things Galen divides into those that are "inevitable and innate," including progressive change in *krasis* with age, the constant flux of bodily material and the accumulation of superfluities, and those that are "external and contingent," like airs, waters, places, and regimen generally.

For a body to be in the best state, it must be *eukratic* throughout, have a normal amount of innate heat, *pneuma* normal in quality and quantity, and a sustained balance between what is lost from the body and what is gained to replace the loss. The points where hygienic measures can be effectively applied are in preserving the *krasis* of the body and its parts within the normal range, maintaining a satisfactory level of innate heat, preserving satisfactory levels of *pneuma,* regulating the amount and nature of the superfluities arising from concoction in the body, aiding

[24] *Const. Art. Med.,* I.296K; Johnston, *On the Constitution,* 120–23.

replacement of lost material, and as far as possible controlling external factors.

Methods

Briefly described below are the measures Galen considers and describes in his *Hygiene*.

Diet: Dietary measures can affect *krasis* and can also facilitate concoction. An example of the former is the use of hot and wet foods to counter the coldness and dryness of old age. An example of the latter is the use of diet to help manage *plethora* and *kakochymia* in the fatigues. Galen considers several common foods and drinks that can be important in maintaining health. These include milk, in infancy with respect to the condition of the nurse providing it, and more generally with respect to its sources; bread, both how it is prepared and what it is taken with; and wine, the various kinds and their respective benefits and harms. There is also the question of how many meals a person should have per day in different circumstances.

Massage: Galen considers two aspects of massage: massage in itself and massage as an adjunct to something else, especially exercise. In Book 2 (chaps. 2, 3, and 6), he gives a detailed account of the kinds of massage, taking issue with Theon's classification. Basically, there is a threefold division in both quantity and quality: much, moderate, and little in the former, and hard/firm, moderate, and soft/gentle in the latter. Galen quotes Hippocrates' summarizing statement: in quantity, much massage reduces flesh while moderate massage enfleshes; in quality, firm massage binds while gentle massage loosens.

Exercise: Galen quotes Hippocrates' brief statement

that exercise / exertion precedes food and expands on this. Variations in exercise involve speed, vigor, and violence. He describes specific exercises, from the simple to the complex. In general, the sequence should be, exercise, apotherapy, bathing, food, rest. Basically, exercise increases innate heat. The drying effect of exercise can be counteracted by apotherapy. In deciding on exercises for an individual, the important factors are environment, constitution of the body, age, and the observable effects.

Baths: These can influence *krasis* among other things. Variable factors are the kind of water used, water temperature, time of immersion, and preparatory measures. A young person with the best constitution has no need of baths for hygiene. Baths should be avoided in wet conditions.

Apotherapy: This can be seen as a part of exercise or as a kind of exercise. The measures are massage, suppression of breath (*pneuma*), and bindings applied by the masseur. Its aims are evacuation of superfluities and the prevention of fatigue after exercise. Galen notes Asclepiades' objection to apotherapy as "filling the head."

Purging: This can be used in fatigues and generally to evacuate superfluities and help correct *plethora* and *kakochymia* if present.

Phlebotomy: This also gets rid of excessive humors in *plethora* and *kakochymia* and is particularly useful for the tensive fatigues. Galen comments on Erasistratus' opposition to phlebotomy.

Medications: These can be used to facilitate concoction in *plethora* and *kakochymia*. Galen describes a number of compound medicinal preparations—oxymel, apomel, the Diospoliticum medication—to help deal with uncon-

cocted humors and to rid the body of superfluities, especially in the aged. He describes the medicinal preparation from quinces that stimulates appetite in the anorexias, improves concoction in the stomach, and strengthens that organ.

Other important hygienic measures are rest, sleep, regulation of sexual activity, attention to environmental factors, and the functions of the soul. At one point, Galen lists the components of a hygienic regimen as four: things to be administered, things to be done, things to be evacuated, and things to be applied externally. In essence, and following Hippocrates, a healthy regimen is based on moderation in all things, particularly exertion / exercise, food, drink, and sexual activity.

GALEN'S WRITINGS ON HYGIENE

Galen has three extant works specifically devoted to hygiene:

Hygiene (On the Preservation of Health—San. Tuend., VI.1–452K)

Thrasybulus, On Whether Hygiene Belongs To Medicine or Gymnastics? (*Thras.*, V.806–98K)

On Exercise with a Small Ball (*Parv. Pil.*, V.899–910K)

In addition, there are several short sections on hygiene in both his *On the Constitution of the Art of Medicine* and his *Art of Medicine,* primarily concerning the aims and subdivisions of the subject. Then there are five works referred to several times in his *Hygiene:* two on the constitution of the body (*On the Best Constitution of Our Body* and *On the Best State*) and three on foods and diet (*The*

Powers of Foods, On the Good and Bad Humors of Nutriments, and *On the Thinning Diet*). Finally, the several works stating his fundamental views on the structure and functions of the body are important for a proper understanding of the concepts and practices of hygiene. Those particularly referred to are *On the Elements according to Hippocrates, On Mixtures (Kraseis), On the Natural Faculties*, and the four treatises on the classification and causes of diseases and symptoms.

The first of the three specific works listed, the treatise entitled *Hygiene,* is Galen's major work on the subject. It is the only substantial work on the subject that has survived from ancient times. In this, although he states his plan for the book and also offers summarizing sections at intervals, the plan is not strictly followed. The plan, as stated, was to set out, in the first five books, the requirements of a regimen for health for each of the stages of life in the case of a person with a faultless constitution. The stages are: infancy and early childhood (0–7), late childhood / early adolescence (8–14), late adolescence / early adulthood (15–21), the prime of life (?22–42), the postprime decline (?43–63), and old age (?64 to death).[25] The final book was intended to consider hygiene for the individual with a less than faultless constitution who is not yet diseased. In fact, while he broadly adheres to his plan in the first two books, books three and four are largely devoted to the fatigues, both following exercise and spontaneous, of which he gives a comprehensive account. Book

[25] Galen does not actually specify the ages for these stages in the *Hygiene.*

five is largely about hygiene in old age, although it does include introductory material on hygiene in the "healthy *dyskrasias.*" The final book completes his consideration of this subject and concludes with two chapters that seem something of an afterthought—one on issues associated with sexual intercourse in men and one on medicinal preparations from quinces.

Interwoven into the basic structure are detailed considerations of the major components of hygiene unrelated to age groups: exercises, massage, baths, and diet. Overall, the treatise provides a detailed and comprehensive account of the subject of hygiene, considering in depth both theoretical and practical matters. Put together with his *Method of Medicine,* and accepting his argument that hygiene is one of the two components of medical practice, the two works together offer a complete account of the practical aspects of that practice. Like a number of Galen's works, his *Hygiene* remained influential and useful over many centuries.

The second work, dedicated to the otherwise unknown Thrasybulus,[26] is of considerably less practical interest and much more of its time, so to speak. Ostensibly, it addresses the issue of whether hygiene falls within the purview of the doctor or the gymnastic trainer. It seems that at the time, such trainers were not only conducting the practice of hygiene (exercise, massage, bathing, etc.) but were also writing on the subject. Theon, whose views on massage are criticized in Galen's *Hygiene,* is a case in point. In the work

---

[26] The pseudo-Galenic treatise, *Opt. Sect.,* I.106–223K, is also dedicated to Thrasybulus.

to Thrasybulus, Galen addresses the stated question and comes to the not unexpected conclusion that not only is hygiene the business of the doctor but also that gymnastic trainers tend to employ methods that are counterproductive as regards health, in pursuit of the particular ends of the athletes under their supervision.

The work comprises forty-seven short sections, which can be grouped, somewhat arbitrarily, as follows:

1–17: These deal with definitions and concepts pertaining to health and related matters. Galen makes this distinction between medicine and gymnastics: medicine is the art of restoring health in those who are diseased and preserving health in those who are healthy; gymnastics is the art of preserving health and producing good condition (*euexia*). He identifies three grades of health: (1) health per se, (2) natural good condition, (3) good condition in athletes. He reiterates his basic definition of health, found in a number of his writings—that is, functions in accord with nature in every part of the body and a constitution in accord with nature. A perfect constitution allows perfect functions.

18–21: These review the things with which the body must "interact": ambient air (surrounding and inhaled), food and drink, rest and movement, sleep and waking. In preserving health, it is necessary to make small corrections in these things, as opposed to the large corrections necessary in treating disease.

22–29: These are predominantly about hygiene as an art, using the arguments regarding the definition of an art in his *On the Constitution of the Art of Medicine*.

30–31: These two long sections express ideas central to

Galen's concept of health and hygiene. In particular, he identifies three components of hygiene: (1) starting with subjects who are *euektic* (in good health / condition) and keeping them so; (2) starting with subjects who are recovering from disease and restoring health in them (analeptic); (3) starting with subjects who are in between health and disease and restoring health in them (hygienic, or "*phylactic*"). He adds, as a fourth, prevention of disease (prophylactic).

32–35: He adds several corollaries to his basic thesis: (1) the *euektic* component is the province of gymnastics and involves regimen and exercise; (2) gymnastics is part of the hygienic art; (3) the materials considered by Hippocrates were airs, water, places, winds, seasons, food, drink, and daily activities.

36–40: In these he deals with further aspects of the original question, in part from a historical perspective. His basic conclusion is that the therapeutic art concerning the body is divided into two parts—medicine (ἰατρική) and hygiene (ὑγιεινή). On the materials of hygiene, he extends his earlier division: substances taken (food, drink, medications), activities performed (ordinary daily activities, exercises), matter voided (urine, feces, sweat), external influences (air, waters, etc.).

41–47: In these final sections Galen addresses some specific issues about gymnastics as one part of the art of hygiene, which in turn is one part of the art of caring for the body, which ultimately is medicine. He makes one of his somewhat intemperate attacks, this time on gymnastic trainers, and particularly their efforts in the pursuit of extreme functions in athletes.

In his concluding statement to the work, he writes:

> It is not, therefore, unreasonable, when asked what
> art hygiene is part of, to answer "the medical art."
> Since the name has been extended further, and no
> longer signifies the part but the whole art concern-
> ing the body, Hippocrates and all the doctors of the
> present time are rightly so named, for they know
> the greatest parts of the art itself are two—thera-
> peutic and hygienic. In turn, they know gymnastics
> is part of hygiene itself, as has also been shown
> before. (V.897–98K)

The third treatise (*On Exercise with a Small Ball*) is
very short (V.899–910K). It is about a specific exercise that
Galen describes as the most beneficial, although there is
no actual description of the exercise itself in the work.[27] It
may involve two people throwing and catching a small ball.
It exerts the body in all parts, is a pleasure to do (delights
the soul), requires no additional equipment, and is essentially without risk. Galen lists what he sees as the three key
requirements of exercise: health of the body, harmony of
the parts, virtue of the soul. The sequence should be: exercise, soft massage with olive oil, and a hot bath.

SPECIFIC ISSUES

There are three specific issues that are given extended
treatment in Galen's *Hygiene:* fatigue, old age, and the
so-called healthy *dyskrasias.* Each is considered below.

---

[27] There is apparently a detailed description of this exercise
in Mercuriale's *De arte gymnastica* of 1569.

Fatigue

Galen's discussion of the various forms of fatigue occupies the major parts of Books 3 and 4. In extant writings from the time prior to Galen, there are the following three significant accounts of the condition:

1. Hippocrates, *Regimen* 2:[28] Three groups are identified as suffering fatigue: those not in training who have overexerted themselves in exercise, those who are trained but have used unaccustomed exercises, and those who have overexerted themselves in their customary exercises. The condition is attributed to σύντηξις (*syntexis,* colliquescence) of flesh. If the products of this breakdown are not evacuated via sweat and breathing, they collect in the fleshy parts, which become hot and fatigued. The treatments include hot baths, purging, gentle walks, weight loss, rest, and sudorific unguents.

2. Aristotle, *Problems* 5:[29] This short chapter considers questions about several aspects of fatigue after exertions such as walking and running on uneven or inclined surfaces, and other gymnastic activities. The mechanism is again identified as σύντηξις (colliquescence) producing superfluities, which, if they remain in the body, cause fatigue. The aims of treatment are elimination of the causative superfluities and symptomatic treatment of the fatigue itself. Measures include further exercise, induced vomiting, baths, and massage. A distinction is made be-

[28] See Hippocrates, *Regimen* 2 (66), trans. W. H. S. Jones, LCL 150, 359–65.

[29] See Aristotle, *Problems* I–XXI, trans. W. S. Hett, LCL 316, 134–65.

tween summer fatigues, best treated by bathing, and winter fatigues, best cured by anointing with oil.

3. Theophrastus, *On Fatigue*:[30] In his short treatise on the subject, Theophrastus adheres to the theory of σύντηξις (colliquescence) as the primary causative factor. If the products of this process are not excreted, or otherwise disposed of, fatigue occurs, with particular involvement of joints and sinews. Theophrastus also mentions *osteokopos*, which Galen includes as a fatigue-like condition. In addition, Theophrastus speaks of fatigue being generated by the body being dried out, corresponding to Galen's "fatigue-like" condition, which the latter adds to his three primary kinds of fatigue. For Theophrastus, management involves moistening agents (drinks, baths) and resolving fatigue with fatigue.

Galen significantly advanced thinking on fatigues, and his teaching remained influential for many centuries. First, he divided fatigues into nonspontaneous (following exercise) and spontaneous, citing a Hippocratic source.[31] Second, he divided fatigues into three kinds—wound-like, tensive, and inflammation-like—based on the presenting symptom. To these he added a fourth, which he called a fatigue-like condition, characterized by thinness and dryness. He also used the term *osteokopos*, which he applied to describe the deep-seated nature of the pain and discomfort. Third, he considered the existence of "combina-

[30] In *Theophrastus of Eresus: On Sweat, On Dizziness and On Fatigue,* ed. W. W. Fortenbaugh, R. W. Sharples, and M. G. Sollenberger (Leiden: E. J. Brill, 2003), 251–308.

[31] Hippocrates, *Aphorisms* 2.5, *Hippocrates* IV, LCL 150, 108–9.

tion" fatigues; if the "fatigue-like condition" is included, there are fifteen possible combinations, which he sets out diagrammatically. In Galen's view, the major causes of fatigues were *plethora* and *kakochymia*. He does not mention σύντηξις (colliquescence) in his *Hygiene*, but the process seems to be essentially the same as that described by earlier writers. There are two basic measures of management: evacuate the superfluity / humor by purging and / or phlebotomy, and change the superfluity / humor by facilitating concoction; this involves medicinal preparations such as oxymel, apomel, and the Diospoliticum medication.

Old Age

Galen's ideas on hygiene for the aged occupy the major part of Book 5 of his *Hygiene*. He, and others before him, saw the basic problem of aging as being an increasing dryness and coldness in *krasis*. Some degree of correction was achievable through heating and moistening agents—foods (using some and avoiding others); drinks, including wine (he gives details on wines); movement and exercise; hot baths; and massage. He also recommended the use of some medications to help get rid of superfluities. Exercise should be moderate and involve customary activities; weak parts should be rested (unlike in the young in whom they should be exercised). In general, care should take into account the condition of the body as a whole, customs, and any disturbing afflictions. Galen provides two interesting case reports of old men—Antiochus the doctor and Telephus the grammarian.

"Healthy" Dyskrasias

Following some initial consideration in the final two chapters of Book 5, Book 6 is basically about hygiene in people who do not have the best constitution of the body or cannot follow the ideal regimen. He deals in succession with the various *dyskrasias* when they fall short of being diseases (i.e., still permit normal function). Both uniform and nonuniform (regular and irregular) *dyskrasias* are covered. The two important causes are *plethora* and *kakochymia*. Basic general measures in management are diet, bathing, massage, moderate exercise, and moderation in sexual activity. More specific measures are downward purging and phlebotomy.

TERMINOLOGY

The four groups of terms considered here are terms that have been transliterated. The basic reason for doing this is the absence of satisfactory terminological equivalents in English and that attempts to devise such are likely to be misleading. These terms are essentially technical terms in Galen's concepts of anatomy, physiology, and pathology and need to be understood in that context. Transliteration helps to identify them as such. In two instances the same term remains in use, albeit with a somewhat different meaning.

Homoiomeres / Organic Parts: There are several places where Galen clearly defines what he means by *homoiomeres*. Thus, in *On the Elements according to Hippocrates,* he describes them as ". . . the primary parts with respect to perception," and lists arteries, veins, sinews,

ligaments, membranes, and flesh as *homoiomeres* in humans (I.493K). In *On the Opinions of Hippocrates and Plato*, the list differs slightly, including "cartilage, bones, sinews, membranes, ligaments and all other such things" (VIII.4.7–14, De Lacy, 2.500). Here he also provides the following definition, identifying the term's biological application as originating with Aristotle: "Therefore, bodies in one outline (*prosgraphe*) are often called *homoiomeres* because all their parts are similar to each other and to the whole, and they are also often called simple or primary." In *On the Differentiae of Diseases*, Galen lists arteries, veins, sinews, bones, cartilage, ligaments, membranes, and flesh as *homoiomerous* structures and clearly states that these are components of organic bodies and are themselves formed from the primary elements (VI.841K). Finally, in *The Method of Medicine*, he writes: "A part is *homoiomerous*, as the name itself clearly shows, that is divisible into similar parts throughout, like the vitreous and the crystalloid and the specific substance of the membranes of the eye" (X.848K). Galen wrote a book on the subject, *On the Differences of Uniform Parts*, which only survives in Arabic.[32]

The basic meaning of "organ" as instrument is retained in Galen's use of the term, which he defines as follows: "I term an organ a part of an animal that carries out a complete function, like the eye with respect to vision, the tongue with respect to speech, and the legs with respect to walking. In this way too, artery, vein and sinew are organs and also parts of the animal. And according to this

[32] CMG, Suppl. O, III, G. Strohmaier, *Galeni De partium homoeomerium differentia libelli* (Berlin: Akademie Verlag, 1970).

usage of terms, at least as defined not only by us but also by the Greeks of old, the eye will be termed a 'constituent part,' a 'part,' and an 'organ'" (X.47K).

Krasis / Eukrasia / Dyskrasia: Krasis has a basic meaning of mixing, combining, or blending. In certain contexts it is translated as "temperament," and in fact Galen's treatise on *krasias, On Mixtures,* is titled in Latin *De temperamentis.* In his system of medicine, *krasis* refers specifically to the mixing or blending of the four elemental qualities (hot, cold, wet, and dry). If these are properly blended (this is a conceptual issue rather than an observational one, although inferences can be drawn from signs), the body or parts thereof are in a state of *eukrasia.* If there is a preponderance of one quality, or one of the four possible combinations of two qualities, this constitutes a *dyskrasia.* There are, then, four possible mono-*dyskrasias* and four compound *dsykrasias* in addition to *eukrasia.* In any given structure (whole body or body part), there can be a single mono-*dyskrasia* or a compound *dyskrasia,* or more than one type of *dyskrasia* can be present, creating an irregular (nonuniform, anomalous) *dyskrasia* in that structure. The other point of particular relevance in *Hygiene* is Galen's identification of what he calls a "healthy *dyskrasia*"—that is, a *dyskrasia* that is compatible with normal function. Once in *Hygiene* Galen uses the term *akrasia,* but the context suggests that this is used here for ἀκράτεια, meaning "debility" or "lack of self-control."

Euchymia / Kakochymia / Plethora: These terms apply to the amounts and states of the four humors in the body. The terms and what they represent are relevant generally to the maintenance of health and specifically to the etiology of fatigues. Thus, a person may be described as *euchy-*

mous if the four humors (blood, yellow bile, black bile, phlegm) are normal in quality and quantity. If there is excess of one or more, the term *kakochymia* applies; if all four are in excess, the term *plethora* applies. The terms *euchymous* and *kakochymous* are also applied to foods;[33] in this context they can be understood as relating to the power of a particular food to produce *kakochymia* in the body. Both *kakochymia* and *plethora* can produce a disturbance of health that falls short of being an actual disease, and then come within the province of hygiene. As mentioned above, an example is the nonspontaneous fatigues. The following quote is from Galen's *Method of Medicine:*

> How you must take care of the whole body when it is in a pathological state is something I have spoken about at length, both throughout what has gone before and also in the work *On Plethora.*[34] Now I shall speak of the chief points of the discussion. When the humors are increased to an equal degree to each other, [doctors] call this "abundance" or *plethora.* On the other hand, when the body is already full of yellow or black bile, or phlegm, or the serous humors, they call such a condition *kakochymia* and not *plethora. Plethora* is treated by the letting of blood, and by numerous baths, exercises and rubbings, as well as by dispersing medications, and in addition by all fastings, which I treated fully in the treatise on hygiene. *Kakochymia,* however, is treated by the

---

[33] The title of Galen's short work on this subject is *On the Good and Bad Humors of Nutriments* (VI.749–815K). The Latin title is given as *De bonis malisque alimentorum sucis.*

[34] *Plenit.,* VII.513–83K.

specific evacuation appropriate for each of the humors in excess. There was also discussion about this in the section on prophylaxis in my work, *Hygiene*. (X.891–92K)

Apepsia / Bradypepsia / Dyspepsia: With a primary meaning of softening, ripening or being changed by heat, πέψις is applied to food before and after ingestion, and to wine. In Galen's considerations of health and disease, the basic term and its derivatives are applied to processes that alter ingested food and allow it to be assimilated by the tissues of the body. These processes are not confined to the stomach, so concoction is preferred as the translation over digestion. The following definition is given in *On the Differentiae of Symptoms:*

> Certainly, with regard to the alterative capacity in the stomach, there is the case of nothing occurring at all, when all such food as is taken in remains as it is in every quality. "Weakly" has acquired the specific term, *bradypepsia,* just as defectively is a change of the food to an unusual quality (*dyspepsia*), so that all three symptoms occur with the one failed function. Concoction (*pepsis*), as this function is called, is an alteration of the food to the quality appropriate for the animal. *Bradypepsia* is a change to the same quality, either over a long time or with difficulty. A change to another quality, one that is not in fact in accord with nature, they call *apepsia.* Privation of function is also referred to by the same term. But it is clearer for this alone to be called *apepsia,* the defective change, *dyspepsia,* and the weak (slow) change, *bradypepsia.* (VII.66K)

TEXTS AND TRANSLATIONS

Three texts were used in the present translation:

> Kühn, VI.1–452 (Greek and Latin)
> Koch's Greek text in the Corpus Medicorum Graecorum series (V.4.2)
> Linacre's Latin translation, 1547 edition

Diels lists Greek manuscripts in Berlin, Leipzig, London, Paris, Rome, and Venice. He also lists numerous Latin manuscripts, particularly in Rome, and several Arabic manuscripts. Koch used the following in the preparation of his 1923 edition:

> the two Venice MSS—Marcianus 276, s. XII, f. 3–68; Marcianus 282, s. XV, f. 101–53
> the one Rome MS—Reginensus 173, s. XV, f. 233–352 and 361
> the Latin translations of Burgundio of Pisa (1110–1193) and Niccolò da Reggio (fl. 1315–1348)
> the works of Aëtius of Amida (Venice, 1534) and Oribasius, including Daremberg's French translation of the latter

Durling lists nine editions of Linacre's Latin translation as follows:[35] Paris, 1517; Venice, 1526; Cologne, 1527; Paris, 1530; Paris, 1538; Tubingen (edited by Fuchsius), 1541; Lyons, 1547; Lyons, 1549; Lyons, 1559. He also lists a Latin translation by G. Tarchagnore, Venice, 1559. Linacre's translation of the *Hygiene* in 1517 was the first of his several valuable translations of Galen's treatises. It is

---

[35] Durling, *"Chronological Census."*

said he had access to only one manuscript.[36] Fuchs says of his edited text: "Linacre's version is often so obscure as to be unintelligible without recourse to the original. Hence, Linacre has been blamed for excessive *severitas* and *gravitas*."[37]

The present translation is based on the Koch edition. This has been compared with the Kühn text and points of significant difference noted (Ko = Koch, Ku = Kühn). Linacre's Latin translation is indicated by L. The most notable difference between Koch and Kühn is the transposition of approximately eight Kühn pages from chapter 3 of Book 5 (line 13, 321K to line 8, 329K) to chapter 10 of the same book, where the fragment is inserted after the penultimate word in line 9, 358K.

SUMMARIES OF THE SIX BOOKS

Book 1

The fifteen chapters of this book can be divided into five groups:

1. Preliminary Theoretical Considerations (1–4)

The difference between hygiene and therapeutics is that the former preserves the state of the body while the latter changes it. For both it is necessary to know, first, the condition of the body; second, that health is a balance of the qualities (hot, cold, dry, wet) in the *homoiomeres* and an

---

36 See R. J. Durling, in Maddison et al., *Essays,* 86.
37 See Maddison et al., *Essays,* editors' introduction, xxiii.

accord with nature, in terms of conformation, size, number, and position, in the organic parts; and third, what harms may befall the body. (chap. 1)

Things that may harm or destroy the body fall into two classes: (1) those that are inevitable and innate, which are essentially the progressive change in *krasis* from that of the primary generative materials (blood and semen) from hot and wet to dry and cold, and the flowing away of bodily substance due to the innate heat; (2) "external" factors that are neither inevitable nor set in motion by ourselves. (chap. 2)

Hygiene is largely about the first class. Galen identifies three primary objectives of a healthy regimen: (1) replacement of things emptied out; (2) separation of superfluities; (3) avoidance of premature aging. (chap. 3)

In chapter 4 he repeats the basic division of causes of deterioration into those that are internal, intrinsic, and inevitable and those that are external and contingent. The former include aging, the constant flux of substance, and the accumulation of superfluities. The latter are basically divided into those that are ever-present (the environment) and those that are due to variably occurring external factors. He concludes this section by posing the question of whether the art of hygiene concerns all these things or only those that affect or change the four basic qualities.

2. Health and Hygiene (5–6)

Obviously, to preserve health, one must be clear what health is. It is a balance. But importantly it is a balance that allows of variation. The assessment of health is based

on an evaluation of the functions of the body. Galen offers another definition of health: the absence of pain and any impediment to the functions of life. There are several digressions on aspects of these basic ideas, and in particular on the range that exists in healthy functions, which is in part dependent on the stage of life. Galen bases this on seven-year periods to include infancy and early childhood, late childhood and early adolescence, late adolescence and early adulthood, the prime of life, the postprime decline, and old age. (chap. 5)

The aims of hygiene are to keep the perfect perfect and to improve the imperfect. He reiterates his basic definition of the best constitution that is fundamental to his considerations of hygiene: "it is perfect *eukrasia* and, at the same time, the conformation of the parts being adapted perfectly to their functions, and in addition, provision in all cases of the number, size, and arrangement with each other of all these that is beneficial for the functions." (chap. 6)

3. Hygiene for the First Stage of Life (7–11)

Galen starts with the first seven years of life and bases his discussion on the infant with the best constitution of the body, dealing successively with the care of the newborn, including the provision of breast milk (chap. 7), movement and exercise (chap. 8), the state of the nurse responsible for the child (chap. 9), feeding more generally, massage and exercise (chap. 10), and drinks, especially water and wine (chap. 11). Although these chapters are ostensibly about the infant / child with the best constitution, there are several digressions.

4. Superfluities and Their Excretion (12–14)

Although Galen begins chapter 12 by announcing his intention to consider hygiene for a child during the second seven-year period, after making two points briefly—that what the child (boy) intends to do in life is relevant to his early regimen and that attention must now be given to the care of the soul—the main focus of this and the next two chapters is on the creation of superfluities and their elimination. Superfluities are the unusable residues of concocted food and drink and are produced in different parts of the body—stomach, liver, arteries, and veins, and the individual parts. There are three points of excretion: the rectum for dry / solid superfluities, the urinary bladder for wet / liquid superfluities, and the skin for wet and vaporous superfluities through sweat and imperceptible transpiration.

He then considers the problem of retention of excretions and its management, identifying various causes of impaired excretion. (chap. 13)

The management of retained superfluities / excretions involves the exhibition of opposites, either by dietary adjustments or medications, baths and exercises. (chap. 14)

5. Synopsis (15)

Galen accepts the term "hygienist" for one who practices the art. Hygiene, like therapeutics, concerns bodies, signs, and causes—bodies that are to be kept healthy, signs that provide diagnostic information, and causes through which health is produced. He makes a fourfold division of causes into: (1) things to be administered; (2) things to be done;

(3) things to be evacuated; (4) things to be applied externally.

Book 2

The twelve chapters in this book can be divided into five groups:

1. Hygiene for the Third Seven-Year Period (1–2)

Galen announces his intention to consider the role of hygiene in the care of a young person with the best constitution, particularly a male, during the third seven-year period (14–21). There is a brief digression on what "best" signifies, and on the differences in bodies and ways of life. "Best" is a person with a body of the best constitution living a life that affords complete freedom for the care of that body. (chap. 1)

The principles of hygiene for this age range are given in two quotes from Hippocrates: (1) exertions should precede food; (2) exertions, food, drink, and sex should all be in moderation. Three additional issues are addressed: (1) on sex—this is really only for those in the prime of life; (2) on the correspondence between exertion, movement, and exercise—this is largely terminological, concerning *ponos;* (3) on exercise—what are its benefits and how to judge the appropriate times and preparations. (chap. 2)

2. Massage (3, 4, and 6)

There is a long section on massage as a preparation for exercise, the varieties of massage, and the views of Theon,

particularly in comparison to Hippocrates' much more concisely expressed views. (chap. 3)

The next chapter continues the discussion on massage, with a focus on Theon, identifying his errors (in Galen's view). Galen provides a table of nine varieties of massage based on the different combinations of the three different levels of quantity and quality. The basic effects of massage were clearly stated by Hippocrates: firm massage binds; soft massage loosens; much massage reduces flesh; moderate massage increases flesh. For a boy in the third seven-year period who is perfectly healthy, there should be no massage at all other than in preparation for and recovery from exercise—the latter is apotherapy. (chap. 4)

After the terminological section (chap. 5), Galen returns to massage, identifying a division between the good effects it has in itself and the benefits derived from it as an aid to other measures, especially exercise (chap. 6).

3. Terminology (5)

The four terms considered are the pairs, *skleros* and *malakos,* and *araios* and *puknos.* The components of the first pair are clear in meaning. The point Galen makes about the second pair, which I have generally rendered "loose-textured" ("rarefied") and "condensed," is that in the present context, they indicate large pores and small pores, respectively.

4. Exercises and Massage (7–11)

The initial chapter considers how to judge the appropriate amount of massage and exercise. Important determining

factors are the ambient air, season, place, bodily constitution, age, and visible effects on the person being massaged. (chap. 7)

Galen then details various exercises and makes a distinction between exercises, activities, and preparation for activity. He stresses the importance of a proper balance between exercises and apotherapy. He speaks of variations in the exercises themselves in terms of speed, vigor, and violence, and the circumstances of the exercises. There is a digression on the difference between the hygienist, the gymnastic trainer, and the doctor—an issue addressed in detail in the work *Thrasybulus*. (chap. 8)

Galen considers the general effects of exercise regardless of the kind—in particular, increase of innate heat. Some further terminological issues are addressed. There are then descriptions of various exercises falling into the category of vigorous. (chap. 9)

Next he deals with rapid and violent exercises, the latter being a combination of vigorous and rapid. (chap. 10)

The final chapter in this group describes exercises specific for different parts of the body. There is a digression on the parts of the body and their ability to move independently as opposed to being dependent on other parts for their movement. He lists three types of movement: (1) initiated internally and intrinsically; (2) initiated externally; (3) compelled by medications. Finally, he considers the effects of various movements on the body. (chap. 11)

5. Conclusions (12)

Galen returns to the young lad in question. The gymnastic trainer must know what is best for him, bearing in mind

the basic principle—what is best for the best constitution is to preserve it and not change it. Also, the choice of exercises must take into account the specific parts to be strengthened. The key for the body of the best constitution is moderation. Galen considers how to achieve this. Finally, there is mention of what should follow exercise: apotherapy, bathing, food, drink, sleep, and ambulation.

Book 3

The thirteen chapters in this book can be divided into six groups:

1. Introduction (1)

Galen presents a recapitulation of Books 1 and 2 and deals with the age ranges 0 to 14 and 14 to 21 concerning aspects of the appropriate regimen.

2. Apotherapy (2–3)

There is a long discussion of apotherapy considered in two ways: as a part of exercise and as a kind of exercise. The basic aims of apotherapy are: (1) to prevent fatigue after exercise; (2) evacuation of superfluities. The components of apotherapy are: (1) massage (Galen considers kinds); (2) suppression of the breath / *pneuma* (Galen considers mechanisms); (3) bindings applied by masseurs (chap. 2).

Galen then offers a defense against criticisms of his descriptions of apotherapy, dealing both with unnamed critics who accuse him of prolixity and with Asclepiades' specific criticism of suppression of the breath, which he claims has the adverse effect of filling the head. (chap. 3)

3. Baths (4)

Here the use of baths after exercise is considered, with attention given to the following: (1) the kinds of water; (2) different water temperatures; (3) time of immersion; (4) preparation, including massage. In hygiene for the person under consideration with the best constitution (14–21), there should be no need for bathing after exercise, but if there is bathing it should be regarded as part of apotherapy. Galen provides details of the method of evaluating the effects of bathing.

4. Fatigues (5–9)

Fatigue which follows exertion is the term given to an abnormal sensation involving the whole body or parts thereof. There are three "simple" fatigues, distinguished by the nature of the abnormal feeling (wound-like, tensive, inflammation-like) and four possible combinations of these, giving a total of seven. In the wound-like fatigue there is an abundance of thin and acrid superfluities in the affected parts. In the tensive fatigue there is excessive stretching of the muscle fibers in direct line with the tension. In the inflammation-like fatigue the muscles are heated by superfluities being drawn in. There is also *ostokopos,* in which there is a deep pain, like bones being broken, and in addition a fatigue-like condition in which the whole body is parched and drawn tight. (chap. 5)

Wound-like fatigue has several causes: (1) *kakochymous* and excrementitious bodies; (2) recent *apepsias;* (3) too long in the sun; (4) too much exercise; (5) massage

with certain oils and movements. Cure is by dispersion of the superfluities, massage with certain oils, and movements. For the tensive fatigue the basis of treatment is relaxation, using soft massage with sweet oil warmed in the sun, *eukratic* baths, and a long period in warm water. (chap. 6)

Inflammation-like fatigue is characterized by the person feeling pain when touched and on movement, and being overly hot. It occurs following exercise in those unaccustomed to exercise. There are three aspects of the cure: (1) evacuation of superfluities; (2) relief of tension; (3) cooling of what seems inflamed. There is a fourth condition, which is fatigue-like but not truly a fatigue. Its features are thinness and dryness. (chap. 7)

In treating the fatigues, a specific diet is required for each kind. Either rest or activity may be appropriate—rest for the tensive and inflammation-like fatigues and further activity for the wound-like (excrementitious) fatigues. There is variation in opinion on the use of baths. There is general controversy about the fatigue-like condition—Galen considers Theon's views. (chap. 8)

There are four possible combinations of the three "simple" fatigues. In evaluating compound conditions, one must gauge the relative strength and importance of the components. Also, the fourth fatigue-like condition may be combined. Galen sets out the possible combinations in diagrammatic form to give fifteen possible conditions in all. Other possible combinations may occur if *stegnosis* (stoppage of the pores) is included. This is considered in the next chapter. (chap. 9)

5. Stoppage of the Pores/*Stegnosis* (10)

Galen defines the term. The causes are thick and viscid superfluities, astringents, and cooling agents. The diagnostic features are listed. The cure is heating by more vigorous exercises, warmer baths, and oils.

6. Exercises (11–13)

Galen advises on exercises following sexual intercourse. Drying is identified as a consequence of both vigorous exertion (exercises) and sexual intercourse. Two approaches are recognized: preparatory exercises prior, and apotherapy following. Both have merit. Galen considers the circumstances that might favor one or the other. (chap. 11)

Brief consideration is given to exercises in other conditions—grief, insomnia, the *apepsias,* anger, lack of drink, and prolonged idleness. Massage should accompany the exercises. (chap. 12)

The final chapter gives brief details of the indications for and the techniques of morning and evening massage. (chap. 13)

Book 4

This book is largely about the spontaneous fatigues—i.e., those not associated with exercise / exertion. Galen here crosses the boundary between hygiene and therapeutics, as he acknowledges (VI.300K). Four groups of chapters are identified:

1. Fatigue in General (1–3)

Before discussing the symptoms of the spontaneous fatigues, Galen questions whether the diagnosis and treatment of these fatigues falls under hygiene or therapeutics, again raising the issue of a category intermediate between health and disease. He states his intention of discussing, in this book, fatigues and related conditions occurring apart from exercise. (chap. 1)

There are three aspects to consider: (1) the symptoms of fatigue; (2) the condition of fatigue; (3) the causes of fatigues. In the wound-like fatigue, an abnormal sensation is the symptom; acridity of thin, warm fluids is the condition; the causes are movement/exercise and *kakochymia*. Spontaneous tensive fatigues are due to *plethora* stretching the parts, the causes being *plethora* and *kakochymia*. *Plethora* and *kakochymia* are also causes of inflammation-like fatigues. (chap. 2)

The two therapeutic options in the fatigues are to evacuate the offending superfluity or to change it. The actual means is specific to each case, but the basic principle is "opposites cure opposites." (chap. 3)

2. Wound-like Fatigues (4–9)

The primary cause of wound-like fatigue is *kakochymia*. Galen goes into considerable detail on the recognition of the kind of abnormal humor, this being the major determining factor in treatment. Purging and phlebotomy are the means of evacuating the abnormal humor; change of the humor is effected by facilitating concoction through measures such as rest / sleep, bathing, massage, fasting,

certain foods, and medications. Details of the last are given. (chap. 4)

Galen considers the dangers of, and contraindications to, phlebotomy, exercise, and bathing. Details of certain compound medications are given, accompanied by an apology for such a detailed digression. (chap. 5)

Further details of the preparation of medications (oxymel, apomel) are given, followed by thoughts on the choice of wines. If these measures bring improvement, treatment can move on to bathing, anointing, massage, and exercise. Other factors causing undesirable movement of humors are considered. (chap. 6)

The essential pathophysiology of wound-like fatigues is excess of unconcocted humors, produced by exercise or heat, moving from the veins into the flesh. Massage and rest are important components of treatment. The preparation of the Diospoliticum medication is described. (chap. 7)

Further details on compound medications are given: from the seeds of the silver fir and from the blossoms of the black poplar. (chap. 8)

Galen describes a variant of the wound-like fatigue in which unconcocted humor is distributed throughout the whole body. Details of treatment are given. (chap. 9)

3. Tensive and Inflammation-like Fatigues (10)

The tensive fatigue apart from exercise is due to excessive stretching of certain parts. Phlebotomy is the treatment of choice if there is excess of blood in the veins. Galen considers Erasistratus' rejection of phlebotomy. The inflammation-like fatigues are characterized by heat

and swelling. Phlebotomy is again the treatment of choice. Nosebleeds and fierce sweats may help. The site of venesection is determined by the location of the symptoms. Adjuncts to phlebotomy are dietary measures, baths, and medications.

4. Some General Thought on Concoction / Digestion (11)

Galen mentions the three phases of concoction / digestion: in the stomach, in the veins, and in the individual parts. He refers to his work *On the Natural Faculties,* which covers these processes in detail.

Book 5

The twelve essays in this book can be divided into three groups:

1. General and Theoretical Considerations (1–2)

Galen begins by apologizing for the length of the whole treatise—it is necessitated by the subject. His approach has been to write about people who have been kept free of disease. He again stresses the variation that exists in healthy people and the need for different regimens for different people and different stages of life. He gives a brief account of his own life and how he benefitted from hygiene. (chap. 1)

He then focuses on the principles of diagnosis and treatment, considering first how to recognize different states of the body. The approach to management involves

three stages: (1) identifying the problem; (2) determining the cause or causes; (3) correction using the measures of hygiene. The last include movement, exercises, foods, drinks, baths, massage, sleep, regulation of sexual activity, the environment, and the functions of the soul. (chap. 2)

2. Hygiene for the Aged (3–10)

The basic problem is a body that is becoming increasingly dry and cold in terms of *krasis*. Correction, as far as it is possible, is through heating and moistening agents—especially appropriate foods, movements and exercises, hot baths and massages, and wine. (chap. 3)

He recognizes the difficulty of maintaining health in an old person and ponders the issue of whether old age should be regarded as a disease, a morbid condition between health and disease, or an unstable state of health. Regardless, it is a stage of life during which people readily become diseased. He makes some specific recommendations on exercise and diet, and describes two cases: Antiochus the doctor and Telephus the grammarian. (chap. 4)

He then considers wine for old people. Beneficial are aged wines of the warmer variety that are thin in consistency and tawny-orange in color. He lists some specific wines. (chap. 5)

Next he considers foods that are likely to cause obstruction, dietary measures to avoid this, and medications to deal with obstruction that has already occurred. (chap. 6)

There is then a chapter on bread and milk specifically, both for old people and more generally. (chap. 7)

He next considers several specific conditions that may afflict old people, the appropriate therapeutic measures, and the particular problems that may occur in the aged. (chap. 8)

Medications to rid the body of superfluities are detailed. Galen mentions premature aging due to disease and refers to his work *On Marasmus*. (chap. 9)

In the final chapter of this major group, exercises are considered, particularly for old people, but also more generally. Determining factors are: (1) the condition of the body as a whole; (2) customs; (3) troublesome afflictions. The particular principles of exercises in old people are that they should involve activities that are customary and that they should be moderate. Moreover, weak parts should be rested, unlike in the young. (chap. 10)

3. Correction of *dyskrasias* Compatible with Health (11–12)

The first of these two chapters focuses on some general therapeutic principles with reference to several Hippocratic statements and two of Galen's own works—*The Art of Medicine* and *On Mixtures*. The recurring theme of the need to recognize the individual variations among people is again raised and exemplified by a detailed case report on Premigenes, the Peripatetic philosopher, in relation to baths. (chap. 11)

The final chapter offers a review of the appropriate measures for different natures / *krasias* and considers also the kinds of disorders that are likely to affect particular natures. (chap. 12)

Book 6

The fifteen chapters in this book can be divided into four groups:

1. General Considerations (1–2)

Galen starts by reviewing the essential points from the first five books. The basic issues are three: (1) what health is; (2) what is damaging to health, including the progressive change in *krasis* toward the dry and cold through the successive stages of life; (3) important components of a healthy regimen. The first five books focus on the person with the best constitution of the body who is free to devote himself to maintaining his health. This, the final book, considers those who do not have the best constitution and / or cannot follow the ideal regimen because of the demands on their time (e.g., civic affairs, work, servitude). (chap. 1)

In considering types of abnormal constitution, Galen returns to his basic division of bodily structure into *homoiomeres,* which may be affected by *dyskrasias,* organic parts, which are subject to disturbances of magnitude, conformation, number, and position, and the body as a whole. He makes some general points about the best *krasis* in relation to the stages of life for both physical and psychical functions. He refers to a third bodily state between health and disease (neutral or neither), considered in detail in his *Art of Medicine.* This state is characterized by freedom from disease but functions that are somewhat deficient. (chap. 2)

2. *Dyskrasias*—Regular and Anomalous (3–8)

In the first of these chapters, Galen considers hot *krasias* and their combination with dryness and wetness. A hot *krasis* is characteristic of the early stages of life and is susceptible to diseases and symptoms caused by yellow bile, and to hepatic diseases. Diet, massage, and medications for preserving health in such *krasias* are considered. (chap. 3)

Next, Galen considers the cold *krasias* and their possible combinations: (1) with *eukrasia* in the other antithesis; (2) with wetness; (3) with dryness. In the first case the aim is a regimen steering a middle course between wetness and dryness. If associated with wetness, there should be avoidance of baths, appropriate exercises and diet, and certain unguents. If associated with dryness, there should be exercises, appropriate foods, hot wines, plenty of sleep, and attention to the evacuation of superfluities. (chap. 4)

Next, after a preamble about variations in regular and irregular (nonuniform, anomalous) *dyskrasias,* a broad definition of health as freedom from pain and unhindered functions, and the causes of deviation from health, Galen focuses on the person with a perfect constitution who is doomed to a life of servitude and can attend to his health only after sundown. (chap. 5)

Galen then considers the two main causes of persistent ill-health: excess / *plethora* and *kakochymia.* The principles of the regimen for the former are restoration of balance through diet, bathing, massage, and moderate exercise. With the latter, there is no single objective—it depends on the kind of *kakochymia.* Common features

are, however, downward purging of the stomach and moderation in sexual activity. He refers to three of his books: *The Powers of Foods, On the Good and Bad Humors of Nutriments,* and *On the Thinning Diet.* (chap. 6)

Next, he deals with ill-health due to a faulty constitution, considering particularly dietary measures: (1) how many meals per day; (2) what kinds of food; (3) measures to effect downward purging. Galen gives the example of himself on a typical day. He also considers *bradypepsias* (abnormally slow or reduced concoction) and their correction. (chap. 7)

Thinness and fatness are then considered. The former may be due to a dry and cold *dyskrasia* throughout the body or failure of the distributive and nutritive capacities. In treatment, pitch plasters are effective, but if this measure is resisted by the patient, alternatives include exercises, massage, and cold baths. In fatness, there is need to reduce the distribution of nutriments and increase evacuations from the body. Measures include downward purging and exercises, particularly running. (chap. 8)

3. Problems Due to Fluxes from the Head (9–13)

At the start of the next chapter, Galen announces his attention of dealing with nonuniform *dyskrasias,* previously defined. The main focus is, however, on the pathological effects of superfluities flowing from the head to structures below. The foundation of the therapeutic approach is to deal with the source, but symptomatic treatment of the secondarily affected structures include baths of medicinal waters and various oils. (chap. 9)

Galen then speaks of pain due to what are obviously

the vagus nerves. This is attributed to the flow of hot ichors from the head to the cardiac orifice of the stomach. He gives details of appropriate dietary measures and medications. (chap. 10)

Secondary effects such as nephrolithiasis and the arthritides are considered and details of the appropriate regimen (particularly diet) are given. (chap. 11)

Galen considers the adverse effects of fluxes to the eyes and ears, and their management. (chap. 12)

There is general consideration of the nature of the problems caused by the flow of superfluities from one part of the body to another—what determines the direction of flow and its ultimate end point, the role of external factors, and the general principles of treatment. (chap. 13)

4. Miscellaneous (14–15)

The first of the final two chapters considers hygiene for men in relation to sexual activity, starting with the description of a particularly debilitating condition associated with laxity of the cardiac orifice of the stomach due to the excessive emission of qualitatively abnormal semen, including nocturnal emissions. Galen also considers variations in the frequency of sexual activity and their relevance for health. (chap. 14)

The final chapter is added as an afterthought following a reminder from a friend. It describes the preparation made from the juice (and possibly the flesh) of quinces, which is useful for increasing the appetite in anorexias, improving concoction in the stomach, and generally strengthening that organ. (chap. 15)

GENERAL BIBLIOGRAPHY

TEXTS AND TRANSLATIONS OF THE
HYGIENE (DE SANITATE TUENDA)

Green, R. M. *A Translation of Galen's Hygiene.* Spring-field, IL: C. C. Thomas, 1951.

Koch, K. *Galeni De sanitate tuenda* in CMG V.4.2, which also contains *Galeni De alimentorum facultatibus* (G. Helmreich), *Galeni De bonis malis sucis* (C. Kalb-fleisch), and *Galeni De victu attenuante* (O. Hartlich). Leipzig and Berlin: Teubner, 1923.

Kühn, C-G. *Claudii Galeni Opera Omnia.* 20 vols. Hildes-heim: Georg Olms Verlag [Leipzig: 1821–1833], 1997 reprint.

Linacre, Thomas. *De sanitate tuenda.* Lyons: G. Rouille, 1547.

TRANSLATIONS OF OTHER
RELEVANT GALENIC WORKS

Boudon, V. *Galien: Exhortation a l'étude de la médicine; Art médical.* Paris: Les Belles Lettres, 2002.

Boudon-Millot, V. *Galien: Introduction general, Sur l'ordre de ses propres livres, Sur ses propres livres, Que l'excellent médicin est aussi philosophe.* Paris: Les Belles Lettres, 2007.

Brain, P. *Galen On Bloodletting.* Cambridge: Cambridge University Press, 1986.

Brock, A. J. *Galen On the Natural Faculties.* LCL 71. Cambridge, MA: Harvard University Press, 1963 [1916].

Daremberg, C. *Oeuvres anatomiques, physiologiques et médicales de Galien.* 2 vols. Paris: J-P Baillière, 1854–1856.

De Lacy, P. H. *Galen On the Doctrines of Hippocrates and Plato.* CMG, V.4.1.2. Berlin: Akademie-Verlag, 1978.

———. *Galen On the Elements According to Hippocrates.* CMG, V.1.2. Berlin: Akademie-Verlag, 1996.

Garcia Novo, Elsa. *Galen: On the Anomalous Dyskrasia.* Berlin: Logos Verlag, 2012.

Grant, M. *Galen on Food and Diet.* London: Routledge, 2000.

Grmek, M. D. *Diseases in the Ancient Greek World.* Baltimore: The Johns Hopkins University Press, 1989.

Harris, H. A. *Sport in Greece and Rome.* Ithaca: Cornell University Press, 1972.

Johnston, I. *Galen: On Diseases and Symptoms.* Cambridge: Cambridge University Press, 2006.

———. *Galen: On the Constitution of the Art of Medicine. The Art of Medicine. A Method of Medicine to Glaucon.* LCL 523. Cambridge, MA: Harvard University Press, 2016.

Johnston, I., and G. H. R. Horsley. *Galen: Method of Medicine.* 3 vols. LCL 516, 517, 518. Cambridge, MA: Harvard University Press, 2011.

May, M. T. *Galen On the Usefulness of the Parts of the Body.* Ithaca: Cornell University Press, 1968.

Meyerhof, M., and J. Schacht. *Galen über die medizinische*

Namen. Berlin: Abhandlungen der Preussischen Akademie der Wissenschaften, 1931.

Nutton, V. *Galen on My Own Opinions.* CMG, V.3.2. Berlin: Akademie-Verlag, 1999.

Powell, O. *Galen: On the Properties of Foodstuffs.* Cambridge: Cambridge University Press, 2002.

Siegel, R. E. *Galen's System of Physiology and Medicine.* Basle: S. Karger, 1968.

———. *Galen on the Affected Parts.* Basle: S. Karger, 1976.

Singer, P. N. *Galen: Selected Works.* Oxford: Oxford University Press, 1997.

———. *Galen. Psychological Writings.* Cambridge: Cambridge University Press, 2014.

Walzer, R., and M. Frede. *Three Treatises on the Nature of Science.* Indianapolis: Hackett Publishing Company, 1985.

GENERAL WORKS

Boudon-Millot, V. *Galien de Pergame.* Paris: Les Belles Lettres, 2012.

Conrad, L. I., M. Neve, V. Nutton, R. Porter, and A. Wear. *The Western Medical Tradition (800 BC to AD 1800).* Cambridge: Cambridge University Press, 1995.

Diels, H. A. *Die Handschriften der antiken Ärzte.* Vol. 1, *Hippocrates and Galen.* Leipzig: Zentralantiquariat, 1970 reprint.

Durling, R. J. "A Chronological Census of Renaissance Editions and Translations of Galen." *Journal of the Warburg and Courtald Institutes* 24 (1961): 230–305.

GENERAL BIBLIOGRAPHY

Garcia-Ballester, L. *Galen and Galenism*. Aldershot: Ashgate Variorum, 2002.

Garofalo, I. *Erasistrati Fragmenta*. Pisa: Giardini Editori e Stampatori, 1988.

Hankinson, R. J., ed. *The Cambridge Companion to Galen*. Cambridge: Cambridge University Press, 2008.

Maddison, F., M. Pelling, C. Webster. *Essays on the Life and Work of Thomas Linacre (c. 1460–1524)*. Oxford: Clarendon Press, 1977.

Mattern, S. P. *Galen and the Rhetoric of Healing*. Baltimore: The Johns Hopkins University Press, 2008.

———. *The Prince of Medicine: Galen in the Roman Empire*. New York: Oxford University Press, 2013.

McGovern, P. E. *Ancient Wine*. Princeton: Princeton University Press, 2003.

Mercuriale, Girolamo. *De arte gymnastica*. Edited by C. Pennuto. Florence: Olschki, 2008.

Nutton, V., ed. *The Unknown Galen*. London: Bulletin of the Institute of Classical Studies, supp. 77, 2002.

———. *Ancient Medicine*. London: Routledge, 2004.

Sarton, G. *Galen of Pergamon*. Lawrence, KA: University of Kansas Press, 1954.

Tissot, S. A. D. *An Essay on Diseases Incident to Literary and Sedentary Persons*. London: J. Nourse, 1769.

Von Staden, H. *Herophilus. The Art of Medicine in Early Alexandria*. Cambridge: Cambridge University Press, 1989.

ABBREVIATIONS

WORKS OF GALEN CITED IN THIS EDITION

Alim. Fac.	*De alimentis facultatibus*	The Powers of Foods
Animi Mores	*Quod animi mores corporis temperamenta sequuntur*	The Soul's Dependence on the Body
Ars M.	*Ars medica*	The Art of Medicine
Bon. Habit.	*De bono habitu*	On Good Condition
Bon. Mal. Suc.	*De bonis malisque alimentorum sucis*	On the Good and Bad Humors of Nutriments
Caus. Puls.	*De causis pulsuum*	The Causes of the Pulses
Comp. Med. Gen.	*De compositione medicamentorum per genera*	On the Composition of Medications according to Kind

Comp. Med. Loc.	*De compositione medicamentorum secundum locos*	On the Composition of Medications according to Places
Const. Art. Med.	*De constitutione artis medicae*	On the Constitution of the Art of Medicine
Cur. Rat. Ven. Sect.	*De curandi ratione per venae sectionem*	On Treatment by Bloodletting
Defin. Med.	*Definitiones medicae*	Medical Definitions
Diagn. Puls.	*De diagnoscendis pulsibus*	Diagnosis by the Pulses
Diff. Puls.	*De differentiis pulsuum*	The *Differentiae* of the Pulses
Diffic. Resp.	*De difficultate respirationis*	Difficulties in Breathing
Elem. Hippocr.	*De elementis secundum Hippocratem*	On the Elements according to Hippocrates
Hipp. Fract.	*In Hippocratis De fracturis commentarii*	On Hippocrates' *Fractures*
Hipp. Off. Med.	*In Hippocratis de officina medici*	On Hippocrates' Surgery
Hist. Phil.	*Historia Philosopha*	History of Philosophy

Hp. Aph.	*Hippocratis aphorismos*	Commentary on Hippocrates' Aphorisms
HVA	*In Hippocratis de acutorum morborum victu*	On Hippocrates' Regimen in Acute Diseases
Inaequal. Intemp.	*De inaequali intemperie*	On Anomalous Dyskrasia
Libr. Propr.	*De libris propriis*	On My Own Books
Marc.	*De marcore*	On Marasmus
Mixt.	*De temperamentis*	On Mixtures (*Kraseis*)
MM	*De methodo medendi*	The Method of Medicine
Morb. Diff.	*De morborum differentiis*	On the *Differentiae* of Diseases
Nat. Fac.	*De naturalibus facultatibus*	On the Natural Faculties
Opt. Const.	*De optima corporis nostri constitutione*	On the Best Consitution of our Bodies
Opt. Sect.	*De optima secta ad Thrasybulum*	On the Best Sect, to Thrasybulus
Ord. Libr. Propr.	*De ordine librorum propriorum*	On the Order of My Own Books

Part. Hom. Diff.	*De partium homoiomerium differentia libellis*	On the Differences of Uniform Parts
Parv. Pil.	*De parvae pilae exercitio*	On Exercise with a Small Ball
Plac. Hippocr. Plat.	*De placitis Hippocratis et Platonis*	On the Opinions of Hippocrates and Plato
Plenit.	*De plenitudine*	On Plethora
Praen.	*De praenotione ad Epigenem*	On Prognosis, for Epigenes
Praesag. Puls.	*De praesagitione ex pulsibus*	Prognosis from the Pulses
Protr.	*Protrepticus*	An Exhortation to the Study of the Arts
Puls. ad Tir.	*De pulsibus ad tirones*	The Pulses for Beginners
San. Tuend.	*De sanitate tuenda*	On the Preservation of Health (Hygiene)
Sect.	*De sectis ad eos introducuntur*	On the Sects
Simpl. Med.	*De simplicium medicamentorum temperamentis et facultatibus*	On the Nature and Powers of Simple Medications

Subf. Emp.	*Subfiguratio Empirica*	Outlines of Empiricism
Sympt. Caus.	*De symptomatum causis*	On the Causes of Symptoms
Sympt. Diff.	*De symptomatum differentiis*	On the *Differentiae* of Symptoms
Syn. Puls.	*Synopsis de pulsibus*	Synopsis of the Pulses
Ther.	*De theriaca ad Pisonem*	On Theriac, to Piso
Ther. Pamph.	*De theriaca ad Pamphilianum*	On Theriac to Pamphilianus
Thras.	*Thrasybulus sive utrum medicinae sit an gymnasticae hygieine*	Thrasybulus, On Whether Hygiene belongs to Medicine or Gymnastics
Trem.	*De tremore, palpitatione, convulsione et rigore*	On Tremor, Palpitation, Convulsion, and Rigor
UPart.	*De usu partium*	On the Use of the Parts
UPuls.	*De usu pulsuum*	The Use of the Pulses
Venae Sect.	*De venae sectione adversus Erasistratum*	On Phlebotomy, against Erasistratus

| Vict. Att. | *De victu attenua-tate* | On the Thinning Diet |
| Voc. | *De voce* | On the Voice |

REFERENCE WORKS

CMG   Corpus Medicorum Graecorum
EANS   *The Encyclopedia of Ancient Natural Scientists.* Edited by P. T. Keyser and G. L. Irby-Massie. London, Routledge, 2008.
LCL   The Loeb Classical Library.
LSJ   *A Greek-English Lexicon.* H. G. Liddell, R. Scott, and H. S. Jones. Oxford, 1990 reprint.
OED   *The Oxford English Dictionary.* 12 vols. Oxford, 1933.
S   *Stedman's Medical Dictionary.* 27th ed. Baltimore, MD: Lippincott, Williams and Wilkins, 2000.

ΓΑΛΗΝΟΥ ΥΓΙΕΙΝΩΝ ΛΟΓΟΣ

HYGIENE

A

1K 1. Τῆς περὶ τὸ σῶμα τἀνθρώπου τέχνης μιᾶς οὔσης, ὡς ἐν ἑτέρῳ δέδεικται γράμματι, δύο ἐστὶ τὰ πρῶτά τε καὶ μέγιστα μόρια· καλεῖται δὲ τὸ μὲν ἕτερον αὐτῶν ὑγιεινόν, τὸ δὲ ἕτερον θεραπευτικόν, ἔμπαλιν ἔχοντα πρὸς ἄλληλα ταῖς ἐνεργείαις, ἐπειδή γε τῷ μὲν φυλάξαι, τῷ δ' ἀλλοιῶσαι πρόκειται τὴν περὶ τὸ σῶμα κατάστασιν. ἐπεὶ δὲ καὶ χρόνῳ καὶ ἀξιώματι πρότερόν ἐστιν ὑγεία νόσου, χρὴ δήπου καὶ ἡμᾶς, ὅπως ἄν τις ταύτην φυλάξειεν, ἐσκέφθαι πρότερον, 2K ἐφεξῆς δὲ καί, ὡς ἄν τις ἄριστα νόσους ἐξιῷτο. κοινὴ δ' ἀμφοτέροις ὁδὸς τῆς εὑρέσεως, εἰ γνοίημεν, ὁποία τίς ἐστιν ἡ διάθεσις τοῦ σώματος, ἣν ὑγείαν ὀνομάζομεν· οὐ γὰρ ἂν οὔτε φυλάττειν αὐτὴν παροῦσαν οὔτ' ἀνακτήσασθαι διαφθειρομένην οἷοί τε ἦμεν ἀγνοοῦντες τὸ παράπαν, ἥτις πότ' ἐστι. γέγραπται δὲ ἡμῖν ἑτέρωθι καὶ περὶ τοῦδε καὶ δέδεικται τῶν μὲν ὁμοιομερῶν ὀναμοζομένων ἡ ὑγεία, ψυχροῦ καὶ θερμοῦ καὶ ξηροῦ καὶ ὑγροῦ, συμμετρία τις ὑπάρχουσα, τῶν δ' ὀργανικῶν ἐκ τῆς τῶν ὁμοιομερῶν συνθέσεώς τε καὶ ποσότητος καὶ πηλικότητος καὶ διαπλάσεως ἀποτε-

BOOK I

1. Although there is one art pertaining to the body of man, as I have shown in another treatise, there are two primary and major parts of this.[1] One of these is called hygiene and the other therapeutics. They are the opposites of each other in terms of functions, since what is proposed on the one hand is to preserve the state of the body, while on the other it is to change it. But since, in both time and importance, health comes before disease, it clearly behooves us to consider first how someone might preserve health and then, in turn, how someone might best cure diseases. There is a common path of discovery for both—it is to know what kind of condition of the body it is that we call health. Should we be altogether ignorant of what this condition is, we could neither preserve health when present nor restore it, if it were being destroyed. I have also written about this elsewhere and have shown that health of the so-called *homoiomeres* is a balance of cold and hot, and dry and wet, while health of the organic parts, whose composition is derived from the *homoiomeres,* is determined by number, magnitude and conformation.[2] As a result,

1K

2K

[1] Both *Ars medica* and *De constitutione artis medicae* deal with this issue—see Johnston, *On the Constitution*, where both are translated.   [2] See particularly *On the Differentiae of Diseases* in Johnston, *Galen: On Diseases and Symptoms.*

3

λουμένη. ὥστε καὶ ὅστις ἂν ἱκανὸς ᾖ φυλάττειν
ταῦτα, φύλαξ οὗτος ἀγαθὸς ὑγείας ἔσται. φυλάξει δὲ
πρότερον ἐξευρὼν ἅπαντας τοὺς τρόπους, καθ᾽ οὓς
διαφθείρεται. ὥσπερ γάρ, εἰ καὶ παντάπασιν ἀπαθὲς
ἦν ἡμῶν τὸ σῶμα, τῆς προνοησομένης αὐτοῦ τέχνης
οὐκ ἂν ἐδεήθημεν, οὕτω νῦν, ἐπειδὴ πάσχει πολυειδῶς, ἀναγκαῖόν ἐστι προστήσασθαί τινα τέχνην, τὴν
ἁπάσας αὐτοῦ τὰς βλάβας γινώσκουσαν καὶ φυλάττεσθαι δυναμένην.

3K   2. Εἰσὶ δὲ βλάβαι τε καὶ διαφθοραὶ τοῦ σώματος
ἡμῶν διτταὶ κατὰ γένος· αἱ μὲν γάρ τινες αὐτῶν
ἀναγκαῖαί τ᾽ εἰσι καὶ σύμφυτοι, τὴν οἷον ῥίζαν ἔχουσαι τὰς ἀρχὰς τῆς γενέσεως, ἔνιαι δ᾽ οὐκ ἀναγκαῖαι
μὲν οὐδὲ ἐξ ἡμῶν αὐτῶν ὁρμώμεναι, διαφθείρουσαι δ᾽
οὐδὲν ἧττον ἐκείνων τὸ σῶμα. διαιρήσομεν δ᾽ ἤδη
χωρὶς ἑκατέρας. αἷμα καὶ σπέρμα τῆς γενέσεως ἡμῶν
εἰσιν αἱ ἀρχαί, τὸ μὲν αἷμα οἷον ὕλη τις εὔρυθμός τε
καὶ εὐπειθὴς εἰς ἅπαν τῷ δημιουργῷ, τὸ δὲ σπέρμα
τὸν τοῦ δημιουργοῦ λόγον ἔχει. κέκραται δὲ ἑκάτερον
μὲν ἐκ τῶν αὐτῶν στοιχείων κατὰ γένος, ὑγροῦ καὶ
ξηροῦ καὶ θερμοῦ καὶ ψυχροῦ, ἤ, εἴπερ ἐθέλει τις οὐκ
ἀπὸ τῶν ποιοτήτων, ἀλλ᾽ ἀπὸ τῆς οὐσίας ὀνομάζειν
αὐτά, γῆς καὶ ὕδατος ἀέρος τε καὶ πυρός· οὕτω γὰρ
ἡμῖν ἐν τῷ Περὶ τῶν καθ᾽ Ἱπποκράτη στοιχείων ἀποδέδεικται, διαφέρουσι δὲ ἐν τῷ ποσῷ τῆς μίξεως. τῷ
μὲν γὰρ σπέρματι πλέον ἐνυπάρχει πυρώδους τε καὶ

---

3 "Impassible" and "passible" are used to render the Greek

whoever is up to the task of preserving these will be a good guardian of health. And he will preserve health, if he first discovers all the ways by which it is vitiated. We would, in fact, have no need of an art protecting health, if our body were altogether impassible,[3] but, since it is passible in many ways, it is necessary for us to establish an art that recognizes all the harms that befall the body and is able to protect it.

2. The harms and destructions of our body are twofold in terms of class: (1) those that are inevitable and innate, having their root, as it were, in the beginnings of formation; (2) those that are neither inevitable nor set in motion by ourselves, but which are no less destructive of the body than the former. I shall now distinguish each of these. Blood and semen are the origins of our genesis: blood is a kind of material that is fitting and well-adjusted in every way for the Demiurge, while the semen has the generative principle of the Demiurge.[4] Each of these has been mixed from the same elements according to class—wet and dry, and hot and cold. Or if someone should not wish to name them from the qualities but from the substance—earth and water, and air and fire. I have demonstrated this in my book *On the Elements according to Hippocrates*.[5] They differ, however, in the amount of the mixture. In the semen, more fiery and airy substances are present, whereas

3K

terms ἀπαθής and παθητός. On the former, see Aristotle, *Metaphysics* 1019a31. The meanings are: "incapable of suffering and feeling pain" (i.e., being affected) and the converse.

[4] On the use of λόγος in "generative principle," see Zeno, SVF, 1.28.   [5] *Elem. Hippocr.*, I.413–508K (English trans., De Lacy, CMG [1994]).

ἀερώδους οὐσίας, τῷ δ' αἵματι γεώδους τε καὶ ὑδατώ-
4K δους· ἐπικρατεῖ γε μὴν ἔτι κἂν τούτῳ τὸ μὲν θερμὸν
τοῦ ψυχροῦ, τὸ δὲ ὑγρὸν τοῦ ξηροῦ, καὶ διὰ τὴν ἐπι-
κράτειαν ταύτην οὐ ξηρόν, ὥσπερ ὀστοῦν καὶ ὄνυξ
καὶ θρίξ, ἀλλ' ὑγρὸν εἶναι λέγεται, τὸ δὲ σπέρμα ξηρότερον μέν ἐστιν ἢ κατὰ τὸ αἷμα, ῥυτὸν μὴν καὶ
ὑγρὸν ὑπάρχει καὶ αὐτό.

καὶ οὕτως ἑκατέρωθεν ἡμῖν ἡ ἀρχὴ τῆς γενέσεως
ἐξ ὑγρᾶς οὐσίας ἐστίν, ἣν οὐχ ὑγρὰν δήπου φυλάττεσθαι προσήκει, μέλλουσάν γε νεῦρα καὶ ἀρτηρίας
καὶ φλέβας καὶ ὀστᾶ καὶ χόνδρους καὶ ὑμένας ὅσα τ'
ἄλλα τοιαῦτα γενήσεσθαι. συγκαταβεβλῆσθαι τοίνυν ἀναγκαῖόν ἐστιν εὐθὺς ἀπὸ τῆς πρώτης γενέσεως
ἰσχυρότερον ἐν τῇ κράσει τὸ ξηραντικὸν στοιχεῖον.
ἔστι δὲ τῇ φύσει τοιοῦτον μάλιστα μὲν τὸ πῦρ, ἤδη
δὲ καὶ ἡ γῆ· ξηρὸν γάρ τι χρῆμα καὶ ἥδε. ἀλλὰ τῆς
μὲν γῆς οὐχ οἷόν τε ἦν μίγνυσθαι πλεῖον, ὑγρῶν εἶναι
δεομένων τῶν ἀρχῶν· τοῦ πυρὸς δ' οὔτε κωλύει τι
πλέον μιχθῆναι καὶ κέκραται τοσούτῳ πλέον ἐν ἀμφοῖν, ὡς μήτε φρύγειν ἤδη καὶ καίειν αὐτάρκως τε
5K ξηραίνειν. καὶ γὰρ αὖ καὶ τὴν πρὸς τὰς κινήσεις ἑτοι-
μότητα τοσοῦτον ὑπάρχον τὸ θερμὸν ἱκανὸν ἦν παρα-
σχεῖν. ὑπὸ τούτου δὴ τὰ μὲν πρῶτα συνίσταταί τε καὶ
βραχεῖάν τινα πῆξιν λαμβάνει τὸ κύημα· μετὰ δὲ
ταῦτα ἐπὶ μᾶλλον ἤδη ξηραινόμενον οἷον ὑπογραφάς
τινας ἴσχει καὶ τύπους ἀμυδροὺς ἑκάστου τῶν μορίων·
εἶτ' ἐπὶ πλέον ξηρανθὲν οὐχ ὑπογραφὰς μόνον οὐδ'
ἀμυδροὺς τοὺς τύπους, ἀλλ' ἀκριβὲς ἑκάστου τὸ εἶδος

6

in the blood there are more earthy and watery substances, although even in the latter, heat prevails over cold and   4K
moisture over dryness. On account of this preponderance, blood is not dry like bone, nails and hair, but is described as being moist, while semen is drier than blood but is itself fluid and moist.

And in this way, on each side, the origin of our genesis is from moist substances, although of course it is not appropriate to preserve the moistness for what will become nerves, arteries, veins, bones, cartilage, membranes and other such things. Accordingly, it is necessary for there to have been laid down in the mixture (*krasis*), right from the first genesis, the element that is stronger in *krasis* in terms of drying capacity. Fire is, in nature, particularly this kind of element, but so too is earth, for this is also dry matter. However, it is not possible for more earth to be mixed, since the origins need to be moist. On the other hand, there is nothing to stop more fire being mixed and prevailing to such an extent in both (i.e., blood and semen), so as not to now parch or burn and to dry sufficiently. In truth,   5K
there should be as much heat as is sufficient to provide readiness for movements. From this, then, the embryo is first formed and takes on a small degree of solidity. After this, when it is already dried out still more, it has as it were the outlines and indistinct forms of each of the parts. Then, when it is dried out even more, it not only has the outlines and indistinct forms but has the exact form of

ἴσχει. καὶ δὴ καὶ ἀποκυηθὲν ἀεὶ καὶ μᾶλλον ἑαυτοῦ
γίνεται ξηρότερόν τε καὶ ῥωμαλεώτερον, ἄχριπερ ἂν
εἰς ἀκμὴν ἀφίκηται. τηνικαῦτα δὲ τά τε τῆς αὐξήσεως
ἵσταται μηκέτι ἐπιδιδόντων τῶν ὀστῶν διὰ τὴν σκληρότητα, καὶ τῶν ἀγγείων δ' ἕκαστον εἰς εὖρος δια-
φυσᾶται, καὶ σύμπανθ' οὕτω τὰ μόρια κρατύνεται
καὶ εἰς τὴν ἀκροτάτην αὐτῶν ἰσχὺν ἀφικνεῖται. τὸ δὲ
ἀπὸ τοῦδε περαιτέρω τοῦ προσήκοντος ἤδη ξηροτέ-
ρων ἁπάντων τῶν ὀργάνων γινομένων αἵ τε ἐνέργειαι
χεῖρον ἀποτελοῦνται καὶ ἀσαρκότερόν τε καὶ ἰσχνότερον ἑαυτοῦ γίνεται τὸ ζῷον. ἐπὶ πλέον δὴ ἀναξηραι-
νόμενον οὐκ ἀσαρκότερόν γε μόνον, ἀλλὰ καὶ ῥυσὸν
6K ἀποτελεῖται, καὶ τὸ κῶλον ἀκρατὲς καὶ σφαλερὸν ἐν
ταῖς κινήσεσι. καὶ καλεῖται μὲν ἡ τοιαύτη διάθεσις
γῆρας, ἀνὰ λόγον δ' ἐστὶ τῇ τῶν φυτῶν αὐάνσει. καὶ
γὰρ αὖ κἀκείνη γῆράς ἐστι φυτῶν, δι' ὑπερβάλλου-
σαν ξηρότητα γινομένη. μία μὲν οὖν ἥδε σύμφυτος
ἀνάγκη φθορᾶς ἅπαντι τῷ γεννητῷ σώματι, δευτέρα
δὲ τοῖς ζῴοις μάλισθ' ὑπάρχουσα τῆς ὅλης αὐτῶν
οὐσίας ἡ ῥύσις ἐκ τῆς ἐμφύτου θερμότητος ἀποτε-
λουμένη.

ταύτας μὲν οὖν τὰς βλάβας οὐδενὶ γεννητῷ σώματι
φυγεῖν ἐγχωρεῖ, τὰς δ' ἄλλας βλάβας, ὅσαι ταύταις
ἕπονται, δυνατὸν φυλάξασθαι προμηθούμενον. ἡ γένεσις δὲ κἀκείνων ἐκ τοῦ τὰς ἀναγκαίας βλάβας ἐπανορθοῦσθαι πειρᾶσθαι. ῥεούσης γὰρ τῆς οὐσίας συ-
στάσεως ἁπάντων τῶν ζῴων, εἰ μή τις ἕτερον ὅμοιον
ἀντεισάγοι τῷ ἀπορρέοντι, διαφορηθήσεταί τε καὶ

8

each part. And now, when it is brought forth, it always becomes drier than it was, and stronger, until it reaches its full development. Then there is a cessation of the increase when the bones no longer grow due to hardness, while each of the vessels is extended in width, and in this way all the parts become strong and reach the highest point of their strength. However, further on from this, when all the organs have already become drier than is appropriate, the functions are made worse and the organism becomes less well-fleshed, and leaner than it was. As it is dried up still more, it is not only leaner, but is also made shriveled and 6K the limbs are weak and unsteady in their movements. Such a condition is called old age and is analogous to the drying up of plants, for that too is the old age of plants, arising from excessive dryness. There is, then, this one innate and inevitable destruction for every body created, while there is a second, existing particularly in animals, which is a flowing away of their whole substance brought about by the innate heat.

It is impossible, then, for any begotten body to escape these harms, whereas those other harms which follow these can be guarded against with proper care. Moreover, the genesis of these latter harms is from attempting to rectify the inevitable harms. For since the existing substance of all animals flows away, unless some other substance like that flowing away is brought in, the whole body

σκεδασθήσεται σύμπαν οὕτω τὸ σῶμα. καὶ διὰ τοῦτο
οἶμαι τὴν φύσιν οὐ τοῖς ζῴοις μόνον, ἀλλὰ καὶ φυτοῖς
εὐθὺς ἐξ ἀρχῆς συμφύτους δοῦναι δυνάμεις ἐφιεμένας
7K τῶν ἀεὶ ἐλλειπόντων. οὔτε γὰρ ἐσθίειν οὔτε πίνειν
οὔτε ἀναπνεῖν διδασκόμεθα πρός τινος ἀλλ' ἐξ ἀρχῆς
ἔχομεν ἁπάντων τούτων ἐν ἡμῖν αὐτοῖς τὰς δυνάμεις
ἄνευ διδαχῆς ἐπιτελούσας ἅπαντα. διὰ μὲν οὖν τῆς
ἐδωδῆς ἀναπληροῦμεν, ὅσον ἀπερρύη τῆς ξηροτέρας
οὐσίας, διὰ δὲ τοῦ πόματος τὸ κενωθὲν[1] τῆς ὑγροτέρας ἀντεισάγομεν, εἰς τὴν ἀρχαίαν ἐπανάγοντες
ἄμφω συμμετρίαν. οὕτω δὲ καὶ τῆς ἀερώδους τε καὶ
πυρώδους οὐσίας τὴν συμμετρίαν ἀναπνοαῖς τε καὶ
σφυγμοῖς διασῴζομεν. ἀποδέδεικται δὲ καὶ περὶ τούτων ἁπάντων ἰδίᾳ καθ' ἕκαστον ἐν ἑτέροις γράμμασι,
καὶ προσήκειν ἡγοῦμαι τῷ λόγῳ τῷ νῦν, ὅσα δέδεικται δι' ἐκείνων ὑποθέσεις ποιησάμενον πρὸς τὴν ὑγιεινὴν πραγματείαν, οὕτως ἔχεσθαι τῶν ἐφεξῆς.

3. Ἐπειδὴ γὰρ ἀπορρεῖ μὲν ἁπάντων τῶν ζῴων
ὁσημέραι πολὺ μέρος τῆς οὐσίας διὰ τὴν ἔμφυτον
θερμότητα, δεόμεθα δὲ ὑπὲρ τοῦ τὴν συμμετρίαν
αὐτῆς διαφυλάττεσθαι σιτίων τε καὶ πομάτων, ἀναπνοῆς τε καὶ σφυγμῶν, ἐξ ἀνάγκης ἀκολουθήσει
τοῖσδε περιττωμάτων γένεσις. εἰ μὲν γάρ, οἷόνπερ ἦν
8K τὸ κενωθέν, ἕτερον ἀκριβῶς τοιοῦτον εἴχομεν αὐτῷ
προσφῦσαι δι' ὅλου, κάλλιστον ἂν ἦν τοῦτο καὶ ὑγι-
εινότατον. ἐπεὶ δὲ τὸ μὲν ἀπορρέον ἑκάστου τῶν

---

[1] τὸ κενωθὲν add. Ko

will be carried away and dispersed in this manner. Because of this, I think, Nature gave animals, right from their beginnings, innate powers that aim at those things that are lacking at any time—and not only to animals but also to plants. For we do not learn from someone else to eat, drink and breathe; from the beginning we have in ourselves capacities that accomplish everything without instruction. Through food, then, we replenish as much of the dry substance as has flowed away and through drink we replenish that which has been emptied out of the moister [substance], restoring both to their original balance. And in the same way, we maintain the balance of airy and fiery substances through respiration and pulses. I have demonstrated all these things separately in other writings.[6] I think it is now germane to the discussion, having shown through those works what creates the foundation of the matter of health, to proceed as follows.

3. Since a great part of the substance of all animals flows away every day due to the innate heat, we need food, drink, respiration and pulses for preserving the balance of this substance, but the creation of superfluities will inevitably follow from these things. If we were able to replace precisely what has been emptied out with something else and assimilate this completely, it would be the best and most healthy situation. However, since what flows away

7K

8K

[6] This is taken as a general reference to a number of Galen's works—perhaps particularly the work on the elements referred to in the previous note, the four treatises on the classification and causation of diseases and symptoms (Johnston, *Galen: On Diseases and Symptoms*) and his *Mixt.*, I.509–694K (English trans., Singer, *Galen: Selected Works*).

GALEN

μορίων τοιοῦτον τὴν φύσιν ἐστίν, οἷόνπερ αὐτὸ τὸ
μόριον, οὐδὲν δὲ τῶν ἐσθιομένων ἢ πινομένων ἀκριβῶς ἐστι τοιοῦτον, ἀναγκαῖον ἐγένετο τῇ φύσει προ-
μεταβάλλειν τε καὶ προπέττειν αὐτὰ καὶ ὡς ἔνι μάλιστα προπαρασκευάζειν ὅμοια τῷ θρεψομένῳ σώματι.
κἂν τούτῳ τὸ μὴ κατεργασθὲν ἀκριβῶς μηδ᾽ ἐξομοι-
ωθὲν οὔτε προσφύεται τῷ σώματι καὶ περιττὸν ὂν
ἀλᾶται κατὰ τὰς ἔνδον εὐρυχωρίας, ὅθενπερ αὐτῷ καὶ
τοὔνομα πρὸς τῶν ἔμπροσθεν ὀρθῶς ἐτέθη περίττωμα. ἐπειδὴ οὖν τὸ μὲν ἐσθίειν τε καὶ πίνειν ἀναγκαῖα τοῖς ζῴοις ὑπάρχει, ἀκολουθεῖ δὲ τούτοις ἡ τῶν
περιττωμάτων γένεσις, ὄργανά τε πρὸς τὴν ἀπόκρισιν αὐτῶν ἡ φύσις παρεσκεύασεν καὶ δυνάμεις αὐτοῖς
ἐνέθηκε, δι᾽ ὧν κινούμενα τὰ μὲν ἕλκει, τὰ δὲ ἐκπαραπέμπει,[2] τὰ δὲ ἐκκρίνει τὰ περιττώματα. καὶ χρὴ δήπου μήτ᾽ ἐμφράττεσθαι κατά τι μήτε ἀρρωστεῖν κατὰ
τὰς ἐνεργείας ὑπὲρ τοῦ καθαρὸν ἀεὶ καὶ ἀπέριττον
9K   διαφυλάττεσθαι τὸ σῶμα.

καί σοι δύο μὲν ἤδη σκοποὺς τούσδε πρὸς δίαιταν
ὑγιεινὴν ὁ λόγος ὑφηγήσατο, τὸν μὲν ἕτερον ἀνα-
πλήρωσιν τῶν κενουμένων, τὸν δ᾽ ἕτερον ἀπόκρισιν
τῶν περιττωμάτων. ὁ γὰρ δὴ τρίτος ὁ περὶ τοῦ μὴ
ταχύγηρον γίνεσθαι τὸ ζῷον ἐξ ἀνάγκης ἕπεται τοῖς
εἰρημένοις. εἰ γὰρ μηδὲν ἁμαρτάνοιτο μήτε ἐν τῷ τὸ
κενούμενον ἀναπληροῦσθαι μήτε ἐν τῷ τὰ περιττώματα ⟨μὴ⟩ μένειν ἔνδον, ὑγιαίνοι τ᾽ ἂν ἐν τῷδε τὸ
ζῷον ἀκμάζοι τε μέχρι παμπόλλου. περὶ μὲν δὴ τοῦδε
καὶ αὖθις εἰρήσεται τοῦ λόγου προϊόντος.

12

from each of the parts is the same in nature as the part itself, whereas nothing that is eaten or drunk is exactly this sort of thing, it naturally becomes necessary to change and digest these things beforehand, and as far as possible to prepare them beforehand to be like the body that will be nourished. And in this, what is not entirely prevailed upon or assimilated is not incorporated into the body, but passes into the internal spaces (cavities) as superfluity. As a consequence, the term "superfluity" was correctly applied to this by our predecessors. Therefore, since eating and drinking are essential for animals, while the generation of superfluities follows these activities, Nature provided organs for the excretion of these things and endowed them with powers through which they are moved to attract, send along and expel the superfluities. And for the purpose of always keeping the body clean and free of superfluities, it is also clearly necessary for these organs to be neither obstructed in any way nor weakened in their functions.   9K

Now the discussion has instructed you in these two objectives pertaining to a healthy regimen: one is the replacement of those things emptied out and the other is the separation of the superfluities. There is, in fact, a third point regarding this—that the animal does not age prematurely—which follows of necessity from the things that have been said. For if nothing goes wrong in the replenishment of what is emptied and none of the superfluities are retained within, the animal will in this way be healthy and will flourish for a long time. I shall speak of this again as the discussion proceeds.

---

2 ἐκπαραπέμπει Ku, παραπέμπει Ko

4. Τὸ δ' ὑπόλοιπον ὧν ἐξ ἀρχῆς διελέσθαι προὐθέμεθα προσθῶμεν ὑπὲρ τοῦ διωρίσθαι σαφῶς ἤδη τοὺς ὑγιεινοὺς σκοπούς, ὁποῖοί τέ εἰσι καὶ ὁπόσοι. ἔφαμεν γάρ, ὡς, εἰ μὲν ἀπαθὲς ἦν ἡμῶν τὸ σῶμα, καθάπερ ἀδάμας ἤ τι τοιοῦτον, οὐδεμιᾶς ἂν ἐδεῖτο τέχνης ἐπιστατούσης αὐτῷ· ἐπειδὴ δὲ διττὰς ἔχει τῆς φθορᾶς αἰτίας, τὰς μὲν ἔνδοθεν καὶ ἐξ ἑαυτοῦ, τὰς δὲ ἐκ τῶν ἔξωθεν προσπιπτόντων, ἀναγκαῖον αὐτὸ δεῖ-
10K σθαι προνοίας οὐ μικρᾶς. ἐξ ἑαυτοῦ μὲν οὖν ἐδείχθη κατὰ διττὸν τρόπον διαφθειρόμενον, ἢ διὰ γῆρας ἐπὶ θάνατον προϊὸν ἢ διὰ τὸ ῥεῖν ἀεὶ τὴν οὐσίαν αὐτοῦ, καθ' ἕτερον δὲ τὸν ἑπόμενον οἷς ἐσθίει τε καὶ πίνει τὸν ἐκ τῆς τῶν περιττωμάτων γενέσεως. ἐξ ἑαυτοῦ μὲν οὖν ὧδέ πως φθείρεται, τῶν δὲ ἔξωθεν αὐτῷ προσπιπτόντων ἓν μὲν ἀχώριστόν τέ ἐστι καὶ διαπαντὸς ὑπάρχον αὐτῷ καὶ ὡς ἂν εἴποι τις σύμφυτον, ὁ περιέχων ἀήρ, τὰ δ' ἄλλα οὔτ' ἀναγκαῖα καὶ κατὰ χρόνους τινὰς ἀτάκτως ὁμιλοῦντα, τὰ μὲν ὥσπερ ὁ περιέχων ἀὴρ[3] ἢ τῷ θερμαίνειν ἀμέτρως ἢ τῷ ψύχειν ἢ τῷ ξηραίνειν ἢ τῷ ὑγραίνειν βλάπτοντα, τὰ δὲ τῷ θλᾶν ἢ διασπᾶν ἢ τιτρώσκειν ἢ ἔξαρθρόν τι ποιεῖν.

ἔστι μὲν οὖν τις ἐνταῦθα λογικὴ ζήτησις εἰς ἑκάτερον ἐπιχειρεῖσθαι δυναμένη, τινῶν μὲν τῆς περὶ τὸ σῶμα τέχνης ἁπάντων τούτων τὴν φυλακὴν εἶναι λεγόντων, τινῶν δὲ τῶν θερμαινόντων τε καὶ ψυχόντων, ὑγραινόντων καὶ ξηραινόντων μόνον· εἰ γὰρ τὸ θλῶν ἢ τὸ τιτρῶσκον ἤ τι τοιοῦθ' ἕτερον ἐργαζόμενον ἐξ-
11K ίστησι τοῦ κατὰ φύσιν ἡμᾶς, οὐδεμιᾶς εἶναι τέχνης

4. Let me now add the rest of those things I proposed to go over from the outset so as to make a clear division in the objectives of hygiene—that is, what kinds and how many there are. I said that, if our body were impassible, like adamant or some such thing, there would be no need of any art for the care of it. But since it has two sorts of causes of deterioration—those which are internal and of themselves and those which are external and befall it— this requires, of necessity, no little forethought. It was 10K shown that destruction "of itself" is twofold in kind: one is to come to death through aging and the constant flux of substance and the other is the creation of superfluities that follows those things eaten or drunk. This is how the body deteriorates "of itself." Of those things that befall it from without, one which is ever-present and everywhere—one might say natural—is the surrounding air, whereas other things are not essential and are irregular associations occurring from time to time. The surrounding air causes harm by either heating, cooling, drying or moistening immoderately, whereas the other things do so by bruising, tearing, wounding or dislocating.

It is, then, possible to attempt a logical inquiry here into whether we say of the art that preserves the body that it concerns all these things or only those that are heating, cooling, moistening and drying. For if bruising, wounding, or anything else of this sort that acts, changes us from an 11K accord with nature, there is no art for either knowing

---

3 τὰ μὲν ὥσπερ ὁ περιέχων ἀὴρ Ko; ὁ μὲν περιέχων ἀὴρ Ku

οὔτε γινώσκειν οὔτε φυλάττειν τὰ τοιαῦτα. ἐμοὶ δὲ
περὶ μὲν τῶν τοιούτων προβλημάτων οὐ πρόκειται
νῦν διαιρεῖν· ὅπερ δ' ὁμολογούμενόν ἐστι παρ' ἀμφοῖν
ἐξ ἑτοίμου λαβὼν ἐπὶ τὸ προκείμενον ἐπάνειμι. τὸ
γὰρ ἅπασι μὲν τοῖς ἀνθρώποις γινώσκεσθαι τὰ τῷ
τιτρώσκειν ἢ θλᾶν ἤ τι τοιοῦτον ἕτερον ἐργάζεσθαι
βλάπτοντά τε καὶ διαφθείροντα τὴν ὑγείαν, οὐχ
ἅπασι δὲ ὅσα τῷ θερμαίνειν ἢ ψύχειν ἢ ξηραίνειν ἢ
ὑγραίνειν, ὡμολόγηται παρ' ἀμφοῖν. οὐκοῦν οὐδ'
ἡμεῖς αὐτοῖς παρὰ μέλος τι πράττειν δόξομεν, εἰ τὰ
γινωσκόμενα ἅπασιν ὑπερβάντες ἐπὶ τὰ μὴ γινωσκόμενα τὸν λόγον ἄγοιμεν. οὐ γάρ μοι πρόκειται τό γε
νῦν εἶναι σοφιστικὰ ζητήματα διελθεῖν, ἀλλ' ὡς ἄν
τις ἥκιστα νοσήσειεν ὑφηγεῖσθαι.

πάλιν οὖν ἐπὶ τὴν οἰκείαν ἀρχὴν ἀναγάγωμεν τὸν
λόγον ἀναμνήσομέν τ' ἀκριβέστερον ἔτι τῶν ὑποθέ-
σεων αὐτοῦ. τό τε γὰρ εἶναι τὴν ὑγείαν οὐχ ἁπλῶς
12K εὐκρασίαν ἢ συμμετρίαν τῶν στοιχείων, ἐξ ὧν ἐγενό-
μεθα, καθάπερ οἱ πρὸ ἡμῶν ὀλίγου δεῖν ἅπαντες ἐνό-
μιζον, ἀλλὰ μόνον τὴν τῶν ὁμοιομερῶν σωμάτων,
ἀποδεδειγμένον ἡμῖν ἐν ἑτέροις, ὑπόθεσις ἔστω πρὸς
τὰ παρόντα· τό τε τῶν ὀργανικῶν σωμάτων τὴν
ὑγείαν ἐν διαπλάσει τε καὶ ἀριθμῷ καὶ πηλικότητι
καὶ συνθέσει τῶν ὁμοιομερῶν συνίστασθαι καὶ τοῦθ'
ὡσαύτως ὑποκείσθω πρὸς τὰ παρόντα δεδειγμένον
ἑτέρωθι.[4] καὶ μὴν καὶ ὅτι ταῖς κατὰ φύσιν ἐνεργείαις

---

[4] πρὸς τὰ παρόντα δεδειγμένον ἑτέρωθι. Ko; προδεδειγμέ-
νον ἑτέρωθι. Ku

about or preventing such things. However, it is not my present task to make a distinction between such proposals, so taking what is accepted by both sides, I shall return forthwith to the task before me. It is agreed by both sides that the things injurious to and destructive of health by wounding, bruising or bringing about something else of this sort are recognized by all men, which is not the case with all those things that heat, cool, dry or moisten. We shall not, therefore, seem to be doing anything inappropriate if we pass over the things known to all and go on to a discussion of those things not known to all. For it is not, in fact, my present purpose to set out in detail a sophistical investigation, but to indicate how we might suffer as little as possible from disease.

So let us return once more to the proper argument, calling to mind more precisely its proposals. Let the hypothesis regarding the present matters be that health is not simply an *eukrasia* or balance of the elements from which we are created, as almost all our predecessors thought; this only applies to the health of the *homoiomeric* bodies, as I have demonstrated in other writings.[7] And let us also assume in similar fashion for our present purposes that the health of the organic bodies lies in the conformation, number, magnitude and composition of the *homoiomeres,* as has been demonstrated elsewhere.[8] And then let us assume that a healthy constitution is judged by func-

12K

---

[7] See particularly Galen's *Ars M.,* 2–4, I.309–18K; Johnston, *On the Constitution,* 164–77.

[8] See, for example, *Morb. Diff.,* 6–10, VI.855–871; Johnston, *Galen: On Diseases and Symptoms*, 144–52.

ἡ ὑγιεινὴ κατασκευὴ κρίνεται, καὶ ὅτι τῆς ὑγείας ἡ
μὲν ἀρίστη τίς ἐστι καὶ ὡς ἂν οὕτω τις εἴποι τελεία
τε καὶ ἀκριβής, ἡ δ᾽ οἷον ἐλλιπής τε καὶ οὐκ ἀκριβὴς
οὐδὲ τελεία, ἥπερ δὴ καὶ πλάτος ἔχειν πάμπολύ φαμεν, ὑποκείσθω καὶ ταῦθ᾽ ἡμῖν ἐν τῷδε, δι᾽ ἑτέρων τε
ἤδη προαποδεδειγμένα καὶ νῦν οὐχ ἥκιστα δειχθη-
σόμενα. μάλιστα δ᾽ ἀνεγνωκέναι βούλομαι τὸν ὁμι-
λήσοντα τοῖσδε τοῖς γράμμασι τὸ βιβλίον, ἐν ᾧ
σκέπτομαι, τίνος ἐστὶ τέχνης μέρος τὸ ὑγιεινόν (ἐπι-
13K   γράφεται δὲ Θρασύβουλος), ἔτι δὲ καὶ τὸ Περὶ τῆς
ἀρίστης κατασκευῆς τοῦ σώματος ἡμῶν, ἔτι δὲ τὸ
Περὶ τῆς εὐεξίας. ἔστι δ᾽ ἄμφω μικρὰ βιβλίδια, ἃ
προαναγνοὺς εἴ τις ἐπὶ τόνδε τὸν λόγον ἀφίκοιτο,
ῥᾷστα ἂν ἀκολουθήσειε τοῖς νῦν λεγομένοις. ὅτι δὲ
καὶ τὸ Περὶ τῶν καθ᾽ Ἱπποκράτη στοιχείων ἀναγκαῖόν ἐστιν εἰς τὰ παρόντα, πρότερον εἴρηται· καὶ
γὰρ δὴ καὶ ἕπεται ἐκείνῳ τὸ Περὶ τῆς ἀρίστης κατασκευῆς καὶ τὸ Περὶ τῆς εὐεξίας.

5. Ἐπὶ τούτοις ὑποκειμένοις ἀρκτέον ἂν εἴη ἐνθένδε
τῆς ὑγιεινῆς πραγματείας. ἐπειδὴ συμμετρία τίς
ἐστιν ἡ ὑγεία, συμμετρία δὲ πᾶσα κατὰ διττὸν ἀποτελεῖται καὶ λέγεται τρόπον, ποτὲ μὲν εἰς ἄκρον
ἥκουσα καὶ ὄντως οὖσα συμμετρία, ποτὲ δὲ ἀπολειπομένη βραχύ τι τῆς ἀκριβείας, εἴη ἂν καὶ ἡ ὑγιεινὴ
συμμετρία διττή τις· ἡ μὲν γὰρ ἀκριβής τε καὶ ἀρί-

───

9 The four works referred to are, in order, *Thras.*, V.806–898K
(English trans., Singer, *Galen: Selected Works*); *Opt. Const.*,

tions that are in accord with nature, and that this is the best of health, or as someone might describe it, perfect and excellent health, while that which is, as it were, defective and is neither perfect nor excellent has, we may say, a very wide range. Let these things also be taken for granted by us in this work, since they have already been shown through other works and will be no less demonstrated now. It is my particular wish that someone reading this book be familiar with those treatises in which I carefully examine what part of the art hygiene is (the one inscribed to Thrasybulus) and *On the Best Constitution of our Bodies* and then *On Good Condition.* Both the latter two are short books that, if someone has read them before coming to this present work, will allow him to easily follow the present arguments. It is also necessary for our present purposes to read beforehand the work *On the Elements according to Hippocrates,* as I said previously, and then to follow that work with *On the Best Constitution of our Bodies* and *On Good Condition.*[9]

13K

5. Hence it is from these foundations we must make a beginning of the study of hygiene. Since health is a kind of balance, and every balance is brought about and expressed in a twofold manner, at one time coming to the highest point and being truly a balance and at another wanting slightly in perfection, hygiene too is a twofold balance. On the one hand, it is exact, optimal, complete

IV.737–749K (English trans., Singer, *Galen: Selected Works*); *Bon. Habit.,* IV.750–756K (English trans., Singer, *Galen: Selected Works*); *Elem. Hippocr.,* I.413–508K (English trans., De Lacy, CMG [1994]).

19

στη καὶ τελέα καὶ ἄκρα, ἡ δὲ ἀπολειπομένη μὲν ταύτης, οὐ μὴν ἤδη γέ πω τοσούτῳ, ὡς λυπεῖσθαι τὸ 14K ζῷον. ἔστι δὲ κἀνταῦθα λογική τις μᾶλλον ἢ κατὰ τὴν χρείαν τῆς τέχνης ζήτησις, οὐ συγχωρούντων ἐνίων ἕτερον ἑτέρου μᾶλλον ὑγιαίνειν οὐδ' εἶναι πλάτος ἱκανὸν ἐν τῇ διαθέσει τοῦ σώματος, ἣν ὑγείαν ὀνομά- ζομεν, ἀλλ' ἕν τι καὶ ἀπηκριβωμένον οὖσαν αὐτὴν ἄτμητον εἰς τὸ μᾶλλόν τε καὶ ἧττον ὑπάρχειν. ἐμοὶ δὲ ὥσπερ τὸ λευκὸν σῶμα τὸ μὲν ἧττον φαίνεται λευκὸν εἶναι, τὸ δὲ μᾶλλον, οὕτω καὶ τὸ ὑγιαῖνον ἧττόν τε καὶ μᾶλλον εἶναι δοκεῖ τοιοῦτον.

διττὴ δὲ ἀπόδειξις τοῦ λόγου· μία μὲν ἐκ τῆς κατὰ τὰς ἡλικίας μεταπτώσεως· ἀφ' οὗ γὰρ ἂν ἀποκυηθῇ τὸ ζῷον, ἀεὶ μεταβάλλειν ἀναγκαῖον αὐτοῦ τὴν κρᾶ- σιν, ὡς ἔμπροσθεν ἐδείκνυμεν· ὥστ', εἴπερ ἐν μὲν τῷ ποιῷ τῆς κράσεως ἡ ὑγεία, τὸ ποιὸν δ' οὐ μένει ταὐτόν, οὐδὲ τὴν ὑγείαν ἐγχωρεῖ τὴν αὐτὴν φυλάττε- σθαι. δευτέρα δ' ἀπόδειξις ἐκ τῆς κατὰ τὰς ἐνεργείας διαφορᾶς· οὔτε γὰρ τοῖς ὀφθαλμοῖς ὡσαύτως ἅπαντες οἱ ὑγιαίνοντες ὁρῶσιν, ἀλλ' οἱ μὲν μᾶλλον, οἱ δ' ἧτ- τον, οὔτε τοῖς ὠσὶν ὁμοίως ἀκούουσιν, ἀλλὰ κἀνταῦθα 15K πάμπολυ τὸ μᾶλλόν τε καὶ ἧττον, οὐ μὴν οὐδὲ τοῖς σκέλεσιν ὡσαύτως θέουσιν οὐδὲ ἀντιλαμβάνονται ταῖς χερσὶν οὐδὲ τοῖς ἄλλοις ἅπασιν ὀργάνοις ὡσαύ- τως ἐνεργοῦσιν, ἀλλ' ὁ μέν τις βέλτιον, ὁ δὲ χεῖρον. εἴπερ οὖν αἱ διαφοραὶ τῶν ἐνεργειῶν ταῖς τῶν κρά- σεων διαφοραῖς ἀκολουθοῦσιν, ἀνάγκη τοσαύτας εἶ- ναι τὰς τῶν κράσεων διαφοράς, ὅσαιπέρ εἰσι καὶ αἱ

and perfect; on the other hand, it is lacking this perfection, although not yet to such a degree that the organism is distressed. And even here the inquiry of the art is more   14K
theoretical than practical, since there are some who do not agree that one person is more healthy than another, or that there is a significant range in the condition of the body, which we call health, claiming instead that health is one exact thing and is not divisible into more or less. But to me, just as a white body seems to be less or more white, so too does health seem to be less or more in just such a way.

The demonstration of the argument is twofold. One component is from the change relating to the time of life. From the time the animal is born, it is of necessity always changing in its *krasis,* as I have shown before, so, if health lies in the quality of the *krasis,* and this quality does not remain the same, it is not possible for health to be kept the same. The second component of the demonstration arises from the difference in the functions, for those who are healthy do not all see in the same way with their eyes— some see more, others see less—nor do they hear equally with their ears, but here too there is more and less to a   15K
significant degree. Nor do they run in like manner with their legs, or grasp in the same manner with their hands, nor function similarly to all the other organs, but one is better while another is worse. Therefore, if the differences of the functions are consequent upon differences of the *krasias,* there are of necessity as many differences of the *krasias* as there are differences of the functions. If, how-

τῶν ἐνεργειῶν. εἰ δὲ μὴ κράσεων ἐθέλοι τις λέγειν,
ἀλλὰ κατασκευῶν, ἵν᾽ ἐπὶ πάσαις ταῖς αἱρέσεσιν ὁ
λόγος ἐκτείνοιτο, συμπεραίνοιτ᾽ ἂν ὡσαύτως. συμμετρία γὰρ δή τις ἡ ὑγεία κατὰ πάσας ἐστὶ τὰς αἱρέσεις, ἀλλὰ καθ᾽ ἡμᾶς μὲν ὑγροῦ καὶ ξηροῦ καὶ θερμοῦ καὶ ψυχροῦ, κατ᾽ ἄλλους δὲ ὄγκων καὶ πόρων,
κατ᾽ ἄλλους δὲ ἀτόμων ἢ ἀνάρμων ἢ ἀμερῶν ἢ ὁμοιομερῶν ἢ ἀνομοιομερῶν[5] ἢ ὅτου δὴ τῶν πρώτων στοιχείων, ἀλλὰ κατὰ πάντας γε διὰ τὴν συμμετρίαν
αὐτῶν ἐνεργοῦμεν τοῖς μορίοις. εἴπερ οὖν διαφόρως
ἐνεργοῦμεν, διάφορός ἐστι καὶ ἡ καθ᾽ ἕκαστον συμμετρία τῶν στοιχείων, ἥπερ ἦν ἡ ὑγεία. καὶ μὴν καὶ
16K   χωρὶς τοῦ τῶν στοιχείων μνημονεύειν ὧδ᾽ ἂν ὁ λόγος
ἐρωτηθείη.

εἴπερ ταῖς κατασκευαῖς τῶν μορίων ἀκολουθοῦσιν
αἱ ἐνέργειαι, ὅσαιπερ ἂν ὦσιν ἐν ταῖς ἐνεργείαις αἱ
διαφοραί, τοσαῦται κἂν ταῖς κατασκευαῖς ἔσονται·
ἀλλὰ μὴν ἀκολουθοῦσιν ταῖς κατασκευαῖς αἱ ἐνέρ-
γειαι· ἀναγκαῖον ἄρα τοσαύτας εἶναι τῶν κατασκευῶν
τὰς διαφοράς, ὅσαιπερ καὶ αἱ τῶν ἐνεργειῶν. εἰσὶ δὲ
αἱ τῶν ἐνεργειῶν πάμπολλαι· τινες ἄρα καὶ τῶν κατασκευῶν εἰσιν. εἴπερ οὖν ἐν ἅπασι τοῖς ὑγιαίνουσιν αἱ
κατασκευαὶ τῶν μορίων ὑπάρχουσι σύμμετροι, διάφοροι δέ εἰσιν αἱ κατασκευαί, διότι καὶ αἱ ἐνέργειαι
διάφοροι, πάμπολλαί τινες ἄρα συμμετρίαι τῶν κατασκευῶν ἔσονται, ὥστε καὶ ὑγεῖαι πάμπολλαι. καὶ μὴν

[5] ἢ ἀνομοιομερῶν add. Ko

ever, someone does not wish to say "of *krasias*" but "of constitutions," so that the argument extends to all the sects,[10] it could be applied in like fashion. For certainly health is a balance, according to all the sects; according to us it is a balance of moist and dry, hot and cold, whereas according to others it is of corpuscles and pores (*onkoi* and *poroi*), and to others again, of atoms, *anarmoi, amereis, homoiomeres* or *anhomoiomeres* (like or unlike parts)[11] or whatever else of the primary elements. But in all cases, we function through a balance of these things in the parts. If, then, we function differently, there is also a difference in the balance in relation to each of the elements, which is what health is. Furthermore, the argument could be pursued quite apart from any mention of elements.

16K

If the functions were to follow the constitutions of the parts, there will be as many differences in the functions as there are in the constitutions. But the functions follow on from the constitutions. Therefore, of necessity, there are going to be as many differences of constitutions as there are of functions. There are many differences of functions, therefore there are many differences of constitutions. But if the constitutions are different because the functions are different, there will be many balanced constitutions, so there will also be many healthy states. And further, if the

[10] The main medical sects or schools in Galen's time were the Dogmatics (Rationalists), Empirics, Methodics, and Pneumaticists. Galen gives a full account of these sects in *Sect.*, I.64–105K (French trans., Daremberg, *Oeuvres anatomiques;* English trans., Walzer and Frede, *Three Treatises*).

[11] For a summary of these several terms see the General Introduction, xliv–xlvi.

εἰ διαφέρουσιν ἀλλήλων αἱ κατὰ μέρος ὑγεῖαι, ἤτοι
κατὰ τὸ κοινὸν ἐν ἁπάσαις εἶδος, ἀφ' οὗπερ ὑγεῖαι
λέγονται, διοίσουσιν ἢ κατὰ τὸ μᾶλλόν τε καὶ ἧττον
ἀλλήλων διαφέρουσιν· ἀλλὰ μὴν οὐ κατὰ τὸ κοινὸν
εἶδος· ἀδιάφοροι γάρ εἰσιν αἱ ὑγεῖαι· κατὰ τὸ μᾶλλον
ἄρα καὶ ἧττον ἀλλήλων διαφέρουσιν. ὥσπερ γὰρ ἡ
17K ἐν τῇ χιόνι λευκότης τῆς ἐν τῷ γάλακτι λευκότητος,
ᾗ μὲν λευκόν ἐστιν, οὐ διαφέρει, τῷ μᾶλλον δὲ καὶ
ἧττον διαφέρει, τὸν αὐτὸν δὴ τρόπον ἡ ἐν τῷ Ἀχιλλεῖ,
φέρε εἰπεῖν, ὑγεία τῆς ἐν τῷ Θερσίτῃ ὑγείας, καθ'
ὅσον μὲν ὑγεία, ταὐτόν ἐστιν, ἑτέρῳ δέ τινι διάφο-
ρος·[6] καὶ τοῦτο τὸ ἕτερον οὐδὲν ἄλλο ἐστὶν ἢ τὸ μᾶλλόν τε καὶ ἧττον. οὔτε γὰρ ὡς οὐ διαφερόντως ἐνεργοῦμεν ἅπαντες ἔνεστιν εἰπεῖν οὔθ' ὡς δι' ἄλλο τι τὴν
ἀνισότητα ταύτην ἔχομεν ἢ διὰ τὴν κατασκευήν, ἀφ'
ἧς ἐνεργοῦμεν. εἰ δέ τις φήσει μόνους μὲν τοὺς ἅπασι
τοῖς μορίοις ἄκρως ἐνεργοῦντας ὑγιαίνειν, ἡμᾶς δὲ
τοὺς ἄλλους, ὅσοι χεῖρον ἐκείνων ἔχομεν, οὐχ ὑγιαίνειν, ἴστω συμπάσης οὗτος τῆς ὑγιεινῆς πραγματείας
ἀνατρέπων τὴν ὑπόθεσιν. εἰ γὰρ δὴ τὸ φυλάττειν ἣν
παρελάβομεν ὑγείαν ὁ σκοπός ἐστιν αὐτῆς, οὐδεὶς δὲ
ἡμῶν ὑγιαίνει, πρόδηλον, ὡς ἐπ' οὐδενὸς ἐνεργοῦσαν
ἕξομεν ἣν νῦν συστῆσαι βουλόμεθα τέχνην ὑγιεινήν·
οὔκουν οὐδὲ ζητητέον αὐτήν, ἀλλὰ σιωπητέον τε καὶ
18K καταπαυστέον ἤδη τὸν λόγον.

[6] διαφέρει Ku

24

healthy states differ from each other individually, they will differ in the form common to all, from which they are termed healthy states, or they will differ from each other in terms of more and less. But they do not differ in a common form because the healthy states are not different. Therefore, they differ from each other in terms of more and less, for just as the whiteness of snow doesn't differ from the whiteness of milk, in that it is white [in both instances], it does differ in terms of more or less. In the same way, let me say, the health in Achilles doesn't differ from the health in Thersites.[12] In being health, it is the same; it differs in something else. And this is no other than in terms of more and less. For it is not possible to say we do not all function differently, or that we have this inequality from anything other than the constitution from which we function. If, however, someone were to say that only those who function perfectly in all their parts are healthy, and that we others who function less well than those people are not healthy, he should realize he is overthrowing the hypothesis of the whole study of hygiene. For if, in truth, the objective of hygiene is for us to take it upon ourselves to preserve health, and yet none of us are healthy, it is quite obvious that the art of hygiene that we now wish to establish will have no sphere of action, and therefore we must not seek it, but must be silent and end the discussion forthwith.

17K

18K

[12] Achilles, son of Peleus and Thetis, as portrayed in Homer's *Iliad,* is the archetype of the mighty warrior. Thersites, by Homer's account, was the ugliest man in Troy—lame, bowlegged, round-shouldered, and bald. He was killed by Achilles because of his supposed love for Penthelea.

ἁπάσας οὖν ἐκκόπτει τὰς τοιαύτας ἀπορίας ἡ τοῦ
ἀληθοῦς γνῶσις· οὐ γὰρ ἡ τελεία μόνον ἥτις ἐστὶν
ἄτμητος ὑγεία λέγεταί τε καὶ ἔστιν, ἀλλὰ καὶ ἡ τῆσδε
μὲν ἀποδέουσα, μηδέπω δὲ τῆς χρείας ἐκπεπτωκυῖα.
χρήζομεν γὰρ ἅπαντες ἄνθρωποι τῆς ὑγείας εἴς τε
τὰς κατὰ τὸν βίον ἐνεργείας, ἃς ἐμποδίζουσί τε ἢ
διακόπτουσι καὶ καταπαύουσιν αἱ νόσοι, καὶ προσέτι
τῆς ἀνοχλησίας ἕνεκεν· ὀχλούμεθα γὰρ ἐν ταῖς ὀδύ-
ναις οὐ σμικρά. τὴν δὲ τοιαύτην κατάστασιν, ἐν ᾗ
μήτε ὀδυνώμεθα μήτε ἐν ταῖς κατὰ τὸν βίον ἐνεργείαις ἐμποδιζόμεθα, καλοῦμεν ὑγίαν, ἣν εἴ τις ἑτέρῳ
προσαγορεύειν ὀνόματι βούλεται, πλέον οὐδὲν ἐκ τούτων σχήσει, καθάπερ οὖν οὐδ᾽ οἱ τὴν 'ἀειπάθειαν'
εἰσάγοντες. εἰ μὲν γὰρ διὰ τοῦτ᾽ εἰσῆγον αὐτήν, ὅτι
πᾶν σῶμα γεννητόν, ὥσπερ τὰς τῆς γενέσεως αἰτίας,
οὕτω καὶ τὰς τῆς φθορᾶς ἔχει συμφύτους ἐξ ἀρχῆς,
ὡς ἡμεῖς ἐπεδείξαμεν ἔμπροσθεν, ἐπηνοῦμεν ἂν αὐτούς, ὡς ἀληθῆ τε ἅμα καὶ παλαιὰ πρεσβεύοντας
δόγματα. ἐπειδὴ δὲ ὁμοειδῆ τὴν τῶν ὑγιαινόντων σω-
19K μάτων κατάστασιν εἶναι βούλονται τοῖς τῶν νοσούν-
των, οὐκέτ᾽ ἐπαινοῦμεν οὐδὲ ἀποδεχόμεθα τὸ δόγμα·
βέλτιον γὰρ ἦν μακρῷ πλάτος ὑποθέσθαι συχνὸν
ἔχειν τὴν ὑγείαν ἤπερ ἅπαντας ἡμᾶς ἀπαύστῳ νο-
σήματι συνέχεσθαι.

καὶ γὰρ εἰ τὰ σπέρματα[7] τῶν νόσων ἐνυπάρχειν
ἡμῖν φασιν, ἀλλά τοι συγχωροῦσί γε καὶ αὐτοὶ διὰ

[7] post σπέρματα add. πασῶν Ku

26

The knowledge of what is true eradicates all such difficulties, for not only is what is complete or indivisible termed health and is so, but so too is what falls short of this but has not yet become incompatible with use. For all men need health for the functions pertaining to life; it is these that diseases impede, or cut through and stop, and over and above this, they need health for the sake of freedom from disturbance. For we are disturbed to no small extent by pains. We call health that state in which we neither feel pain nor are impeded in the functions pertaining to life. If someone should wish to call it by another name, he will accomplish nothing more by doing this, just as those who introduce the term "perpetual affection"[13] do not. For if they introduce it for this reason—that every created body, just as it has the causes of genesis, in the same way also has the causes of destruction innately from the beginning, as I showed previously—we would commend them as giving first importance to true and ancient doctrines. However, when they wish the state of healthy 19K bodies to be of the same kind as that of diseased bodies, we no longer praise them or accept their doctrine. It would be better by far to assume that a latitude exists in health than for all of us to be continuously beset by unending disease.

People say the seeds of diseases are present in us. But they themselves also in fact agree that these seeds escape

[13] On the term ἀειπάθεια, see Galen's *Ars M.* (I.317K). LSJ, which refers to this usage, has "perpetual passivity," which is clearly not the precise meaning here—see Johnston, *On the Constitution*, 175, where Boudon's note on the term is given.

τὴν σμικρότητα τὴν αἴσθησιν ἡμῶν ἐκφεύγειν αὐτά.
ἔστω τοίνυν, εἰ βούλονται, καὶ ὀδυνηρά τις ἐν ἡμῖν
διάθεσις, ἀλλ' οὕτω σμικρὰ καὶ ἀναίσθητος, ὡς μὴ
λυπεῖν τοὺς ἔχοντας. ἔστωσαν, εἰ βούλονται καὶ πυρε-
τοί, ἀλλ' οὕτω σμικροί, ὡς μήτ' αἴσθησιν ἀπ' αὐτῶν
ἡμῖν γίνεσθαι μηδεμίαν ἐξεῖναί τε καὶ πολιτεύεσθαι
καὶ λούεσθαι καὶ πίνειν ἐσθίειν τε καὶ τἆλλα πράττειν, ὧν δεόμεθα. τὸ γὰρ τῆς χρείας ἀπαρεμπόδιστον
ὁρίζει μᾶλλον τὴν ὑγείαν. οὐδὲ γὰρ ἡ τῶν ἐνεργειῶν
ἀσθένεια νόσου γνώρισμά ἐστιν, οὕτως ἁπλῶς εἰπού-
σιν, ἀλλὰ ἡ παρὰ τὴν ἑκάστου φύσιν. ὡς ἅπαντές γε
κακῶς ὁρῶμεν, εἰ τοῖς ἀετοῖς τε καὶ Λυγκεῖ παραβαλ-
20K λοίμεθα, καὶ δὴ καὶ ἀκούομεν οὐκ ὀρθῶς, εἰ Μελάμ-
ποδι, καὶ τοῖς ποσὶν ἀρρωστοῦμεν, εἴ τις ἡμᾶς Ἰφίκλῳ
παραβάλλοι, καὶ ταῖς χερσίν, εἰ Μίλωνι, καὶ καθ'
ἕκαστον δὴ μόριον ἐγγὺς ἂν ἥκειν νομισθείημεν πη-
ρώσεως, εἰ τοῖς πρωτεύσασι κατά τι παραβαλλοί-
μεθα.

τίς γοῦν ἡμῶν φαύλως ἔχειν οἴεται τῶν ὀφθαλμῶν,
εἰ μὴ βλέποι τοὺς ἀπὸ δυοῖν σταδίων μύρμηκας; ἢ τίς
τῶν ὤτων, εἰ μὴ κατακούοι τῶν ἀφ' ἑξήκοντα σταδίων; ἀλλ' εἰ ταῦτα τὰ γράμματα, τὰ κατὰ τουτὶ τὸ
βιβλίον ἐγγεγραμμένα, μὴ βλέποι τις ὀρθῶς, εὐλόγως οὗτος ἂν ἤδη μέμψαιτο τὰς ὄψεις· οὐ μὴν οὐδ' εἰ

---

14 Lynceus: son of Aphareus and Arene, brother of Idas, was
one of the Argonauts and famous for his keen sight; Melampus is
presumably the mythical seer and ancestor of the Melampodids;

our perception of them due to their small size. Let it be so, then, if they wish, that there is some distressing condition in us, but one so small and imperceptible as not to disturb those who have it. Let there also be, if they wish, fevers, but ones so slight that there is not any perception of them in us, so we can go out and conduct our business, bathe, drink, eat and do the other things we need to do. It is the absence of interference with use that especially defines health. Nor is weakness of the functions a sign of disease, strictly speaking, but [only] what is contrary to the nature of each function. In fact, we all see badly if we compare ourselves to eagles and to Lynceus, and we do not hear properly if we compare ourselves to Melampus.   20K Furthermore, we are weak in the feet when compared to Iphicles, and in the hands when compared to Milo.[14] Indeed, in each part we would be deemed to come close to being disabled, if we were to compare ourselves to those who are preeminent in respect of that part.

    Anyway, which of us would consider himself defective in the eyes, if he couldn't see ants from two *stadia,* or in the ears, if he couldn't hear something from sixty *stadia?* But if someone could not properly see the letters I have written in this book, he would now reasonably blame his eyes, although he would not be right to blame them, if he

Iphicles was the mythical twin brother of Heracles and is said to have been involved in some of the latter's exploits; Milo of Croton was an athlete and wrestler famed for his strength, particularly in his hands. Galen gives a disparaging (and somewhat amusing) account of some of his most famous feats—see his *Protr.,* I.34–35K (French trans., Boudon, *Galien;* English trans., Singer, *Galen: Selected Works*).

ταῦτα τέσσαρας ἀποστήσας πήχεις μὴ βλέποι, δικαίως ἂν μέμφοιτο, πλὴν εἰ τῶν οὕτω τις ὀξυωπε-
στάτων εἴη τὴν φύσιν, ὡς καὶ ταῦτα ἐμβλέπειν. οὕτω
γάρ, οἶμαι, καὶ μέμψεται καὶ δικαίως φήσει, ὥσπερ
ἅπαντες ἄνθρωποι λέγουσιν, ὡς τόδε τι κατὰ τὸν ἔμπροσθεν χρόνον ἐνεργῶν εἶτα νῦν οὐκ ἐνεργεῖ. τὸν
μὲν γὰρ τοιοῦτον ἐν νόσῳ τινὶ φήσομεν ὑπάρχειν,
εἴπερ μὴ διὰ γῆρας ταῦτα πάσχοι· καίτοι καὶ τοῦτο
νόσον εἶναι λέγουσιν ἔνιοι. τοὺς δ᾽ ἄλλους ἅπαντας,
21K οἷς φύσει μήτ᾽ ὀξὺ βλέπειν μήτ᾽ ἀκούειν ὑπάρχει
μήτε θέειν ὠκέως ἤ τι τοιοῦτον ἕτερον ἐνεργεῖν ἰσχυρῶς, οὔτε νοσεῖν οὔθ᾽ ὅλως παρὰ φύσιν ἔχειν ὑπολη-
ψόμεθα. πᾶσαι μὲν γὰρ αἱ νόσοι παρὰ φύσιν, οὐκ
ἔχουσι δ᾽ οἱ τοιοῦτοι παρὰ φύσιν, ὥσπερ οὐδ᾽ οἱ γέ-
ροντες. οὔκουν ἁπλῶς γε τῶν ἐνεργειῶν εὐρωστίᾳ τε
καὶ ἀρρωστίᾳ κριτέον ἐστὶ τοὺς ὑγιαίνοντάς τε καὶ
νοσοῦντας, ἀλλὰ τὸ κατὰ φύσιν μὲν τοῖς ὑγιαίνουσι,
τὸ παρὰ φύσιν δὲ τοῖς νοσοῦσι προσθετέον, ὡς εἶναι
τὴν μὲν ὑγείαν διάθεσιν κατὰ φύσιν ἐνεργείας ποιη-
τικήν, τὴν δὲ νόσον διάθεσιν παρὰ φύσιν ἐνεργείας
βλαπτικήν.

οὔτε γὰρ ἡ κατὰ φύσιν διάθεσις ἤδη καὶ ὑγεία·
διάθεσις γάρ τίς ἐστι κατὰ φύσιν ἥ τε τῶν Αἰγυπτίων
μελανότης ἥ τε τῶν Κελτῶν λευκότης ἥ τε τῶν Σκυ-
θῶν πυρρότης· ἀλλ᾽ οὐδὲν τῶν τοιούτων ὑγείας δηλωτικόν,[8] διότι μηδ᾽ ἐν χρώμασιν ὅλως ἡ ὑγεία· οὔτ᾽ εἰ
παρὰ φύσιν, ἤδη καὶ νόσος· ὡς εἴη γ᾽ ἂν οὕτω νόσος
ἥ τ᾽ ἐξ ἡλίου μελανότης ἥ τ᾽ ἐκ μακρᾶς σκιατροφίας

could not see the letters from four *cubits* away, unless he was one of those with very sharp vision by nature, who could also see them. For in this case, I think he will both blame them and will rightly say what all men say—that at a previous time he could do this, but cannot do it now. We shall say in response that such a person is in the grip of some disease, unless he is affected in this way through old age. And indeed, some say old age is a disease too. However, we shall not assume all others in whom vision is not sharp or hearing is not acute by nature, or who cannot run   21K
swiftly, or function strongly in some other such thing, are diseased or altogether contrary to nature. All diseases are contrary to nature, but such people are not contrary to nature, just as the aged are not. One must not, therefore, determine those who are healthy and those who are diseased simply on the basis of the strength or weakness of the functions; one must apply the term "in accord with nature" to those who are healthy and the term "contrary to nature" to those who are diseased, as health is a condition in accord with nature productive of functions and disease a condition contrary to nature injurious to functions.

A condition in accord with nature is not thereby also health, for the swarthiness of Egyptians, the paleness of Celts and the ruddiness of Scythians are conditions in accord with nature but none of these conditions is indicative of health because health is not in colors at all. Also, a condition contrary to nature is not already a disease, for in this way darkness from the sun or pallor from a long time in the shade would be diseases. Rather, it is necessary

---

8 ὑγείας δηλωτικόν Ko; ὑγίεια Ku

22K λευκότης· ἀλλὰ προσθεῖναι χρὴ τῇ μὲν τῆς ὑγείας
ἐννοίᾳ τὸ λόγον αἰτίας ἔχειν αὐτὴν πρὸς τὴν ἐνέργειαν, τῇ δὲ τῆς νόσου τὸ καὶ ταύτην τὴν ἐνέργειαν
βλάπτειν. ἀλλὰ περὶ μὲν τούτων ἐν ἑτέροις εἴρηται
διὰ πλειόνων, εἰς δὲ τὰ παρόντα τοσοῦτον ἀποχρήσει
μόνον ἐξ αὐτῶν εἰλῆφθαι, τὸ πλάτος ἱκανὸν εἶναι τῆς
ὑγείας, καὶ μὴ πᾶσιν ἡμῖν ὑπάρχειν ἴσον ἀκριβῶς.

εἰ δέ τῳ δοκεῖ βίαιον εἶναι καὶ τὴν μὴ παντάπασιν
ἠκριβωμένην εὐκρασίαν ὅμως ἔτι καὶ αὐτὴν ὀνομάζεσθαι, οὗτος ἀναμνησθήτω τῶν κατὰ τὸν βίον ἁπάντων ὀνομάτων. εὔκρατον οὖν τι καὶ πόμα φαμὲν εἶναι
καὶ βαλανεῖον, οὐ μόνον ὅτι τὸ μὲν ἄλλῳ, τὸ δὲ ἄλλῳ
τοιοῦτόν ἐστιν, ἀλλὰ ὅτι καὶ πρὸς τὸν αὐτὸν ἄνθρωπον ἐν πλάτει τοιοῦτον ὑπάρχει· ἀποστραφέντος γοῦν
τοῦ πίνοντος, ἐμβαλὼν εἰς τὸ ποτήριον ἤτοι θερμὸν ἢ
ψυχρὸν βραχὺ λάθοις ἄν.[9] καίτοι γ᾽, εἴπερ ἦν οὕτως
ἀπηκριβωμένον τὸ εὔκρατον, ὡς ἓν εἶναι καὶ ἄτμητον,
οὐκ ἂν ἐπιβαλόντος θερμὸν ἢ ψυχρὸν ἔτι εὔκρατον
23K ἐφαίνετο. κατὰ δὲ τὸν αὐτὸν τρόπον, οὐδ᾽ εἰ βραχύ τις
εἰς τὴν εὔκρατον κολυμβήθραν ἐμβάλλοι ψυχροῦ,
διαφθερεῖ παραχρῆμα τὴν εὐκρασίαν αὐτῆς. οὕτω δὲ
καὶ τὸ περιέχον εὔκρατον εἶναί φαμεν, εἰ καὶ βραχείας ἐφ᾽ ἑκάτερα τροπὰς λαμβάνει. καὶ τί θαυμα-
στόν, εἰ τὴν εὐκρασίαν εἰς ἱκανὸν ἐκτείνουσι πλάτος

[9] post ἄν: εὔκρατον ποιήσας Ku

to add to the concept of health that this has a causative   22K
relation to function, and to the concept of disease, that this
is damaging to function. But I have spoken about these
matters at length in other works.[15] It will be sufficient for
our present purposes to take this much only from these—
that the range of health is very wide and is not exactly the
same for all of us.

If it seems to someone to be doing violence to the term
to call health what is not perfectly *eukratic*[16] in all respects, let him call to mind all the terms pertaining to life.
Thus we say a drink is *eukratic* and a bath, and not only
that it is such to one person and such to another, but that
also it is so in range to the same person. At any rate, if the
one drinking were to turn aside, you could add a little hot
or cold to the drink unseen, and it would still be *eukratic*.[17]
However, if it were exactly *eukratic* in such a way as to be
one and indivisible, and you were to add hot or cold, it
would obviously not still be *eukratic*. In the same way, if   23K
someone were to add a little cold to a *eukratic* pool, it
would not make an immediate difference to its *eukrasia*.
So too, we say the ambient air is *eukratic*, even if it takes
a small turn in either direction. And what would be surprising about everyone extending *eukrasia* to a significant

[15] See particularly Galen's *MM*, Books 1 and 2, and the opening section of his *Morb. Diff.*   [16] The two terms, *eukrasia*
and *dyskrasia*, are retained in the Greek form as fundamental
technical terms in Galen's theory on health and disease. In simple
terms, the former means a good balance of the four elemental
substances, qualities, or humors, and the latter some disturbance
of this balance—see the General Introduction, xlvi–xlviii.

[17] The translation here follows the Kühn text.

33

ἅπαντες, ὅπου γε καὶ τὴν ἐν ταῖς λύραις εὐαρμοστίαν
τὴν μὲν ἀκριβεστάτην δήπου μίαν καὶ ἄτμητον ὑπάρχειν εἰκός, τὴν μέντοι γ᾽ εἰς χρείαν ἰοῦσαν πλάτος
ἔχειν; πολλάκις γὰρ ἡρμόσθαι δοκοῦσαν ἄριστα λύ-
ραν ἕτερος μουσικὸς ἀκριβέστερον ἐφηρμόσατο. παν-
ταχοῦ γὰρ ἡ αἴσθησις ἡμῖν ἐστι κριτήριον ὡς πρὸς
τὰς ἐν τῷ βίῳ χρείας· ὥστε καὶ τὴν εὐκρασίαν δήπου
καὶ τὴν δυσκρασίαν αἰσθήσει κρινοῦμεν. ὡσαύτως δὲ
καὶ τὴν τῆς ἐνεργείας βλάβην ἑκάστου τῶν βεβλαμ-
μένων παρὰ τὸ κατὰ φύσιν, ὅταν εἰς αἰσθητὸν ἥκῃ
μέγεθος, ἤδη νόσον ἡμῖν εἶναι νομιστέον, οὐδὲν ὡς
πρὸς τὰ παρόντα διαφέροντος οὐδ᾽ ἐνταῦθα, πότερον
αὐτὰς ταύτας τὰς βεβλαμμένας ἐνεργείας τὸ νόσημα
24K εἶναι λέγει τις ἢ τὰς διαθέσεις, ὑφ᾽ ὧν βλάπτονται,
ὥσπερ οὐδ᾽ εἰ διαθέσεις τις ἢ κατασκευὰς ὀνομάζειν
ἐθέλοι.

διῄρηται γὰρ ἡμῖν ἑτέρωθι καὶ περὶ τῶνδε καὶ
δέδεικται κατὰ τὰς διαθέσεις τε καὶ κατασκευὰς τοῦ
σώματος ἥ θ᾽ ὑγεία καὶ ἡ νόσος, οὐ κατὰ τὰς ἐνεργείας τε καὶ βλάβας αὐτῶν συνιστάμεναι. ἀλλὰ πρός
γε τὸ φυλάττειν ὑγείαν ἢ ἰᾶσθαι τὰς νόσους οὐδὲν ἐκ
τῆς τούτων ἀκριβείας ὀνινάμεθα. μόνον γὰρ ἀρκεῖ
γινώσκειν, ὡς ἡ μὲν κατασκευὴ τοῦ σώματος, αἰτίας
λόγον ἔχουσα ὡς πρὸς τὴν ἐνέργειαν, ὁ σκοπὸς τῆς
ὑγιεινῆς τε καὶ θεραπευτικῆς ἐστι τέχνης· ταύτην γὰρ
ἡμῖν φυλάττειν μὲν ὑπάρχουσαν πρόκειται, δημιουρ-
γεῖν δὲ ἀπολλυμένην· αἱ δὲ ἐνέργειαι κατ᾽ ἀνάγκην
ἕπονται, ταῖς μὲν χρησταῖς κατασκευαῖς ἄμεμπτοι,

range where even in harps there is, I suppose, the possibility of an harmoniousness that is absolutely exactly one and indivisible, and yet, when it comes to using them, they have a range? In fact, often when a harp seems to be well-tuned, another musician tunes it more accurately. Always our perception is a criterion when it comes to uses in life. As a result, we shall judge both *eukrasia* and *dyskrasia* by perception. Similarly, we must regard as already a disease the damage of function of each of those things damaged in terms of an accord with nature when it reaches a perceptible size—it would make no difference here for our present considerations whether we say the disease is the damaged functions themselves or the conditions by which they are damaged, just as it does not if someone should   24K
wish to term these conditions or constitutions.

I have defined these things for you elsewhere[18] and have shown that both health and disease relate to the conditions and constitutions of the body and do not exist in the functions and their injuries. But in respect of preserving health or treating diseases, we derive no benefit from the exactness of these things. It is alone sufficient to recognize that the constitution of the body, which has the ground of cause in relation to function, is the objective of both the hygienic and therapeutic arts. For the task before us is to preserve health when it is present and to create it when it is being destroyed. The functions follow of necessity; those that are faultless follow good constitutions while

[18] The key works referred to here are the four treatises on diseases and symptoms (Johnston, *Galen: On Diseases and Symptoms*), *MM* (Johnston and Horsley, *Galen: Method of Medicine*), *Opt. Const.* (Singer, *Galen: Selected Works*), and *Mixt.* (Singer, *Galen: Selected Works*).

ταῖς δ' οὐ τοιαύταις μοχθηραί. ὥστ' ἐπειδήπερ ὃ φυ-
λάττομεν καὶ δημιουργοῦμεν αὐτοὶ διάθεσίς τίς ἐστι
καὶ κατασκευὴ τοῦ σώματος, ἕπεται δ' ἐξ ἀνάγκης
αὐτῇ ἡ τῶν ἐνεργειῶν,[10] οὐδὲν ἔτι χρὴ πρὸς τὰ παρ-
25K όντα, πότερον ἐν τῷ τῶν ἐνεργειῶν ἢ τῶν κατασκευῶν
γένει θετέον ἐστὶν ὑγείαν τε καὶ νόσον, ἐπισκοπεῖσθαι.

λαβόντες δ' ἐξ ὑποθέσεως δέον εἶναι φυλάττειν τῆς
κατὰ φύσιν ἡμῶν κατασκευῆς τοῦ σώματος ἐκεῖνα, δι'
ὧν ἐνεργοῦμεν, ἀναμνήσαντές τε πάλιν, ὡς ἡ τῶν
ὁμοιομερῶν εὐκρασία τε καὶ τῶν ὀργανικῶν διάπλα-
σις καὶ θέσις ἀριθμός τε καὶ πηλικότης τῶν ἐνεργειῶν ἐστιν αἴτια καὶ ὡς ἐν πλάτει πάντα ταῦτ' ἐστὶ
καὶ ὡς καθ' ἕκαστον ἄνθρωπον ἴδια, τῶν ἐφεξῆς ἐχώ-
μεθα. μεγάλην δ' εἰς αὐτὰ χρείαν ἡ γνῶσις τοῦ πλά-
τους αὐτῶν παρέχεται. τῆς γὰρ εὐκρασίας διττῆς
οὔσης, τῆς μὲν ὡς πρὸς ἐπίνοιαν μᾶλλον ἢ μόνιμον
ὕπαρξιν, ὡς ἐν ζῴου σώματι, τῆς δὲ ὑπαρχούσης ἅμα
καὶ φαινομένης ἐν ἅπασιν τοῖς ὑγιαίνουσιν, αὐτὴν
πάλιν ἡμᾶς χρὴ τέμνειν τὴν φαινομένην· εὑρεθήσεται
γὰρ οὐ σμικρά τις ἐν αὐτῇ διαφορά. μάθοις δ' ἂν ὡς
ἔστιν ἀληθὲς τὸ λεγόμενον ἐκ τῶν ἡλικιῶν μάλιστα.

τῆς γὰρ τῶν μειρακίων ἡλικίας ἀρίστης οὔσης
πρὸς τὰς καθ' ὁρμὴν ἐνεργείας, ἡ τῶν βρεφῶν δι'
26K ὑγρότητα χείρων ἐστίν, ἡ δὲ τῶν γερόντων διὰ ξηρότητά τε καὶ ψύξιν. ἐν μέντοι ταῖς ἄλλαις ἐνεργείαις
ταῖς φυσικαῖς ὀνομαζομέναις, οἷον αὐξήσεσί τε καὶ
πέψεσι καὶ ἀναδόσεσι καὶ θρέψει, τὰ βρέφη τῶν

those that are disordered follow constitutions that are not good. As a result, since what we ourselves preserve or create is some condition or constitution of the body, while the perfection of the functions follows this of necessity, it is not still necessary for present purposes to consider whether we must place health and disease in the class of functions or constitutions.   25K

Accepting, *ex hypothesi,* that what is needed is to preserve those aspects of the constitution of the body in accord with nature, through which we function, and calling to mind again that the *eukrasia* of the *homoiomeres* and the conformation, position, number and size of the organic bodies are the causes of the functions, let us come next to the fact that all these exist as a range and are specific to each person. The knowledge of the range of these things provides a great advantage in these matters. Because *eukrasia* is twofold—in one respect being more related to a concept than existing as a stable state, as in the body of the animal, and in another respect, as it exists and appears in all those who are healthy—it behooves us to subdivide the actual appearance, for we shall discover no little diversity in this. You may learn that what I have said is true especially with respect to the stages of life.

Thus, the age of youths is the best with regard to the voluntary functions, while that of infants is inferior due to   26K
moistness and that of the aged due to dryness and cold. However, in the other functions termed physical, such as growth, concoction (digestion), distribution and nutrition,

---

[10] *post* ἐνεργειῶν: τελειότης Ku

κατὰ τὰς ἄλλας ἡλικίας ἁπάσας ἀμείνω. ἀλλ᾽ ὅμως
οὐδὲν κωλύονται πάντες οἱ κατὰ πάσας τὰς ἡλικίας
ὑγιαίνειν. ὡς οὖν ἐπὶ τῶν ἡλικιῶν ἔχει, κατὰ τὸν
αὐτὸν τρόπον εὑρήσεις καὶ ἐπὶ τῶν φύσεων αὐτῶν
ἀμήχανον οὖσαν τὴν διαφορὰν ἐν ταῖς κράσεσιν·
ὥστε, εἰ οὕτως ἔτυχε, δυοῖν παίδοιν τὴν αὐτὴν ἡλικίαν
ἀγόντων τὸν μὲν ὑγρότερον εἶναι θατέρου πολλῷ, τὸν
δὲ ξηρότερον, ὡσαύτως δὲ τὸν μὲν θερμότερον, τὸν δὲ
ψυχρότερον. ἐν ὅσοις δὲ σώμασιν ὑπάρχει τὸ πολὺ
τοῦ προσήκοντος ἤτοι θερμοτέροις ἢ ψυχροτέροις ἢ
ὑγροτέροις ἢ ξηροτέροις εἶναι, ἐν τούτοις οὐκ ἔστιν
ἄμεμπτος ἡ κρᾶσις· ἐν ὅσοις δὲ ὑπάρχει μέν τις διαφορὰ παρὰ τὸ κάλλιστα κεκραμένον, οὐ μὴν αἰσθητή
γε διὰ σμικρότητα, τοῦθ᾽ ὡς πρὸς τὴν χρείαν ἐν ἴσῃ
χώρᾳ τιθέμεθα τοῖς ἀρίστοις· ὥστ᾽ εἶναι τῆς κατὰ
πλάτος ὑγείας τὴν μὲν εὔκρατόν τε καὶ ἄμεμπτον ὡς
27K πρὸς αἴσθησιν, τὴν δ᾽ οἷον δύσκρατόν τε καὶ μεμ-
πτήν.

ἐναργέστατα δὲ τοῦ λεγομένου γνωρίσματα παρ-
έχουσιν αἱ κατ᾽ ἰσχνότητα καὶ πολυσαρκίαν διαφοραὶ
τῶν σωμάτων· ἀναγκαῖον γάρ που τὰς ἐναντίας ἕξεις
ἐναντίαις ἕπεσθαι κράσεσιν. ὥσπερ οὖν οὐκ ἐπαινοῦμεν οὔτε τὸ ἄγαν ἰσχνὸν οὔτε τὸ λίαν παχὺ σῶμα,
κατὰ τὸν αὐτὸν τρόπον οὔτε τὰς κράσεις αὐτῶν ἐπαι-
νέσομεν, εἰ καὶ ὅτι μάλιστα βλέπομεν ἀμφοτέρους
ὑγιαίνοντας. αἱ δ᾽ ἐν τῷ μέσῳ τούτων ἕξεις τῶν εὐ-
σάρκων ὀνομαζομένων, ὥσπερ αὐταὶ σύμμετροί τε
καὶ ἄμεμπτοι τὴν ἰδέαν εἰσίν, ἕπονται συμμέτροις τε

infants are better than all the other ages. Nonetheless, there is nothing preventing those of all the ages being healthy. What applies in the case of ages, you will also find to be the same in the case of their natures—the difference in the *krasias* (temperaments) is enormous. So it may come about that there are two children of the same age, one of whom is much moister than the other, or drier, and likewise one is hotter or colder. In those bodies that are much hotter, colder, moister or drier than they should be, the *krasis* is not faultless, while in those bodies in which there is a deviation from the best mixing, but one that is too small to be perceptible, when it comes to use, we put this in the same category as the best. As a result, in terms of the range of health, there is that which is *eukratic* and without fault as far as can be detected and that which is in fact *dyskratic* and faulty.

27K

The differences of bodies in terms of leanness and plumpness provide the clearest signs of what has been said, for it is inevitable to some degree that opposite states follow opposite *krasias*. Therefore, just as we do not praise either the very thin or the very fat body, in the same way we shall not praise their *krasias*, even if we see both are particularly healthy. The states in between these—states of those called well-fleshed—just as they are in due proportion and faultless in kind, follow moderate and faultless

καὶ ἀμέμπτοις κράσεσι. τὰς τοίνυν τοιαύτας φύσεις
ὡς μὲν πρὸς τὴν ἀκριβεστάτην ἀλήθειαν οὐδ' αὐτὰς
εὐκράτους ἄν τις εἴποι κατά γε τὴν ἁπλῆν καὶ τελέαν
εὐκρασίαν, ὡς μέντοι πρὸς τὴν αἴσθησίν τε καὶ πρὸς
τὴν χρείαν ἀμέμπτους τε καὶ ἀρίστας χρὴ τίθεσθαι.
τεκμήριον δ' ἐναργέστατον, ὅτι μηδ' αὐταὶ τὴν ἀκρι-
28K βεστάτην ἔχουσιν εὐκρασίαν, ἐκ τοῦ μηδέποτε μένειν
αὐτὰς ὡσαύτως ἐχούσας, ἀλλὰ πρῶτον μὲν δέχεσθαι
τὴν καθ' ἡλικίαν μεταβολήν, οὐδέποτε ἐν ταὐτῷ με-
νούσης οὐδεμιᾶς ἡλικίας, ἀλλ' ἀεὶ πρὸς τὸ ξηρότερον
ἰούσης, δευτέραν δὲ τὴν καθ' ὕπνον τε καὶ ἐγρήγορ-
σιν, ἡσυχίαν τε καὶ κίνησιν, αὐτῶν τε τῶν κινήσεων
τὰς διαφοράς, ἔτι τε πρὸς ταύταις τὴν τοῦ πεινῆν ἢ
διψῆν ἢ ἐσθίειν ἢ πίνειν ἢ ἐμπεπλῆσθαι σίτου ἢ
πόματος προσδεῖσθαι· καὶ πρὸς τούτοις ἔτι καὶ λουτρὰ καὶ θυμοὶ καὶ φροντίδες καὶ λῦπαι καὶ πάνθ' ὅσα
τοιαῦτα μονονουχὶ καθ' ἑκάστην ῥοπὴν ὑπαλλάτ-
τοντα τὴν κρᾶσιν.

οὔκουν χρὴ ζητεῖν ἐν τοσαύτῃ μεταβολῇ τὴν ἀκριβῶς ἀρίστην κρᾶσιν. εἰ γὰρ καὶ συνέπεσέ ποτε καθ'
ἡντιναοῦν ἀρίστην φύσιν, ἀλλ' οὐκ ἔμεινέ γε οὐδ' ἐν
ἀκαρεῖ, ὥστε γ' ἐμοὶ καὶ θαυμάζειν ἐπέρχεται τὴν
δόξαν τῶν ἀνδρῶν, ὅσοι τὴν ὑγείαν τε καὶ τὴν εὐκρασίαν ἁπλῆ τε καὶ μίαν εἶναι νομίζουσιν, εἰ δέ τι
παρὰ ταύτην ἐστίν, οὐχ ὑγείαν εἶναί φασιν· τήν τε
29K γὰρ ἀειπάθειαν ἐσφέροντες οὐκ αἰσθάνονται περὶ
πράγματος ἢ μηδέποτε γεγονότος ἐν ζῴου σώματι

40

krasias. And so, someone might term such natures, according to the strictest truth, not actually *eukratic,* in respect of an absolute and perfect *eukrasia,* but when it comes to perception and use, we must place them among the faultless and best. However, the clearest proof that these *krasias* do not have the most exact *eukrasia* is from   28K
their never remaining the same as they were. First, they yield to the change relating to the age, since no age ever remains the same in itself but always goes toward greater dryness. Second, there are the changes relating to sleep and wakefulness, rest and movement, and the differences of the movements themselves. And further, in addition to these, there are the changes of hunger or thirst, eating or drinking, and satiety or desire for food or drink, and as well as these also, baths, anger, anxiety and grief, and all such things as change the *krasis* merely related to each decisive influence.

It is not, therefore, necessary to seek, in such change, the best *krasis* precisely, for even if it did occur in the best nature in any way at all, it would not remain for a moment. Consequently, it falls to me to wonder at the opinion of men who think that health and *eukrasia* are without latitude and single things. They say that whatever is contrary to this is not health, introducing the idea of "perpetual affection" without realizing they are making an argument   29K
about a matter that has either never existed in the body of

ποιούμενοι τὸν λόγον ἢ μηδ᾽ ἐλάχιστον ὑπομένοντος,
εἰ καί ποτε γένοιτο.

6. Τούτοις μὲν οὖν, ἢν ὀνειρώττουσιν ὑγείαν, ἀπο-
λείπωμεν φυλάττειν, ἡμεῖς δ᾽ ἐπὶ τὰς φαινομένας ἀφι-
κώμεθα καὶ διττὴν αὐτῶν θέμενοι τὴν οὐσίαν, ἣν νῦν
δὴ πέπαυμαι λέγων, ἴδιον ἑκατέρας σκοπὸν ἀποδῶ-
μεν, ἐπὶ μὲν τῆς ἀμέμπτου τὴν ἀκριβῆ φυλακὴν ὡς
πρὸς αἴσθησιν, ἐπὶ δὲ τῆς μεμπτῆς τὴν οὐκ ἀκριβῆ.
πειρᾶσθαι γὰρ χρὴ τὰς ὑγιεινὰς δυσκρασίας ἐπανορθοῦσθαι, τὰς μὲν ξηροτέρας τοῦ δέοντος φύσεις ὑγροτέρας ἐργαζομένους, ὅσαι δὲ ὑγρότεραι, ξηραίνοντας,
οὕτω δὲ καὶ τῶν μὲν θερμοτέρων καθαιροῦντας τὴν
ὑπερβολήν, τῶν δὲ ψυχροτέρων καὶ τούτων κολάζον-
τας τὴν ἀμετρίαν. ὁποίοις δ᾽ ἄν τις ὑγιεινοῖς διαι-
τήμασιν ἐργάζοιτο ταῦτα, προϊὼν ὁ λόγος ἐπιδείξει.

πρότερον γάρ με χρὴ διελθεῖν, ὅπως ἄν τις τῆς
ἀρίστης φύσεως διαφυλάττοι τὴν ὑγείαν, ἔτι δὲ τούτου πρότερον, ἥτις ποτέ ἐστιν ἡ ἀρίστη κατασκευὴ
τοῦ σώματος, ἀναμνήσωμεν. ἔστι δὲ δήπου κατὰ μὲν
τὴν οὐσίαν αὐτὴν τοῦ πράγματος ἐξηγουμένοις ἡ
εὐκρατοτάτη τε ἅμα καὶ τὴν διάπλασιν τῶν μορίων
ἀκριβῶς ἁρμόττουσαν ταῖς ἐνεργείαις αὐτῶν ἔχουσα
καὶ πρὸς τούτοις ἔτι τόν τε ἀριθμὸν ἅπαντα καὶ τὰ
μεγέθη καὶ τὴν πρὸς ἄλληλα σύνταξιν ἁπάντων αὐτῶν χρηστὴν ταῖς ἐνεργείαις παρεχομένη.

κατὰ δὲ τὰ γνωρίσματα τὸ ἀκριβῶς εὔσαρκον
σῶμα τοιοῦτόν ἐστιν, ὃ μέσον ἔφην ὑπάρχειν ἰσχνοῦ
τε καὶ πολυσάρκου· πολύσαρκον δὲ ἢ παχὺ λέγειν οὐ

an animal or never remains for any length of time, even if it were to exist at some time.

6. Let us, then, leave it to them to maintain what they dream to be health while we come to those things that are apparent. And assuming the substance of these to be two-fold, which I have just now ceased speaking about, let us assign a specific objective to each—in the case of the fault-less, perfect preservation according to perception, and in the case of the faulty, the imperfect preservation. For we must attempt to correct the healthy *dyskrasias,* making natures that are drier than they should be, more moist, and those that are more moist than they should be, drier. In the same way too, we should reduce the excess of those who are too hot and we should correct the imbalance of those who are too cold. As the discussion proceeds, I shall show by what kind of healthy regimens someone might do these things.

First, I must go over in detail how someone might preserve the health of the best nature, but even before this, let us remind ourselves of what at any time the best constitution of the body is. For those expounding the actual substance of the matter, it is perfect *eukrasia* and, at the same time, the conformation of the parts being adapted perfectly to their functions, and in addition, provision in all cases of the number, size, and arrangement with each other of all these that is beneficial for the functions.

According to the signs, the perfectly well-fleshed body is such a thing—I said it is the midpoint between being lean and having excessive flesh. It makes no difference

30K

43

διοίσει. οὕτω δὲ καὶ τῶν γε ἄλλων ὑπερβολῶν ἀκριβῶς μέσον ἐστὶ τὸ τοιοῦτον, ὥστε μήτε λάσιον αὐτὸ
δύνασθαι εἰπεῖν τινα ἢ ψιλὸν τριχῶν, ἀλλὰ μήτε μαλακὸν μήτε σκληρὸν ἢ λευκὸν ἢ μέλαν ἢ ἄφλεβον ἢ
εὐρύφλεβον ἢ θυμικὸν ἢ ἄθυμον ἢ ὑπνῶδες ἢ ἀγρυ-
πνητικὸν ἢ ἀμβλὺ τὴν διάνοιαν ἢ πανοῦργον ἢ ἀφρο-
δισιαστικὸν ἢ τοὐναντίον. εἰ δὲ καὶ πᾶσιν εἴη τοῖς
μέρεσιν ἀκριβῶς μέσον ἁπασῶν τῶν ὑπερβολῶν,
31K   ἔστι μὲν δήπου καὶ κάλλιστον ὀφθῆναι τὸ τοιοῦτον,
ὡς ἂν σύμμετρον ὑπάρχον εἰς ἅπαντάς τε τοὺς πόνους ἐπιτήδειον. ἕξει δὲ καὶ τὰ ἄλλα πάντα γνωρίσματα τῆς καθ᾽ ἕκαστον μόριον εὐκρασίας, ἅπερ ἐν
τῷ δευτέρῳ Περὶ κράσεων εἴρηται γράμματι. πολλὰ
γὰρ τῶν σωμάτων εὔκρατα μέν, εἰ τύχοι, ταῖς κεφα-
λαῖς ἐστι, δύσκρατα δὲ τοῖς θώραξι ἢ τοῖς κατὰ γαστέρα τε καὶ τὰ γεννητικὰ μόρια. τινῶν δὲ ἐν τοῖς
κώλοις ἐστὶν ἡ δυσκρασία καὶ πολλοῖς καθ᾽ ἕν τι τῶν
σπλάγχνων ἤ τι μόριον ἕτερον ἓν ἢ πλείω, καθάπερ
οὖν καὶ περὶ πλείονα σπλάγχνα ἐνίοις ἐστὶν ἡ δυσκρασία. πολλοῖς δὲ ἐφώρασα καὶ κατά τι τῶν ὀργανικῶν μορίων δύο κράσεις ὑπαρχούσας, ὥστε, εἰ τύχοι, τὸ μὲν ἄλλο κύτος τῆς γαστρὸς ἅπαν ἑτέρας
εἶναι κράσεως, ἤτοι χρηστῆς ἢ μοχθηρᾶς, ἑτέρας δὲ
μόνον αὐτῆς τὸ στόμα. περὶ μὲν δὴ τούτων ἐν ταῖς
μοχθηραῖς κατασκευαῖς τῶν σωμάτων εἰρήσεται.

7. Περὶ δὲ τῆς ἀρίστης ἤδη λέγωμεν, ἧς ἕκαστον
μόριον ἄμεμπτον ἔχει τὴν σύμπασαν οὐσίαν. ὁ δὴ

whether we say "having excess flesh" or "fat." The same applies to the precise midpoint of the other excesses, so we are able to say someone is bald or hairy, but neither soft nor hard, neither pale nor dark, neither with collapsed nor distended veins, neither passionate nor passionless, neither sleepy nor wakeful, neither dull-witted nor clever, neither lecherous nor the opposite. If, for all the parts, there were to be a precise midpoint between the two extremes, this is clearly the best such state to be seen, as it would be in balance and serviceable for all its tasks. And it will have all the other signs of the *eukrasia* pertaining to each part—these are described in the second book of *On Mixtures*.[19] For many bodies are *eukratic* in the head, as may happen, but *dyskratic* in the chest, or in the abdomen, or in the generative parts. In some the *dyskrasia* is in the limbs, and in many people in one of the internal organs, or some other single part, or several, just as in some the *dyskrasia* is in a larger number of the internal organs. Also, in many cases, I have observed two *krasias* existing in one of the organic parts, so that, as may happen, in another hollow organ of the abdomen, there is a completely different *krasis,* either good or bad, and another one in the cardiac orifice[20] of the stomach. I shall speak about these in relation to pathological (abnormal) constitutions of bodies.

7. Let us speak now about the best [constitution] in which each part is without fault in its whole substance.

31K

[19] *Mixt.*, I.509–694K (English trans., Singer, *Galen: Selected Works*).

[20] Galen's use of terms in relation to the stomach, its parts, and the abdomen generally seems to be somewhat variable.

32K τοιοῦτος ἄνθρωπος ὑπὸ τὴν ὑγιεινὴν ἀγόμενος τέχνην
εὐτυχὴς μὲν ἂν εἴη, εἰ μετὰ τὴν πρώτην ἀποκύησιν
ἐπιστατοῖτο πρὸς αὐτῆς· οὕτω γὰρ ἄν τι καὶ εἰς τὴν
ψυχὴν ὀνίναιτο, τῆς χρηστῆς διαίτης ἤθη χρηστὰ
παρασκευαζούσης· οὐ μὴν ἀλλὰ καὶ εἰ κατά τινα τῶν
ἑξῆς ἡλικιῶν εἰς χρείαν τῆς τέχνης ἀφίκοιτο, καὶ οὕτως ὀνήσεται τὰ μέγιστα.

εἰρήσεται δὲ πρῶτον μέν, ὡς ἄν τις ἐξ ἀρχῆς παραλαβὼν ἄνθρωπον τοιοῦτον ὑγιαίνοντα διὰ παντὸς
ἀποδείξειε τοῦ βίου, πλὴν εἴ τι τῶν ἔξωθεν αὐτῷ συμπίπτει βίαιον, οὐδὲν γὰρ τοῦτό γε πρὸς τὸν τῆς ὑγιεινῆς τέχνης ἐπιστήμονα· δεύτερον δέ, ὅπως ἄν τις, εἰ
καὶ μὴ νεογενὲς εἴη τὸ παιδίον, ἀλλ᾽ ἤδη παιδεύεσθαι
δυνάμενον, ἐπιστατήσειεν αὐτοῦ· καὶ οὕτω καθ᾽ ἑκάστην τῶν ἄλλων ἡλικιῶν. τὸ τοίνυν νεογενὲς παιδίον,
τοῦτο δὴ τὸ ἄμεμπτον ἁπάσῃ τῇ κατασκευῇ, πρῶτον
μὲν σπαργανούσθω, συμμέτροις ἁλσὶν περιπαττόμε-
νον, ὅπως αὐτῷ στερρότερον καὶ πυκνότερον εἴη τὸ
33K δέρμα τῶν ἔνδον μορίων. ἐν γὰρ τῷ κυΐσκεσθαι πάνθ᾽
ὁμοίως ἦν μαλακὰ μήτε ψαύοντος αὐτοῦ τινος ἔξωθεν
σκληροτέρου σώματος μήτ᾽ ἀέρος ψυχροῦ προσπεσόν-
τος, ὑφ᾽ ὧν συναγόμενόν τε καὶ πιλούμενον γένοιτ᾽ ἂν
ἑαυτοῦ τε καὶ τῶν ἄλλων μορίων σκληρότερόν τε καὶ
πυκνότερον. ἐπειδὰν δ᾽ ἀποκυηθῇ, ἐξ ἀνάγκης ὁμιλεῖν
μέλλει καὶ κρύει καὶ θάλπει καὶ πολλοῖς σκληρο-
τέροις ἑαυτοῦ σώμασι. προσήκει διὰ ταῦτα τὸ σύμφυτον αὐτῷ σκέπασμα παρασκευασθῆναί πως ὑφ᾽

Certainly, such a person would be fortunate if, after the   32K
time of his birth, he came to the art of hygiene and were
to have the support of this. In this way, he would also
derive some benefit for the *psyche,* since a good regimen
is a preparation for good habits. But even if, during one of
the subsequent stages of life, he were to come to the use
of the art, he would also derive the greatest benefit in this
way.

What I shall say first is that, if someone were to take
charge of such a person from the beginning, he would
make that person healthy throughout his whole life, unless
some violence were to befall him from without, for this
has nothing to do with being versed in the art of hygiene.
Second, I shall relate how someone would support a child,
even if he were not newborn but already able to be trained;
and the same with each of the other stages of life. Moreover, in the case of the newborn child who is faultless in
his whole constitution, let us first wrap him in swaddling
clothes, after sprinkling moderate amounts of salt over
him, as a result of which the skin would be firmer and
thicker than the internal parts. For when the baby is *in*   33K
utero everything is similarly soft since it is neither touched
by any harder body from without nor does cold air strike
it. It is due to these things that the skin becomes contracted and compressed—both harder and more dense
than it was before, and more than the other parts. When
it is born, the infant is inevitably going to come into contact with both cold and heat, and with many bodies harder
than itself. Because of these things, it is appropriate for us
to prepare its natural covering in some way, so it is best

ἡμῶν ἄριστον εἰς δυσπάθειαν. ἱκανὴ δὲ ἡ διὰ μόνων
τῶν ἁλῶν παρασκευὴ τοῖς γε κατὰ φύσιν ἔχουσι βρέφεσιν. ὅσα γὰρ ἤτοι μυρίνης φύλλων ξηρῶν περιπατ-
τομένων ἤ τινος ἑτέρου τοιούτου δεῖται, μοχθηρῶς δή
που διάκειται. πρόκειται δ᾽ ἡμῖν τό γε νῦν εἶναι περὶ
τῶν ἄριστα κατεσκευασμένων τὸν λόγον ποιεῖσθαι.
ταῦτ᾽ οὖν, ὡς εἴρηται, σπαργανωθέντα γάλακτί τε
χρήσθω τροφῇ καὶ λουτροῖς ὑδάτων χρηστῶν· ὑγρᾶς
γὰρ χρῄζει τῆς συμπάσης διαίτης, ἅτε καὶ τὴν κρᾶσιν ὑγροτέραν ἔχοντα τῶν ἐν ταῖς ἄλλαις ἡλικίαις.

34K   ἔοικε δὲ τοῦτο πρῶτον εὐθὺς ἥκειν σκέμμα τῶν
ἀναγκαίων εἰς δίαιταν ὑγιεινήν. εἰσὶ γὰρ οἳ νομίζουσιν ἀεὶ δεῖσθαι ξηραίνεσθαι τὰς ὑγροτέρας φύσεις,
ὥσπερ γε καὶ θερμαίνεσθαι μὲν τὰς ψυχροτέρας
ὑγραίνεσθαι δὲ τὰς ξηροτέρας, ψύχεσθαι δὲ τὰς θερ-
μοτέρας· ὑπὸ μὲν γὰρ τῶν ὁμοίων ἑκάστην τῶν ἀμετριῶν αὐξάνεσθαι, κολάζεσθαι δὲ καὶ καθαιρεῖσθαι
πρὸς τῶν ἐναντίων, ἑνὶ δὲ λόγῳ "τὰ ἐναντία τῶν ἐναντίων" ὑπάρχειν "ἰάματα." ἐχρῆν δὲ αὐτοὺς μὴ τοῦτο
μόνον Ἱπποκράτους ἀνεγνωκέναι τε καὶ μνημονεύειν,
ὡς τὰ ἐναντία τῶν ἐναντίων ἐστὶν ἰάματα, ἀλλὰ κἀ-
κεῖνα, δι᾽ ὧν φησιν· "αἱ ὑγραὶ δίαιται πᾶσι τοῖσι
πυρετταίνουσι ξυμφέρουσι, μάλιστα δὲ παιδίοισι καὶ

---

21 The Greek term is *dyspatheia,* for which LSJ gives four
somewhat disparate meanings: deep affliction, firmness in resist-
ing, and the capabilities of endurance and insensitivity; there is
no specific medical meaning given. In the OED it is taken as the

able to resist being affected [dyspathic].[21] In those infants who are in accord with nature, an adequate preparation is through salts alone. Those who are abnormal to some degree need to be covered over with the dried leaves of myrtle or some other such thing. However, what lies before us now is to make the discussion about those with the best constitutions. These, as I said, having been wrapped in swaddling clothes, are to be provided with milk as nourishment and baths of beneficial waters; the whole regimen needs to be moist, inasmuch as the *krasis* is more moist than in the other stages of life.

This seems to bring us immediately to the first question of what is the essential for a healthy regimen. For there 34K are those who think natures that are too moist always need to be dried, just as natures that are too cold always need to be heated, natures that are too dry to be moistened, and those that are too hot to be cooled, for each of the imbalances is increased by like things but corrected and reduced through the opposites—in short, "opposites are the cures of opposites."[22] But they must not only know and remember Hippocrates' statement, that "opposites are the cures of opposites," but also that in which he says, "moist regimens are beneficial to all who are febrile, particularly children, and others who are accustomed to being so

opposite of sympathy, with the meaning of antipathy, aversion, and dislike. It is described as a rare term; reference is made to Galen's *MM*. In that work it is described as a confusing term associated with Thessalus (X.267K). In *Opt. Const.* it is used in a similar sense to that above (IV.743K).

[22] Hippocrates, *Aphorisms* 2.22, *Hippocrates* IV, LCL 150, 112–13, and *Breaths* 1, *Hippocrates* III, LCL 147, 228–29.

τοῖσιν ἄλλοισι τοῖσιν οὕτως εἰθισμένοισι διαιτᾶσθαι." φαίνεται γὰρ ἐνταῦθα παράλληλα θεὶς ἐφεξῆς
τὰ τρία, νόσημά τε καὶ ἡλικίαν καὶ ἔθος. ἀπὸ μὲν τοῦ
νοσήματος ἔνδειξιν λαμβάνων τῶν ἐναντίων, ἀπὸ δὲ
τῆς ἡλικίας τε καὶ τοῦ ἔθους τῶν ὁμοίων. τῷ μὲν γὰρ
πυρετῷ (νόσημα δέ ἐστι τοῦτο θερμὸν καὶ αὐχμῶδες)
αἱ ὑγραὶ δίαιται χρησταί· τοῖς δὲ παιδίοις (οὐ γὰρ
35K νόσημα τούτοις γε, ἀλλὰ κατὰ φύσιν ἢ ἡλικίαν) τὸ
ὁμοιότατον ὠφελιμώτατον. οὕτω δὲ καὶ τοῖς ἔθεσιν,
ὡς ἂν καὶ αὐτοῖς ἐπικτήτους τινὰς ἐν τοῖς σώμασι
φύσεις ἐργαζομένοις, ἡ τῶν ἐναντίων προσφορὰ βλα-
βερωτάτη. καὶ δεόντως τοῖς μὲν κατὰ φύσιν ἔχουσι
σώμασι φυλάττεσθαι χρὴ τὴν οἰκείαν ἕξιν, τοῖς δὲ
νοσοῦσιν ἀλλοιοῦσθαί τε καὶ πρὸς τοὐναντίον ἐπάγε-
σθαι. φυλάττεται μὲν οὖν ἕκαστον ὑπὸ τῶν ὁμοίων,
ἀλλοιοῦται δὲ ὑπὸ τῶν ἐναντίων.

οὔκουν ξηραίνειν χρὴ τοὺς παῖδας, ὅτι μὴ παρὰ
φύσιν αὐτοῖς ἡ ὑγρότης, ὥσπερ ἐν βράγχοις τε καὶ
κορύζαις καὶ κατάρροις [καὶ ὑδέροις],[11] ἀλλ' ἐν τοῖς
φύσει διαιτᾶν ὑγραίνουσι τὸ κατὰ φύσιν φυλάττοντα
ταῖς διαίταις ὑγραινούσαις λουτροῖς τε ποτίμων ὑδάτων (ὅσα γὰρ ἐμφαίνει τινὰ φαρμακώδη ποιότητα,
ξηραίνει πάντα, καθάπερ τὰ θειώδη καὶ ἀσφαλτώδη
καὶ νιτρώδη[12] καὶ στυπτηριώδη) καὶ τροφὴν καὶ ποτὸν
παρέχειν ὅτι μάλιστα κράσεως ὑγροτάτης. οὕτω δὲ
καὶ ἡ φύσις αὐτὴ προὐνοήσατο τῶν παίδων τροφὴν

---

[11] καὶ ὑδέροις *add.* Ko   [12] καὶ νιτρώδη *add.* Ko

treated."[23] Here he has obviously placed three things in juxtaposition—disease, stage of life and custom. From the disease, he takes the indication of opposites while from the stage of life and custom, he takes the indication of similars. For in a fever (which is a hot and dry disease), the moist regimens are beneficial, whereas in children (for in them it is not in fact a disease but a stage of life and accords with nature), what is most similar is most helpful.   35K It is also the same with customs, as these produce certain acquired natures in bodies for which the exhibition of opposites is very harmful. And what is right for bodies in accord with nature, is that one must preserve the proper (i.e., preexisting) state, whereas in those who are diseased, one must set in motion a change to the opposite. Therefore, each of the two is preserved by similars but changed by opposites.

One must not, then, dry children because the moistness in them is not contrary to nature, as it is in coughs, coryzas, catarrhs and dropsies.[24] Rather, one must feed them with things that are moist in nature, maintaining an accord with nature by means of moist diets and baths of potable waters (for those that display any medicinal quality, such as those that are sulfurous, full of asphalt or contain sodium carbonate or alum, are all drying), and most of all provide food and drink that is particularly moist in *krasis*. And in this way too, Nature herself gave forethought to children, preparing moist nutriment for

[23] Hippocrates, *Aphorisms* 1.16, *Hippocrates* IV, LCL 150, 106–7.

[24] One might question Koch's addition of "dropsies" here as being somewhat incongruous.

αὐτοῖς ὑγρὰν παρασκευάζουσα τὸ γάλα τῆς μητρός.
36K ἄριστον μὲν οὖν ἴσως καὶ τοῖς ἄλλοις ἅπασι βρέφεσι
τὸ γάλα τῆς μητρός, πλὴν εἰ μὴ τύχοι νενοσηκός,
οὐχ ἥκιστα δὲ καὶ τῷ τῆς ἀρίστης κράσεως, ὑπὲρ οὗ
νῦν ὁ λόγος ἐστί. εἰκὸς γάρ που τῆς τούτου μητρὸς
ἄμεμπτον εἶναι τό τε σύμπαν σῶμα καὶ τὸ γάλα.

ἐξ αἵματος μὲν οὖν ἔτι κυουμένοις ἡμῖν ἡ τροφή·
ἐξ αἵματος δὲ καὶ ἡ τοῦ γάλακτος γένεσις ὀλιγίστην
μεταβολὴν ἐν τοῖς μαστοῖς προσλαβόντος. ὥσθ᾽ ὅσα
παιδία τῷ τῆς μητρὸς γάλακτι τρέφεται, συνηθε-
στάτῃ τε ἅμα καὶ οἰκειοτάτῃ χρῆται τροφῇ. φαίνεται
δὲ οὐ μόνον παρασκευάσασα τὴν τοιαύτην τροφὴν ἡ
φύσις τοῖς βρέφεσιν, ἀλλὰ καὶ δυνάμεις αὐτοῖς εὐθὺς
ἐξ ἀρχῆς ἐμφύτους παρασχοῦσα, ἕνεκα τῆς χρήσεως
αὐτοῦ. καὶ γὰρ καὶ γεννηθεῖσιν εἴ τις ἐνθείη παραχρῆμα τῷ στόματι τὴν θηλὴν τοῦ μαστοῦ, βδάλλει
τε τὸ γάλα καὶ καταπίνει προθυμότατα· καὶ ἢν ἀνιώ-
μενά τε καὶ κλαυθμυριζόμενα τύχῃ, τῆς λύπης οὐκ
ἐλάχιστον ἴαμα αὐτοῖς ὁ τιτθὸς τῆς τρεφούσης ἐστὶν
ἐντιθέμενος τῷ στόματι.

τρία γὰρ οὖν δὴ ταῦτα ταῖς τροφοῖς ἐξεύρηται τῆς
37K λύπης τῶν παιδίων ἰάματα τῇ πείρᾳ διδαχθείσαις, ἐν
μὲν τὸ νῦν δὴ λελεγμένον, ἕτερα δὲ δύο, κίνησίς τις
μετρία καὶ φωνῆς εὐμέλειά τις, οἷς χρώμεναι διαπαντὸς οὐ καταπραΰνουσι μόνον, ἀλλὰ καὶ εἰς ὕπνον
ἀπάγουσιν αὐτὰ δηλούσης αὖ κἀν τούτῳ τῆς φύσεως,
ὅτι πρὸς μουσικὴν καὶ γυμναστικὴν οἰκείως διάκειται. καὶ ὅστις οὖν ἱκανός ἐστι καλῶς χρήσασθαι ταῖς

them—that is, mother's milk. Thus mother's milk is very
likely best for all infants other than those who happen to   36K
be diseased, and not least for the infant with the best
krasis, about whom the present discussion is. Anyway, it is
likely that the mother of such a child is faultless both in
respect of her body as a whole and her milk.

Our nourishment, while still *in utero,* is from blood,
while the genesis of milk is also from blood that has un-
dergone the least change in the breasts. As a consequence,
those children who are nourished by mother's milk make
use of the most customary and fitting nutriment. So it is
clear that Nature has not only prepared such a nutriment
for infants, but has also provided innate powers for them,
right from the beginning, for the purpose of using this
nutriment. For also, when the infant has been born, if
someone immediately places the nipple of the breast in its
mouth, it sucks the milk and drinks it down very eagerly.
And if infants should happen to be distressed and crying,
the remedy for their grief is not least the breast of the
nurse placed in their mouths.

There are, then, these three remedies for the distress
of children discovered by nurses who have learned from
experience. One is that which we have now discussed; the   37K
other two are moderate movement and a certain modula-
tion of the voice. Those who use these things continually
not only settle the infant but also put it to sleep, revealing
even in this an innate predisposition toward music and
exercise. Therefore, someone who is effective in using

τέχναις ταύταις, οὗτος καὶ σῶμα καὶ ψυχὴν παιδεύσει κάλλιστα.

8. Ταῖς γοῦν τροφοῖς αἱ τῶν παιδίων κινήσεις ἔν τε λίκνοις καὶ σκίμποσι καὶ ταῖς σφῶν αὐτῶν ἀγκάλαις ἐξεύρηνται. καί πως ἂν καὶ τοῦθ᾽ ἕτερον ἡμῖν ἥκοι σκέμμα πρὸς ὑγείας τήρησιν ἀναγκαιότατον, Ἀσκληπιάδου μὲν ἄντικρύς τε κἀκ τοῦ φανερωτάτου κατεγνωκότος γυμνασίων, Ἐρασιστράτου δὲ ἀτολμότερον μὲν ἀποφηναμένου, τὴν δ᾽ αὐτὴν Ἀσκληπιάδη γνώμην ἀποδεικνυμένου, τῶν δ᾽ ἄλλων σχεδὸν ἁπάντων ἰατρῶν ἐπαινούντων οὐ πρὸς εὐεξίαν μόνον, ἀλλὰ καὶ πρὸς ὑγείαν αὐτά. τριττὰ δ᾽ ἐστὶ γένη τά γε πρῶτα τῶν γυμνασίων, ὅσαιπερ καὶ αἱ τῶν κινήσεων

38K διαφοραί· ἢ γὰρ ἐξ ἑαυτῶν ἢ ὑφ᾽ ἑτέρων ἢ διὰ φαρμάκων κινούμεθα. τὸ μὲν δὴ τρίτον εἶδος τῆς κινήσεως οὐδαμῶς ὑγιαίνουσι πρέπον· ἡ δ᾽ ὑφ᾽ ἑτέρου κίνησις ἔν τε τῷ πλεῖν καὶ ἱππεύειν καὶ ὀχεῖσθαι καί, ὡς ἀρτίως λέλεκται, διά τε λίκνων καὶ σκιμπόδων καὶ ἀγκαλῶν γίνεται. τοῖς μὲν οὖν νεογενέσι παιδίοις οὔπω δεῖ κινήσεως τηλικαύτης, ἡλίκη διά τε τῶν ὀχημάτων καὶ τῶν πλοίων καὶ τοῖς ἱππαζομένοις γίνεται· τοῖς δ᾽ ἤδη τὸ τρίτον ἢ τὸ τέταρτον ἔτος ἀπὸ τῆς πρώτης γενέσεως ἄγουσιν ἐγχωρεῖ καὶ δι᾽ ὀχημάτων καὶ πλοίων κινεῖσθαι τὰ μέτρια· ἑπταετῆ δὲ γενόμενα τὰ παιδία καὶ τῶν ἰσχυροτέρων ἀνέχεται κινήσεων, ὥστε καὶ ἱππεύειν ἐθίζεσθαι. ἐξ ἑαυτῶν δὲ κινεῖσθαι τὰ παιδία τηνικαῦτα δύναται πρῶτον, ὅταν ἕρπειν ἀπάρξηται, καὶ μᾶλλον, ἐπειδὰν ἀπάρξηται[13] βαδί-

these arts well is also someone who will be best at nurturing both body and mind.

8. Anyway, the movements in cradles, cots and their own bent arms were discovered by the nursemaids of children. And we might also come in some way to the other issue of what is most essential for the maintenance of health, although Asclepiades opposed and gave the clearest condemnation of exercises, while Erasistratus spoke less vehemently but displayed the same opinion as Asclepiades. But almost all other doctors praise them, not only in regard to a good bodily condition but also in regard to health. There are three primary classes of exercises, and as many differences of movements—those by ourselves, those by other things and those when we are moved 38K through medications. Certainly, the third kind of movement in no way is fitting for those who are healthy. Movement by something else occurs in sailing, horse riding and driving a carriage and, as I said just now, through cradles, cots and the bent arms [of nurses]. For newborn babies there is not yet need, at such a young age, of the movement that arises through carriages, boats and horses. For those who are already three to four years old, it is permissible to bring them to a moderate amount of movement through carriages and boats. When children are seven, they tolerate stronger movements, so they can also accustom themselves to horse riding. Children are first able to move by themselves at a young age when they begin to crawl, and more so when they begin to walk. But they

---

13 ἀπάρξηται add. Ko

GALEN

ζειν· μὴ βιάζεσθαι δὲ αὐτὰ πρὸ τοῦ δέοντος, ὅπως μὴ
διαστραφῇ τὰ κῶλα. δηλοῖ μὲν κἂν τῷδε τῆς ἡλικίας,
εἰς ὅσον ἡμῶν ἡ φύσις ᾠκείωται γυμνασίοις. οὐ γὰρ
ἄν, οὐδ᾽ εἰ κατακλείσαις παιδία, οἷός τε εἴης κωλῦσαι
διαθέειν τε καὶ σκιρτᾶν ὡσαύτως γε τοῖς πώλοις τε
39K καὶ μόσχοις. ἱκανὴ γὰρ ἡ φύσις ἐν ἅπασι τοῖς ζῴοις
ὁρμὰς οἰκείας ἐνθεῖναι πρὸς ὑγείαν τε καὶ σωτηρίαν.

ἀλλ᾽ οἱ περὶ τὸν Ἀσκληπιάδην οὐδὲν τῶν τοιούτων
ἐννοήσαντες ἐπὶ πολλῆς σχολῆς σοφίσματα πλέκουσιν, ἐπιδείκνυσθαι πειρώμενοι τὰ γυμνάσια μηδὲν εἰς
ὑγείαν συντελοῦντα. πρὸς ἐκείνους μὲν οὖν δὴ καὶ
αὖθις εἰρήσεται τὰ εἰκότα, ὡς τῆς ἀδολεσχίας ἐπὶ
τέλους εἰρησομένης· πρόκειται γάρ μοι νῦν οὐ σοφιστῶν ἀδολεσχίας ἐλέγχειν, ἀλλ᾽ αὐτὸ τὸ χρήσιμον
εἰς ὑγείαν ἐκδιδάσκειν.

ἐπὶ δὲ τοὺς παῖδας ἐπάνειμι τοὺς ἄριστα κατεσκευ-
ασμένους τὸ σῶμα. τούτους δ᾽ εἰκὸς εἶναι καὶ τὸ τῆς
ψυχῆς ἦθος ἀμέμπτους· ὡς ὅσοι γε θυμικώτεροι τοῦ
δέοντός εἰσιν ἢ ἀθυμότεροί[14] τινες ἢ ἀναισθητότεροι
ἢ λιχνότεροι τοῦ προσήκοντος, ἀνάγκη τούτους οὐκ
ὀρθῶς κεκρᾶσθαι τοῖς μέρεσιν ἐκείνοις τοῦ σώματος,
οἷς ἐνεργοῦμεν ἕκαστα, τῶν εἰρημένων. γέγραπται δ᾽
ὑπὲρ αὐτῶν ἐπὶ πλέον ἐν τοῖς Περὶ τῶν Ἱπποκράτους
καὶ Πλάτωνος δογμάτων ὑπομνήμασιν. ἀλλ᾽ ὅ γε νῦν
ἡμῖν προκείμενος ἐν τῷ λόγῳ παῖς ἄριστός ἐστι τὰ

14 post ἀθυμότεροί: ἢ εὐαισθητότεροι, ἢ ἀναισθητότεροι
τινες, Ku; τινες ἢ ἀναισθητότεροι Ko

should not be forced into walking beyond what is needed lest their legs become distorted. And it is clear, even at this age, to what degree our nature has been made fit for exercises. For even if you were to shut children in, you would not be able to prevent them running about and frolicking like foals and calves. Nature is sufficient to instill in all 39K animals the proper impulses toward health and preservation.

But the followers of Asclepiades, give no consideration to such things and in their great idleness weave sophisms attempting to show that exercises contribute nothing to health. Now to those men I shall say again what is appropriate—let them pursue their idle talk to completion. My task now is not to refute the idle talk of Sophists but to teach what is actually useful for health.

I shall return to the children who are best constituted with respect to the body. It is also right for children that the ethos of the soul is faultless. As there are some who are more high-spirited than they should be, or too faint-hearted, some who are excessively sensitive, or too insensitive or greedy than is appropriate,[25] of necessity these have not been properly mixed (i.e., do not have a proper *krasis*) in those parts of the body through which we carry out each of the things mentioned. I have written at greater length about these matters in the treatises *On the Opinions of Hippocrates and Plato*.[26] But, in fact, the child now before us in the discussion is the best in all these things.

[25] The Kühn text has been followed here.
[26] See Galen's *Plac. Hippocr. Plat.*, V.181–805K (English trans., De Lacy, *Galen on the Doctrines of Hippocrates and Plato*, CMG V.4.1.2). See particularly V.181K ff.

40K πάντα. τούτου τοίνυν ἐπανορθοῦσθαι μὲν οὐδὲν χρὴ
τῶν τῆς ψυχῆς ἠθῶν, φυλάττειν δ᾽, ὅπως μὴ δια-
φθαρῇ. φυλάττεται δ᾽ ἅπαν ὑπὸ τῶν αὐτῶν κατὰ γένος, ὑφ᾽ ὧνπερ καὶ διαφθείρεται. διαφθείρεται δὲ τὸ
τῆς ψυχῆς ἦθος ὑπὸ μοχθηρῶν ἐθισμῶν ἐν ἐδέσμασί
τε καὶ πόμασι καὶ γυμνασίοις καὶ θεάμασι καὶ ἀκού-
σμασι καὶ τῇ συμπάσῃ μουσικῇ. τούτων τοίνυν
ἁπάντων ἔμπειρον εἶναι χρὴ τὸν τὴν ὑγιεινὴν τέχνην
μετιόντα καὶ μὴ νομίζειν, ὡς φιλοσόφῳ μόνῳ προσήκει πλάττειν ἦθος ψυχῆς· ἐκείνῳ μὲν γὰρ δι᾽ ἕτερόν
τι μεῖζον τὴν τῆς ψυχῆς αὐτῆς ὑγείαν, ἰατρῷ δὲ ὑπὲρ
τοῦ μὴ ῥᾳδίως εἰς νόσους ὑπομεταφέρεσθαι τὸ σῶμα.
καὶ γὰρ θυμὸς καὶ κλαυθμὸς καὶ ὀργὴ καὶ λύπη καὶ
πλεῖον τοῦ δέοντος φροντὶς ἀγρυπνία τε πολλὴ ἐπ᾽
αὐτοῖς γενομένη πυρετοὺς ἀνάπτουσι καὶ νοσημάτων
μεγάλων ἀρχαὶ καθίστανται, ὥσπερ καὶ τοὐναντίον
ἀργὴ διάνοια καὶ ἄνοια καὶ ψυχὴ παντάπασιν ἄθυμος
ἀχροίας καὶ ἀτροφίας ἐργάζεται πολλάκις ἀρρωστίᾳ
τῆς ἐμφύτου θερμότητος. χρὴ μὲν γὰρ φυλάττειν
41K ἅπαντος μᾶλλον ἐν ὅροις ὑγιεινοῖς τὴν σύμφυτον
ἡμῖν θερμότητα. φυλάττεται δὲ ὑπὸ τῶν συμμέτρων
γυμνασίων οὐ κατὰ τὸ σῶμα μόνον, ἀλλὰ κατὰ τὴν
ψυχὴν γινομένων. αἱ δ᾽ ἄμετροι κινήσεις ἐν ἐπιθυ-
μίαις τε καὶ διαλογισμοῖς καὶ θυμοῖς, αἱ μὲν ὑπερ-
βάλλουσαι χολωδέστερον ἀποφαίνουσι τὸ ζῷον, αἱ δ᾽
ἐλλείπουσαι φλεγματικώτερον καὶ ψυχρότερον. καὶ
δὴ καὶ ταῖς μὲν προτέραις ἕξεσιν οἵ τε πυρετοὶ καὶ
ὅσα θερμότερα πάθη, ταῖς δ᾽ ἑτέραις ἐμφράξεις καθ᾽

Accordingly, in him it is not necessary to restore the dis- 40K
positions of the soul, but to preserve them so they are not
corrupted. However, everything is preserved by those
same things in respect of class by which it is also destroyed.
The ethos of the soul is destroyed by bad habits in foods,
drinks and exercises, by sights and sounds, and by music
in general. Therefore, it is necessary for someone who
practices the art of hygiene to be experienced in all these
things and not to think it appropriate to mold the ethos of
the soul by philosophy alone, for he more than anyone else
is responsible for the health of the soul itself. The doctor,
on the other hand, is responsible for not allowing the body
to slip easily into diseases. For passion, weeping, anger,
grief, unnecessary anxiety, and severe insomnia arising
from these things kindle fevers and the origins of major
diseases are established, just as conversely, an idle mind,
folly and all in all a spiritless soul often produce pallor and
atrophies through a weakness of the innate heat. For it is
necessary above all to preserve our innate heat within 41K
healthy limits. It is preserved by the occurrence of moderate exercise, not only of the body but also of the mind
(soul). Immoderate movements occur in passions, arguments and anger. Movements that are excessive render the
animal more bilious; those that are deficient render the
animal more phlegmatic and cold. And furthermore, in
the former states, the fevers and those affections that are
hotter all occur, while in the latter there are obstructions

ἧπάρ τε καὶ σπλάγχνα, ἐπιληψίαι τε καὶ ἀποπληξίαι
ἤ τι τοιοῦτον ἄλλο,[15] καὶ συνελόντα φάναι, τὰ καταρ-
ροϊκά τε καὶ ῥευματικὰ νοσήματα, συμπίπτει πάντα.
καὶ οὐκ ὀλίγους ἡμεῖς ἀνθρώπους νοσοῦντας ὅσα ἔτη
διὰ τὸ τῆς ψυχῆς ἦθος ὑγιεινοὺς ἀπεδείξαμεν, ἐπανορ-
θωσάμενοι τὴν ἀμετρίαν τῶν κινήσεων.

 οὐ σμικρὸς δὲ τοῦ λόγου μάρτυς καὶ ὁ πάτριος
ἡμῶν θεὸς Ἀσκληπιός, οὐκ ὀλίγας μὲν ᾠδάς τε γρά-
φεσθαι καὶ μίμους γελοίων καὶ μέλη τινὰ ποιεῖν ἐπι-
τάξας, οἷς αἱ τοῦ θυμοειδοῦς κινήσεις σφοδρότεραι
γενόμεναι θερμοτέραν τοῦ δέοντος ἀπειργάζοντο τὴν
42K κρᾶσιν τοῦ σώματος, ἑτέροις δέ τισιν, οὐκ ὀλίγοις
οὐδὲ τούτοις, κυνηγετεῖν καὶ ἱππάζεσθαι καὶ ὁπλομαχεῖν. εὐθὺς δὲ τούτοις διώρισε τό τε τῶν κυνηγεσίων
εἶδος, οἷς τοῦτο προσέταξε, τό τε τῆς ὁπλίσεως, οἷς
δι' ὅπλων ἐκέλευσε τὰ γυμνάσια ποιεῖσθαι. οὐ γὰρ
μόνον ἐπεγείρειν αὐτῶν τὸ θυμοειδὲς ἐβουλήθη, ἄρ-
ρωστον ὑπάρχον, ἀλλὰ καὶ μέτρον ὡρίσατο τῇ τῶν
γυμνασίων ἰδέᾳ. οὐ γὰρ ὡσαύτως θήγεται τὸ θυμοειδὲς εἰς ἀγρίους ὗς ἢ ἄρκτους ἢ ταύρους ἤ τι τῶν
οὕτως ἀλκίμων θηρίων ἢ ἐπὶ λαγωοὺς ἢ δορκάδας ἢ
τι τῶν οὕτω δειλῶν, οὐδ' ὡσαύτως ἐπί τε τῆς κούφης
ὁπλίσεως καὶ τῆς βαρείας, ὥσπερ οὐδὲ ἐν τῷ θέειν
ὠκέως ἢ μετρίως κινεῖσθαι καὶ μετὰ τοῦ φιλονεικεῖν
ἑτέροις ἢ καθ' ἑαυτόν. οὕτω δὲ καὶ τῶν μέγα κεκρα-

15 ἤ τι τοιοῦτον ἄλλο, add. Ko

involving the liver and internal organs, epilepsies and apoplexies or something else of this sort, and generally speaking, catarrhal and rheumatic diseases all happen. And every year we make many people, who are diseased in terms of the ethos of the soul, healthy when we correct the imbalance of movements.

A significant witness of this argument is our ancestral god, Asclepius,[27] who ordered a number of odes to be written and humorous mimes and certain lyrics to be created, in which more violent movements of someone who was passionate made the *krasis* of the body hotter than it should be, while for certain others, and there were quite a few of them, [there was] hunting, horse riding and practicing the use of arms [that did the same]. He immediately determined the kind of hunting for those whom he ordered to hunt, and the arms for those whom he ordered to carry out the military exercises. For not only did he wish to stir up their passion when it was weak, but also to establish a limit of moderation in the kind of exercises. For passion is not aroused in the same way against wild boars, bears, bulls, and other similarly strong wild animals as it is in the case of hares and deer, and some other timid animals. Nor is it similar in the case of light and heavy arms, just as it is not in running swiftly or moving moderately, nor with liking to contend against others or against one-

42K

[27] Asclepius (Aesculapius in Latin) was the legendary god of healing, featuring first in the works of Homer and Hesiod. The two accounts in the OCD (under the Greek and the Latin names) give a good account of his significance and alleged activities. See also E. J. and L. Edelstein, *Asclepius: A Collection and Interpretation of the Testimonies* (1943).

γότων ἢ ἐγκελευομένων τε καὶ παροξυνόντων ἐπὶ τοὺς
πόνους ἢ σιωπώντων οὐκ ὀλίγον διαφέρει. ἀλλὰ περὶ
μὲν τούτων ἐν τοῖς ἔπειτα λόγοις ἐπὶ πλέον εἰρήσεται.

τὰ δὲ σμικρὰ παιδία τὰ τὴν ἀρίστην ἔχοντα κρᾶσιν
(ὑπὲρ τούτων γὰρ ἦν ὁ λόγος) οὐκ ὀλίγης ἐπιμελείας
δεῖται πρὸς τὸ μηδεμίαν ἐν αὐτοῖς τῆς ψυχῆς ἄμετρον
43K γίνεσθαι κίνησιν· ἅτε γὰρ οὐδέπω λόγῳ χρώμενα τῷ
κλαίειν τε καὶ κεκραγέναι καὶ θυμοῦσθαι καὶ κινεῖν
ἀτάκτως ἑαυτὰ διασημαίνει τὴν ἀνίαν. ἡμᾶς οὖν χρὴ
στοχαζομένους, ὅτου δεῖται, παρέχειν ἑκάστοτε τοῦτο,
πρὶν αὐξηθεῖσαν αὐτῶν τὴν λύπην εἰς σφοδροτέραν
καὶ ἄτακτον κίνησιν ὅλην ἐμβαλεῖν τὴν ψυχὴν ἅμα
τῷ σώματι· ἤτοι γὰρ ἐξ ἑαυτῶν ὀδαξούμενα ἢ πρός
τινος ἔξωθεν ἀνιώμενα ἢ ἀποπατεῖν ἢ οὐρεῖν ἢ ἐσθίειν
ἢ πίνειν ἐθέλοντα κλαίει τε καὶ κινεῖται πλημμελῶς,
ὥσπερ σφαδάζοντα. γένοιτο δ' ἄν ποτε καὶ θάλπους
ἐπιθυμεῖν αὐτὰ κρύει ταλαιπωρούμενα καὶ ἀναψύξεώς
τινος ὑπὸ θάλπους ἀμέτρου διοχλούμενα καί ποτε μὴ
φέροντα τὸ πλῆθος τῶν ἐπιβεβλημένων ἱματίων· ἀνιᾷ
γὰρ δὴ καὶ τοῦτο πολλάκις οὐ σμικρά, καὶ μάλιστα
κατὰ τὰς ὑποστροφὰς ὅλου τοῦ σώματος καὶ τὰς τῶν
κώλων κινήσεις. ἀλλὰ καὶ αὐτὸ τὸ ἡσυχάζειν ἐπὶ
πλέον οὐ σμικρῶς λυπηρόν· οὐδενὸς γὰρ ἀμετρίᾳ
χαίρει ζῷον οὐδέν, ἀλλ' ἀεὶ τοῦ συμμέτρου χρῄζει.
44K σύμμετρον δὲ οὐχ ἓν ἅπασιν, ἀλλ' ἐν τῷ πρός τι
πᾶσα συμμετρία. διὸ χρὴ τὸν ἐπιμελούμενον ἀνατρο-
φῆς παιδίων, στοχαστικὸν ἀκριβῶς ὑπάρχοντα τοῦ
συμμέτρου τε καὶ οἰκείου, παρέχειν ἑκάστοτε τοῦτο,

self. And it makes no little difference to those exercising whether the onlookers shout loudly, urging and spurring them on or are silent. But I shall say more about these things in the discussions that follow.

Small children, who have the best *krasis* (for the argument was about them), need no little care so that no immoderate activity of the mind (soul) occurs in them. Inasmuch as they do not yet have the use of speech, they show their distress by crying, calling out, becoming restive, and moving themselves in a disorderly fashion. It behooves us, then, to work out what they need and provide it on each occasion before their distress is increased to the point of throwing the entire soul along with the body into more violent and disorderly movement. For when they either feel irritation from something within themselves, or are distressed by something external, or wish to defecate, urinate, eat or drink, they cry and move in a disorderly way, as if they are struggling. It also happens sometimes that, being distressed by cold, they desire heat, or troubled greatly by excessive heat, they desire some cooling, and on occasion cannot bear the number of coverings put on them. For this often distresses them to no small degree, and particularly in relation to turning the body as a whole or moving the limbs. But also actual silence, if it is excessive, is distressing to no small degree, for no animal takes pleasure in anything immoderate—there is always the need for moderation. However, moderation is not one and the same for all things; every moderation is in relation to something. Accordingly it is necessary for someone who has charge of rearing children to calculate precisely what is moderate and fitting, and provide this on each occasion

43K

44K

πρὶν αὐξηθεῖσαν αὐτῷ τὴν λύπην εἰς ἀμετρίαν κινήσεως ἐμβαλεῖν τό τε σῶμα καὶ τὴν ψυχήν· εἰ δ' ἄρα
ποτὲ καὶ λάθοι τὸ λυποῦν αὐξηθέν, ἐπανορθοῦσθαι
πειρᾶσθαι τὴν λύπην αὐτῷ τε τῷ παρέχειν αὐτίκα τὸ
ἐπιθυμηθὲν ἢ ἐκκόπτειν τὸ ἀνιῶν ἔτι τε τῇ κινήσει τῇ
διὰ τῶν ἀγκαλῶν καὶ τοῖς μέλεσι τῆς φωνῆς, οἷς εἰώ-
θασιν αἱ σοφώτεραι τῶν τροφῶν χρῆσθαι.

ἐγὼ γοῦν ποτε δι' ὅλης ἡμέρας παιδίου κλαίοντός
τε καὶ θυμουμένου καὶ σφοδρῶς καὶ ἀτάκτως ἑαυτὸ
μεταβάλλοντος ἐξεῦρον τὸ λυποῦν ἀπορουμένης τῆς
τροφοῦ· ὡς γὰρ οὔτε πρὸς τὸν τιτθὸν ἐντεθειμένον
οὔτε προϊσχομένης αὐτὸ τῆς τρεφούσης, εἰ ἀποπατεῖν
ἢ οὐρεῖν ἐθέλοι, καθίστατο, παρηγορεῖτο δ' οὐδέν,
οὐδ' ὁπόταν ἐν ταῖς ἀγκάλαις ἐνθεμένη κατακλίνειν
ἐπιχειρήσειεν, ἐθεασάμην δ' ἐγὼ[16] τὴν στρωμνὴν αὐ-
45K τοῦ καὶ τὰ περιβλήματά τε καὶ ἀμφιέσματα ῥυπαρώ-
τερα καὶ αὐτὸ τὸ παιδίον ἤδη ῥυπῶν τε καὶ ἄλουτον,
ἐκέλευσα λοῦσαί τε καὶ ἀπορρύψαι καὶ τὴν στρωμνὴν
ὑπαλλάξαι, καὶ πᾶσαν τὴν ἐσθῆτα καθαρωτέραν ἐρ-
γάσασθαι· καὶ τούτων γενομένων, αὐτίκα μὲν ἐπαύ-
σατο τῶν ἀμέτρων κινήσεων, αὐτίκα δὲ καθύπνωσεν
ἥδιστόν τε καὶ μακρότατον ὕπνον. εἰς δὲ τὸ καλῶς
ἐστοχάσθαι πάντων τῶν ἀνιώντων τὸ παιδίον οὐκ ἀγ-
χινοίας μόνον, ἀλλὰ καὶ τῆς περὶ τὸ τρεφόμενον αὐτὸ
συνεχοῦς ἐμπειρίας ἐστὶ χρεία.

9. Ταῦτ' οὖν ἅπαντα περὶ τὸ παιδίον εἰς τρίτον ἔτος
ἀπὸ τῆς πρώτης γενέσεως ἀξιῶ πραγματεύεσθαι καὶ
πρὸς τούτοις ἔτι τῆς τρεφούσης αὐτὸ οὐ σμικρὰν ποι-

before the distress in the child is increased to the point of throwing the body and soul into an imbalance of movement. And if the increase in distress has for a time escaped notice, attempt to correct the distress in the child by immediately providing what is desired or removing what is distressing, and as well as this, with movement by rocking in the bent arms and modulations of the voice, which the more skilled nurses are accustomed to use.

At all events, on one occasion, when a baby was crying, restive and turning itself about in a violent and disorderly fashion for a whole day, and the nurse was at a loss, I myself discovered what was distressing it. For when he didn't settle when the nipple was put in his mouth or when the nurse held him out in front of her to see if he wanted to defecate or urinate, and nothing consoled him, not even when she attempted to lay him down after rocking him in her arms, I saw that his bed and its coverings, as well as 45K his clothes, were rather soiled, and the baby himself was dirty and unwashed. I directed her to wash and clean him thoroughly, change the bed, and make all the clothing cleaner. Once these things had happened, he immediately stopped the excessive movements and straightway fell into a very sweet and very prolonged sleep. In properly evaluating all the things that distress a young child, not only is there need of wisdom but also of a long-continuing experience in nursing itself.

9. I think it is important to take the trouble to do all these things concerning the child up to the third year from birth, and in addition to these, the nurse should give no

---

[16] ἐγὼ *add.* Ko

εἶσθαι πρόνοιαν ἐδεσμάτων τε πέρι καὶ πομάτων,
ὕπνων τε καὶ ἀφροδισίων καὶ γυμνασίων, ὡς ἂν ἄριστον εἴη τὴν κρᾶσιν τὸ γάλα. γίνοιτο δ᾽ ἂν τοιοῦτον,
εἰ τὸ αἷμα χρηστότατον εἴη. ἔστι δὲ χρηστότατον τὸ
μήτε πικρόχολον μήτε μελαγχολικὸν μήτε φλεγματῶδες μήτ᾽ ὀρρώδει τινὶ μήθ᾽ ὑδατώδει συμμιγὲς
46K ὑγρότητι. γεννᾶται δὲ τοιοῦτον ἐπί τε τοῖς συμμέτροις
γυμνάσμασι καὶ τροφαῖς εὐχύμοις τε ἅμα καὶ κατὰ
καιρὸν τὸν προσήκοντα καὶ κατὰ μέτρα τὰ δέοντα
λαμβανομέναις, ὥσπερ οὖν καὶ ἐπὶ πόμασιν εὐκαίροις
τε καὶ μετρίοις· ὑπὲρ ὧν ἁπάντων ἐν τοῖς ἔπειτα λόγοις ἀκριβῶς διορισθήσεται.

ἀφροδισίων δὲ παντάπασιν ἀπέχεσθαι κελεύω τὰς
θηλαζούσας παιδία γυναῖκας. αἵ τε γὰρ ἐπιμήνιοι
καθάρσεις αὐταῖς ἐρεθίζονται μιγνυμέναις ἀνδρί, καὶ
οὐκ εὐῶδες ἔτι μένει τὸ γάλα. καί τινες αὐτῶν ἐν γαστρὶ λαμβάνουσιν, οὗ βλαβερώτερον οὐδὲν ἂν εἴη
παιδίῳ γάλακτι τρεφομένῳ. δαπανᾶται γὰρ ἐν τῷδε
τὸ χρηστότατον τοῦ αἵματος εἰς τὸ κυούμενον. ἅτε
γὰρ ἀρχὴν ζωῆς ἐν ἑαυτῷ ἰδίαν περιέχον, ὑπὸ ταύτης
τε διοικεῖται καὶ διὰ παντὸς ἐπισπᾶται τὴν οἰκείαν
τροφὴν ὡσανεὶ ἐνερριζωμένον τε τῇ μήτρᾳ καὶ ἀχώριστον ὑπάρχον ἀεὶ νύκτωρ τε ἅμα καὶ μεθ᾽ ἡμέραν
ἔλαττόν τε εἰκότως ἐν τῷδε καὶ φαυλότερον ἀποτελεῖται τὸ τῆς κυούσης αἷμα, καὶ διὰ τοῦτο καὶ τὸ
γάλα μοχθηρόν τε καὶ ὀλίγον ἐν τοῖς τιτθοῖς ἀθροί-
47K ζεται. ὥστε ἔγωγε βουλεύσαιμ᾽ ἄν, εἰ κινήσειεν ἡ θη-
λάζουσα τὸ παιδίον, ἑτέραν ἐξευρίσκειν τροφόν, ἐπι-

little forethought to her own food and drink, her sleep, sexual activity and exercises, so that her milk is the best in terms of *krasis*. Such a thing would occur if her blood were at its very best—that is, neither picrocholic, melancholic, phlegmatic, whey-like, nor mixed with some other watery fluid. Such blood is produced by moderate exercises, nu-   46K triments that are *euchymous* and taken at the appropriate time and in the necessary amount, just as also by drinks that are timely and moderate. All these things will be precisely defined in the discussion that follows.

I direct women who are nursing (i.e., providing milk for) little children to abstain from sexual activity altogether. The menstrual flow is stirred up by sexual intercourse and the milk no longer remains sweet. And some women become pregnant; nothing is more harmful than this for the infant being nourished by their milk. In this case, the best of the blood is used up in the fetus, inasmuch as the origin of life is specifically contained in the blood itself and is provided for by this. Throughout, [the fetus] draws the proper nutrition as if it is rooted in the uterus and always inseparable, both night and day. Because of this, the blood of the pregnant woman becomes less suitable and of poorer quality, and also on this account, the milk is in a bad state and little is collected in the breasts. As a consequence, I myself would recommend that, if a woman nursing an infant should become pregnant, another nurse   47K

σκεπτομένους τε καὶ δοκιμάζοντας αὐτῆς ἀκριβῶς τὸ
γάλα γεύσει καὶ ὄψει καὶ ὀσφρήσει. καὶ γὰρ γευομένοις καὶ ὀσμωμένοις ἡδὺ καὶ θεωμένοις[17] λευκόν τε
καὶ ὁμαλὸν καὶ μέσως ἔχον ὑγρότητός τε καὶ παχύτη-
τος ὀφθήσεται τὸ ἄριστον γάλα· τὸ δὲ μοχθηρὸν ἤτοι
παχὺ καὶ τυρωδέστατον ἢ ὑγρὸν καὶ ὀρρῶδες καὶ
πελιδνὸν καὶ ἀνώμαλον ἐν συστάσει καὶ χροιᾷ καὶ
γευομένοις πικρότατον καὶ ἄλμης ἤ τινος ἑτέρας ἀλ-
λοκότου ποιότητος ἔμφασιν παρέξει· τὸ δὲ τοιοῦτον
οὐδὲ πρὸς τὴν ὀσμὴν ἡδύ. ταῦτα μὲν ἔστω σοι γνωρίσματα μοχθηροῦ τε καὶ χρηστοῦ γάλακτος, οἷς
τεκμαιρόμενος, ὅταν ᾖ διὰ κύησιν ἢ καὶ νόσημά τι
περὶ τὴν μαῖαν[18] γενόμενον ἐφ᾽ ἑτέραν ἀνάγκη τροφὸν
ἰέναι, τὴν κρίσιν τε καὶ τὴν αἵρεσιν αὐτῆς ποιεῖσθαι.

10. Τρέφειν δὲ τὸ παιδίον τὰ μὲν πρῶτα γάλακτι
μόνῳ· ἐπειδὰν δ᾽ ἐκφύσῃ τοὺς πρόσθεν ὀδόντας, ἐθίζειν ἤδη πως αὐτὸ καὶ τῆς παχυτέρας ἀνέχεσθαι τρο-
48K φῆς, ὥσπερ οὖν καὶ τοῦτο αὐτὸ τῇ πείρᾳ διδαχθεῖσαι
ποιοῦσιν αἱ γυναῖκες, ἄρτου μέντοι πρῶτον, ἔπειτα δ᾽
ὀσπρίων τε καὶ κρεῶν ὅσα τ᾽ ἄλλα τοιαῦτα προμασώ-
μεναι καὶ ἐντιθεῖσαι τοῖς στόμασι τῶν παιδίων. ἀνατρίβειν δὲ χρὴ τὸ σῶμα τῶν βρεφῶν ἐλαίῳ γλυκεῖ,
καθάπερ τοῦτ᾽ αὐτὸ ποιοῦσιν ἐπιτηδείως αἱ πλεῖσται
τῶν τροφῶν εὐθὺς ῥυθμίζουσαί τε καὶ διαπλάττουσαι
τὰ μόρια. ἀλλ᾽ ἐπί γε τοῦ νῦν ὑποκειμένου κατὰ τὸν
λόγον παιδίου, τὴν κατασκευὴν τοῦ σώματος ἀμέμπτως ἔχοντος, οὐδὲν χρὴ περιεργάζεσθαι τὴν τροφὸν
εἴς γε τὴν τῶν μελῶν εὐρυθμίαν, ἀλλ᾽ ἀνατρίβειν τε

should be found, considering and assuming her milk
would be altogether better in taste, appearance and odor.
And for those tasting and smelling the sweetness and looking at it, the best milk will be seen as white and uniform,
and is midway between watery and thick. Poor milk is either thick and very cheesy, or watery and whey-like, livid
and variable in consistency and color, and is very bitter to
those tasting it. And it will give the impression of saltiness
or some other unusual quality. Such milk is not sweet to
the smell. Take these as your signs of bad and good milk
and base your judgment on them. Whenever either preg-
nancy or some disease is present involving the nurse and
it is necessary to go on to another nurse, make the decision
and choice on this.

10. Nourish the infant with milk alone at first. When
he cuts his first teeth, accustom him to some degree to go
on to tolerate thicker food, just as women, taught by expe-   48K
rience, do—bread first, then pulses and meat and other
such things, chewing them beforehand, then putting them
into the mouths of the infants. One ought to rub the body
of infants with sweet oil, just as the majority of nurses
deliberately do, immediately shaping and molding the
parts. But in the case of the child now under discussion—
that is, one who has a faultless constitution of the body—
the nurse must not be overly officious regarding orderly
movements of the limbs, but should rub them moderately

---

[17] καὶ θεωμένοις add. Ko

[18] post νόσημά: τι περὶ τὴν μαῖαν Ko; τί ἐστι περὶ τὴν
μητέρα Ku

τὰ μέτρια καὶ λούειν ὁσημέραι, καθόσον οἷόν τε μὴ
περιεχομένου γάλακτος ἀπέπτου κατὰ τὴν γαστέρα·
κίνδυνος γὰρ ἀναληφθῆναι τοῦτο πρὶν πεφθῆναι καλῶς εἰς ὅλον τὸ σῶμα τοῦ παιδίου. πολὺ δὲ δὴ μᾶλλον, εἰ καὶ τὴν γαστέρα τις αὐτὴν ἀνατρίβοι γάλακτος μεστήν, ἐμπλήσει τε τὸ σῶμα τροφῆς ἀπέπτου
πληρώσει τε τὴν κεφαλήν. διὸ χρὴ πολλὴν πρόνοιαν
πεποιῆσθαι τοῦ μὴ λαμβάνειν τὴν τροφὴν τὸ παιδίον
μήτε πρὸ λουτρῶν μήτε πρὸ τῶν ἀνατρύψεων.

49K   γένοιτο δ' ἂν τοῦτο, παραφυλαττούσης ἀκριβῶς
τῆς τροφοῦ τὸν ἐπὶ τοῖς μακροτέροις ὕπνοις καιρόν·
ἐν τούτῳ γὰρ μάλιστα τὴν κοιλίαν ἤτοι παντάπασιν
κενὴν ἢ πεπεμμένην ἤδη τὴν τροφὴν περιέχουσαν
εὑρεῖν ἔστιν, οὐδ',[19] ὥσπερ νῦν ποιοῦσι[20] ἔνιαι μὲν τῶν
τροφῶν, ἕνα τινὰ χρόνον ἀφορίσασαι τῆς ἡμέρας,
ἔνιαι δ',[21] ὅταν αὐταὶ σχολάσωσι, [τοῦ] τηνικαῦτα
προνοούμεναι· διότι πολλάκις ἀναγκαῖόν ἐστι βλάπτεσθαι τὰ παιδία ἤπερ ὠφελεῖσθαι. ὁ γὰρ ὑφ' ἡμῶν
ἀφοριζόμενος καιρὸς ἄλλοτε εἰς ἄλλον ἐμπίπτει χρόνον ἤτοι τῆς ἡμέρας ἢ τῆς νυκτός. ἐπὶ μέντοι τῶν
μειζόνων ἤδη παιδίων, ὅσα καὶ πληγαῖς καὶ ἀπειλαῖς
ἐπιπλήξεσί τε καὶ νουθετήσεσι πείθεται, καιρὸς ἂν
εἴη διττὸς εἰς ἀνάτριψίν τε καὶ λουτρόν· ὁ μὲν πρό-
τερός τε καὶ ἄριστος, ἐπειδὰν ἐξαναστάντα τῶν ἑωθινῶν ὕπνων, εἶτα παίξαντα τροφὴν αἰτῇ. τότε γὰρ ἐπι-
τίθεσθαι μάλιστα αὐτοῖς χρή, τὸ μὲν σῶμα πρὸς
ὑγίειαν τε ἅμα καὶ εὐεξίαν ἀσκοῦντας, τὴν δὲ ψυχὴν
εἰς εὐπείθειάν τε καὶ σωφροσύνην, οὐκ ἄλλως παρ-

and bathe [the infant] every day, as far as possible when
no unconcocted milk is contained in the stomach, for
there is a danger of this being taken up into the whole body
of the infant before it is properly concocted. And much
more, if someone were to rub the stomach itself while it
is full of milk, this will fill the body with unconcocted food
and fill up the head. Accordingly, it is necessary to have
given considerable care to the infant not taking nourishment either before bathing or before massage.

This will occur if the nurse pays very close attention to   49K
the time after longer sleeps. At this time particularly, the
belly is found to be either entirely empty or to contain
food that has already been concocted. Do not, as some
nurses now do, designate one particular time of day [for
feeding], or as others do, who only provide nutriment
when they have time. Accordingly, often and inevitably,
the infants are harmed rather than helped. The appropriate time I am defining may fall at one time or another,
either during the day or during the night. However, in
children who are already bigger—those who are prevailed
upon by blows, threats, rebukes and admonitions—there
may be two different times for rubbing and bathing. The
first and best is when children, after getting up from their
early morning sleep and then playing, want food. For at
that time we should apply ourselves to them especially,
and we can train the body toward health and, at the
same time, a good state, and the mind to obedience and

---

19 οὐδ᾽ Ko; εἰ δ᾽ Ku

20 ποιοῦσι Ko; ποιοῦσι, βλάπτουσιν Ku

21 post ἔνιαι δ᾽,: ὅταν αὐταὶ σχολάσωσι, [τοῦ] τηνικαῦτα
Ko; ὅταν σχολάσωσι, τὸ τηνικαῦτα Ku

50K ἕξειν αὐτοῖς τὴν τροφὴν φάσκοντας, εἰ μὴ προθύμως
ἐπακούσαιεν, εἰς ὅσον ἂν ἐθέλωμεν ἡμεῖς, ἀνατρίψεσί
τε καὶ λουτροῖς. οὗτος μὲν οὖν ὁ ἄριστος καιρός.

εἰ δέ τις ἀσχολία τὸν τρέφοντα τὸ παιδίον ἀπάγει,
μέτριον ἄρτου δόντα παίζειν ἐπιτρέπειν, εἰς ὅσον ἂν
βούληται, κἄπειτ᾽ αὖθις αἰτῆσαν αὐτὸ τηνικαῦτα τρίβειν τε καὶ λούειν. οὐ μὴν πίνειν γε ἐπιτρεπτέον ποτὲ
αὐτοὺς πρὸ τῶν λουτρῶν ἐπὶ τοῖς σιτίοις· ἀθροωτέρα
γὰρ ἂν οὕτως ἡ ἀνάδοσις εἰς τὸ σῶμα τῶν ἐν τῇ
γαστρὶ περιεχομένων γένοιτο. χρὴ δὲ ἐπὶ τῶν ἀμέμπτως ὑγιαινόντων σωμάτων φυλάττεσθαι τοῦτο. μεμ-
πτέαι γὰρ αἱ τοιαῦται διαθέσεις τε καὶ κατασκευαὶ
τῶν σωμάτων εἰσίν, ἐφ᾽ ὧν ἄμεινόν ἐστι πρὸ τῶν λουτρῶν διδόναι σιτία. καὶ διορισθήσεται περὶ αὐτῶν ἐν
τοῖς ἔπειτα λόγοις· ἀλλὰ νῦν γε τῆς ὑποθέσεως μνη-
μονευτέον ἐστίν, ὡς τὸν ἄριστα σώματος ἔχοντα
παῖδα διαιτῶμεν, ὅπως ἡμῖν φυλάττοιτο τοιοῦτος. ἐφ᾽
οὗ βέλτιόν ἐστιν ἡγεῖσθαι τὰ λουτρὰ τῶν σιτίων· εἰ
δ᾽ ἐν τοιούτῳ χωρίῳ τρέφοιτο ὁ παῖς, ἐν ᾧ βαλανεῖον
51K οὐκ ἔνεστιν (ἴσως μὲν οὖν οὐδ᾽ ὁμιλήσουσι τοῖσδε
τοῖς γράμμασιν οἱ τοιοίδε), λούουσι μὲν ἐν σκάφαις
αἱ τροφοὶ κἀνταῦθα τοὺς παῖδας, ἕως ἂν εἰς τὸ δεύτερον ἢ καὶ τρίτον ἔτος ἀπὸ γενετῆς ἐξίκωνται· μείζο-
νας δὲ γενομένους, εἰ καὶ μὴ καθ᾽ ἑκάστην ἡμέραν,
ἀλλὰ καὶ διὰ τρίτης γέ που καὶ τετάρτης ἀλείφουσί
τε καὶ ἀνατρίβουσιν· εἰ δὲ μὴ κωλύοι τὰ τῆς ὥρας, αἱ
λίμναι τε καὶ οἱ ποταμοὶ λουτρὸν αὐτοῖς εἰσιν, οἷόνπερ ἡμῖν τὸ βαλανεῖον.

72

moderation, saying we will not otherwise provide food for   50K
them unless they readily consent to rubbing and bathing
to the extent we might wish. This, then, is the best time.

If, however, some business calls the nurse who is
attending the child away, after giving him a moderate
amount of bread, he should be allowed to play as much as
he wants, and then rub and bathe him at such a time as he
wants food again. One must not permit children to drink
after eating before their bath, for in this way the distribution to the body of the things contained in the stomach
would occur in too concentrated a fashion. It is necessary
to keep to this in bodies that are perfectly healthy. There
are those conditions and constitutions of bodies that are
faulty; in these, it is better to give food before a bath. I
shall distinguish these in the discussions that follow. But
now we must recall what the proposal is—that we should
provide a regimen for the child having the best body so
that we might preserve it as such. In this case it is better
for the bath to precede food. If, however, the child is being
brought up in the sort of place in which there is not a
bathing room, (although perhaps such people would not
come into contact with these writings), nurses in this situ-   51K
ation bathe the children in basins until they reach the
second or third year from birth. When the children become bigger, they anoint and massage them, if not every
day, at least every three or four days. And unless the time
of year prevents it, pools and rivers are a bath for them,
just as the bathing room is to us.

παρὰ μέν γε τοῖς Γερμανοῖς οὐ καλῶς τρέφεται τὰ παιδία. ἀλλ' ἡμεῖς γε νῦν οὔτε Γερμανοῖς οὔτε ἄλλοις τισὶν ἀγρίοις ἢ βαρβάροις ἀνθρώποις ταῦτα γράφομεν, οὐ μᾶλλον ἢ ἄρκτοις ἢ λέουσιν ἢ κάπροις ἤ τισι τῶν ἄλλων θηρίων, ἀλλ' Ἕλλησι καὶ ὅσοι τῷ γένει μὲν ἔφυσαν βάρβαροι, ζηλοῦσι δὲ τὰ τῶν Ἑλλήνων ἐπιτηδεύματα. τίς γὰρ ἂν ὑπομείνειε τῶν παρ' ἡμῖν ἀνθρώπων εὐθὺς ἅμα τῷ γεννηθῆναι τὸ βρέφος ἔτι θερμὸν ἐπὶ τὰ τῶν ποταμῶν φέρειν ῥεύματα, κἀνταῦθα, καθάπερ φασὶ τοὺς Γερμανούς, ἅμα τε πεῖραν αὐτοῦ ποιεῖσθαι τῆς φύσεως ἅμα τε κρατύνειν τὰ σώματα, βάπτοντας εἰς τὸ ψυχρὸν ὕδωρ ὥσπερ τὸν διάπυρον σίδηρον; ὅτι μὲν γάρ, ἐὰν ὑπομείνῃ τε καὶ
52K μὴ βλαβῇ, καὶ τὴν ἐκ τῆς οἰκείας φύσεως ἐπεδείξατο ῥώμην καὶ τὴν ἐκ τῆς πρὸς τὸ ψυχρὸν ὁμιλίας ἐπεκτήσατο, πρόδηλον πάντως· ὅτι δ', εἰ νικηθείη πρὸς τῆς ἔξωθεν ψύξεως ἡ ἔμφυτος αὐτοῦ θερμότης, ἀναγκαῖον αὐτίκα τεθνάναι, καὶ τοῦτ' οὐδεὶς ἀγνοεῖ. τίς οὖν ἂν ἕλοιτο νοῦν ἔχων καὶ μὴ παντάπασιν ἄγριος ὢν καὶ Σκύθης εἰς τὴν τοιαύτην πεῖραν ἀγαγεῖν αὐτοῦ τὸ παιδίον, ἐν ᾗ θάνατός ἐστιν ἡ ἀποτυχία, καὶ ταῦτα μηδὲν μέγα τι μέλλων ἐκ τῆς πείρας κερδανεῖν; ὄνῳ μὲν γὰρ ἴσως ἤ τινι τῶν ἀλόγων ζῴων ἀγαθὸν ἂν εἴη μέγιστον, οὕτω πυκνὸν καὶ σκληρὸν ἔχειν δέρμα, ὡς ἀλύπως φέρειν τὸ κρύος· ἀνθρώπῳ δέ, λογικῷ ζῴῳ, τί ἂν εἴη μέγα τὸ τοιοῦτον; οὐδὲ γὰρ εἰς ὑγείαν ἁπλῶς οὑτωσὶ λέγων ἄν τις ἐπιτήδειον ὑπάρχειν τὸ πυκνότατον καὶ σκληρότατον δέρμα δεόντως ἂν εἴποι. διττῆς

Among the Germans, the little children are not nurtured well. But I am not now writing these things for Germans or other wild and barbaric people, any more than I am for bears, lions or wild boars, or any other wild animals, but for Greeks and for those born barbarians in race who emulate the practices of the Greeks. For who, of those dwelling among us, would tolerate an infant, still warm from the birth, being immediately carried to the flowing waters of a river, and there, as they say the Germans do, in an attempt to test its nature and at the same time strengthen its body by dipping it into cold water, like red-hot iron? Because, on the one hand, if it survives and suffers no harm, it is clear at any rate that it has both 52K demonstrated the strength of its own nature and has gained in addition further strength from the contact with the cold. On the other hand, no one is unaware of the fact that, if the innate heat of the infant were to be overcome by the external cold, it will inevitably die immediately. Who then, in his right mind, who was not a total savage and a Scythian, would choose to subject his own infant to such a test in which failure means death and nothing of significance will be gained from surviving the test? Perhaps it might be a great good for an ass or some of the other irrational animals to have a thick and hard skin so as to be able to bear the cold without distress, but for man, a rational animal, what would be great about such a thing? Speaking purely and simply of health, no one would be obliged to say that the thickest and hardest skin is needed.

γὰρ οὔσης βλάβης τοῖς τῶν ζῴων σώμασιν, ἑτέρας
μὲν ἀπὸ τῶν ἔξωθεν αἰτίων, ἑτέρας δὲ ἀπὸ τῶν ἔνδοθεν, εὐάλωτόν ἐστι τοῖς μὲν ἔξωθεν πᾶσιν, ὧν μαλα-
53K   κόν ἐστι καὶ ἀραιὸν τὸ δέρμα, τοῖς δὲ ἔνδοθεν, ὧν
πυκνόν τε καὶ σκληρόν.

καὶ διὰ τοῦθ᾿ Ἱπποκράτης περὶ τῶν ἀπὸ τροφῆς ἐν
ἡμῖν ἐπ᾿ ὠφελείᾳ τε καὶ βλάβῃ γινομένων διδάσκων
ἐπ᾿ ἄλλοις πολλοῖς καὶ τοῦτ᾿ ἔγραψεν· "ἀραιότης σώματος ἐς διαπνοήν, οἷσι πλέον ἀφαιρέεται, ὑγιεινό-
τερον, οἷσι δ᾿ ἔλασσον, νοσερώτερον." βέλτιον οὖν
ἑκατέρας πεφυλάχθαι τὰς ὑπερβολὰς καὶ μήτ᾿ εἰς
τοσοῦτον πυκνὸν τὸ δέρμα παρασκευάζειν, ὡς κωλύειν διαπνεῖσθαι καλῶς, μήθ᾿ οὕτως ἀραιόν, ὡς ὑπὸ
παντὸς αἰτίου τῶν ἔξωθεν αὐτῷ προσπιπτόντων ἑτοίμως βλάπτεσθαι. τοιοῦτον δὲ καὶ φύσει ἐστὶ τὸ σῶμα
τοῦ νῦν ἡμῖν προκειμένου τῷ λόγῳ παιδίου, μέσον
ἁπασῶν τῶν ὑπερβολῶν. οὕτως οὖν αὐτὸ διαιτητέον,
ὡς φυλάττειν ἀεὶ τῆς κατασκευῆς τὴν ἀρετήν. φυλα-
χθήσεται δέ, κατὰ μὲν τὴν πρώτην ἡλικίαν ἐν γάλακτι τρεφομένῳ καὶ λουτροῖς γλυκέων ὑδάτων θερμῶν λουομένῳ, ὅπως ὅτι μάλιστα μέχρι πλείστου
μαλακὸν αὐτῷ διαμένον τὸ σῶμα πολλὴν ἐπίδοσιν εἰς
τὴν αὔξησιν ποιοῖτο. μετὰ δὲ ταῦτα, καθ᾿ ὃν ἂν ἤδη
54K   χρόνον εἰς διδασκάλους δύναιτο φοιτᾶν, οὐκ ἀναγκαῖον ἔτι λουτροῖς χρῆσθαι συνεχέσι, ἀλλ᾿ ἀρκεῖ δια-
παλαίειν μανθάνοντα σύμμετρά τε πονεῖν ἐνταῦθα
πρὸ τῶν σιτίων, ἀλουτεῖν δὲ ἤδη τὰ πλείω. τὸ δ᾿ ὑπερ-
πονεῖν, ὥσπερ ἔνιοι τῶν παιδοτριβῶν ἀναγκάζουσι

For since there is a twofold harm to the bodies of animals—one from external causes and the other from internal causes—those in whom the skin is soft and fine-textured are readily susceptible to all external causes, 53K while those in whom the skin is thick and hard are readily susceptible to internal causes.

And because of this, Hippocrates, teaching about what occurs in us due to the benefit and harm from foods, also wrote this, in addition to many other things: "In regard to transpiration, thinness of the body is more healthy for those from whom more is eliminated and more morbid in those from whom less is eliminated."[28] Therefore, it is better that each excess is guarded against and not to provide skin so thick that it hinders good transpiration, nor so thin that it is readily injured by every cause that befalls it from without. And such, in nature, is the body of the infant now before us in the discussion, midway between all the excesses. One must, then, treat this body in such a way as to always preserve the excellence of its constitution. And this will be preserved in the first stage of life by nourishment with milk and by bathing with baths of warm, sweet waters so that in particular the body will remain soft for the longest time and so make great progress in growth. After these things, at the time the child is able to visit teachers, it is no longer essential to use baths continually; 54K it is sufficient, when he is learning wrestling, to train moderately there before food and remain unwashed for the most part. But to overtrain, as some wrestling teachers

[28] Hippocrates, *On Nutriment* 28, *Hippocrates* I, LCL 147, 352–53.

τοὺς παῖδας, οὐδαμῶς ἀγαθόν· ἀναυξῆ γὰρ ὑπὸ τῆς
παρὰ καιρὸν σκληρότητος ἀποτελεῖται τὰ σώματα,
κἂν πλείστην ὁρμὴν ἐκ φύσεως εἰς τὴν αὔξησιν ἔχῃ.
11. Οἴνου δὲ τὸν οὕτω πεφυκότα παῖδα μέχρι πλείστου μηδ᾽ ὅλως γεύειν. ὑγραίνει τε γὰρ ἱκανῶς καὶ
θερμαίνει τὸ σῶμα πινόμενος οἶνος ἐμπίπλησί τε τὴν
κεφαλὴν ἀτμῶν ἐν ταῖς ὑγραῖς καὶ θερμαῖς κράσεσι,
οἷάπέρ ἐστι καὶ ἡ τῶν τοιούτων παιδίων. ἀλλ᾽ οὔτε
ἐμπίπλασθαι καλὸν αὐτοῖς τὴν κεφαλὴν οὔτε ὑγραίνεσθαι καὶ θερμαίνεσθαι περαιτέρω τοῦ προσήκοντος. εἰς τοσοῦτον γὰρ ἥκουσιν ὑγρότητός τε καὶ θερμότητος, ὡς, εἰ καὶ βραχὺ παραυξήσειέ τις ὁπότερον
αὐτῶν, ἐν ἀμετρίᾳ καθίσταται. οὐσῶν δὲ πασῶν τῶν
ἀμετριῶν φευκτῶν, ἡ τοιαύτη μάλιστ᾽ ἂν εἴη φευκτή,
καθ᾽ ἣν οὐκ εἰς τὸ σῶμα μόνον, ἀλλὰ καὶ εἰς τὴν
55K   ψυχὴν ἡ βλάβη διικνεῖται. διόπερ οὐδὲ τοῖς ἤδη τε-
λείοις ἄνευ τοῦ προσήκοντος μέτρου πινόμενος οἶνος
ἀγαθός· εἰς θυμούς τε γὰρ ἐκκαλεῖται προπετεῖς καὶ
ὕβριν ἀσελγῆ καὶ τὸ λογιζόμενον τῆς ψυχῆς ἀμβλύ
τε καὶ τεθολωμένον ἐργάζεται. ἀλλὰ τούτοις μὲν ἑτοίμως εἰς τὴν τῶν χολωδῶν περιττωμάτων ἐπίκρασίν τε
ἅμα καὶ κένωσιν ἐπιτήδειος. οὐχ ἥκιστα δὲ καὶ εἰς
τὴν ἐν αὐτοῖς τοῖς στερεοῖς ὀργάνοις τοῦ ζῴου γινο-
μένην ξηρότητα διά τε πόνους ὑπερβάλλοντας, ἔστιν
ὅτε δὲ καὶ διὰ τὴν οἰκείαν τῆς ἡλικίας κρᾶσιν, ἐπιτήδειος ὁ οἶνος, ὑγραίνων μὲν καὶ ἀνατρέφων, ὅσον
ἀμέτρως ἐξήρανται, πραΰνων δὲ τὸ δριμὺ τοῦ πικροῦ
χυμοῦ καὶ δι᾽ ἱδρώτων καὶ οὔρων ἐκκενῶν. οἱ δὲ παῖ-

compel boys to do, is in no way good. For bodies are rendered unable to grow by hardness contrary to the stage of
life, even if they have a considerable natural impulse toward growth.

11. A child that has been brought up in this way should
not taste wine at all for as long as possible. Since drinking
wine moistens significantly and heats the body, it fills the
head with vapors in the moist and hot *krasias,* which is the
krasis of such children. But it isn't good for them for the
head to be filled, or to be moistened and heated beyond
what is appropriate. They have come to such a degree of
moisture and heat that, if either one of these two should
increase even a little, they are in a state of imbalance.
Although all these imbalances are to be avoided, one such
as this should be particularly avoided, since in it harm
penetrates not only into the body but also into the mind   55K
(soul). For this reason, drinking wine in an inappropriate
amount is not good, even for those who are already fully
grown. It elicits a proneness to anger, insolence and lewd-
ness, dulls the working of the mind, and creates befuddle-
ment. But wine is useful in adults for tempering the bilious
superfluities and, at the same time, is also useful for their
evacuation. Not least, wine is also useful for dryness occurring in the actual solid organs of the organism due to
excessive labors, and sometimes also due to the specific
krasis of the stage of life, since it moistens and nourishes
to the extent that they have become immoderately dry,
mollifies the sharpness of the bitter humor, and effects
elimination by sweat and urine. Children, inasmuch as

δες, ἅτε μήτε τὸν τοιοῦτον ἀθροίζοντες χυμὸν οἰκείαν
τε παμπόλλην ἔχοντες ὑγρότητα, τῶν μὲν ἐξ οἴνου
γενομένων ἀγαθῶν οὐδενὸς προσδέονται, μόνης δὲ
ἀπολαύουσι τῆς αὐτοῦ βλάβης. οὐκοῦν νοῦν ἔχων οὐδεὶς ἐπιτρέψει τοιούτῳ πόματι χρῆσθαι τοὺς παῖδας,
ὃ πρὸς τῷ μηδὲν ἀγαθὸν ἐργάζεσθαι βλάβην ἐξαί-
σιον ἐφεδρεύουσαν ἔχει.

56K   οὐ μὴν οὐδὲ ψυχροῦ πόματος εἰς τὸ παντελὲς εἴργειν κελεύω τοὺς τοιούτους παῖδας, ὥσπερ ἔνιοι ποιοῦσιν, ἀλλ' ἐπί τε σιτίοις τὰ πολλὰ καὶ κατὰ τὰς
θερμοτάτας ὥρας, ὅταν ἥξωσιν αὐτοὶ πρὸς τὸ ψυχρόν, ἐπιτρέπω χρῆσθαι, μάλιστα μέν, εἰ οἷόν τε, πηγαίῳ προσφάτῳ, μηδεμίαν ἐπίκτητον ἔχοντι ποιότητα
μοχθηράν, μὴ παρόντος δὲ τοῦ τοιούτου τοῖς ἄλλοις.
φυλάττεσθαι δὲ τὰ λιμναῖα καὶ θολερὰ καὶ δυσώδη
καὶ ἁλυκὰ καὶ ἁπλῶς εἰπεῖν ὅσα τινὰ ποιότητα κατὰ
τὴν γεῦσιν ἐνδείκνυται· χρὴ γὰρ ἀποιότατον αὐτοῖς
τὸ κάλλιστον ὕδωρ φαίνεσθαι, οὐ πρὸς τὴν γεῦσιν
μόνον, ἀλλὰ καὶ πρὸς τὴν ὀσμήν. εἴη δ' ἂν τὸ τοιοῦτον
ἥδιστόν τε ἅμα πίνοντι καὶ ἀκριβῶς καθαρόν. εἰ δὲ
καὶ ταχέως ὑποχωρεῖ τῶν ὑποχονδρίων, μηδὲν ζητεῖν
ἕτερον βέλτιον, ὡς, ὅσα γε καθαρὰ μέν ἐστι καὶ λαμ-
πρὰ καὶ οὐκ ἀηδῆ πινόμενα, παραμένει δὲ ἐπὶ πλέον
ἐν τοῖς ὑποχονδρίοις, ἡμιμόχθηρα νομιστέον. ἀπέχεσθαι δὲ τούτων κελεύω τῶν πάμπαν ψυχρῶν, οὐ μὴν
θερμοῖς γε χρῆσθαι κωλύω. ἀσφαλέστατον μὲν οὖν
57K τῇ πείρᾳ κεκρίσθαι τὸ τοιοῦτον ὕδωρ· εἰ δὲ καὶ διὰ
γνωρισμάτων τις ἐθέλοι προγινώσκειν αὐτοῦ τὴν δύ-

they do not collect such a humor and have a very considerable natural moistness, require none of the benefits occurring from wine and only enjoy the harm of it. So then, no
one with sense will allow children to use such a drink—a
drink which, in addition to doing nothing good, has an
egregious harm lying in wait.

In respect of such children, I do not direct keeping   56K
them away from cold drinks completely, as some do, but
permit their use after food in many instances, and in the
hottest seasons, whenever they themselves are inclined
toward what is cold, and particularly, if possible, from a
fresh spring to which no bad quality has been added. But
if such a spring is not available, use other drinks. One must
be on guard against pools of stagnant water, and muddy,
foul smelling or salty water, or in short, those waters that
display to the taste some particular quality. For the best
water must seem to them to be without qualities (i.e.,
pure)—and not only to taste but also to smell. Such water
should seem sweet and, at the same time, also perfectly
pure to the one drinking it. And if it passes through the
hypochondria quickly, nothing else better could be sought,
since in fact all waters that are pure and clear, and not
unpleasant to drink, but that stay rather a long time in the
hypochondria, must be considered partially bad. I direct
abstention from these [waters] when they are completely
cold, but I do not prevent the use of those that are hot. It   57K
is safest for such water to have been judged by experience.
If, however, someone should wish to know beforehand its

ναμιν, ὅσων αἱ πηγαὶ πρὸς ἄρκτον ἐρρυήκασιν ἐκ
πετρῶν θλιβόμεναι τὸν ἥλιον ἀπεστραμμένον ἔχουσαι, ἀτέραμνά τε καὶ βραδύπορα τὰ τοιαῦτα χρὴ νομίζειν ἅπαντα. εὐθὺς δ᾽ αὐτοῖς ὑπάρχει καὶ τὸ θερμαί-
νεσθαί τε καὶ ψύχεσθαι βραδέως· ὡς ὅσων γε πρός
τε τὰς ἀνατολὰς ἐρρώγασιν αἱ πηγαὶ καὶ διὰ πόρου
τινὸς ἢ γῆς διηθεῖται καθαρᾶς θερμαίνεταί τε καὶ ψύχεται τάχιστα, ταῦτ᾽ ἐλπίζειν εἶναι κάλλιστα πάσαις
ταῖς ἡλικίαις.

 οὐ γὰρ δή, καθάπερ οἴνων τε καὶ σιτίων καὶ γυμνασίων ἐγρηγόρσεώς τε καὶ ὕπνων καὶ ἀφροδισίων ἄλλον ἄλλως ἀπολαύειν προσήκει κατὰ τὰς διαφερού-
σας ἡλικίας, οὕτω καὶ ὕδατος, ἀλλ᾽, ὅπερ ἄριστον
εἴρηται νῦν, τούτῳ χρῆσθαι καὶ παῖδα καὶ νεανίσκον
καὶ πρεσβύτην, ὥσπερ γε καὶ ἀέρα τὸν ἄριστον εἰσ-
πνεῖν ἐν[22] ἅπασιν ὁμοίως χρηστόν. ἄριστον δὲ ἀέρα
λέγω τὸν ἀκριβῶς καθαρόν· εἴη δ᾽ ἂν τοιοῦτος ὁ μήτ᾽
ἐκ λιμνῶν ἢ ἑλῶν ἀναθυμιάσεως ἐπιθολούμενος μήτ᾽
58K ἐκ βαράθρων δηλητήριον αὔραν ἀποπνεόντων, ὁποία
περί τε Σάρδεις ἐστὶ καὶ Ἱερὰν πόλιν ἑτέρωθί τε πολ-
λαχόθι τῆς γῆς. οὕτω δὲ καὶ ὅστις ἔκ τινος ὀχετοῦ
τῶν καθαιρόντων ἢ μεγάλην τινὰ πόλιν ἢ πολυάν-
θρωπον στρατόπεδον ἐπιθολοῦται, μοχθηρὸς ἱκανῶς
ἐστι. μοχθηρὸς δὲ καὶ ὃς ἂν ἔκ τινος σηπεδόνος ἢ
ζῴων ἢ λαχάνων ἢ ὀσπρίων ἢ κόπρου μιαίνηται. καὶ
μὴν καὶ ὅστις ὁμιχλώδης ἐστὶ διὰ ποταμὸν ἢ λίμνην

----

[22] ἐν add. Ku

potency from signs, those waters whose sources flow toward the north, or are squeezed out from rocks and are turned away from the sun must all be considered unsoftened and slow of passage. In them, both the heating and cooling are immediately slow. On the other hand, in those whose sources face the rising sun and are filtered through some channel or pure earth, the heating and cooling are very quick—one should expect these to be the best for all the stages of life.

Water is not like wine, food, exercise, wakefulness and sleep, and sexual activity where it is appropriate for the different stages of life to enjoy the benefits of these things differently. Rather, what I just now said is best is what children, adolescents and old people should use, just as the best air to breathe in is similarly good in all ages. I call the "best air" that which is absolutely pure. Such air should not be from marshy lakes or from the exhalation of marshy ground, or the noxious air that comes forth from pits of 58K the kind there are around Sardis and Hierapolis and many other places in the world.[29] Likewise also, the air made turbid by some aqueduct that drains a great city or a populous military camp is very bad. Bad too is air tainted by putrefaction, either of animals, vegetables, pulses or dung. And also air that is cloudy due to a neighboring river or

[29] Sardis, the chief city of Lydia, was situated under a steep fortified hill in the Hermus valley—see G. M. A Hanfman, *Sardis from Prehistoric to Roman Times* (1983). Hierapolis was an ancient city located on hot springs in Phrygia in southwestern Anatolia. It was noted for its hot baths.

γειτνιῶσαν, οὐκ ἀγαθός, ὥσπερ γε καὶ ὅστις ἂν ἐν
κοίλῳ χωρίῳ πανταχόθεν ὄρεσιν ὑψηλοῖς περιεχο-
μένῳ μηδεμίαν πνοὴν δέχηται· πνιγώδης γὰρ ὅδε καὶ
σηπεδονώδης ἐστὶν ἀνὰ λόγον τοῖς ἀποκεκλεισμένοις
ἐν οἴκοις τισίν, ἐν οἷς εὐρὼς ὑπὸ σηπεδόνος τε καὶ
ἀπνοίας ἀθροίζεται. οἱ μὲν δὴ τοιοῦτοι πάσαις ταῖς
ἡλικίαις λυμαίνονται, ὥσπερ γε καὶ ὁ καθαρὸς ἀκριβῶς ἁπάσαις ταῖς ἡλικίαις ἀγαθός. ἡ δὲ κατὰ θερμότητα καὶ ψυχρότητα καὶ προσέτι ξηρότητα καὶ
ὑγρότητα διαφορὰ τῶν ἀέρων οὐχ ὁμοίως ἔχει πρὸς
ἅπαντας, ἀλλὰ τοῖς μὲν εὐκράτοις σώμασιν ὁ εὔκρατος ἀὴρ[23] ἄριστος, ὅσα δ᾽ ἂν ὑπό τινος ἐξεχούσης
59K   ποιότητος δυναστεύηται, τούτοις ἄριστος ὁ ἐναντιώ-
τατος τῇ κρατούσῃ, ψυχρὸς μὲν τῇ θερμῇ, θερμὸς δὲ
τῇ ψυχρᾷ, καὶ δὴ καὶ τῇ μὲν ὑγροτέρᾳ ξηρός, τῇ δὲ
αὐχμηροτέρᾳ τοῦ προσήκοντος εἰς τοσοῦτον ὑγρότε-
ρος, εἰς ὅσον κἀκείνη τοῦ συμμέτρου ξηροτέρα.

ταυτὶ μὲν ἐν τῷδε τῷ λόγῳ γινώσκειν ἱκανά· ὥσπερ
δ᾽ ἄν τις ἐπανορθοῖτο τὰς ἐκ τῶν μοχθηρῶν ὑδάτων
τε καὶ ἀέρων βλάβας, ἐν ἑτέρῳ βιβλίῳ ῥηθήσεται.
νυνὶ γὰρ οἷον σκοπόν τινα καὶ κανόνα τὴν ἀρίστην
κατασκευὴν τοῦ σώματος ἐπὶ τοῖς ἀρίστοις διαιτή-
μασι διελθεῖν ἔγνωκα· τὰς δὲ τῶν ἡμαρτημένων κατά
τι σωμάτων ἅμα καὶ διαιτημάτων ἐπαλλάξεις ἐν τοῖς
μετὰ ταῦτα γράμμασιν ἁπάσας διαιρήσομεν.

12. Πάλιν οὖν ὁ λόγος ἐπὶ τὸν ἄριστα κατεσκευασμένον παῖδα ἐπανελθὼν τὴν ἀπὸ τῆς πρώτης ἑβδομάδος ἡλικίαν αὐτοῦ μέχρι τῆς δευτέρας ἐκδιηγεί-

marshy lake is not good, just as that in a hollow place enclosed on all sides by lofty mountains, receiving no breeze is not. This air is stifling and foul, analogous to the air in houses that are shut up in which mold collects due to putrefaction and lack of ventilation. Such airs are harmful for all the stages of life, just as absolutely pure air is good for all the stages of life. In relation to heating and cooling, and besides these, drying and moistening, the difference of the airs is not the same for all *krasias.* For *eukratic* bodies, *eukratic* air is best, whereas for those in which some outstanding quality prevails, the most opposite to the prevailing quality is best—cold for hot, hot for cold, and further, dry for the too moist and moist for the too dry, and for that which is drier than is appropriate, what is more moist to the extent to which it is drier than a proper balance.

59K

This is enough to understand in the present discussion. I shall speak in another work about how someone might correct the harms from bad waters and airs.[30] For the time being, I have decided to go over only the objective and rule in the case of the best regimes in regard to the best constitution of the body. I shall break down all the varieties of the faults pertaining to bodies along with the regimens in the books following these.

12. Since the discussion is returning again to the child of the best constitution, let me speak in detail about its stage of life from the first seven years to the second, in

---

[30] See, for example, Book 2 of *Mixt.*, I.572–645K.

---

[23] ἀήρ *add.* Ko

σθω, κατά τε τὴν κρᾶσιν ὁποία τίς ἐστι καὶ ὧντινων
χρῄζει διαιτημάτων. ἡ μὲν δὴ κρᾶσις, ὡς κἂν τοῖς
Περὶ κράσεων ὑπομνήμασι δέδεικται, θερμὴ μέν
60K ἐστιν ὁμοίως, ὑγρὰ δὲ οὐχ ὁμοίως. ἀεὶ γὰρ ἀπὸ τῆς
πρώτης γενέσεως ἅπαν ζῷον ὁσημέραι γίνεται ξηρότερον, οὐ μὴν θερμότερόν γε ἢ ψυχρότερον ὡσαύτως
ἐπὶ πάσης[24] ἡλικίας· ἀλλ᾽ ὅσα μὲν ἄριστα κατεσκεύασται σώματα, παραπλήσιά πως ἐπὶ τούτων ἄχρι τῆς
ἀκμῆς ἡ θερμότης παραμένει· ὅσα δὲ ὑγρότερα καὶ,
ψυχρότερα τῶν ἀρίστων ἐστίν, αὐξάνεται τούτων ἡ
θερμότης. ἀλλ᾽ οὐχ ὅ γε νῦν λόγος ὑπὲρ ἐκείνων
ἐστίν.

ὁ δὲ ἄριστα κατεσκευασμένος ἄνθρωπος ἄχρι τῆς
τεσσαρεσκαιδεκαέτιδος ἡλικίας ἐν τῇ προειρημένῃ
διαίτῃ φυλαττέσθω, γυμναζόμενος μήτε πάνυ πολλὰ
μήτε βίαια, μή πως αὐτοῦ τὴν αὔξησιν ἐπίσχωμεν,
καὶ λουόμενος ἐν θερμοῖς μᾶλλον ἢ ψυχροῖς λουτροῖς·
οὔπω γὰρ οὐδὲ τούτων ἀλύπως ἀνέχεσθαι δυνήσεται.
πλαττέσθω δὲ καὶ τὴν ψυχὴν ἐν τῷδε τῆς ἡλικίας καὶ
μάλιστα δι᾽ ἐθισμῶν τε σεμνῶν καὶ μαθημάτων, ὅσα
μάλιστα ψυχὴν ἐργάζεσθαι κοσμίαν ἱκανά· πρὸς γὰρ
τὰ μέλλοντα κατὰ τὴν ἑξῆς ἡλικίαν αὐτῷ περὶ τὸ
σῶμα πραχθήσεσθαι μέγιστον ἐφόδιόν ἐστιν ἡ εὐ-
κοσμία τε καὶ εὐπείθεια.

61K μετὰ δὲ τὴν δευτέραν ἑβδομάδα μέχρι τῆς τρίτης
εἰ μὲν εἰς τὴν ἄκραν εὐεξίαν ἄγειν αὐτὸν ἐθέλοις, ἤτοι
στρατιώτην τινὰ γενναῖον ἢ παλαιστρικὸν ἢ ὁπωσοῦν
ἰσχυρὸν ἀπεργάσασθαι βουλόμενος, ἧττον τῶν τῆς

respect to what the *krasis* is and what kind of regimens it
needs. Now the *krasis,* as I have shown in the treatises *On
Mixtures,*[31] is similarly hot but not similarly moist. For   60K
always, right from the time of birth, every animal becomes
drier every day but not hotter or colder in the same manner in every stage of life. But in the case of those bodies
that are best constituted, the heat remains about the same
up to full development, whereas in the case of those that
are moister and colder than the best, their heat increases.
However, the discussion is not now about those.

Let a person with the best constitution be kept to the
previously mentioned regimen up to the fourteenth year,
exercising neither very much nor violently, lest in some
way we hold back his growth, and bathing in warm rather
than cold baths, for he will not yet be able to tolerate the
latter without harm. And in this stage of life, let him be
molded with respect to his soul, and particularly through
serious habits and studies that are especially able to make
the soul well-ordered. For regarding those things that are
going to be brought about concerning the body in the
stage of life to follow this one, the greatest support comes
from good conduct and compliance.

However, after the second seven years and up to the   61K
third, if you wish to bring the child to a peak of good
health, wanting to make him either a noble soldier, or a
wrestler, or strong in any other way whatsoever, you may

[31] *Mixt.,* I.572–645K (English trans., Singer, *Galen: Selected
Works*). See particularly I.578K ff.

---

[24] πάσης Ko; πάσαις Ku

ψυχῆς ἀγαθῶν, ὅσα γε εἰς ἐπιστήμην τινὰ καὶ σο-
φίαν ἄγει, προνοήσῃ· τὰ μὲν γὰρ εἰς ἦθος ἐν τῷδε
μάλιστα τῆς ἡλικίας ἀκριβωθῆναι προσήκει. εἰ δὲ τὰ
μὲν κατὰ τὸ σῶμα μέχρι τοῦ κρατυνθῆναι τὰ μόρια
καὶ περιποιῆσαί τιν' ἕξιν ὑγιεινὴν καὶ αὐξῆσαι προ-
αιροῖο, τὸ δὲ λογικὸν τῆς ψυχῆς τοῦ μειρακίου κοσμῆ-
σαι σπουδάζοις, οὐ τῆς αὐτῆς ἐπ' ἀμφοῖν διαίτης
δεηθήσῃ. καίτοι καὶ τρίτον ἂν καὶ τέταρτον εἶδος
βίου εὑρεθείη ποτέ, τῶν μὲν ἐπί τινα βάναυσον ἀφιγμένων τέχνην, καὶ ταύτην ἤτοι γυμνάζουσαν ἢ ἀγύ-
μναστον φυλάττουσαν τὸ σῶμα, τῶν δὲ ἐπὶ γεωργίαν
ἢ ἐμπορίαν ἤ τι τοιοῦτον ἕτερον. ὥστε καὶ χαλεπὸν
εἶναι δοκεῖ ἀριθμῷ τινι περιλαβεῖν πάσας τῶν βίων
τὰς ἰδέας. τῆς μὲν γὰρ ὑγιεινῆς τέχνης ἐπάγγελμά
ἐστιν ἅπασιν ἀνθρώποις ὑποθήκας διδόναι πρὸς
62K ὑγείαν, ἤτοι καθ' ἕκαστον ἰδίας ἢ κοινῇ σύμπασιν
ἁρμοττούσας ἢ τὰς μέν τινας ἰδίας αὐτῶν, τὰς δὲ κοινάς. οὐ μὴν ἐγχωρεῖ γε περὶ πάντων ἅμα διελθεῖν,
ἀλλὰ πρῶτον μὲν ὡς ἄν τις ἐπὶ μήκιστον ἐκτείνων τὴν
ζωὴν ὑγιαίνῃ τὰ πάντα· χρὴ δ', οἶμαι, τὸν τοιοῦτον
βίον ἁπάσης ἀναγκαίας πράξεως ἀποκεχωρηκέναι,
μόνῳ σχολάζοντα τῷ σώματι· δεύτερον δὲ μεθ' ὑποθέσεως ἢ τέχνης ἢ πράξεως ἢ ἐπιτηδεύματος ἢ ὑπηρε-
σίας τινὸς ἤτοι πολιτικῆς ἢ ἰδιωτικῆς ἢ ὅλως ἀναγκαίας ἀσχολίας. οὐδὲ γὰρ ἄλλως ἂν σαφὴς ὁ λόγος
οὔτ' εὐμνημόνευτος οὔτε μεθόδῳ περαινόμενος ἡμῖν
γένοιτο χωρὶς τῆς εἰρημένης ἄρτι τάξεως.

ἐπὶ δὲ τὴν πρώτην ἐπανέλθωμεν ὑπόθεσιν ἐπιδεί-

give less forethought to the good qualities of the soul such as lead to knowledge and wisdom. For it is appropriate for those things pertaining to ethos to be perfected during this time of life particularly. If, however, you prefer the parts pertaining to the body to be strengthened up to a certain point and to attain a certain healthy state and to grow, or if you hope to adorn the rational part of the soul of the boy, you will not need the same regimen in both cases. And indeed, even a third or fourth kind of life may be found at some time or other for those devoted to one of the practical arts that maintains the body either with or without exercise—such arts are farming, commerce or something else of this sort. As a result, it seems difficult to put a number on all the kinds of lives. For the profession of the art of hygiene is to give instructions regarding health to all 62K men, which are suitable either to each person individually or to all people in common, or some of them individually and some in common. It is not possible to go over all these at the same time, but the first point is how someone, extending his life for the longest time possible, may be healthy in all respects. It is, I think, necessary for such a life to be free from all necessary activity, leaving time for the body alone. Second, in the subject under discussion, there is the issue of an art, activity, pursuit, some service (either civic or private), or some wholly necessary occupation. Otherwise our discussion would not be clear or easy to remember or accompanied by method, apart from the order spoken of just now.

Let me return to the first proposal and show how some-

ξωμέν τε, ὅπως ἄν τις[25] ἀρίστην κατασκευὴν σώματος
ἔχων ἀποχωρήσας ἁπάντων τῶν κατὰ τὸν βίον εἰς τὸ
κοινὸν συντελούντων, ἑαυτῷ μόνῳ ζήσειε, μήτε νοσή-
σας μηδέποτε, καθόσον οἷόν τε, μήτε ἀποθανὼν ἔμ-
63K προσθεν τοῦ μηκίστου χρόνου τῆς ζωῆς. ἄφθαρτον
μὲν γὰρ ποιῆσαι τὸ γεννητὸν οὐχ οἷόν τε, κἂν ὅτι
μάλιστα τῶν καθ᾽ ἡμᾶς τις νῦν ἀνὴρ φιλόσοφος ἐπει-
ρᾶτο δεικνύναι τοῦτο διὰ τοῦ θαυμασίου τούτου συγ-
γράμματος, ἐν ᾧ διδάσκει τὴν ὁδὸν τῆς ἀθανασίας.
ἐπὶ πλεῖστον δὲ χρόνον προήκειν ἐγχωρεῖ ποιῆσαι
ζῴου σῶμα, καὶ μάλιστα τοῦ κάλλιστα πεφυκότος.
ἔνια γὰρ οὕτως εὐθὺς ἐξ ἀρχῆς κατεσκεύασται κακῶς,
ὡς μηδ᾽ εἰς ἑξηκοστὸν ἔτος ἀφικέσθαι δύνασθαι, κἂν
αὐτὸν ἐπιστήσῃς αὐτοῖς τὸν Ἀσκληπιόν. ἀλλ᾽ οὐ νῦν
περὶ ἐκείνων ὁ λόγος.

ἐπὶ δὲ τὸν κατεσκευασμένον ἄριστα πάλιν ἐπανελ-
θόντες ἀναμνήσωμεν ὧν ἐν ἀρχαῖς ἀπεδείξαμεν, ὡς
ἐσθίειν μὲν καὶ πίνειν ἀναγκαῖον ἡμῖν ἐστιν, ἐπειδὴ
διαπαντὸς ἀπορρεῖ τι τοῦ σώματος ἡμῶν, ἐπεὶ δὲ
ἐσθίομέν τε καὶ πίνομεν, ἀναγκαῖον αὖθίς ἐστι τῆς
τῶν περιττωμάτων προνοεῖσθαι κενώσεως. ἐπεὶ δὲ
τούτων ἐστὶν εἴδη πολλά, τὰ μὲν τῆς ἐν τῇ γαστρὶ
πεττομένης τροφῆς, τὰ δὲ τῆς ἐν ἥπατι καὶ ἀρτηρίαις
64K καὶ φλεψί, τὰ δὲ τῆς καθ᾽ ἕκαστον μόριόν ἐστι περιττώματα, χρὴ δήπου καὶ τὴν κένωσιν αὐτῶν ἰδίαν εἶ-

[25] post τις: ἀρίστην κατασκευὴν σώματος ἔχων Ko; ἀρίστης τυχὼν κατασκευῆς σώματος Ku

one who has the best constitution of the body, withdrawing from all those things that contribute in life to the general good, might live for himself alone and never at any time be sick, as far as possible, nor die before the longest time of life. For it is not possible to make immortal that which   63K
is generated, even though, particularly, one of our contemporaries who is a philosopher attempted to show this through his remarkable treatise in which he teaches the road to immortality.[32] However, it is possible to make the body of an animal last for a very long time, and especially one that is best in terms of nature. For some are constituted so badly right from the beginning that they are not able to reach even the sixtieth year, if you were to set Asclepius himself over them. However, the discussion about those people is not for today.

So let me return once more to the best constitution, and call to mind those things I showed at the start—that it is necessary for us to eat and drink since something of our body is continually flowing away. When we eat and drink, it is again necessary to give forethought to the evacuation of the superfluities. And since there are many kinds of these—those from food concocted in the stomach, those from what is concocted in the liver, arteries and veins, and those from what is concocted in each part—it   64K
must be clear also that the evacuation of these is specific

[32] It is not clear to whom Galen is referring here. The same person is mentioned again in Book 6 (399K) and there identified as a Sophist.

ναι καθ' ἕκαστον, ὥσπερ γε καὶ ἡ φύσις αὐτὴ φαίνεται τοῦτο ἐξ ἀρχῆς ἐργασαμένη. παρεσκεύασε γὰρ
τοῖς ζῴοις ὄργανα πολλά, τὰ μὲν ἐκκαθαίροντά τε καὶ
διακρίνοντα ταυτὶ τὰ περιττώματα, τὰ δὲ παράγοντα,
τὰ δὲ ἀθροίζοντα, τὰ δὲ ἐκκρίνοντα. καὶ λέλεκται μὲν
ὑπὲρ ἁπάντων ἐπὶ πλέον ἔν τε τοῖς τῶν φυσικῶν δυνάμεων ὑπομνήμασι κἀν τοῖς Περὶ χρείας μορίων· εἰς
δὲ τὸ παρὸν ἔσται καὶ ταῦθ' ἡμῖν ὑπόθεσις τῷ λόγῳ.

το μὲν γὰρ πρῶτον περίττωμα διακρίνεται ἅμα καὶ
προπέμπεται κατὰ βραχὺ διὰ πάντων τῶν ἐντέρων
ἄχρι τῆς κατὰ τὸ καλούμενον ἀπευθυσμένον ἀξιολό-
γου κοιλότητος, ἧς κατὰ τὸ πέρας ἐπίκεινται μύες,
εἴργοντές τε καὶ κατέχοντες ἔνδον αὐτὸ καὶ κωλύοντες
ἀκαίρως ἐκρεῖν· ἐπειδὰν δ' ἱκανῶς ἀθροισθὲν ἀνιαρὸν
ᾖ τῷ ζῴῳ, τηνικαῦτα παριᾶσιν ἔξω φέρεσθαι, συντελούντων τι πρὸς τὸ τάχος τῆς ἀφόδου τῶν κατ' ἐπιγάστριον μυῶν ἅμα τῷ διαφράγματι. τὸ δ' ἐν ἥπατι
65K   περίττωμα, τὸ μὲν οἷόνπερ τὸ καλούμενον ἄνθος ἐν
τοῖς οἴνοις ἐστί, τὸ δ' οἷόνπερ ἡ τρύξ. ἕλκεται δὲ τὸ
μὲν ἕτερον ὑπὸ τῆς ἐπικειμένης τῷ σπλάγχνῳ κύστεως, τὸ δ' ἕτερον ὑπὸ τοῦ σπληνός· ἀθροισθέντα δ'
ἐν τούτοις ἐκκρίνεται, τὸ μὲν εἰς τὴν ἀρχὴν τῶν λεπτῶν ἐντέρων, τὸ δ' εἰς αὐτὴν τὴν γαστέρα, καὶ ἀπὸ
τούτων ἤδη διὰ πάντων τῶν ἐντέρων ἅμα τῷ ξηρο-
τέρῳ περιττώματι τῆς τροφῆς διεξέρχεται. τὸ δὲ ἐν
φλεψὶ καὶ ἀρτηρίαις περίττωμα τοιοῦτόν ἐστιν, οἷον

---

33 *Nat. Fac.*, II.1–205K (English trans., Brock, *On the Natural*

in each case, just as Nature herself has obviously brought about from the beginning. Nature has provided for animals many organs: those that purify and separate the superfluities, those that pass them along, those that gather them together, and those that evacuate them. More has been said about all these in the treatises *On the Natural Faculties* and in those, *On the Use of the Parts*.[33] In the present circumstances, these will be the foundation of our discussion.

The first superfluity is separated and sent onward gradually through all the intestines as far as the substantial cavity termed the rectum; around the outlet of this muscles are placed that close it off and retain it within, preventing the superfluity flowing out in an untimely fashion. But when what is collected is sufficiently distressing to the animal, then the muscles relax and allow it to be carried outward. The muscles in the epigastrium along with the diaphragm contribute jointly to the speed of expulsion. In the case of the superfluity in the liver, in part it is like the 65K so-called flower in wines, while in part it is like the lees. The former is drawn by the bladder underlying the viscus (i.e., the gallbladder) while the other is attracted by the spleen. When it is collected in these places, it is expelled— the former to the beginning of the thin intestines and the latter to the stomach itself. From these it now proceeds through all the intestines, along with the drier superfluity of food. The superfluity in the veins and arteries is such

Faculties; French trans., Daremberg, *Oeuvres anatomiques*—see particularly II.22K ff). *UPart.,* III.1–913 and IV.1–366K (English trans., May, *On the Usefulness;* French trans., Daremberg, *Oeuvres anatomiques*). See particularly III.333K ff.

ὀρρὸς ἐν τῷ πηγνυμένῳ γάλακτι, καθαίροντες δὲ καὶ
τοῦτο οἱ νεφροὶ τῇ κύστει παραπέμπουσιν· ἡ δ'
ἀθροίζει τρόπον ὁμοιότατον τῷ πρόσθεν εἰρημένῳ
περὶ τοῦ ξηροῦ περιττώματος· ἐπίκειται γάρ τις καὶ
τῇδε κατὰ τὸν ἔκρουν ἐπικάρσιος μῦς, κλείων οὕτως
ἀκριβῶς τὸ στόμιον, ὡς μηδὲν ἔξω παραρρεῖν. ἐπειδὰν δὲ καὶ τοῦθ' ἱκανὸν ἤδη γινόμενον ἀνιᾷ τὸ ζῷον,
ἀφίσταται μὲν ὁ μῦς τῆς φρουρᾶς ἀνιείς τε καὶ χαλῶν
ἑαυτόν, ἐκκρίνει δὲ τὸ περιττὸν ἅπαν ἡ κύστις, ἐπιβο-
ηθούντων αὖ καὶ τῇδε πρὸς τὸ τάχος ὥσπερ τῇ τῶν
ξηρῶν περιττωμάτων ἐξόδῳ τῶν μυῶν τῶν κατ' ἐπιγάστριον.

66K   τὸ δ' ὑπόλοιπον γένος τῶν περιττωμάτων γίνεται
μὲν ἐν ἑκάστῳ μορίῳ τοῦ τρέφοντος αὐτὰ χυμοῦ, τὸ
μὲν οἷον ἡμίπεπτόν τι λείψανον, ἀδυνατῆσαν ἐξομοι-
ωθῆναι τῷ τρεφομένῳ, τὸ δ', ὅπερ ἦν ἔμπροσθεν ἀνα-
δόσεως ὄχημα, πληρῶσαν τὴν χρείαν, ὑγρὸν καὶ λεπτὸν ὄν, οἷόνπερ τὸ προειρημένον ὀρρῶδες, ἐκ τῶν
ἀγγείων εἰς τὴν κύστιν συρρεῖ.[26] τούτῳ τῷ περιττώματι πόρος μὲν οὐδείς ἐστιν ἀποτεταγμένος ὑπὸ τῆς
φύσεως, ἐκκρίνεται δὲ διά τε τῶν μαλακῶν σωμάτων
φερόμενον, εἰκόντων αὐτοῦ τῇ ῥύμῃ, καὶ μάλισθ' ὅταν
ὑπὸ πνεύματος ἀθροώτερον ὁρμήσαντος ὠθῆται, καὶ
μέντοι καὶ διὰ τῶν σμικρῶν ἁπάντων πόρων, ὧν ἐστι
πλῆρες ὅλον τε τὸ σῶμα καὶ σύμπαν τὸ δέρμα. λέλεκται δ' ὑπὲρ τῆς γενέσεως αὐτῶν ἐν τοῖς Περὶ κράσεων.

τὸ μὲν δὴ λεπτότερον περίττωμα ῥᾳδίως ἐκκρίνεται

that it is like the whey in curdled milk, and when the kidneys purify this, they send it on to the bladder. And this collects it in a manner very similar to that previously described concerning the dry superfluity. For there is also a transverse muscle lying at the outlet in this, which closes the opening so precisely that nothing flows past to the outside. Whenever this [collection] has already occurred and distresses the animal, the muscle guarding the orifice relaxes, releasing and relaxing itself, and the bladder expels all the superfluity. The muscles of the epigastrium again come to its aid in effecting the swift outward passage, as in the case of the dry superfluities.

The remaining class of superfluities arises in each part 66K from the humor nourishing these. One is a kind of half concocted remnant that cannot be assimilated by what is being nourished and the other is what was formerly a vehicle of distribution that, having fulfilled its use, is moist and thin like the previously mentioned whey and flows from the vessels to the bladder. No channel has been assigned specifically to this superfluity by Nature; after being carried through the soft tissues that yield to its force, it is expelled, and particularly when it is impelled by *pneuma*, rushing in a more concentrated manner, it is forced onward, and indeed, continues through all the small channels of which the whole of the body and the entire skin are full. I have spoken about the genesis of these in the work *On Mixtures*.[34]

Now the thinner superfluity is easily expelled, being

---

[34] See note 30 above—the relevant section is 2.5 (I.614K).

---

[26] συρρεῖ Ko; περιρρεῖ Ku

πρός τε τῆς ἐμφύτου θερμότητος εἰς ἰδέαν ἀτμοῦ λυό-
μενον ὑπό τε βιαίας κινήσεως ἀθρόως ἐκρηγνύμενον·
ὀνομάζεται δὲ τὸ μὲν οὕτως ἐκκριθὲν ἱδρώς· τὸ δὲ ἕτε-
67K ρον, οὐδὲν ἔχον ὄνομα, διότι οὐδὲ γινώσκεται τοῖς
πολλοῖς, ἅτε τὴν ὄψιν ἐκφεῦγον ὑπὸ λεπτότητος,
ἄδηλος αἰσθήσει διαπνοὴ κέκληται πρὸς αὐτῶν τῶν
φωρασάντων αὐτὸ τῷ λογισμῷ. κατὰ δὲ τήνδε τὴν
ἄδηλον αἰσθήσει διαπνοὴν ἐκκρίνεταί τι καὶ τοῦ παχυτέρου περιττώματος· ἰσχυροτέρας δὲ δεῖ τῷδε καὶ
τῆς ἀπαγούσης θερμότητος καὶ τῆς ὠθούσης ῥύμης,
ἢ κίνδυνος αὐτῷ στῆναι κατά γε τὸ δέρμα, πρὶν ἀφικέσθαι πρὸς τὸ πέρας. ἐκ τούτου τοῦ περιττώματος
ἥ τε τῶν τριχῶν ἐδείκνυτο γένεσις, οὐχ ἥκιστα δὲ καὶ
ὁ περὶ τοῖς δέρμασιν ἀθροιζόμενος ἅπασι τοῖς ζῴοις[27]
ῥύπος.

εἴρηταί μοι σχεδὸν ἅπαντα τὰ ἀναγκαῖα τοῦ λόγου
κεφάλαια τῆς τῶν περιττωμάτων γενέσεώς τε ἅμα καὶ
κενώσεως, ἀποδεδειγμένα μὲν ἐν ἑτέραις πραγματείαις, ὧν ὀλίγον ἔμπροσθεν ἐμνημόνευσα, μελλήσοντα
δ᾽ ὑποθέσεις ἀναγκαῖαι γενήσεσθαι τοῖς νῦν ἐνεστη-
κόσι λόγοις. ἐπειδὴ γὰρ ἐκκενοῦσθαι χρὴ ταῦτα, μο-
χθηρὰ ταῖς ποιότησιν ὑπάρχοντα, κἂν Ἀσκληπιάδης
μὴ βούληται, χρὴ πρῶτον μὲν ἐπίστασθαι τὰς αἰτίας
αὐτῶν τῆς ἐπισχέσεως, ἐφεξῆς δὲ πειρᾶσθαι μήτε
68K περιπίπτειν αὐταῖς, εἰ δὲ καὶ περιπέσοιμέν ποτε, διὰ
ταχέων ἐπανορθοῦσθαι τὸ σφάλμα πειρᾶσθαι. τὸ μὲν

[27] τοῖς ζῴοις add. Ko

96

released by the innate heat in the form of breath. This breaks out and is collected together due to vigorous movement. What is expelled in this way is called sweat. However, the other, which has no name because it is not known 67K to the majority in that it escapes visual perception due to thinness has been called "transpiration imperceptible to sensation" by those who discovered it by reason. And in this transpiration imperceptible to sensation, some of the thicker superfluity is also expelled. For this there is need for greater strength, both of the driving heat and of the impelling force, the danger from it being that it may stay in the skin before coming to the end. The genesis of hair has been shown to be from this superfluity, no less also the dirt collected on the skin in all animals.

I have mentioned almost all the essential points of the discussion of the genesis of the superfluities along with their evacuation; these have been shown in other treatises that I mentioned a little earlier and that are going to become the foundation essential for the discussions now being set in place. Since these things must be evacuated, being harmful due to their qualities, even if Asclepiades does not wish it to be so,[35] it is first necessary to know the causes of their being held back, and then next to attempt not to encounter these, or if we are going to encounter 68K them at some time, to attempt to correct the fault as

[35] On Asclepiades' theories, foundational for Methodism and based on a system of *poroi* and *onkoi,* see J. Vallance, *The Lost Theory of Asclepiades of Bithynia* (1990).

δὴ μὴ περιπίπτειν ἐκ τοῦ γινώσκειν, πότερον αὐτάρκως ἀποκρίνεται ἢ μή, περιγίνεται· τὸ δ' ἐπανορθοῦσθαι μεθόδου τινὸς ἑτέρας προσδεῖται.

13. Λεγέσθω δὴ πρῶτον μέν, ὑφ' ὧν ἴσχεται τῶν
εἰρημένων ἐκκρίσεων ἑκάστη· δεύτερον δέ, ὅπως ἄν
τις ἐπεσχημένην αὐτὴν προτρέψειεν. ἡ μὲν δὴ τῶν
περὶ τὴν γαστέρα περιττωμάτων ἐπίσχεσις ἤτοι διὰ
τὰ λαμβανόμενα σιτία τε καὶ ποτὰ γίνοιτ' ἂν ἢ διὰ
τὴν γαστέρα μετὰ τῶν ἐντέρων, διὰ μὲν τὰ σιτία καὶ
ποτὰ παρά τε τὴν ποιότητα καὶ ποσότητα τῶν ληφθέντων ἔτι τε πρὸς τούτοις τάξιν τε καὶ τὸν τρόπον
τῆς χρήσεως, παρὰ μὲν τὴν ποιότητα, στρυφνῶν ἢ
αὐστηρῶν ἢ ξηρῶν ταῖς συστάσεσιν ὑπαρχόντων,
παρὰ δὲ τὴν ποσότητα, τοῦ προσήκοντος ἢ πλειόνων
ἢ ἐλαττόνων, παρὰ δὲ τὴν τάξιν, εἰ τὰ μὲν ξηρὰ καὶ
στύφοντα πρότερον, τὰ δὲ ὑγρὰ καὶ λιπαρὰ καὶ γλυκέα δεύτερα προσενέγκαιτο, παρὰ δὲ τὸν τρόπον τῆς
69K χρήσεως, εἰ δέον δὶς σιτεῖσθαι, πᾶσαν εἰς ἅπαξ
προσενέγκαιτο τὴν τροφήν.

ἡ δὲ περὶ τὴν γαστέρα τε καὶ τὰ ἔντερα τῆς ἐπισχέσεως τῶν περιττωμάτων αἰτία διά τε τὴν φύσιν
αὐτῶν καὶ διὰ τὴν ἐπίκτητον γίνεται διάθεσιν. αἱ μὲν
δὴ περὶ τὴν φύσιν αὐτῶν αἰτίαι τῶν μοχθηρῶν εἰσι
τοῦ σώματος κατασκευῶν, ὡς ἐν τῷ περὶ ἐκείνων εἰρήσεται λόγῳ· περὶ δὲ τῶν ἐπικτήτων ἐν τῷδε ῥητέον.
ὀκτὼ διαφοραὶ τῶν προσφάτων εἰσὶ τῆς γαστρὸς διαθέσεων, ἅπασαι δυσκρασίαι κατὰ γένος ὑπάρχουσαι,
τέσσαρες μὲν ἁπλαῖ, θερμότης καὶ ψυχρότης, ξηρό-

quickly as possible. Ensuring that they are not encoun-
tered comes from the knowledge of whether they are sufficiently separated or not; correction, however, requires
some other method.

13. So then, let me state first those things by which each
of the previously mentioned excretions is held back, and
second, how someone might urge it on when it is held
back. Now delay of the superfluities involving the stomach
may arise either through the foods and drinks taken in or
through the stomach together with the intestines, being
due to the foods and drinks on account of the quality and
quantity of what was taken in, and in addition to these, the
order and manner of use. It is on account of the quality,
when they are astringent, bitter or dry in composition; it
is on account of the quantity, when this is more or less than
appropriate; it is on account of the order, if those that are
dry and astringent are presented first and those that are
moist, fatty and sweet are presented second; it is on account of the manner of use, if the nutriment that ought to   69K
be eaten as two meals is presented all at once.

The cause of the delay of superfluities involving the
stomach and intestines is due to the nature of these and
to an acquired condition. Now the causes involving the
nature of these [structures] are those of bad constitutions
of the body, as will be described in the discussion about
those. I must, however, speak here about the acquired
[conditions]. There are eight *differentiae* of the newly acquired conditions of the stomach; all are *dyskrasias* according to class. Four are simple: hot, cold, dry and moist,

της καὶ ὑγρότης, τέσσαρες δὲ ἄλλαι σύνθετοι, θερμότης τε ἅμα καὶ ξηρότης, καὶ θερμότης ἅμα ὑγρότητι,
καὶ ψυχρότης ἅμα καὶ ξηρότης, καὶ ψυχρότης ἅμα
ὑγρότητι. χρὴ δ᾽ εἰς τοσοῦτον ἥκειν μεγέθους ἑκάστην τῶν δυσκρασιῶν, ὡς ἀσθενῆ φανερῶς ἐργάσασθαι τὴν προωστικὴν δύναμιν, ἤτοι τῆς γαστρὸς
μόνης ἢ τῶν λεπτῶν ἐντέρων ἢ τῶν παχέων, ἢ καὶ
συμπάντων ἅμα τῶν εἰρημένων ἤ τινων ἐν αὐτοῖς.
συνίστανται δ᾽ αἱ τοιαῦται δυσκρασίαι ποτὲ μὲν ἀπὸ
τῶν εἴσω τοῦ σώματος λαμβανομένων, ἔστι δ᾽ ὅτε καὶ
ἀπὸ τῶν ἔξωθεν προσπιπτόντων, ἀπὸ μὲν τῶν εἴσω
70K τοῦ σώματος, ὁπόταν ἐν ταῖς τροφαῖς ἢ τοῖς πόμασι
φαρμακωδεστέρα τις ᾖ δύναμις, ἢ θερμαινόντων ἢ
ψυχόντων ἢ ξηραινόντων ἢ ὑγραινόντων ἢ θερμαινόντων τε ἅμα καὶ ξηραινόντων ἢ καὶ κατ᾽ ἄλλην τινὰ
συζυγίαν ἐνεργούντων, ἀπὸ δὲ τῶν ἔξωθεν προσπιπτόντων, ἤτοι τοῦ περιέχοντος ἡμᾶς ἀέρος ἀμέτρως
θερμαίνοντος ἢ ψύχοντος ἢ ξηραίνοντος ἢ ὑγραίνον-
τος ἢ κατὰ συζυγίαν τινὰ τούτων ἐνεργοῦντος ἢ ἀπό
τινος ὕδατος, ἐν ᾧ τις ἔτυχε λουσάμενος, ἢ ἀλείμμα-
τός τινος ἢ ἁπλῶς ὅτου δή τινος ἑτέρου προσπεσόν-
τος ἔξωθεν τῇ γαστρὶ θερμαίνειν ἢ ψύχειν ἢ ξηραίνειν ἢ ὑγραίνειν ἀμέτρως δυναμένου. διὰ ταύτας μὲν
δὴ τὰς αἰτίας ἐπέχεται ἡ γαστήρ.

τὸ δὲ πικρόχολον περίττωμα διά τε τὴν ἀρρωστίαν
τῆς ἑλκούσης ἢ ἐκκρινούσης αὐτὸ δυνάμεως καὶ διὰ
στενοχωρίαν τῶν παραγόντων τε καὶ ἐκκρινόντων ὀργάνων. ἀλλ᾽ ἡ μὲν ἀρρωστία τῆς τε κύστεως ὅλης τῆς

while the four others are compound: hot and dry, hot and moist, cold and dry and cold and moist. It is necessary for each of the *dyskrasias* to come to such a magnitude that it renders the expulsive capacity obviously weak, either of the stomach alone, or of the thin intestines, or of the thick intestines, or of all those mentioned at the same time, or of some among them. Such *dyskrasias* sometimes arise from those things taken into the body, and sometimes from those things befalling it from without. It is from those things taken into the body whenever, in the nutriments 70K and drinks, there is some more medicinal potency operating, either heating, cooling, drying or moistening, or heating and drying at the same time, or also acting according to one of the other conjunctions. It is from those things befalling externally when either the ambient air is immoderately heating, cooling, drying or moistening, or according to some conjunction of these operating, or from some water in which someone happened to bathe, or some unguent, or simply from anything else whatsoever that befalls the stomach from without that is capable of heating, cooling, drying or moistening immoderately. Due to these causes, then, the stomach is hindered.

The picrocholic superfluity [is retarded][36] by the weakness of the capacity attracting or expelling it and by the narrowness of the conducting and expelling organs. But a weakness of the whole bladder on the liver (gallbladder)

[36] Added from Linacre's Latin translation (*moratur*).

ἐπὶ τῷ ἥπατι καὶ τῶν ἀπ' αὐτῆς εἰς τὸ σπλάγχνον
ἡκόντων στομάτων καὶ τῶν εἰς τὸ ἔντερον ἐξερευγο-
71K μένων πόρων ἐπὶ προσφάτου δυσκρασίας γένοιτ' ἄν,
ἐφ' ἧσπερ καὶ ἡ προωστικὴ δύναμις ἐλέγετο βλάπτεσθαι τῶν ἐντέρων τε καὶ τῆς γαστρός· ἡ δὲ στενοχωρία ἢ διὰ φλεγμονὴν ἢ σκίρρον ἢ ἔμφραξιν ἢ τὴν ἐκ
τῶν περιεχόντων αὐτὰ θλῖψιν ἢ μύσιν τῶν στομάτων.
αὗται δ' αἱ θλίψεις πάλιν ἐκ τῶν περιεχόντων ἢ διὰ
πλῆθος ἄμετρον τῶν ἐν αὐτοῖς περιεχομένων ἢ διὰ
φλεγμονὴν ἢ σκίρρον, ὥσπερ γε καὶ ἡ μύσις ἢ διά
τι τούτων ἢ διὰ ξηρότητα. τῆς ξηρότητος δ' αὐτῆς
αἴτια τά τε στύφοντα σφοδρῶς ἐστι καὶ τὰ θερμαίνοντα μετὰ τοῦ ξηραίνειν. τὰ μὲν γὰρ ἐκθλίβοντά τε
τὴν ὑγρότητα καὶ αὐτὰ τὰ <συνεστῶτα>[28] συνάγοντα
τε καὶ σφίγγοντα καὶ πιλοῦντα, τὰ δὲ διαφοροῦντα[29]
τὴν ξηρότητα ἐργάζεται. φλεγμονὴ δὲ καὶ σκίρρος
ἤδη γε νοσήματα φανερῶς ἐστιν· ὥστ' ἐκπέπτωκεν
τῆς ὑγιεινῆς πραγματείας, αὖθίς τε καὶ περὶ αὐτῶν
εὐκαιρότερον εἰρήσεται.

κατὰ δὲ τὸν αὐτὸν τρόπον οὐδὲ τὸ μελαγχολικὸν
ἐκκαθαρθήσεταί ποτε περίττωμα, τοῦ μὲν σπληνὸς
72K ἀνὰ λόγον ἔχοντος τῇ χοληδόχῳ κύστει, τῆς δ' εἰς
αὐτὸν τεταμένης φλεβὸς ἀπὸ τῶν πυλῶν τοῦ ἥπατος
τοῖς ἕλκουσιν ἀγγείοις τὸ χολῶδες περίττωμα, τῆς δ'
ἐκ τοῦ σπληνὸς εἰς τὴν γαστέρα φερομένης φλεβὸς
τῷ τὴν χολὴν ἐξερευγομένῳ πόρῳ. τὸ δὲ καθ' ἕκαστον
τῶν τρεφομένων τοῦ ζῴου μορίων περίττωμα διά τε

and of the openings leading away from this to the viscus and from the channels emptying themselves into the intestines may occur from a recently acquired *dyskrasia,* in which case also the expulsive capacity of the intestines and stomach is said to be harmed. Narrowness may be due to inflammation, induration, obstruction, compression by the structures surrounding it, or occlusion of the openings. These compressions may, in turn, arise from those surrounding structures due either to an excessive amount of things contained in them, or due to inflammation or induration, just as occlusion may occur due to one of these things or to dryness. Causes of the dryness itself are things that are strongly astringent and things that are heating along with drying. The former bring about dryness by squeezing out the moisture and by bringing together the things that have been opened, and by compressing and thickening, whereas the latter create dryness by dispersing the fluids. Inflammation and induration are clearly already diseases and so fall outside the matter of hygiene. It will be more opportune to speak about these subsequently.

71K

In the same way the melancholic superfluity will not be cleared out when the spleen is in a state analogous to the gallbladder, and when the vein that extends to this from the portal fissure of the liver is in a state analogous to the vessels drawing the biliary superfluity, and when the vein leading from the spleen to the stomach is analogous to the bile duct. However, the superfluity in each of the parts of the animal being nourished will be hindered by its amount

72K

---

28 ⟨συνεστῶτα⟩ Ko; κεχηνότα Ku
29 post διαφοροῦντα: τὰς ὑγρότητας Ku

πλῆθος καὶ πάχος αὐτοῦ καὶ προσέτι καὶ γλισχρότητα
ἢ δι᾽ ἀρρωστίαν τῆς ἀλλοιούσης αὐτὸ θερμότητος καὶ
διὰ στενοχωρίαν τῆς διεξόδου κωλυθήσεται. πλῆθος
μὲν δὴ καὶ πάχος καὶ γλισχρότης ἤτοι γε ἐκ τῆς τῶν
πομάτων καὶ ἐδεσμάτων γίνεται φύσεως ἢ ἐκ προσφάτου τινὸς ἀρρωστίας τῆς ἀλλοιωτικῆς ἐν τῷ τρεφομένῳ μορίῳ δυνάμεως· ἡ δ᾽ ἀσθένεια τῆς ἀλλοιούσης[30] αὐτὸ θερμότητος ἀγυμνασίας ἔκγονος ὑπάρχει·
καὶ ἡ στενοχωρία δὲ τῆς διεξόδου διὰ σκίρρον καὶ
φλεγμονὴν ἔμφραξίν τε καὶ θλίψιν καὶ μύσιν γίνεται·
τούτων δὲ ἑκάστων τὴν γένεσιν ὀλίγον πρόσθεν εἰρή-
καμεν.

ἔνια μέντοι μόρια πρὸς τοῖς ἀδήλοις τούτοις πόροις
ἑτέρους τινὰς ἐκροὰς ἔχει σαφεῖς καὶ αἰσθητούς,
73K ὥσπερ ἐγκέφαλός τε καὶ ὀφθαλμός. καὶ τοῦτο γίνεται
πρὸς τῆς φύσεως ἢ διὰ τὸ κύριον τοῦ μέρους ἢ διὰ
τὴν ἀκρίβειαν τῆς ἐνεργείας ἢ διὰ πυκνότητα τῶν
περιεχόντων σωμάτων. ὁ μὲν γὰρ ἐγκέφαλος οἶκός
τίς ἐστι τῆς λογικῆς ψυχῆς καὶ στεγανῷ περιλαμ-
βανόμενος ὀστῷ διὰ μεγίστων τε καὶ πλείστων ὀχε-
τῶν ἐκκαθαίρεται, πρῶτον μὲν τῶν κατὰ τὰς ῥῖνάς τε
καὶ τὴν ὑπερῴαν, δεύτερον δὲ τῶν καθ᾽ ἑκάτερον οὖς
καὶ τρίτον τῶν κατὰ τὰς τοῦ κρανίου ῥαφάς· οὐκ ἀπει-
κὸς δὲ καὶ εἰς τοὺς ὀφθαλμούς τι συρρεῖν ἐξ αὐτοῦ
περιττόν. ὁ δὲ ὀφθαλμὸς οὐχ ὡς κύριος ἔτι οὗτός γε,
ἀλλ᾽ ὡς ἀκριβῶς καθαρὸς εἶναι δεόμενος εἰς τὴν τῆς
ἐνεργείας ἀκρίβειαν, αἰσθητοῖς τε ἐκκενοῦται πόροις

and thickness, and in addition, viscidity, or due to the
weakness of the heat changing it, and due to the narrowness of the outflow passage. Amount certainly, thickness
and viscidity arise either from the nature of the drinks and
foods or from some recent weakness of the alterative capacity in the part being nourished, whereas the weakness
of the heat changing it is born of lack of exercise. The
narrowness of the outflow passage arises due to induration, inflammation, obstruction, compression and occlusion. I have spoken about the genesis of each of these a
little earlier.

Some parts have, however, in addition to these imperceptible channels, certain other obvious and perceptible
channels, like the brain and eye. And this occurs from 73K
Nature, due either to the importance of the part, the precision of its function, or the density of the surrounding
structures. Thus the brain is a sort of dwelling of the rational soul, and being surrounded by a bony covering, is
purified by very large and numerous channels. The first of
these are in the nostrils and palate, the second are those
in each ear, and the third is in the sutures of the cranium.
And it is not unlikely that some of the superfluity from
the brain flows to the eyes. Although the eyes are not as
important as the brain, they are, in fact, still important,
and need to be completely pure for the precision of their

---

30 ἀλλοιούσης Ko; οὐ λυούσης Ku

ἅπαν ὅσον ἐν αὐτῷ περίττωμα γεννᾶται κατά τε τὴν ῥῖνα καὶ τὰ βλέφαρα.,

14. Τὰ μὲν οὖν τῶν περιττωμάτων αἴτιά τε καὶ ὄργανα λέλεκται. ὅπως δ' ἄν τις ταῦτα κατεσχημένα κενώσειεν, ἐφεξῆς λεκτέον, ἀρξαμένους αὖθις ἀπὸ τῶν κατὰ τὴν γαστέρα. κοινὸν μὲν οὖν ἐπὶ πάντων παράγ-
74K γελμα τὴν ἐναντίαν αἰτίαν τῇ τὴν βλάβην ἐργασα-
μένῃ προσάγειν, ἴδιον δὲ καθ' ἑκάστην, εἰ μὲν ὀλι-
γώτερα καὶ ξηρότερα προσαράμενος ἐπισχεθείη τὴν γαστέρα, πλείω τε ἅμα προσφέρειν καὶ ὑγρότερα, εἰ δὲ ξηρότερα, πλείω μὲν μὴ προσφέρειν, ὑγρότερα δέ, εἰ δ' αὐστηρὰ καὶ στρυφνά, γλυκέσι τε καὶ λιπαροῖς εὐωχεῖν, εἰ δὲ τῇ τάξει πλημμελῶς, εἰς τὸ δέον ἐπανάγειν, εἰ δὲ ἅπαξ ἀντὶ τοῦ δίς, οὐ μόνον δίς, ἀλλὰ καὶ πολλάκις προσφέρειν. κατὰ δὲ τὸν αὐτὸν τρόπον τὰς προσφάτους δυσκρασίας ἐξιᾶσθαι τοῖς ἐναντίοις, ὑγραίνοντα μέν, εἰ ξηρανθείη, θερμαίνοντα δέ, εἰ ψυ-
χθείη, κἀπὶ τῶν ἄλλων ποιοτήτων ἀναλόγως. αἱ δ' ὗλαι τούτων ἐν ταῖς περὶ τῶν φαρμάκων γεγραμμέ-
ναις ἡμῖν πραγματείαις εἴρηνται.

ἐπισχεθείσης δὲ τῆς ξανθῆς χολῆς, ἐπὶ μὲν ἐμφρά-
ξει τῇ λεπτυνούσῃ διαίτῃ χρηστέον εἴρηται δ' ἡ ὕλη τῆς τοιαύτης διαίτης ἑτέρωθι δι' ἑνὸς γράμματος· ἐπὶ δὲ θλίψεσι, ταῖς μὲν διὰ τὴν τῶν ὁμιλούντων τοῖς τῆς χολοδόχου πόροις σωμάτων ἄμετρον πλήρωσιν, εἰ

---

37 This is taken to be a general reference to the three major works on materia medica: *Simpl. Med.*, XI.379–982K and XII.1–

function. They clear out all the superfluity generated in them through perceptible channels in the nose and eyelids.

14. I have, then, spoken of the causes and organs of the superfluities. Next I must say how someone might evacuate those that are retained, starting again from those in the stomach. There is a common precept applicable to all cases—apply the opposite cause to the one that created   74K the harm, specific in each case. If the stomach is hindered because what is consumed is too little or too dry, provide more and moister things. If what is consumed is too dry, do not provide more; instead, provide what is more moist. If, however, what is consumed is bitter and astringent, feed well with things that are sweet and fatty. If the defect is in the order, correct what you need to correct—provide food once instead of twice, or not only twice but frequently. In the same way, treat the recently acquired *dyskrasias* with opposites—moistening agents, if the stomach has been dried, heating agents, if it has been cooled, and similarly in the case of the other qualities. The materials of these are described in the works I have written on medications.[37]

If the yellow bile is retained due to obstruction you must use a thinning diet—the material of such a diet is described elsewhere in one book.[38] If retention is due to compression caused by excessive fullness of the bodies

377K; *Comp. Med. Loc.,* XII.378–1003K and XIII.1–361K; *Comp. Med. Gen.,* XIII.362–1058. None of these has yet been translated into a modern language.

[38] *Vict. Att.,* CMG V.4.2 (English trans., Singer, *Galen: Selected Works*).

75K　μὲν διὰ πάχος χυμῶν, τῇ λεπτυνούσῃ διαίτῃ χρηστέον, εἰ δὲ διὰ πλῆθος, τῇ κενούσῃ· εἰ δὲ διὰ φλεγμονὴν ἢ σκίρρον, ἐκπέπτωκεν ἤδη τὰ τοιαῦτα τῆς
ὑγιεινῆς πραγματείας· εἰ δὲ διὰ δυσκρασίαν ὑπό-
γυιον, ἀντεισάγοντα τὴν ἡττημένην ποιότητα. κατὰ
δὲ τὸν αὐτὸν τρόπον ἰᾶσθαι χρὴ τὴν μύσιν τῶν στο-
μίων, ἐπὶ μὲν τοῖς αὐστηροῖς γενομένην, τὰ λιπαρὰ
καὶ γλυκέα κελεύοντα λαμβάνειν, ἐπὶ δὲ τοῖς θερμαί-
νουσι καὶ ξηραίνουσιν, ὅσα ψύχει καὶ ὑγραίνει. λεχθήσεται δ' ἡ τῶν τοιούτων σιτίων ὕλη ἐν τοῖς ἑξῆς
ὑπομνήμασιν.

　ὁ αὐτὸς δὲ τρόπος ἐστὶ τῆς καθάρσεως ἐπισχεθέντι
τῷ τρίτῳ γένει τῶν περιττωμάτων, ὃ καθ' ἕκαστον τοῦ
ζῴου μόριον ἔφαμεν συνίστασθαι. εἰ μὲν γὰρ μύσει
ἔτι τὰ στόματα τῶν πόρων, ἐπανορθοῦσθαι χρὴ διὰ
τῶν ἐναντίων τοῖς βλάψασι, τὰς μὲν ἐπὶ τοῖς ψύχου-
σιν αἰτίοις στεγνώσεις ἐκθερμαίνοντα, καθάπερ ὅσαι
πυκνωθέντων ἡμῶν ὑπὸ κρύους γίνονται, τὰς δὲ ἐπὶ
θερμότητι καὶ ξηρότητι ἐμψύχοντά τε καὶ ἀνυγραί-
76K　νοντα, καθάπερ ὅσαι δι' ἐκκαύσεις· οὕτω δὲ καὶ ὅσαι
διά τι τῶν στυφόντων προσέπεσον, ὥσπερ καὶ ὅσα
τῶν ὑδάτων ἐστὶ στυπτηριώδη, λιπαραῖς τε καὶ μαλακαῖς τρίψεσιν ἅμα τοῖς τῶν γλυκέων ὑδάτων λουτροῖς. εἰ δὲ ἐμφραχθεῖεν οἱ πόροι διὰ πάχος ἢ πλῆθος
ἢ γλισχρότητα περιττωμάτων, ἥ τε λεπτύνουσα δίαιτα
τούτοις ἁρμόσει καὶ ὅσα τέμνει καὶ θερμαίνει φάρμακα, τὰ μὲν εἴσω τοῦ σώματος λαμβανόμενα, τὰ δὲ
ἔξωθεν ἐπιτιθέμενα, καὶ πρὸ τούτων ἁπάντων γυμνά-

adjacent to the bile-containing channels, or due to thickness of the humors, you must use the thinning diet, but if 75K it is due to an excess in amount, you must use a diet that is emptying. If due to inflammation or induration, such things already fall outside the matter of hygiene. If due to an acute *dyskrasia,* restore the quality that has been overcome. In the same way, you must cure occlusion of the orifices, occurring due to bitter things, by directing [the patient] to take fatty and sweet things, and in those occluded due to heating and drying agents, those things that cool and moisten. The material of such foods will be described in the books that follow.[39]

There is the same method of purification in the third class of retention of superfluities that I said arises in each part of the animal. For if the orifices of the channels are closed, you must correct this through the opposites to the injuring agents, heating those occluded by cooling causes, such as those occurring in us when we are contracted (thickened) by icy cold, while cooling and moistening those occluded due to heating and drying causes, such as those due to heatstrokes. And in the same way too, you 76K must treat those occurring due to one of the astringents, such as the sulfurous waters, having had an impact, with oils and soft massage along with baths of sweet waters. If, however, the channels are obstructed due to the thickness, abundance or viscidity of the superfluities, the thinning diet will be suitable for these, as are those medications that cut and heat, whether taken into the body or applied

[39] Books 5 and 6.

σια. καὶ γὰρ καὶ ταῦτα λύειν τε δύναται τὰ περιττώματα καὶ διὰ τῶν πόρων ἐκκενοῦν, καὶ τοσούτῳ
πλεονεκτεῖ τῶν λεπτυνόντων ἐδεσμάτων τε καὶ φαρμάκων, ὅσῳ βέλτιόν ἐστι μηδὲν βλαπτόμενον εἰς τὴν
τῶν σωμάτων ἕξιν ἐκκενοῦσθαι τὰ περιττώματα τοῦ
σὺν τῷ τάς τε σάρκας συντήκεσθαι καὶ ἰσχνοῦσθαι
τὰ στερεά. αὗται μὲν γὰρ αἱ βλάβαι τοῖς θερμοῖς τε
καὶ λεπτύνουσιν ἐφεδρεύουσι φαρμάκοις· ἐπὶ δὲ τοῖς
γυμνασίοις οὐ μόνον οὐδὲν τοιοῦτον, ἀλλὰ καὶ ῥώμη
τοῖς ὀργάνοις ἐγγίνεται, τῆς θερμότητος ἀναζωπυ-
ρουμένης αὐτοῖς κἀκ τῆς πρὸς ἄλληλα τῶν σωμάτων
77K   παρατρίψεως σκληρότητός τέ τινος καὶ δυσπαθείας
ἐγγιγνομένης.

ὡς δ᾽ ἄν τις ἐν καιρῷ γυμνάζοιτο καὶ μέτρῳ τῷ
προσήκοντι χρῷτο καὶ τάξει καὶ ποιότητι τῶν κατὰ
μέρος ἐνεργειῶν τῇ δεούσῃ, νυνὶ μὲν οὐ πρόκειται
λέγειν, ὥσπερ οὐδὲ περὶ τροφῆς καὶ καιροῦ καὶ
μέτρου καὶ τάξεως καὶ ποιότητος, οὐδὲ περὶ τῶν λεπτυνόντων ἐδεσμάτων τε καὶ πομάτων, οὐδὲ περὶ τῶν
ἀλλοιούντων κατὰ ποιότητα φαρμάκων· οὐδενὸς γὰρ
αὐτῶν οὐδέπω τὴν κατὰ μέρος εἴπομεν χρῆσιν, ἀλλ᾽
ἠρκέσθημεν ἐπὶ κεφαλαίων μόνον διελθεῖν· ἐν δὲ τοῖς
ἑξῆς ὑπομνήμασιν ὑπὲρ ἁπάντων ἐπὶ πλέον εἰρήσεται.

15. Νυνὶ μὲν γὰρ εἰς σύνοψιν ἀγαγεῖν ἠβουλήθην
ἅπασαν τὴν πραγματείαν, ὡς μηδεμίαν ὕλην ὑγιεινὴν[31] λαθεῖν, ἧς ἔμπειρον εἶναι χρὴ τὸν ὑγιεινόν. οὐδὲν γὰρ χεῖρον ὑγιεινὸν ὀνομάζειν τὸν ἐπιστήμονα

externally, and in preference to all these, exercises. For truly, these are also able to dissolve the superfluities and evacuate them through the channels. This has the advantage over thinning foods and medications to the extent that it is better to evacuate the superfluities of the body with nothing harmful to the state of the body in which the fleshes are melted away and the solid structures reduced. For these harms lie in wait with heating and thinning medications. Not only does no such thing happen due to exercises, but also strength arises in the organs, since the heat is restored in them, and even from the rubbing of the bodies against each other, a certain hardness and resis-   77K tance to affection (*dyspatheia*) arises.

How someone should exercise in a timely fashion and use an appropriate measure, with the required order and quality of the actions individually, it is not my task to discuss now, just as it is not to discuss, regarding food, the timing, amount, order and quality of nutriments, or thinning foods and drinks, or medications that alter qualities. For I have not yet said anything about the use of any of these, but have been satisfied with going through the chief points only. I shall say rather more about all these things in the books that follow.

15. Now I wished to provide a synopsis of the whole matter, so that no material that it is necessary for the hygienist to be practiced in escapes notice. For it is no bad thing to call someone skilled in the whole art of health, a

---

31 ὑγιεινὴν *add.* Ko

καὶ ὑγιεινῆς ἁπάσης τέχνης, ὥσπερ τὸν μόνης τῆς
περὶ τὰ γυμνάσια γυμναστήν· ἀτὰρ οὖν καὶ ὠνόμα-
σεν Ἐρασίστρατος οὕτως αὐτόν. ἐν δὲ τοῖς ἑξῆς ὑπο-
78K μνήμασιν ἑκάστης τῶν εἰρημένων ὑλῶν ἐπισκεψό-
μεθα τόν τε καιρὸν καὶ τὴν ποιότητα καὶ τὴν ποσότητα
καὶ τὸν τρόπον τῆς χρήσεως, ὡς μηκέθ᾽ ὕλην μόνον,
ἀλλ᾽ ὑγιεινὸν αἴτιον γίνεσθαι.

ἐν τρισὶ γὰρ τούτοις γένεσι πρώτοις ἐστὶν ἡ ὑγιεινὴ πραγματεία, καθάπερ καὶ ἡ θεραπευτική, σώμασί
τε καὶ αἰτίοις καὶ σημείοις· σώμασι μὲν αὐτοῖς τοῖς
ὑγιαίνουσιν, ἃ χρὴ φυλάττεσθαι τοιαῦτα, σημείοις δὲ
τοῖς συμβεβηκόσιν αὐτοῖς, ἐξ ὧν διαγινώσκεται,
αἰτίοις δέ, ὑφ᾽ ὧν ἡ φυλακὴ τῆς ὑγείας γίνεται. τέτταρας δὲ τῆς τούτων ὕλης τὰς διαφορὰς οἱ δοκιμώτατοι τῶν νεωτέρων ἰατρῶν ἔθεντο, προσφερόμενα καὶ
ποιούμενα καὶ κενούμενα καὶ ἔξωθεν προσπίπτοντα,
προσφερόμενα μὲν ἐδέσματά τε καὶ πόματα καὶ εἴδη
τινὰ τῶν φαρμάκων εἴσω τοῦ σώματος λαμβανόμενα
καὶ τὸν εἰσπνεόμενον ἀέρα, ποιούμενα δὲ τρίψεις τε
καὶ περιπάτους καὶ ὀχήσεις καὶ ἱππασίαν καὶ σύμπασαν κίνησιν. εἰ δ᾽ οὐ πᾶσα κίνησίς ἐστι γυμνάσιον,
ἀλλ᾽ ἡ σφοδροτέρα μόνον, προσκείσθω ὧδε τῇ κινήσει τὸ γυμνάσιον, ὡς εἶναι τὰ ποιούμενα κινήσεις τε
καὶ γυμνάσια. συγκαταριθμοῦνται δὲ τῷ γένει τῶν
79K αἰτίων τούτων καὶ ἐγρηγόρσεις καὶ ὕπνοι καὶ ἀφροδίσια. τὰ δὲ ἔξωθεν προσπίπτοντα, πρῶτος μὲν ὁ περι-

----

40 Erasistratus (3rd c. BC) contributed much to and wrote

"hygienist," just as it is not to call someone skilled only in the art of exercise, a "gymnastic trainer." Indeed, even Erasistratus[40] also named such a person in this way. In the books that follow, I shall consider each of the previously mentioned materials in respect of the timing, quality, 78K quantity and manner of use, so that it is no longer just a material but becomes a cause of health.

Hygiene as a subject lies in these three primary classes, just as therapeutics also does; that is, in bodies, causes and signs—in actual bodies that are healthy and must be maintained so, in the signs that have occurred in them from which we will make a diagnosis, and in the causes through which the preservation of health arises. The most notable of the younger doctors established four *differentiae* of the material of these [causes]: things to be administered, things to be done, things to be evacuated and things that befall from without. Things to be administered are foods, drinks, certain kinds of medications taken into the body and the inspired air. Things to be done are massage, walking, driving, horse riding and all movement. Since not every movement is gymnastics, but only what is more vigorous, let me thus add exercise to movement, so the things to be done are movements and exercises. Also included in the class of these causes are wakefulness, sleep and sexual 79K activity. Things that befall from without are first, the ambi-

extensively on medical theory and practice. On a number of critical points he and Galen held opposing views. No extant works by Erasistratus survive. I. Garofalo has prepared a collection of fragments (*Erasistrata Fragmenta*); fragments 153–67 are headed "hygiene"; fragments 115–67 may all be from the lost work *On Hygiene*.

113

ἔχων ἡμᾶς ἀήρ ἐστιν, ἔπειθ᾽ ὅσα λουομένοις ἢ ἀλειφομένοις[32] ἢ διαπαλαίουσιν ἐν κόνει προσπίπτει τῷ
δέρματι, καὶ εἰ δή τι φάρμακόν ἐστιν οὐκ ἐκβαῖνον
ὅρους ὑγιεινούς, ὥσπερ ἅλες[33] ἢ νίτρον ἢ ἀφρόνιτρον
ἤ τι τῶν αὐτοφυῶν ὑδάτων θερμῶν. ἡ δὲ τῶν κενουμένων ὕλη προείρηται μὲν ὀλίγον ἔμπροσθεν.

εἰ δ᾽ ὀρθῶς ἀντιδιήρηται τοῖς προειρημένοις τρισὶ
γένεσι τῶν αἰτίων, οὐ ῥᾴδιον ἀποφήνασθαι. τάχα γὰρ
ἂν ἦν βέλτιον ὑπὸ μὲν τῶν προσφερομένων καὶ ποιουμένων καὶ προσπιπτόντων ἔξωθεν ἀλλοιοῦσθαί τε καὶ
μετακοσμεῖσθαι φάναι τὰ κατὰ τὸ σῶμα, τὴν μεταβολὴν δὲ αὐτῶν γίνεσθαι κατά τε τὸ ποιὸν καὶ κατὰ
τὸ ποσόν, καὶ κατὰ μὲν τὸ ποιὸν ἐν τῷ θερμαίνεσθαι
καὶ ψύχεσθαι καὶ ξηραίνεσθαι καὶ ὑγραίνεσθαι, κατὰ
δὲ τὸ ποσὸν ἐν τῷ τρέφεσθαι καὶ κενοῦσθαι, καὶ
αὐτήν γε τὴν κένωσιν εἶναι διττήν, ἑτέραν μὲν τῶν
περιττωμάτων, ὑπὲρ ὧν ὀλίγον ἔμπροσθεν ἐλέγομεν,
ἑτέραν δὲ τὴν αὐτῆς τῆς οἰκείας ἡμῶν οὐσίας ἀπορ-
ροήν, ἥτις ἀντίκειται τῇ θρέψει. ἐπισημήνασθαι δὲ
80K χρὴ κἀνταῦθα τὴν ὁμωνυμίαν, ἣν Ἱπποκράτης ἐν τῷ
περὶ τροφῆς ἡμῶν συγγράμματι διείλετο, φάμενος
ὧδε· "τροφὴ δὲ τὸ τρέφον, τροφὴ καὶ τὸ οἷον, τροφὴ
καὶ τὸ μέλλον."

τῇ μὲν γὰρ κατὰ τὸ πρῶτον σημαινόμενον τροφῇ
τε καὶ πέψει τὴν ἀπορροὴν τῆς οὐσίας ἀντιδιαιρεῖ-
σθαι χρή, τῇ δὲ κατὰ τὸ δεύτερον σημαινόμενον[34]

---

[32] post ἢ ἀλειφομένοις: add. ἢ διαπαλαίουσιν ἐν κόνει Ko

ent air, then those things that contact the skin in bathing, anointing with oil, and wrestling in dust, and finally, some medication not falling outside the limits of hygiene, such as salt, myrtle, niter, native sodium carbonate or one of the natural waters that are hot. The material of the things being evacuated was spoken of a little earlier.

Whether the differentiation into the three previously mentioned classes of causes has been made correctly is not easily demonstrated. Perhaps it would be better to say that those things pertaining to the body are changed and rearranged by things that are administered, done or befall from without, while the change of these occurs in both quality and quantity. In quality, in being heated, cooled, dried and moistened, and in quantity, in being nourished and evacuated, and that the actual emptying is twofold—the one of superfluities that I spoke about a little earlier and the other when the intrinsic substance of our nature flows away, which is the antithesis of being nourished. Here it is necessary to indicate the homonymy, which Hippocrates, in his work *On Nutriment,* distinguished 80K when he said: "Nutriment is that which nourishes; nutriment is also that which is fit to nourish; and nutriment is that which is about to nourish."[41]

In relation to the first signification, it is necessary to oppose the outflow of substance to nutriment and concoction. In relation to the second signification, it is necessary

[41] See Hippocrates, *On Nutriment* 8, *Hippocrates* 1, LCL 147, 344–45.

---

[33] *post* ἅλες: *add.* ἢ μύρτα Ku
[34] *post* τὸ δεύτερον: *add.* σημαινόμενον Ko

αἱμορραγίαν τε καὶ ἁπλῶς εἰπεῖν αἵματος ἅπασαν κένωσιν, τῇ δὲ κατὰ τὸ τρίτον ἔμετόν τε καὶ λειεντερίαν. ἀλλὰ περὶ μὲν τῶν τοιούτων διαιρέσεων, ὅπως ἄν τις ἐθέλῃ, τιθέσθω. τῶν δ' ὑλῶν ἁπασῶν τῶν ὑγιεινῶν ἐπίστασθαι ἀναγκαῖον τὰς δυνάμεις τῷ τὴν ὑγιεινὴν τέχνην μετιόντι. καὶ γὰρ καὶ ἡ ἐπιδέξιος αὐτῶν χρῆσις ἐντεῦθεν ὥρμηται. γίνεται δ' ἡ ἐπιδέξιος αὐτῶν χρῆσις, ἐὰν τόν γε καιρὸν ἑκάστου καὶ τὸ μέτρον εὕρωμεν. ὥστε ἐπὶ ταῦτα χρὴ προϊέναι μᾶλλον, οὐ τὰς μοχθηρὰς αἱρέσεις διεξελέγχειν. ἀλλὰ ἐπειδὴ μέγεθος ἱκανὸν ὁ πρῶτός μοι λόγος ἔχει, τοῦτον μὲν ἐνταῦθα καταπαύσω, τὰ δ' ὑπόλοιπα τῆς πραγματείας ἐν τοῖς ἑξῆς διηγήσομαι.

to oppose hemorrhage and, to speak simply, every evacuation of blood to nutriment and concoction. In relation to the third signification, it is necessary to oppose vomiting and diarrhea [to nutriment and concoction]. But on the matter of such divisions, let each person establish those he wishes to. It is necessary for the one going about the art of hygiene to know the powers of all the materials that are health-producing, for the skillful use of them arises here, if we are to discover what is timely and measured in each case. Consequently, it is necessary to proceed to these rather than delay by refuting mistaken sects.[42] But since my first discussion has already gone on long enough, I shall stop this here, and set out what remains of the matter in the books to follow.

---

[42] See note 10 above. Galen was particularly critical of the Methodics; less so of the Empirics.

B

81K 1. Τὰ μὲν δὴ κεφάλαια καὶ τοὺς σκοποὺς τῆς ὑγιεινῆς τέχνης ὁ πρόσθεν λόγος προείρηκεν· τὰ δὲ κατὰ μέρος ἅπαντα πειρᾶσθαι χρὴ διελθεῖν, ἀρξαμένους αὖθις ἀπ᾽ ἐκείνων, εἰς ἅπερ ἐτελεύτα τὸ πρῶτον γράμμα. ὑποκείσθω δή τις ἡμῖν τῷ λόγῳ παῖς ὑγιεινότατος φύσει, τῆς τρίτης ἑβδομάδος ἐτῶν ἀρχόμενος, ἐφ᾽ οὗ πλάττειν τε καὶ κοσμεῖν τὸ σῶμα προκείσθω καθ᾽ ὅσον οἷόν τε κάλλιστα.

καὶ πρῶτον τοῦτ᾽ αὐτὸ διοριστέον, τί ποτε βούλεται τῷ λόγῳ τὸ κάλλιστα προσκείμενον. βούλεται δὲ
82K τόδε. ὥσπερ αὐτῶν τῶν σωμάτων ἐδείχθη παμπόλλη τις οὖσα διαφορά, κατὰ τὸν αὐτὸν τρόπον καὶ τῶν βίων, οὓς βιοῦμεν, εἴδη πάμπολλά ἐστιν. οὔκουν ἐγχωρεῖ τὴν ἀρίστην τοῦ σώματος ἐπιμέλειαν ἐν ἅπαντι τῷ προχειρισθέντι βίῳ συστήσασθαι ἀλλὰ τὴν μὲν ὡς ἐν ἑκάστῳ βελτίστην οἷόν τε, τὴν δ᾽ ἁπλῶς ἀρίστην οὐκ ἐγχωρεῖ κατὰ πάντας τοὺς βίους ποιήσασθαι. πολλοῖς γὰρ τῶν ἀνθρώπων μετὰ περιστάσεως πραγμάτων ὁ βίος ἐστί. καὶ βλάπτεσθαι μὲν ἀναγκαῖόν ἐστιν αὐτοῖς ἐξ ὧν πράττουσιν, ἀποστῆναι δ᾽ ἀδύνατον. ἔνιοι μὲν γὰρ ὑπὸ πτωχείας εἰς τοὺς τοιού-

118

BOOK II

1. The preceding discussion has stated the chief points and 81K
objectives of the art of hygiene. We must attempt to go
over all these individually, beginning again from those at
which the first book ended. Let us, then, assume in our
discussion some child who is very healthy in nature, starting the third seven-year period of life. In the case of such
a child let us propose to form and prepare the body in the
best way possible.

And we must first distinguish this itself—what we
mean by attaching the term "the best" at this time. What
we mean is this: just as it was shown that there are very 82K
many differences of the bodies themselves, in the same
way there are also very many kinds of lives we lead. Therefore it is not possible for the best care of the body to be
established in every life that is undertaken, but the best
that is possible in each case, for what is absolutely the best
is impossible to achieve in all lives. For many people, life
is bound up with the circumstances of their activities, and
for them harm is inevitable from what they do—this cannot be avoided. Some happen upon such lives through

τους ἐμπίπτουσι βίους, ἔνιοι δ' ὑπὸ δουλείας, ἤτοι
πατρόθεν εἰς αὐτοὺς καθηκούσης ἢ αἰχμαλώτοις ληφθεῖσιν ἢ ἁρπαχθεῖσιν, ἅσπερ καὶ μόνας δουλείας
ὀνομάζουσιν οἱ πολλοὶ τῶν ἀνθρώπων. ἐμοὶ δὲ δο-
κοῦσι καὶ ὅσοι διὰ φιλοτιμίαν ἢ δι' ἐπιθυμίαν ἥν-
τιναοῦν εἵλοντο βίον ἐν περιστάσεσι πραγμάτων, ὡς
ὀλίγιστα δύνασθαι σχολάζειν τῇ τοῦ σώματος ἐπιμελείᾳ, καὶ οὗτοι δουλεύειν ἑκόντες οὐκ ἀγαθαῖς δε-
σποίναις. ὥστε τούτοις μὲν οὐκ ἐγχωρεῖ γράψαι τὴν
83K ἁπλῶς ἀρίστην ἐπιμέλειαν τοῦ σώματος· ὅστις δὲ
ἀκριβῶς ἐλεύθερος ὑπάρχει καὶ τύχῃ καὶ προαιρέσει,
δυνατὸν ὑποθέσθαι τῷδε, ὡς ἂν ὑγιαίνοι τε μάλιστα
καὶ ἥκιστα νοσήσειε καὶ γηράσειεν ἄριστα.

καὶ μέν γε καὶ ἡ ὑγιεινὴ μέθοδος, καθάπερ οὖν καὶ
ἄλλη πᾶσα μέθοδος, ἀρχὴν διδασκαλίας τοιαύτην
ἐπιζητεῖ. τὸ γὰρ ἁπλοῦν καὶ ἄμεμπτον ἐν ἑκάστῳ γένει, καθάπερ τις κανών, ἁπάντων ἐθέλει προτετάχθαι
τῶν οὐχ ἁπλῶν οὐδ' ἀμέμπτων. ἁπλοῦν δὲ καὶ ἄμεμπτον ἐν μὲν τοῖς σώμασι τὸ κατεσκευασμένον ἄριστά
ἐστιν, ἐν δὲ τοῖς βίοις τὸ ἀκριβῶς ἐλεύθερον. ταῦτ'
οὖν ἄμφω πρῶτα συζευγνύσθω κατὰ τόνδε τὸν λόγον·
εἶθ' ἑξῆς ἑκάστῃ κατασκευῇ σώματος μοχθηρᾷ βίος
ἐλεύθερος μιγνύσθω· κἄπειθ' ἑξῆς ἀρίστῃ κατασκευῇ
σώματος ἕκαστος τῶν ἐν δουλείᾳ τινὶ βίων· ἐπὶ δὲ
τοῖσδε τὰς μοχθηρὰς τῶν σωμάτων κατασκευὰς
ἐπαλλάξωμεν τοῖς μοχθηροῖς βίοις, εἰ μέλλει τέλειος
ἡμῖν ὁ λόγος ἔσεσθαι.

poverty and some through slavery, either coming down to them from their fathers, or by being taken away as prisoners, or being snatched away, which the majority of people call the only slavery. However, there are also those who seem to me to have chosen a life caught up in the circumstances of their activities, either through ambition or whatever kind of desire, so they are least able to spend time on the care of their bodies; these men are willing slaves to bad mistresses. As a result, it is impossible to set down what, for these men, is the absolute best care of the body. But in the case of someone who is completely free, whether by chance or by choice, it is possible to lay down the following: how he might be most healthy and least diseased and how he might grow old in the best way.

83K

In fact, the method of hygiene seeks just such a principle of instruction, as does every other method, for in each class that which is simple and without fault, is what someone might wish to be established as a standard for all that is neither simple nor without fault. In bodies, that which is simple and without fault is the best constitution; in lives, it is complete freedom. First, let us link together both these things in this discussion. Next, let a life that is free be combined with each defective constitution of the body. And then, in turn, let each of the lives spent in some kind of enslavement be combined with the best constitution of the body. After these, let us join the defective constitutions of bodies with defective lives, so our discussion will come to completion.

84K 2. Τίνα ποτ' οὖν προσήκει τίθεσθαι ἀρχὴν τῆς ὑγιεινῆς πραγματείας τῷ κάλλιστά τε κατεσκευασμένῳ
τὸ σῶμα καὶ τρίτης ἑβδομάδος ἐτῶν ἀρχομένῳ καὶ
μόνῃ σχολάζοντι τῇ τοῦ σώματος ἐπιμελείᾳ, τὰ δ'
ἄλλα πάντα πάρεργα τεθειμένῳ· ἐμοὶ μὲν οὖν δοκεῖ
τὴν ὑφ' Ἱπποκράτους εἰρημένην ἔν τε τοῖς Ἀφο-
ρισμοῖς, ἔνθα γράφει "πόνοι σιτίων ἡγείσθωσαν," ἔν
τε τῷ τῶν Ἐπιδημιῶν ἕκτῳ κατὰ τήνδε τὴν ῥῆσιν·
"πόνοι, σῖτα, ποτά, ὕπνοι, ἀφροδίσια, πάντα μέτρια."
καὶ γὰρ καὶ τὸ ποσὸν ἀφώρισεν ἅπασι, προσθεὶς τὸ
κατὰ τὸ τέλος τῆς ῥήσεως ὄνομα τὸ "μέτρια," καὶ τὸν
καιρὸν ἐδίδαξε τῇ τάξει τοῦ λόγου. εἰς γὰρ τὴν τῆς
ὑγείας φυλακὴν ἄρχειν μὲν χρὴ τοὺς πόνους, ἕπεσθαι
δὲ σιτία τε καὶ ποτά, εἶθ' ἑξῆς ὕπνους, εἶτα ἀφροδίσια
τοῖς γε δὴ μέλλουσιν ἀφροδισίοις χρῆσθαι. τὰ μὲν
γὰρ ἄλλα πάντα κοινὰ πάσης ἡλικίας ἐστί, τὰ δ'
ἀφροδίσια μόνης τῆς τῶν ἀκμαζόντων, ἡνίκαπερ
αὐτῶν καὶ ἡ χρεία, ὡς ταῖς γε πρόσθεν τε καὶ ὄπισθεν ἡλικίαις ἢ οὐδ' ὅλως σπερμαίνειν ἢ οὐ γόνιμον
σπερμαίνειν ἢ μοχθηρῶς γόνιμον ὑπάρχειν. ἀλλὰ
85K γὰρ ὁ περὶ μὲν τῶν ἀφροδισίων λόγος εἰς τὴν οἰκείαν
ἀναβεβλήσθω τάξιν.

ἀπὸ δὲ τῶν πόνων ἀρκτέον, αὐτὸ τοῦτο πρῶτον ἐν
αὐτοῖς διελομένους, εἴτε ταὐτόν ἐστι πόνος τε καὶ
κίνησις καὶ γυμνάσιον, εἴτε πόνος μὲν καὶ κίνησις
ταὐτόν, ἕτερον δέ τι τὸ γυμνάσιον, εἴτε κίνησις μὲν

---

[1] Hippocrates, *Epidemics* 6.4(23), LCL 477, 254–55.

2. What in the world, then, is it appropriate to establish as the principle of the matter of hygiene for someone with 84K
the best constitution of the body entering the third seven-year period of life—someone who has the time to devote himself solely to the care of the body and to set aside all other things as secondary? It seems to me it is the statement by Hippocrates in the *Aphorisms,* where he writes: "Let exertions precede food."[1] In relation to this statement, in the sixth book of *Epidemics* [there is]: "Exertions, food, drink, sleep and sexual activity all in moderation."[2] Indeed, he also set the amount for all these things, placing the term "moderation" at the end of the statement, and he taught the appropriate time in the order of the statement. For the preservation of health, exertions[3] must come first, followed by food and drink, and next in order, sleep, and then sexual intercourse for those who intend to engage in this. All the other things are common to all the stages of life, whereas sexual activity is only for those in their prime, at which time there is need of this, since, in the stages before and after, insemination is altogether out of the question or is unfruitful or bad in terms of fertility. But let 85K
me put off the discussion about sexual activity to its proper place.

We must start from exertions, first making this distinction among these things: whether exertion (work), movement and exercise are the same, or exertion and movement the same but exercise something different, or

[2] Hippocrates, *Epidemics* 6.6(2), LCL 477, 262–63.

[3] The Greek term πόνος has various meanings. LSJ includes "work" (especially hard work), "toil," "labor," "bodily exertion," and "exercise" as well as "distress" and "trouble," among other terms. "Exertion" is predominantly used here. Galen himself expands somewhat on the term.

123

ἕτερον, οὐδὲν δ᾽ ὁ πόνος τοῦ γυμνασίου διαφέρει. ἐμοὶ
μὲν δὴ δοκεῖ μὴ πᾶσα κίνησις εἶναι γυμνάσιον, ἀλλ᾽
ἡ σφοδροτέρα μόνη. ἐπεὶ δ᾽ ἐν τῷ πρός τι τὸ σφο-
δρόν, εἴη ἂν ἡ αὐτὴ κίνησις ἑτέρῳ μὲν γυμνάσιον,
ἑτέρῳ δ᾽ οὐ γυμνάσιον. ὅρος δὲ τῆς σφοδρότητος ἡ
τῆς ἀναπνοῆς ἀλλοίωσις· ὡς, ὅσαι γε κινήσεις οὐκ
ἀλλοιοῦσι τὴν ἀναπνοήν, οὔπω ταύτας ὀνομάζουσι
γυμνάσια· εἰ δ᾽ ἤτοι μεῖζον ἢ ἔλαττον ἢ θᾶττον ἢ
πυκνότερον ἀναγκασθείη τις ἀναπνεῖν ἐπὶ κινήσει
τινί, γυμνάσιον ἡ τηλικαύτη κίνησις ἐκείνῳ γενήσεται. τοῦτο μὲν δὴ κοινῇ γυμνάσιον ὀνομάζεται, ἰδίᾳ
δέ, ἀφ᾽ οὗπερ καὶ τὰ γυμνάσια προσαγορεύουσιν
ἅπαντες, ἔν τινι κοινῷ τῆς πόλεως οἰκοδομησάμενοι
χωρίῳ, εἰς ὅπερ καὶ ἀλειψόμενοί τε καὶ διατριψόμενοι
86K καὶ διαπαλαίσοντες ἢ δισκεύσοντες ἤ τι τοιοῦτον
ἄλλο πράξοντες ἥκουσιν.

ἡ δὲ τοῦ πόνου προσηγορία ταὐτόν μοι δοκεῖ ση-
μαίνειν θατέρῳ τῷ ὑπὸ τοῦ γυμνασίου ὀνόματος εἰρημένῳ δηλοῦσθαι τῷ κοινῷ. καὶ γὰρ καὶ οἱ σκάπτοντες
καὶ οἱ θερίζοντες καὶ ἱππαζόμενοι πονοῦσί τε καὶ
γυμνάζονται κατὰ τὸ κοινὸν τοῦ γυμνασίου σημαινόμενον. ἐμοὶ μὲν οὖν οὕτω διῃρήσθω περὶ τῶν ὀνομάτων, καὶ κατὰ ταῦτα τὰ σημαινόμενα πᾶς ὁ ἐφεξῆς
λόγος ἀκουέσθω· εἰ δέ τις ἑτέρως βούλεται χρῆσθαι,
συγχωρῶ· οὐδὲ γὰρ ὑπὲρ ὀνομάτων ὀρθότητος ἥκω
σκεψόμενος, ἀλλ᾽ ὡς ἄν τις ὑγιαίνοι μάλιστα· καὶ
πρὸς τοῦτ᾽ αὐτὸ χρήσιμον ὑπάρχον μοι περί τε τῶν
γυμνασίων καὶ πόνων καὶ ξυλλήβδην εἰπεῖν ἁπάσης

movement is different whereas exertion is no different from exercise. It certainly seems to me that not every movement is exercise but only that which is quite vigorous. However, since vigor is a relative term, it may be that the same movement is exercise for one person but not for another. The defining feature of vigor is the change of breathing, inasmuch as those movements that do not change the breathing, people do not yet term exercises, whereas if someone is forced to breathe more or less, or faster or slower after some movement, such a movement will become an exercise for that person. This is called "exercise" by common consent, whereas separately, among those which all term exercises, are those done in some public place they have built in the city to which they come to be anointed or massaged, and wrestle or throw the discus, or do something else of this sort. 86K

The term *ponos* (exertion, work, exercise) seems to me to signify the same thing as the other thing which has been said to be signified to people in general by the term "exercise." For those who dig, reap and ride horses both work and exercise according to the general signification of exercise. Let me make the distinction about the terms in this way, and let the whole discussion that follows be understood according to these significations. If, however, someone should wish to use [the term] otherwise, I agree, for I have not come to examine the correctness of terms but how someone might be most healthy. And since, for this very purpose, it is useful for me to make a division con-

κινήσεως διελέσθαι, τὰ σημαινόμενα τῶν ὀνομάτων
ἠναγκάσθην ἀφορίσασθαι. αἱ μὲν δὴ τῶν γυμνασίων
χρεῖαι καὶ διὰ τοῦ πρώτου μὲν εἴρηνται λόγου, βέλτιον δ' ἂν εἴη καὶ νῦν ἐπανελθεῖν αὐτὰς διὰ βραχέων,
ἐπειδὴ σκοπός τε ἅμα καὶ κριτήριον αὗται τυγχάνου-
87K σιν οὖσαι πάντων τῶν κατὰ μέρος ἐν τῇ περὶ τὰ γυμνάσια τέχνῃ πραττομένων.

ἦσαν δέ, ὡς οἶμαι, διτταὶ κατὰ γένος, αἱ μέν τινες
εἰς τὴν τῶν περιττωμάτων κένωσιν, αἱ δὲ εἰς αὐτὴν
τῶν στερεῶν σωμάτων τὴν εὐεξίαν διαφέρουσαι.
ἐπειδὴ γάρ ἐστι κίνησις σφοδρὰ τὸ γυμνάσιον,
ἀνάγκη τρία μὲν πρῶτα[1] ταῦτα γίνεσθαι πρὸς αὐτοῦ
κατὰ τὸ γυμναζόμενον σῶμα, τήν τε σκληρότητα τῶν
ὀργάνων ἀλλήλοις παρατριβομένων, τήν τε τῆς ἐμφύτου θερμότητος αὔξησιν, τήν τε τοῦ πνεύματος
κίνησιν βιαιοτέραν, ἕπεσθαι δὲ τούτοις τἆλλα σύμπαντα τὰ κατὰ μέρος ἀγαθὰ τοῖς σώμασιν ἐκ γυμνασίων γινόμενα, διὰ μὲν τὴν σκληρότητα τῶν ὀργάνων
τήν τε δυσπάθειαν αὐτῶν καὶ τὴν πρὸς τὰς ἐνεργείας
εὐτονίαν, διὰ δὲ τὴν θερμότητα τήν τε τῶν ἀναδιδο-
μένων ὁλκὴν ἰσχυρὰν καὶ τὴν ἀλλοίωσιν ἑτοιμοτέραν
καὶ τὴν θρέψιν βελτίονα καὶ χύσιν ἁπάντων τῶν σωμάτων, ἐφ' ᾗ χύσει τὰ μὲν στερεὰ μαλάττεσθαι, τὰ
δὲ ὑγρὰ λεπτύνεσθαι, τοὺς πόρους δ' εὔρους γίνεσθαι
συμβαίνει, διὰ δὲ τὴν τοῦ πνεύματος ἰσχυρὰν κίνησιν ἐκκαθαίρεσθαί τε τοὺς πόρους ἀναγκαῖόν ἐστι καὶ
κενοῦσθαι τὰ περιττώματα.

88K ἀλλ' εἴπερ ταῦτα ποιεῖ τὸ γυμνάσιον, οὐ χαλεπὸν

126

cerning exercises and exertions, and in a word every movement, I have been compelled to distinguish the significations of the terms. Although the uses of exercise have also been discussed in the first book, it would be better to recapitulate these briefly now, since these happen to be both the objective and the criterion of all the things done individually in the art of exercise.   87K

These were, I think, twofold in terms of class, drawing a distinction between those that pertain to the evacuation of superfluities and those that pertain to the actual good state of the solid bodies. Since exercise is vigorous movement, of necessity these three primary things arise from it in respect of the exercising body—hardness of the organs when they rub together with each other, an increase of the innate heat, and the more forceful movement of the *pneuma*—while all the other things good for bodies that arise in turn from exercises follow these things: (1) due to the hardness of the organs, their resistance to affection and vigor in regard to their functions; (2) due to the heat and the strong attraction of those things being distributed, there is a greater readiness to change, better nutrition and diffusion to all the bodies; what happens due to this diffusion is that the solid bodies are softened, liquids are thinned and channels become wider; (3) due to the strong movement of the *pneuma* the channels are inevitably cleaned out and the superfluities evacuated.

But if exercise does these things, it is no longer difficult   88K

---

1 μὲν πρῶτα Ko; μόνον Ku

ἔτι τὸν καιρὸν τῆς χρήσεως ἐξευρεῖν. διότι μὲν γὰρ
ἀναδόσεσι συνεργεῖ, οὐ χρὴ πλῆθος ὠμῶν καὶ
ἀπέπτων οὔτε σιτίων οὔτε χυμῶν ἢ κατὰ τὴν κοιλίαν
ἢ ἐν τοῖς ἀγγείοις περιέχεσθαι· κίνδυνος γὰρ αὐτοῖς
ἑλχθῆναι πρὸς ἅπαντα τοῦ ζῴου τὰ μόρια, πρὶν χρη-
στοῖς γενέσθαι πεφθεῖσι σχολή. διότι δὲ ἐκκαθαίρει
τοὺς πόρους καὶ κενοῖ τὰ περιττώματα, κάλλιον αὐτὸ
πρὸ τῶν σιτίων ἀναλαμβάνεσθαι. "τὰ" μὲν γὰρ "μὴ
καθαρὰ τῶν σωμάτων, ὁκόσον ἂν θρέψῃς, μᾶλλον
βλάψεις." ὥστε ἐκ τῶν εἰρημένων εὔδηλον, ὡς οὗτος
ἄριστός ἐστι γυμνασίων καιρός, ἡνίκα ἡ μὲν χθιζινὴ
τροφὴ τελέως ᾖ κατειργασμένη τε καὶ πεπεμμένη τὰς
δύο πέψεις, τήν τε ἐν τῇ γαστρὶ καὶ τὴν ἐν τοῖς ἀγγείοις, ἑτέρας δ' ἐφεδρεύῃ τροφῆς καιρός. εἰ δὴ τοῦδε
πρόσθεν ἢ ὄπισθεν γυμνάζοις, ἢ χυμῶν ἀπέπτων
ἐμπλήσεις τὸ ζῷον ἢ τὴν ὠχρὰν χολὴν ἐπιτρέψεις
γεννηθῆναι πλείονα.

89K   γνώρισμα δὲ τοῦ τοιούτου καιροῦ τῶν οὔρων ἡ
χροιά· τὸ μὲν οὖν ὑδατῶδες ἄπεπτον ἔτι σημαίνει τὸν
ἐκ τῆς γαστρὸς ἀναδοθέντα χυμὸν ἐν τοῖς ἀγγείοις
περιέχεσθαι, τὸ δὲ πυρρὸν καὶ χολῶδες ἐκ πολλοῦ
κατειργάσθαι, τὸ δὲ μετρίως ὠχρὸν ἄρτι τῆς δευτέρας
πέψεως γεγενημένης ἐστὶ σημεῖον. ὅταν γὰρ μηδέπω
χρώζηται τῇ χολῇ τὸ οὖρον, ὑδατῶδές τε καὶ λευκὸν
φαίνεται, ὅταν δὲ πλέον ἀναδέξηται τοῦ προσήκοντος,
πυρρόν. ἐπειδὰν δὲ συμμέτρως ᾖ πυρρὸν ἢ μετρίως
ὠχρόν, τηνικαῦτα τοίνυν ἄγειν ἐπὶ τὰ γυμνάσια, προ-
αποθέμενον ὅσον ἂν ἐν τῇ κύστει καὶ τοῖς ἐντέροις

to discover the appropriate time for its use. Since it assists distribution, there must not be an abundance of raw and unconcocted foods or humors contained in either the stomach or the vessels, otherwise there is a danger of these being drawn to all parts of the animal before there is time for them to become useful as a result of concoction. And because it clears out the channels and evacuates the superfluities, it is better to partake of it before food, for with "unpurged bodies, the more you nourish them, the more you harm them."[4] Consequently, it is clear from what has been said that the best time for exercises is when the nourishment from the previous day has been completely worked upon and digested by the two digestions—that in the stomach and that in the vessels—and the time for further nourishment is near. If you exercise either before or after this, you will fill the organism with unconcocted humors and allow yellow bile to become abundant.

The color of the urine is a sign of such a time. If it is 89K watery, this signifies that there is still unconcocted humor distributed from the stomach and contained in the vessels. However, urine that is tawny and bile-containing indicates that digestion is, for the most part, accomplished. Urine that is moderately pale is a sign of the second digestion having just now taken place. For whenever the urine is not yet tinged by bile, it appears watery and pale. However, when it receives more bile than is appropriate, it is tawny. Whenever it is moderately tawny or moderately pale, proceed to exercises under these circumstances after excretion of as much of the superfluity as is contained in the

[4] Hippocrates, *Aphorisms* 2.10, *Hippocrates* IV, LCL 150, 110–11. See also Galen, *In Hp. Aph.*, XVIIB.414K.

τοῖς κάτω περιεχόμενον ἢ περίττωμα· κίνδυνος γὰρ
κἀκ τῶν τοιούτων εἰς τὴν ἕξιν τοῦ σώματος ἐνεχθῆναί
τι τῇ ῥύμῃ τῆς ἐν τοῖς γυμνασίοις θερμότητος ἀναρ-
πασθέν. εἰ μὲν οὖν εὐθέως ἀποδυσάμενός τις ἐπὶ τὰς
ἰσχυροτέρας ἔρχοιτο κινήσεις, πρὶν μαλάξαι σύμπαν
τὸ σῶμα καὶ λεπτῦναι τὰ περιττώματα καὶ τοὺς πόρους εὐρῦναι, κίνδυνος μὲν καὶ ῥῆξαί τι καὶ σπάσαι
τῶν στερεῶν σωμάτων, κίνδυνος δὲ καὶ τὰ περιττώ-
90K ματα τῇ τοῦ πνεύματος ῥύμῃ κινήσαντος τοὺς πόρους
ἐμφράξαι.

εἰ δὲ κατὰ μικρὸν προθερμήνας προμαλάξειε μὲν
τὰ στερεά, προλεπτύνειε δὲ τὰ ὑγρὰ καὶ τοὺς πόρους
εὐρύνειε, κίνδυνος οὐδεὶς ἔτι οὔτε τοῦ ῥῆξαί τι μόριον
οὔτε τοῦ τοὺς πόρους ἐμφράξαι ἂν καταλάβοι τὸν
γυμναζόμενον. ὅπως οὖν ταῦτα γίνοιτο, χρὴ προθερ-
μάναντα μετρίως ἀνατρίψαντά τε σινδόνι τὸ σύμπαν
σῶμα κἄπειτα δι᾽ ἐλαίου τρίβειν. οὐ γὰρ δὴ εὐθέως
γε χρῆσθαι τῷ λίπει συμβουλεύω πρὶν θερμανθῆναί
τε τὸ δέρμα καὶ τοὺς πόρους εὐρυνθῆναι καὶ συλλήβδην εἰπεῖν εὐτρεπισθῆναι τὸ σῶμα πρὸς τὸ κατα-
δέξασθαι τὸ ἔλαιον. ἱκαναὶ δ᾽ εἰς τοῦτο παντάπασιν
ὀλίγαι περιαγωγαὶ τῶν χειρῶν ἄθλιπτοί[2] τε καὶ μετρίως ταχεῖαι σκοπὸν ἔχουσαι θερμῆναι τὸ σῶμα
χωρὶς τοῦ θλῖψαι· καὶ γὰρ δὴ καὶ φανεῖταί σοι τούτων
γινομένων ἔρευθος εὐανθὲς ἐπιτρέχον ἅπαντι τῷ δέρματι. τότ᾽ οὖν ἤδη τὸ λίπος ἐπάγειν αὐτῷ καὶ τρίβειν
γυμναῖς ταῖς χερσὶ συμμέτρως ἐχούσαις σκληρότητός τε καὶ μαλακότητος, ὅπως μήτε συνάγηται

bladder and the lower intestines, for there is a danger of
such things being carried to the substance of the body by
the strength of the heat carried off in the exercises. If,
then, immediately after stripping off, someone were to go
to overly strong movements before softening the whole
body, thinning the superfluities and dilating the channels,
there is a danger of breaking and tearing apart the solid
bodies, and also a danger that the superfluities, moved by   90K
the force of the *pneuma,* will block up the channels.

But if, by warming a little beforehand, one were to
soften the solid parts, thin the fluids and dilate the channels, there would no longer be a danger of breakage of any
part or obstruction of the channels, if it were to catch the
person exercising. Therefore, in order for these things to
occur, it is necessary to warm the whole body beforehand
and to massage it moderately with a fine linen cloth, and
then to rub it with oil. I certainly do not recommend the
immediate use of the oil before the skin has been warmed,
the pores dilated, and, in a word, the body made ready to
receive the oil. All in all, a few strokes of the hands that
are painless and moderately quick are sufficient for the
purpose and have the objective of warming the body apart
from the rubbing. Furthermore, when these things occur,
a florid redness will seem to you to be spreading over the
whole skin. At that time, introduce the oil to it and massage with bare hands which are midway between hard and
soft, so the body is neither constricted and compressed,

---

² ἄθλιπτοί Ko; ἄλυπτοί Ku

91K　καὶ σφίγγηται τὸ σῶμα μήτε ἐκλύηται καὶ χαλᾶται
περαιτέρω τοῦ προσήκοντος, ἀλλ᾽ ἐν τῇ φύσει φυλάττηται.

τρίβειν δὲ κατὰ μὲν τὰς πρώτας ἐπιβολὰς ἀτρέμα,
τοὐντεῦθεν δ᾽ ἤδη κατὰ βραχὺ παραύξοντα καὶ μέχρι
γε τοσούτου τὴν τρίψιν ἐπὶ τὸ ῥωμαλεώτερον μεταγαγεῖν, ὡς θλίβεσθαι μὲν ἤδη σαφῶς τὴν σάρκα, μὴ
θλᾶσθαι δέ· μὴ πολλῷ δὲ χρόνῳ τὴν οὕτως ἰσχυρὰν
τρίψιν ἐπάγειν, ἀλλ᾽ ἅπαξ ἢ δὶς ἐφ᾽ ἑκάστου μέρους.
οὐ γὰρ ὥστε σκληρῦναι τὸ σῶμα τοῦ παιδὸς οὕτω
τρίβομεν, ὅταν ἤδη τοῖς πόνοις προσάγωμεν, ἀλλ᾽
ὑπὲρ τοῦ προτρέψαι τε εἰς τὰς ἐνεργείας καὶ συστρέψαι τὸν τόνον καὶ τὴν ἐκ τῆς μαλακῆς τρίψεως ἀραιότητα σφίγξαι. σύμμετρον γὰρ αὐτοῦ τὸ σῶμα φυλάττεσθαι χρὴ καὶ οὐδαμῶς οὔτε σκληρὸν οὔτε ξηρὸν
ἀποτελεῖσθαι, μή πως ἐπίσχωμέν τι τῆς κατὰ φύσιν
αὐξήσεως. τοῦ δὲ χρόνου προϊόντος, ὅταν ἤδη μειράκιον ἡμῖν γίνηται, τότε καὶ τῇ σκληροτέρᾳ τρίψει
χρησόμεθα καὶ ταῖς μετὰ τὰ γυμνάσια ψυχρολουσίαις. ἀλλὰ περὶ μὲν τούτων αὖθις εἰρήσεται.

92K　3. Ἐν τῇ δὲ τρίψει ὡς πρὸς τὰ γυμνάσια παρασκευαζούσῃ, σκοπὸν ἐχούσῃ μαλάξαι τὰ σώματα, τὴν
μέσην σκληρᾶς καὶ μαλακῆς ἐπικρατεῖν χρὴ ποιότητα καὶ κατ᾽ ἐκείνην τυποῦσθαι τὸ σύμπαν. πολυειδεῖς δὲ ταῖς ἐπιβολαῖς τε καὶ περιαγωγαῖς τῶν χειρῶν αἱ ἀνατρίψεις γιγνέσθωσαν, οὐκ ἄνωθεν κάτω
μόνον ἢ κάτωθεν ἄνω φερομένων αὐτῶν, ἀλλὰ καὶ
πλαγίων καὶ λοξῶν ἐγκαρσίων τε καὶ σιμῶν. καλῶ δὲ

nor loosened and relaxed beyond what is fitting, but is   91K
maintained within natural limits.

In the first applications the massage should be gentle.
Thereafter, increase it gradually until in fact it comes to
the point of being quite strong so as to clearly compress
the flesh without bruising it. However, do not apply strong
massage for a long period of time, but once or twice to
each part. We should not massage in such a way that we
harden the body of the boy whom we are now preparing
for his exertions. Rather, we should use it to prepare for
the actions, bring together the tone, and compress the
loose texture resulting from the soft massage. We must
maintain his body in a moderate state and in no way make
it hard or dry lest somehow we inhibit its natural growth.
As time proceeds and he becomes a young lad, we shall
also at that time use harder massage and cold baths after
the exercises. But I shall speak about these things again.

3. In the preparatory massage before exercises, which   92K
has the objective of softening the bodies, it is necessary
for a quality midway between hard and soft to prevail, and
to model everything in relation to that. The rubbings
should be of various kinds with laying on and moving
around of the hands, not only moving them from above
downward and from below upward, but also sideways,
crosswise, obliquely and circularly. I call "oblique" the op-

ἐγκάρσιον μὲν τὸ ἐναντίον τῷ εὐθεῖ, σιμὸν δὲ τὸ
βραχὺ τούτου παρεγκλῖνον ἐφ' ἑκάτερα, καθάπερ γε
καὶ τὸ τῆς εὐθύτητος ἑκατέρωσε πρὸς ὀλίγον ἐκτρε-
πόμενον ὀνομάζω πλάγιον· ὅσον δ' ἀκριβῶς μέσον
ἐστὶν ἐγκαρσίας τε καὶ εὐθείας φορᾶς, λοξὸν τοῦτο
προσαγορεύω. καὶ μὲν δὴ καὶ τρίψιν τε καὶ ἀνάτριψιν
οὐ διοίσει λέγειν, εἰδότας ὅτι τὸ μὲν τῆς ἀνατρίψεως
ὄνομα συνηθέστερον τοῖς παλαιοῖς ἐστι, τὸ δὲ τῆς
τρίψεως τοῖς νεωτέροις. πολυειδεῖς δὲ κελεύω γίνεσθαι
τὰς ἐπιβολάς τε καὶ περιαγωγὰς τῶν χειρῶν ἕνεκα
93K τοῦ συμπάσας ὡς οἷόν τε τῶν μυῶν τὰς ἶνας ἐκ παντὸς μέρους ἀνατρίβεσθαι.

το γὰρ οἴεσθαι τὴν μὲν ἐγκάρσιον ἀνάτριψιν, ἣν
δὴ καὶ στρογγύλην ὀνομάζουσιν ἔνιοι, σκληρύνειν
καὶ πυκνοῦν καὶ σφίγγειν καὶ συνδεῖν τὰ σώματα,
τὴν δὲ εὐθεῖαν ἀραιοῦν τε καὶ χαλᾶν καὶ μαλάττειν
καὶ λύειν ἐκ τῆς αὐτῆς ἐστιν ἀγνοίας, ἐξ ἧσπερ καὶ
τὰ ἄλλα, ἃ περὶ τῆς τρίψεως εἴρηται τοῖς πλείστοις
τῶν γυμναστῶν. πλεῖον γὰρ οὐδὲν οὐδεὶς ἔχει περὶ
δυνάμεων τρίψεως εἰπεῖν ὧν Ἱπποκράτης ἔγραψεν ἐν
τῷ Κατ' ἰητρεῖον, ὑπὸ μὲν τῆς σκληρᾶς δεῖσθαι τὰ
σώματα φάσκων, ὑπὸ δὲ τῆς μαλακῆς λύεσθαι, καὶ
ὑπὸ μὲν τῆς πολλῆς ἰσχναίνεσθαι, σαρκοῦσθαι δὲ
ὑπὸ τῆς μετρίας. ἔχει δὲ ἡ ῥῆσις ὧδε· "ἀνάτριψις δύναται λῦσαι, δῆσαι, σαρκῶσαι, μινυθῆσαι· ἡ σκληρὰ
δῆσαι, ἡ μαλακὴ λῦσαι, ἡ πολλὴ μινυθῆσαι, ἡ με-

posite to straight, and "circular" a little deviation from this
to each side, just as I also, in fact, term "sideways" the
deviation of the straight, turning a little to the side. What
is to a precise extent midway between oblique and straight,
I call "crosswise." And indeed, it will make no difference
whether we say massage or rubbing,[5] realizing that the
term rubbing was more customary among the ancients
while massage is more customary among those of recent
times. I direct the applications and moving around of the
hands to be of various kinds for the purpose of all the fi-   93K
bers from every part of the muscles being rubbed, as far
as possible.

The idea that oblique rubbing, which some also call
circular, hardens, thickens, compresses and binds together
the bodies, but that straight rubbing thins, relaxes, softens
and loosens, arises from that ignorance which is also the
source of the other things said about massage by the majority of gymnastic trainers. For no one has anything more
to say about the powers of massage than those things Hippocrates wrote in his treatise, *In the Surgery,* when he said
that bodies are bound up by hard rubbing, loosened by
soft rubbing, reduced by much rubbing and enfleshed by
moderate rubbing. His statement is as follows: "Rubbing
is able to loosen, bind, enflesh and reduce. Hard rubbing
binds; soft rubbing loosens; much rubbing reduces; and

---

[5] The two terms are τρῖψις and ἀνάτριψις, both of which have
the meaning of "massage," "rubbing," or "friction." The former is
given a wider range in LSJ. Galen essentially uses them inter-
changeably, the only distinction being that made here.

τρίη σαρκῶσαι." τέσσαρες γὰρ αὗται διαφοραὶ κατὰ
γένος ἐπὶ τέσσαρσι δυνάμεσί τε καὶ χρείαις τῶν τρίψεων ἁπασῶν εἰσιν. εἰ δὲ καὶ τὰς μέσας αὐτῶν προσ-
λογιζοίμεθα συνεμφαινομένας ταῖς εἰρημέναις, ἐξ αἱ
94K πᾶσαι διαφοραὶ γενήσονται. πόθεν οὖν ἐπῆλθε τοῖς
πλείστοις τῶν νεωτέρων γυμναστῶν οὕτω πολλὰς
διαφορὰς γράψαι τρίψεων, ὡς μηδὲ ἀριθμῆσαι ῥᾳδίως αὐτὰς δύνασθαι; πόθεν ἄλλοθεν ἢ ὅτι λογικῆς
θεωρίας ἀγύμναστοι παντάπασιν ὄντες οὐ συνεῖδον
ἅμα ταῖς οἰκείαις τῆς τρίψεως διαφοραῖς ἐνίοτε μὲν
καὶ τῶν ἔξωθέν τινος μνημονεύοντες, ἐνίοτε δὲ καὶ τὰς
ἀπεργαζομένας ἑκάστην τρίψιν αἰτίας ἀναγράφοντες,
ἔστιν ὅτε δὲ καὶ πρὸς ἀλλήλας ἐπιπλέκοντες αὐτάς τε
τὰς γνησίας διαφορὰς καὶ ὅσας οὐκ ὀρθῶς αὐταῖς
προσέθεσαν.

ὅταν μὲν γὰρ λέγωσι, τὰς τρίψεις ἀλλήλων διαφέρειν τῷ τὰς μὲν ἐν ὑπαίθρῳ γίγνεσθαι, τὰς δὲ ἐν
καταστέγῳ, τὰς δὲ ἐν ὑποσυμμιγεῖ σκιᾷ, καὶ τὰς μὲν
ἐν ἀνεμώδει χωρίῳ, τὰς δὲ ἐν γαληνῷ, καὶ τὰς μὲν ἐν
θερμῷ, τὰς δὲ ἐν ψυχρῷ, καὶ τὰς μὲν ἐν ἡλίῳ, τὰς δὲ
ἐν προστάδι βαλανείου, τὰς δὲ ἐν βαλανείῳ,[3] τὰς δὲ
ἐν παλαίστρᾳ, καὶ τοιοῦτόν τινα ποιοῦντες κατάλο-
γον, οὐκ οἰκείας διαφορὰς τρίψεων λέγουσιν, ἀλλ᾽ ὧν
οὐκ ἄνευ τινός ὁ τριβόμενός ἐστιν. ἀνάγκη γὰρ πάν-
95K τως αὐτὸν ἔν τινι χωρίῳ τῶν κατὰ τὴν οἰκουμένην

[3] τὰς δὲ ἐν προστάδι βαλανείου, τὰς δὲ ἐν βαλανείῳ, Ko;
τὰς δὲ ἐν βαλανείῳ, τὰς δὲ πρὸ βαλανείου, Ku

136

moderate rubbing enfleshes."[6] These are the four differences in terms of class for the four powers and uses of all massages. And if we count in addition the intermediates of these, indicating them at the same time as those mentioned, there will be six differences in all. From where, 94K then, did it come to the majority of younger gymnastic trainers to write that there are so many differences of massages that they cannot easily enumerate them? From where else than being wholly unpracticed in logical theorizing, do they not realize they sometimes mention, at the same time with the intrinsic differences of massage, one of the external factors, and sometimes describe the causes which effect each massage? Sometimes they also intermingle with each other the actual legitimate differences and those not correctly assigned to these.

Whenever they say that massages differ from one another by virtue of the fact that some occur in the open air, some indoors, some in partial shade, some in a windy place, some in a calm place, some in a hot place, some in a cold place, some in the sun, some in the vestibule of a bathing room, some in a bathing room and some in the wrestling school, in making some such catalog, they are not describing specific differences of massage but those things which the person being massaged is not without. For it is altogether necessary for him to be in some place in the inhabited world. And in fact, in addition to the 95K

[6] Hippocrates, *In the Surgery* 17, *Hippocrates* III, LCL 149, 76–77.

ὑπάρχειν, καὶ πρός γε τῷ χωρίῳ χειμῶνος ἢ θέρους
ἢ κατά τινα τῶν ἄλλων ὡρῶν. ἐπειδὰν δὲ τὰς μὲν τῷ
μετὰ πλείονος ἐλαίου, τὰς δὲ τῷ μετὰ ἐλάττονος ἢ
παντάπασιν ἐλαίου χωρίς, ἤτοι διὰ τῶν χειρῶν μόνον
ἢ μετὰ κόνεως ἢ διὰ σινδόνων, καὶ τούτων ἤτοι σκληρῶν ἢ μαλακῶν γίνεσθαι διαφέρειν ἀλλήλων λέγωσι
τὰς τρίψεις, αἰτίων καταρίθμησιν ποιοῦνται τῶν ἤτοι
σκληρὰν ἢ μαλακὴν ἀπεργαζομένων τὴν τρίψιν. ἐξ
οὗ γένους τῶν αἰτίων ἐστὶ καὶ τὸ τὰς χεῖρας τῶν τρι-
βόντων ἤτοι σκληρὰς ἢ μαλακὰς εἶναι καὶ ἤτοι πιέ-
ζειν σφοδρῶς ἢ πράως ἐφάπτεσθαι.

τὸ δὲ τρίτον εἶδός ἐστι τῶν πολλὰς τῶν τρίψεων
οἰομένων εἶναι διαφορὰς ἐκ τοῦ κατὰ συζυγίας τινὰς
ἐπιπλέκειν ἀλλήλοις ἅπαντα τὰ νῦν εἰρημένα. ὅσοι
μὲν οὖν αὐτῶν ἢ τὰς ἔξωθεν περιστάσεις τῶν πραγμάτων ἢ τὰς αἰτίας τῶν οἰκείων διαφορῶν ἀλλήλαις ἐπι-
πλέκουσιν, εὐφωρατότεροι γίνονται μὴ γινώσκοντες
ὀρθῶς· ὅσοι δὲ κατὰ τὰς οἰκείας διαφορὰς ποιοῦνται
τὰς συζυγίας, ἧττον οὗτοι γνωρίζονται σφαλλόμενοι.
96K εἰσὶ δὲ οἳ καὶ σοφίας δόξαν ἀπηνέγκαντο καὶ δοκοῦσί
τι πλέον εὑρηκέναι τῶν ὑφ᾽ Ἱπποκράτους εἰρημένων.
ὧνπερ οὖν καὶ Θέων ὁ γυμναστής ἐστιν, ὅστις ἔδοξε
βέλτιον Ἱπποκράτους ἐγνωκέναι περὶ τρίψεως. ἀφορι-
σαμένου γὰρ ἐκείνου κατὰ τὴν προγεγραμμένην ῥῆσιν ἐν μὲν τῇ κατὰ ποιότητα διαφορᾷ τήν τε μαλακὴν
καὶ τὴν σκληράν, ἐν δὲ τῇ κατὰ ποσότητα τήν τε
πολλὴν καὶ τὴν μετρίαν, ὁ Θέων οὐκ ἀξιοῖ μνημο-

place, it must be in winter or summer, or one of the other seasons. Whenever they say that massages differ from one another by being in some cases with more oil, in some with less oil, and in some without any oil at all, or with hands alone, or with powder, or with a fine linen cloth, and of these either hard or soft, they are making a computation of the causes which make the massage either hard or soft. From this class of causes, there are also the hands of those massaging—whether they are hard or soft, and whether they press strongly or are laid on gently.

The third kind is of those who think there are many differences of massages from the fact that they weave all the things just said with each other according to certain conjunctions. Those of them who interweave with each other either the external circumstances of the matters, or the actual matters themselves, or the causes of the specific differences are easily recognized as not understanding correctly. However, those who do make conjunctions of the specific differences are less easily recognized as falling into error. And there are those who have gained the repu- 96K tation of wisdom, and seem to have discovered something more than was said by Hippocrates. One of these is Theon, the gymnastic trainer, who seemed to have known about massage better than Hippocrates. For the latter, in the previously quoted statement, makes a distinction in terms of quality in the differences between hard and soft, and in terms of quantity, between much and moderate. Theon, however, doesn't think it worthwhile to mention either

νεύειν οὔτε ποιότητος οὔτε ποσότητος ἰδίᾳ, γράφων
ἐν ἄλλοις τισὶ κἂν τῷ τρίτῳ τῶν Γυμναστικῶν ὧδε·

ἀρέσκει περὶ τρίψεως παραγγέλλοντας δεῖν ἀεὶ
συναρμόζειν ταῖς ποιότησι τὰς ποσότητας.
καθ᾽ ἑαυτὰς μὲν γὰρ ἀτελεῖς εἶναι πρὸς τὴν ἐν
τοῖς ἔργοις κατόρθωσιν. τὴν γοῦν μαλακὴν τρίψιν παρὰ τὴν ποσότητα τριῶν ἀποτελεσμάτων
ποιητικὴν γίνεσθαι. τὴν μὲν γὰρ ὀλίγην ἀνιέναι
ποσῶς τὴν σάρκα καὶ εὐαφῆ ποιεῖν, τὴν δὲ πολλὴν διαφορεῖν καὶ τήκειν, τὴν δ᾽ αὐτάρκη σαρκοῦν τὸ σῶμα πλαδαρᾷ καὶ συγκεχυμένῃ⁴
σαρκί. ὁμοίως δὲ καὶ τὴν σκληρὰν τρίψιν παρὰ
τὴν ποσότητα τὸν ἴσον ἀριθμὸν τῶν ἀποτελε-
97K σμάτων ποιεῖν· πολλὴν μὲν γὰρ προσαχθεῖσαν
σφίγγειν τὰ σώματα καὶ συνδεῖν καὶ φλεγμονῇ
τι παραπλήσιον ἀπεργάζεσθαι, τὴν δὲ αὐτάρκη
σαρκοῦν μεμνωμένῃ⁵ καὶ εὐπεριγράπτῳ σαρκί,
τὴν μέντοι γε ὀλίγην ἐνερευθῆ πρὸς ὀλίγον
χρόνον τὴν ἐπιφάνειαν ποιεῖν.

οὐκ ἀξιοῖ δὲ περὶ τῆς σκληρᾶς τρίψεως ἰδίᾳ καθ᾽
ἑαυτὴν οὐδὲν παραγγέλλειν τὸν γυμναστήν, ἀλλὰ
συναρμόζειν αὐτῇ τὸ ποσόν, εἴπερ τι μέλλοι ποτὲ
κατόρθωμα τέλεον⁶ ἐν τοῖς ἔργοις τῆς τέχνης ἔσεσθαι.
κατὰ δὲ τὸν αὐτὸν τρόπον οὐδὲ περὶ τῆς μαλακῆς ἰδίᾳ

---

⁴ συγκεχυμένῃ Ko; κεχυμένῃ Ku
⁵ μεμνωμένῃ Ko; μεμειωμένῃ Ku; imminuta L

quality or quantity specifically, writing in certain other works, and in the third book of his *Gymnastics* as follows:

> In instructions about massage it is accepted that it is necessary to always combine the quantities with the qualities, in that they are incomplete in themselves in terms of the successful accomplishment in the actions. In fact, soft massage is three times more effective in a comparable amount because a little of it relaxes the flesh and makes it soft to the touch, whereas a lot disperses and liquefies, while an adequate quantity enfleshes the body with moist and liquid flesh. Similarly also, hard massage produces effects in proportion to amount. For a large amount brings about compression of bodies, binds them together and creates something resembling inflammation, whereas an adequate amount enfleshes with reduced flesh of good contour, but a small amount makes the surface red for a short time.[7]

97K

He does not, however, consider it worthwhile to give the gymnastic trainer any specific instructions about hard massage in itself, other than to fit together the quantity with it, if at any time there is going to be a successful outcome that will follow from the actions of the art. In the same way, he thinks that nothing need now be suggested

[7] For Theon, see EANS, 795, and the General Introduction to the present work, xix–xx.

---

[6] τέλεον Ko; πλέον Ku

καθ' ἑαυτὴν οὐδὲν ἡγεῖται δεῖν[7] νῦν ὑποτίθεσθαι· μὴ
γὰρ δύνασθαί ποτε γίνεσθαι μαλακὴν τρίψιν αὐτὴν
καθ' ἑαυτὴν μόνην ἄνευ τοῦ πολλὴν ἢ ὀλίγην ἢ σύμμετρον ὑπάρχειν. εἶθ' ἑξῆς διηγεῖται κατὰ συζυγίαν,
ὅσα περὶ τοῖς σώμασιν ἡμῶν ἐργάζεσθαι πεφύκασιν·
τὴν μὲν ὀλίγην τε ἅμα καὶ μαλακὴν ἀνιέναι ποσῶς
τὴν σάρκα καὶ εὐαφῆ ποιεῖν, ἀποφαινόμενος οὐδὲν
ἄλλ' ἢ τὸ "λῦσαι" πρὸς Ἱπποκράτους εἰρημένον ἑτέροις ὀνόμασιν ἑρμηνεύων· τὸ γὰρ ἀνιέναι τὴν σάρκα
καὶ εὐαφῆ ποιεῖν τί ἄλλο ἢ λύειν ἐστὶ τὰ συνδεδεμένα
98K τε καὶ συνηγμένα; προσέθηκε δὲ τῷ λόγῳ τὸ ποσῶς,
οὐ τὸ γένος τῆς ἐνεργείας ὑπαλλάττων, ἀλλὰ τὸ ποσὸν ἐν αὐτῷ διορίζων. μαλάττει γὰρ ἐπ' ὀλίγον ἡ
τοιαύτη τρίψις, ὡς, εἴ γ' ἐπὶ πλεῖον γίνοιτο, μαλάξει
μὲν ἔτι καὶ νῦν, ἀλλὰ μειζόνως ἢ πρόσθεν. ὅτι τοίνυν
οὐ μεγάλως, οὐδ' ἱκανῶς, ἀλλὰ καὶ βραχέως ἡ ὀλίγη
καὶ μαλακὴ τρίψις ἀνίησί τε καὶ μαλάττει τὰ σώματα,
διὰ τῆς τοῦ ποσῶς προσθήκης ἐδήλωσεν, οὐδὲν[8] οὐδέπω κατά γε τοῦτο τῶν ὑφ' Ἱπποκράτους εἰρημένων
διδάσκων περιττότερον, ὥσπερ οὐδ' ἐν τῷ φάναι, "τὴν
αὐτάρκη" καὶ μαλακὴν "σαρκοῦν τὸ σῶμα πλαδαρᾷ
καὶ συγκεχυμένῃ σαρκί." διότι μὲν γὰρ αὐτάρκης,
σαρκώσει, διότι δὲ μαλακή, λύσει, τοῦτ' ἔστι μαλάξει, ὅπερ ἴσον ἐστὶ τῷ πλαδαρὰν καὶ κεχυμένην ἐργάσασθαι τὴν σάρκα.

ἐχρῆν δ' αὐτόν, ὥσπερ ὑπὲρ τούτων εἶπεν ὀρθῶς,
οὕτω καὶ ὁπότε περὶ τῆς πολλῆς τε ἅμα καὶ μαλακῆς
διαλέγεται, μὴ τοῦτο μόνον εἰπεῖν, ὅτι διαφορεῖν καὶ

about soft massage specifically in itself, in that it is not possible at any time for there to be soft massage by itself alone, without there being much, little or a moderate amount. Then, next in order, he sets out in detail, according to the conjunction, those things naturally effected in our bodies—massage that is small in amount, gently slackens the flesh and makes it soft to the touch—which is nothing other than explaining what Hippocrates meant by saying "loosens," expressing the matter in other terms, for what else is slackening the flesh and making it soft to the touch than releasing what is bound and held together? He added to the discussion the amount, not changing the class of the action but determining the amount in it. For such massage softens to a small extent, just as, if it is still more, it will now still soften, but more than before. Accordingly, he showed it is not massage that is prolonged and large in amount that softens bodies, but that which is brief and small in amount, through adding the amount. In doing this, he is not adding anything more to the teaching set out by Hippocrates, just as he is not when he says sufficient and soft massage enfleshes the body with moist and liquefied flesh. Because it is sufficient, it will enflesh; because it is soft, it will loosen (that is, it will soften), which is equivalent to making the flesh moist and liquefied.

98K

It behooved him, just as he spoke correctly about these things, to do the same when he explains much and soft, and not only say that it naturally disperses and dissolves,

---

7 δεῖν om. Ku
8 οὐδὲν add. Ko

τήκειν πέφυκεν, ἀλλὰ καὶ ὁποίαν τινὰ τὴν ὑπόλοιπον
99K ἐργάζεται σάρκα. οὐ γὰρ δὴ πᾶσαν τὴν οὐσίαν διαφορεῖ καὶ τήκει, καθάπερ τὸ πῦρ, ἀλλά τι καὶ κατα-
λείπει πάντως. τοῦτο οὖν τὸ καταλειπόμενον ὁποῖόν τι
τὴν ἰδέαν ἐστὶν ἐχρῆν, οἶμαι, προσκεῖσθαι τῷ λόγῳ
καὶ μὴ μόνον τὸ τῆς πολλῆς τρίψεως ἴδιον ἔργον
εἰπόντα τὸ τῆς μαλακῆς παραλιπεῖν. ὅτι μὲν γὰρ
ἡ πολλὴ διαφορεῖ, καὶ πρὸς Ἱπποκράτους εἴρηται
πρόσθεν· ἀλλ᾽ οὐχ ὑπὲρ τῆς πολλῆς ἁπλῶς ἐνεστή-
σατο τὸν λόγον ὁ Θέων, ἀλλὰ κατὰ συζυγίαν ἠξίωσε
διδάσκειν, ἀπὸ τῆς πρώτης μὲν ἀρξάμενος τῆς τε ὀλίγης τε ἅμα καὶ μαλακῆς, ἐπὶ δευτέραν δὲ μεταβὰς
τὴν μαλακήν τε ἅμα καὶ πολλήν, εἶθ᾽ ἑξῆς τρίτης
μνημονεύσας τῆς μαλακῆς τε ἅμα καὶ συμμέτρου.
ὥσπερ οὖν ἐπί τε τῆς πρώτης καὶ τρίτης συζυγίας
οὐκ ἐσιώπησε τὸ τῆς ποιότητος ἔργον, οὕτως ἐχρῆν
αὐτὸν οὐδὲ ἐπὶ τῆς δευτέρας παραλιπεῖν, ἀλλὰ κἀνταῦθα φάναι τὴν πολλήν τε ἅμα καὶ μαλακὴν τρίψιν
διαφορεῖν τε ἅμα καὶ μαλακὴν ἀπεργάζεσθαι τὴν
σάρκα· πλὴν εἰ τὸ τήκειν ἀντὶ τοῦ μαλάττειν χρὴ
δέξασθαι, καὶ οὕτως ἀληθὲς μὲν εἴη τὸ εἰρημένον. οὐχ
100K ὅπως δὲ διαβάλλει τὴν Ἱπποκράτους διδασκαλίαν,
ἀλλὰ καὶ παντὸς μᾶλλον ὁμολογεῖ αὐτήν.

εἴπερ γὰρ ἡ μαλακὴ τρίψις ἀεὶ μὲν ἁπαλὸν ἐργάζεται τὸ σῶμα, κἂν ὀλίγη κἂν πολλὴ κἂν σύμμετρος
ὑπάρχῃ, προσέρχεται δ᾽ αὐτῇ οὐδέν τι παρὰ τῆς
ποσότητος ἕτερον, ἀχώριστον ἔσται τῆς μαλακῆς τρίψεως τὸ μαλάττειν, ὥσπερ, οἶμαι, καὶ τῆς σκληρᾶς τὸ

144

but also what it makes the remaining flesh like. For this
certainly does not disperse and dissolve the whole sub-   99K
stance, as fire does, but always leaves something remaining. Therefore, it is necessary, I think, to add to the discussion what sort of flesh is left behind in respect of kind, and
not to only to state the specific action of much massage,
omitting that of soft massage. Hippocrates said earlier that
much massage disperses, but Theon did not make his discussion simply about much; he also thought it worthwhile
to teach about combination, beginning from the first—
that is, slight and soft, then proceeding to the second,
which is soft and much, and then next making mention of
the third, which is soft and moderate. Then, just as in the
case of the first and third conjunctions, he did not remain
silent on the action of quality, so it was incumbent upon
him not to leave it out in the case of the second, but to say
here too that much and soft massage disperses and, at the
same time, makes the flesh soft. Even if we must accept
"to liquefy" instead of "to soften," the statement would
thus be true. In just such a manner, Theon not only does
not slander the teaching of Hippocrates, but above all   100K
agrees with it.

　　If soft massage always makes the body soft, whether it
is little, much, or moderate, and nothing else is added to
this in terms of quantity, then softening will be inseparable
from soft massage, just as, I think, hardening will be from

σκληρύνειν ἢ συνάγειν ἢ σφίγγειν ἢ δεῖν ἢ πυκνοῦν
ἢ δυσαφῆ τὴν σάρκα ποιεῖν ἢ ὡς ἄν τις ἑτέρως ἑρμηνεύειν ἐθέλῃ ταὐτὸν πρᾶγμα. πρῶτον μὲν γὰρ καὶ
κύριον ὄνομα τῆς οὕτω διατιθεμένης σαρκός ἐστι τὸ
σκληρόν, ὥσπερ καὶ τῆς ἐναντίας αὐτῇ τὸ μαλακόν.
οὐ μὴν ἀλλὰ καὶ τῶν ἄλλων ἕκαστον, ὡς εἶπον, ἐγχωρεῖ λέγειν. καὶ διὰ τίνα μὲν τὴν αἰτίαν οὕτω πολλοῖς ὀνόμασιν οἷόν τε χρῆσθαι καθ᾽ ἑνὸς πράγματος,
ὀλίγον ὕστερον εἰρήσεται· νυνὶ δέ (ὡμολόγηται γὰρ
τοῦτο) τὴν μὲν αἰτίαν αὐτοῦ λέγειν ἔν γε τῷ παρόντι
παραλίπωμεν.

ὅτι δ᾽ ἐξ ἀνάγκης ἕπεται τῇ μαλακῇ τρίψει τὸ μαλακὰ ποιεῖν τὰ σώματα, πρόδηλον ἐκ τῶν εἰρημένων.
101K εἰ γὰρ μήτε πολλὴν μήτ᾽ ὀλίγην μήτε σύμμετρον τὴν
τοιαύτην τρίψιν παραλαβὼν σκληρὸν ἐργάσασθαί
ποτε δυνήσῃ τὸ σῶμα, δῆλον, ὡς ἀχώριστον αὐτῆς
ἐστι τὸ μαλάττειν, ὥσπερ γε καὶ τῆς σκληρᾶς τὸ συνδεῖν τε καὶ σκληρύνειν. καὶ γὰρ καὶ ταύτην ἄν τ᾽ ὀλίγην ἄν τε πολλὴν ἄν τε σύμμετρον παραλάβῃς, οὐδέποτε μαλάξεις τὸ σῶμα κατ᾽ οὐδεμίαν ποσότητα,
διὰ παντὸς δὲ σκληρὸν ἀπεργάσῃ μᾶλλον ἢ ἧττον,
ἐπὶ πλέον μὲν τρίβων μᾶλλον, ἐπ᾽ ἔλαττον δὲ ἧττον.
εἰ δὲ καὶ παντελῶς ὀλίγας τὰς σκληρὰς ἐπιβολὰς
ποιησάμενος ἀρκεσθείης, ἀνὰ λόγον τῇ τούτων βρα-
χύτητι καὶ σκληρότης ἀπαντήσεται. καθάπερ γὰρ καὶ
ὁ πυρὶ πλησιάζων ἀεὶ μὲν θερμαίνεται, κἂν ἐπὶ πολὺν
χρόνον αὐτῷ τύχῃ πλησιάζων κἂν ἐπ᾽ ὀλίγον, ἀλλὰ
μᾶλλον μὲν τοῦθ᾽ ὁ πολυχρονίως ὁμιλῶν, ἧττον δὲ ὁ

hard massage, or drawing together, compressing, binding, condensing, or making the flesh hard to the touch, or however else someone might wish to interpret the same matter. For the first and principal term for flesh so disposed is "hard," just as the opposite to this is "soft," although of the rest, as I said, it is possible to use whatever name one pleases.[8] And the reason why it is possible in this way to use many names for one matter, I shall speak about a little later. For the present, (for this is what is agreed upon), let us postpone speaking about the reason for this for the time being.

That making bodies soft necessarily follows soft massage is clear from what has been said. For if by undertaking either much, little or a moderate amount of such massage, you will never be able to make the body hard at any time, it is obvious that softening is inseparable from it, just as binding together and hardening are inseparable from hard massage. For truly, whether you undertake this either little, much or moderately, you will never soften the body, regardless of the quantity. You will always make it hard to a greater or lesser degree—more hard by more massage and less hard by less. Also, if you were to be entirely satisfied with making few hard strokes, the hardness you encounter will be in proportion to their fewness, just as when you come near to a fire, you are always heated, whether you happen to approach it for a long or short time, but more if the association is of long duration, less if

101K

[8] The translation of this sentence, difficult to understand in the Greek, follows the Kühn Latin.

μέχρι βραχέος, ἐπ' ἐλάχιστον δὲ ὁ ψαύσας μόνον,
οὕτω κἀν ταῖς τρίψεσιν ὁμοιοῦται διαπαντὸς τὸ σῶμα,
πρὸς μὲν τῆς μαλακῆς μαλαττόμενον, ὑπὸ δὲ τῆς
σκληρᾶς σκληρυνόμενον, οὐ μὴν ἴσῳ γε τῷ μέτρῳ
διαπαντός, ἀλλ' ὑπὸ μὲν τῆς πλείονος μᾶλλον, ὑπὸ δὲ
τῆς ἐλάττονος ἧττον, ὥσπερ γε καὶ ὑπὸ μὲν τῆς πλεί-
102K στης μάλιστα, ὑπὸ δὲ τῆς ἐλαχίστης ἥκιστα.

ὁ τοίνυν Θέων, ὁπότε περὶ τῆς μαλακῆς τε καὶ
πολλῆς τρίψεως διαλεγόμενος διαφορεῖν καὶ τήκειν
εἶπεν αὐτήν, εἰ μὲν τὸ κενοῦν διὰ τοῦ τήκειν δηλοῖ,
πλέον οὐδὲν σημαίνει τοῦ διαφορεῖν· ὥστε δὶς μὲν
ἂν εἴη ταὐτὸν εἰρηκώς, παραλελοιπὼς δὲ προσθήκην
ἀναγκαίαν τῆς γινομένης περὶ τὸ σῶμα διαθέσεως ἐκ
τῆς μαλακῆς τρίψεως· εἰ δὲ τὸ μαλάττειν ἢ λύειν ἢ
χαλᾶν ἢ ὅπως ἂν ἑτέρως ὀνομάζειν ἐθέλῃ, παραλε-
λείψεται μὲν οὕτως οὐδὲν αὐτῷ, τὰ δ' Ἱπποκράτους
πάντα λέγων φωραθήσεται διὰ μοχθηροτέρου τρόπου
διδασκαλίας. ὅτι μὲν οὖν ὁ τοιοῦτος τρόπος τῆς διδασκαλίας μοχθηρότερός ἐστιν ἧς Ἱπποκράτης ἔγραψεν, ὀλίγον ὕστερον ἐπιδείξομεν· ὅτι δέ, εἴπερ τὸ
τήκειν ἀντὶ τοῦ μαλάττειν εἴρηκεν, οὐδὲν τῶν Ἱπποκράτους λέγει περιττότερον, ἄντικρυς δῆλον. εἰπόντος
γὰρ ἐκείνου, τὴν μὲν μαλακὴν λύειν, τὴν δὲ πολλὴν
ἰσχναίνειν, εὔδηλον ἂν εἴη παντί γε τῷ νοῦν ἔχοντι
συλλογίσασθαι, τὴν ἐκ τῆς μαλακῆς τε ἅμα καὶ πολ-
103K λῆς σύνθετον ἰσχναίνειν τε ἅμα καὶ μαλάττειν.

ἀλλ' οὐχ ὁ Θέων ἔοικεν οὕτω γινώσκειν, ἀλλ' ὅπερ
ἐν τῷ πρώτῳ τῶν Γυμναστικῶν ἰσχναίνειν εἶπε δι'

it is of short duration, and least if you only just reach it. It is always like this also with massaging the body—it is softened by soft massage and hardened by hard massage—although not always in equal measure, but more by more massage and less by less, just as [it is affected] most by the most massage and least by the least.   102K

Accordingly, when Theon, discussing soft and much (prolonged) massage, said it disperses and liquefies, if by liquefying he is indicating emptying, he is signifying nothing more than dispersing, so he would have said the same thing twice while omitting what must be added, which is the condition occurring in the body from the soft massage. If, however, he says to soften, loosen or relax, or whatever else he might wish to term it, he will in this way leave out nothing from it, but he will be detected as saying all the things Hippocrates said through a more burdensome kind of teaching. That such a manner of teaching is inferior to what Hippocrates wrote, I shall demonstrate a little later. But if he says "to liquefy" instead of "to soften," it is perfectly clear he is saying nothing more than Hippocrates said. For when the latter says soft massage dissolves and much massage reduces, it would be quite clear, at least to anyone of sense, who would conclude [from the information given] that massage which combines softness and much, reduces and at the same time also softens.   103K

But Theon doesn't seem to see things in this way. Rather, while in the first book of his *Gymnastics*, he speaks

ἑνὸς ῥήματος, τοῦτ᾽ ἐν τῷ τρίτῳ διὰ δυοῖν, τοῦ τε
διαφορεῖν καὶ τήκειν. ἔχει γὰρ οὖν δὴ καὶ ἡ ἐν τῷ
πρώτῳ ῥῆσις ὧδε· "ἐκ δὲ τῶν ἐναντίων τὴν μαλακὴν
τρίψιν πολλὴν μὲν γενομένην ἰσχναίνειν τὰ σώματα,
αὐτάρκη δὲ σαρκοῦν τρυφερᾷ καὶ κεχυμένῃ σαρκί."
φανερῶς γὰρ ἐνταῦθα τὴν μαλακήν τε ἅμα καὶ πολλὴν τρίψιν ἰσχναίνειν ἔφη τὰ σώματα. ὅτι δ᾽ ἀναγκαῖον ἦν οὐ μόνον τὸ τῆς πολλῆς ἴδιον, ἀλλὰ καὶ τὸ
τῆς μαλακῆς εἰπεῖν, αὐτὸς ὁ Θέων ἐνεδείξατο διὰ τῆς
ἑξῆς ῥήσεως, ἐν ᾗ περὶ τῆς σκληρᾶς τρίψεως διδάσκων τὴν πολλὴν σφίγγειν τὰ σώματα καὶ συνδεῖν
καὶ φλεγμονῇ τι παραπλήσιον ἐργάζεσθαί φησιν. ὃ
γὰρ εἶχε συμμέτρως ὑπάρχουσα τῇ ποσότητι, τοῦτο
πλεονασθεῖσα μᾶλλον ἐπεκτήσατο. πάντες οὖν ὁμο-
λογοῦσι τὴν σύμμετρον σκληρὰν σαρκοῦν τὸ σῶμα
σκληρᾷ σαρκί, καθάπερ γε καὶ τὴν σύμμετρον μαλακὴν σαρκοῦν μὲν καὶ αὐτήν, ἀλλὰ μαλακῇ τῇ σαρκί.

104K   κινδυνεύει δὲ καὶ ὁ Θέων, ἡνίκα ὑπὲρ τῆς μαλακῆς
τε ἅμα καὶ πολλῆς τρίψεως ὁ λόγος ἦν αὐτῷ, μηδὲν
τῆς μαλακῆς ἔργον εἰρηκέναι, ὁπότε δὲ περὶ τῆς
σκληρᾶς τε ἅμα καὶ πολλῆς, μηδὲν τῆς πολλῆς. εἰ
γὰρ τὸ τήκειν ταὐτὸν τῷ κενοῦν καὶ διαφορεῖν ὑπολά-
βοιμεν, ὑπὲρ μὲν τῆς πολλῆς ἔσται τι λελεγμένον
αὐτῷ, περὶ δὲ τῆς μαλακῆς οὐδέν, ὥσπερ γε κἂν τῷ
περὶ τῆς σκληρᾶς τε καὶ πολλῆς ὑπὲρ μὲν τῆς σκληρᾶς εἴρηται τὸ σφίγγειν καὶ συνδεῖν καὶ φλεγμονῇ τι
παραπλήσιον ἀπεργάζεσθαι, περὶ δὲ τοῦ πλήθους

of "reducing" using a single word, in the third book he uses two words—"dispersing" and "liquefying." Then he also has the following statement in the first book: "In contrary fashion, soft massage in large amount reduces bodies whereas moderate massage enfleshes them with delicate and liquefied flesh."[9] Clearly here he said that massage which is soft and at the same time large in amount reduces bodies. That it was necessary to state specifically not only the large amount but also the softness, Theon himself showed by the following statement in which, teaching about hard massage, he says that much compresses bodies and binds them together, and creates something similar to inflammation. For what it has when it is moderate in amount, it has in abundance when it is more protracted. All, then, agree that a moderate amount of hard massage enfleshes the body with hard flesh, just as a moderate amount of soft massage enfleshes the same, but with soft flesh.

However, when his discussion was about soft and long-continued massage, Theon was in danger of not saying anything on the action of soft [massage], whereas when his statement was about hard and long-continued massage, he was in danger of not saying anything about the large amount. If we take liquefying to be the same as evacuating and dispersing, he will have said something about the large amount, but there will be nothing about the softness, just as in [his statement] about hardness and large amount, he speaks about the hardness compressing, binding and bringing about something like inflammation, but says

104K

9 The present quotes are from Theon's lost work *Gymnastrion* referred to in note 7 above.

151

οὐδέν· καίτοι δίκαιον ἦν τι καὶ περὶ τῆς ποσότητος εἰπεῖν. ἀπεφήνατο γοῦν, ὡς, εἰ σύμμετρος εἴη τῇ ποσότητι, παχύνει.

τί ποτ᾽ οὖν ἡ πολλὴ ποιήσει μετὰ τὴν σύμμετρον ὑπάρχουσα; πάντως γὰρ ἤτοι φυλάξει τὸ ταύτης ἔργον ἢ ἀλλοιώσει. ἀλλ᾽ εἰ μὲν φυλάξει, τὸ πλῆθος μετὰ τὴν συμμετρίαν τῶν τρίψεων οὐδὲν ἐργάσεται περιττότερον· εἰ δέ τι δράσει πλέον, ἤτοι καθαιρήσει τῆς σαρκώσεως ἢ προσθήσει. καθαιροῦσα μὲν οὖν λεπτυνεῖ, προστιθεῖσα δὲ σαρκώσει. ἀλλὰ μὴν οὐ σαρκοῖ· λεπτύνειν γοῦν ἀναγκαῖον αὐτήν. οὐ μὴν εἰπέ γε ὁ Θέων οὐδὲν ὑπὲρ τῆς κατὰ τοῦτο διαφορᾶς, ἀλλ᾽ ὅλως ἐσιώπησε, μήτ᾽ εἰ λεπτύνει μήτ᾽ εἰ παχύνει ἢ φυλάττει τὴν ἐκ τῆς συμμέτρου τρίψεως σάρκωσιν ἡ σκληρὰ πλεονασθεῖσα μηδὲν ὅλως ἀποφηνάμενος, ἀλλὰ μόνον ὅτι σφίγγει καὶ συνδεῖ παραπλησίως φλεγμονῇ. ἐχρῆν δ᾽ οὐ ταῦτα μόνον εἰπεῖν, ἀλλ᾽ ὅτι καὶ λεπτύνει.

4. Φαίνεται τοίνυν[9] ἢ μὴ δυνηθεὶς συνιέναι τῆς τοῦ Ἱπποκράτους ἐν τοῖσδε τέχνης ἢ μὴ βουληθεὶς ἐπαινεῖν τὸν ἄνδρα διὰ βραχείας οὕτω ῥήσεως ἁπάσας τε τὰς διαφορὰς τῶν τρίψεων εἰπόντα καὶ πρὸς ταῖς διαφοραῖς ἑκάστης αὐτῶν ἐνέργειάν τε καὶ δύναμιν. ὡς μὲν γὰρ ἄν τις οἰηθείη κατὰ τὴν πρόχειρον οὑτωσὶ φαντασίαν, ὑπὲρ τεττάρων εἴρηκε μόνον· ἔχει δὲ οὐχ οὕτω τἀληθές, ἀλλ᾽ ἑτέρας ἐνεδείξατο δύο ταῖς εἰρημέναις ἐξ ἀνάγκης συνεπινοουμένας, ἐν μὲν τῇ κατὰ ποιότητα διαφορᾷ τὴν μέσην τῆς σκληρᾶς τε καὶ μα-

nothing about the amount—and indeed, it would have been right to also speak about the quantity. Anyway, he did declare that, if there is moderation in quantity, it thickens.

What in the world, then, will "large amount" do if it is after "moderate"? At all events, it will either preserve the action of the massage or it will change it. But if it preserves it, the quantity of massage after moderation will bring about nothing more. If it does do something more, it will either reduce the growth of flesh or it will increase it. If it reduces, then it thins, whereas, if it increases, it enfleshes. But it doesn't enflesh, so, of necessity, it thins. In fact, Theon said nothing about the difference in this, but was altogether silent, making no declaration at all about whether it thins or thickens, or whether it preserves the flesh from the moderate massage or increases the hardness. He only said that it compresses and binds together, [and creates something] similar to inflammation. He should not only say this, but also that it thins.

105K

4. It is obvious, then, that Theon was either unable to understand the art of Hippocrates in these matters or that he did not wish to praise the man because he spoke so briefly in this way on all the differences of massage, and as well as the differences, on the function and potency of each of them. For someone might be made to think, from the image presented in this way, that he spoke about four kinds only. But this is not true; he did show there were two other kinds which must be included with those spoken of. These are in the difference in terms of quality between

---

9 *post* τοίνυν *add.* ὁ Θέων Ku

λακῆς, ἥπερ δὴ καὶ σύμμετρός ἐστιν, ἐν δὲ τῇ κατὰ
ποσότητα τὴν ὀλίγην· ἀναγκαῖον γὰρ τὴν ὀλίγην
106K ἐναντίαν εἶναί τινα τῇ πολλῇ. τὸ δὲ τῆς παλαιᾶς ἑρμηνείας εἶδος οὕτως ἐστὶ βραχυλόγον, ὡς πολλὰ
πολλάκις ὑπερβαίνειν δοκεῖν τῇ λέξει τῶν ἐξ ἀνάγκης
ἑπομένων τοῖς λεγομένοις. καὶ διὰ τοῦτ᾽, οἶμαι, γράφομεν αὐτῶν ὑπομνήματα, ποδηγοῦντες τοὺς δι᾽
ἀγυμνασίαν ἀδυνάτους ἕπεσθαι τάχει λέξεως παλαιᾶς, καθάπερ κἂν τῷδε τῷ λόγῳ ποιοῦμεν. εἰ γὰρ ἡ
μὲν σκληρὰ δύναται δεῖν, ἡ δὲ μαλακὴ λύειν, ὅσα μὲν
ἐκλέλυται σώματα πέρα τοῦ μετρίου, σκληρῶς ἀνατριπτέον, ὅσα δ᾽ ἔσφιγκται, μαλακῶς. εἰ δέ τι συμμέτρως ἔχει, τοῦτ᾽ εὔδηλον ὡς οὔτε μαλακῶς οὔτε
σκληρῶς, ἀλλ᾽ ὅσον οἷόν τε τὰς ὑπερβολὰς ἑκατέρας
φυλαττόμενον. ὁ δὲ θαυμάσιος Θέων εὐθὺς τοῦτο
πρῶτον ἔσφαλται, μήτε τὴν δύναμιν εἰπών ποτε τῆς
συμμέτρου κατὰ ποιότητα τρίψεως μήτε τὴν χρείαν,
ἀλλ᾽ ἀεὶ παρερχόμενος αὐτὴν ὥσπερ οὐκ οὖσαν. περὶ
μὲν δὴ τοῦδε καὶ μικρὸν ὕστερον ἐροῦμεν. ὥσπερ δ᾽
ἐν ταῖς κατὰ ποιότητα διαφοραῖς οὐ σκληρὰ καὶ μαλακὴ μόνον ἐστίν, ἀλλὰ καὶ σύμμετρος, οὕτω κἂν
ταῖς κατὰ τὸ ποσὸν οὐ πολλὴ καὶ ὀλίγη μόνον, ἀλλὰ
καὶ μετρία.

107K   τί ποτ᾽ οὖν ἐν μὲν ταῖς κατὰ ποιότητα τὴν μέσην
παρέλιπεν ὁ Ἱπποκράτης, ἐν δὲ ταῖς κατὰ ποσότητα
τὴν ὀλίγην; ἴσως γὰρ ἄν τινι δόξειεν ἀλόγως τοῦτο
πρᾶξαι· χρῆναι γὰρ ἐν ταῖς βραχυλόγοις διδασκα-
λίαις ἀφορίζεσθαι μὲν ταῖς ἄκραις ἐναντιότησι τὰ

hard and soft (firm and gentle), which is also moderate, and in the difference in terms of quantity regarding little, where little is of necessity opposite to much. But the form 106K of the ancient explanation expresses things so briefly that it often seems to pass over, in the statement, many things that necessarily follow the things stated. And it is because of this, I think, that we write commentaries about them as guides for those who, being unpracticed, are unable to follow the brevity of the ancient statement, just as I am doing in this discussion. Thus, if hard (firm) massage can bind, while soft (gentle) massage can loosen, one must firmly massage those bodies that have been loosened beyond moderation and gently massage those that are bound tightly. If something is moderate, it is clear it should be massaged neither gently nor firmly, but as far as possible protected against each extreme. This is the first point on which the admirable Theon was immediately mistaken, neither speaking at any time of the power of massage that is moderate in quality, nor the use, but always disregarding this, as if it didn't exist. I shall also speak about this a little later. Just as in the differences pertaining to quality, there is not only hard and soft, but also moderate, so too in the differences pertaining to quantity, there is not only much and little, but also moderate.

Why then, in the differences pertaining to quality, did 107K Hippocrates leave out moderate, and in the differences pertaining to quantity, little? Perhaps it might seem to someone that he did this unwittingly because, in the briefly expressed teachings, it is necessary to define mat-

πράγματα, παραλείπεσθαι δὲ τὸ μέσον τε καὶ σύμμετρον ἐν αὐταῖς, ὡς ἐξ ἀνάγκης τοῖς ἄκροις συνεπινο-
ούμενον. ἐγὼ τοίνυν καὶ τοῦτο πειράσομαι διελθεῖν
καὶ σαφηνίσαι τὴν τοῦ παλαιοῦ γνώμην, οὐκ ἀναγκαῖον μὲν ὂν ἐνταῦθα (τὰ γὰρ τοιαῦτα ζητήματα διὰ
τῶν ἐξηγητικῶν ὑπομνημάτων εἰθίσμεθα λύειν), ἀλλ'
ἐπειδὴ ἅπαξ ἐν τῷδε τοῦ λόγου κατέστην, ὡς ἀπολο-
γεῖσθαι τοῖς ἀγυμνάστοις ὑπὲρ Ἱπποκράτους, ἐν οἷς
αὐτοὶ σφαλλόμενοι κατηγοροῦσι τοῦ κατορθοῦντος,
οὐκ ὀκνήσω προσθεῖναι καὶ τοῦτο.

δυοῖν οὖν ὄντων πραγμάτων ὅλῳ τῷ γένει κεχω-
ρισμένων, εἴ γε δὴ τὸ ποιοῦν τοῦ ποιουμένου τῷ γένει
διενήνοχεν, αἱ μὲν τρίψεις ἐκ τῶν ποιούντων εἰσίν, αἱ
δ' ὑπ' αὐτῶν ἀποτελούμεναι κατὰ τὰ σώμαθ' ἡμῶν
διαθέσεις ἐκ τῶν ποιουμένων. ὥστε καὶ τὰς ἐναντιότη-
108K τας ἀναγκαῖον ἑτέρας μὲν ἐν τῷ τῶν τρίψεων, ἑτέρας
δ' ἐν τῷ τῶν διαθέσεων ὑπάρχειν γένει, ἐν μὲν τῷ τῶν
τρίψεων τήν τε μαλακὴν καὶ τὴν σκληρὰν καὶ τὴν
πολλὴν καὶ τὴν ὀλίγην, ἐν δὲ τῷ τῶν διαθέσεων τὴν
οἷον δέσιν τε καὶ λύσιν τῶν σωμάτων καὶ τὴν ἰσχνότητα καὶ τὴν σάρκωσιν. ἡ μὲν οὖν προτέρα τῶν διαθέσεων ἐναντίωσις ὑπὸ τῆς προτέρας κατὰ τὰς τρίψεις
ἐναντιώσεως γίγνεται, ἡ δὲ δευτέρα οὐκέτι. συμβαίνει
γὰρ ἐπ' αὐτῆς τὴν μὲν ἰσχνότητα πρὸς τῆς πολλῆς
γενέσθαι τρίψεως, τὴν δ' ἀνάθρεψιν ὑπὸ τῆς μετρίας.
ἡ γὰρ ὀλίγη σαρκοῦν οὐδέπω δυνατή, διότι δεῖται μὲν
τὸ σαρκωθησόμενον ἀεὶ αἵματός τε παραθέσεως συμμέτρου καὶ δυνάμεως εὐρώστου. καὶ ταῦτα ἄμφω κα-

ters by their opposite extremes, leaving out the median and the moderate in those as being, of necessity, jointly known from the extremes. Therefore, I shall also attempt to go over this and make clear the opinion of the ancient (Hippocrates), although not necessarily here, for I am accustomed to leave such questions to the exegetical commentaries. But since I have been appointed once and for all in the present discussion to speak in his defense to those untrained in Hippocratic medicine, among whom are those who erroneously speak against the man who was successful, I shall not hesitate to also add this.

Since there are two matters wholly separated in class, if in fact what acts is differentiated in class from what is acted upon, the massages are separated into those who do them and the conditions brought about through them in our bodies by the things that are done. As a consequence, it is also necessary that the opposites are those in the class of the massages and those in the class of the conditions. In the class of the massages, there are soft and hard, and much and little; in the class of conditions there are, as it were, the binding together and the loosening of bodies, and the reducing and enfleshing. In the former, the opposition of the conditions arises through the prior opposition in the massages, whereas in the second case this is no longer so. What happens in this is that the reduction arises from the large amount of massage, whereas the restoration arises from the moderate amount. A small amount of massage is not yet able to enflesh because what is going to be enfleshed always needs in addition the joint presence of a moderate amount of blood and a strong capacity, and both these things properly arise in it from moderate massage,

λῶς αὐτῷ πρὸς τῆς συμμέτρου γίνεται τρίψεως, οὐ-
δέτερον δ᾽ ἱκανῶς οὐδ᾽ αὐτάρκως ἐπὶ τῆς ὀλίγης. ἐπεὶ
τοίνυν οὐ συνέβαινον ἐς ταὐτὸν αἱ τῶν τρίψεων ἐναν-
τιώσεις ταῖς τῶν διαθέσεων, ἠνάγκαζε δ᾽ αὐτὸν ἡ τῆς
βραχυλόγου διδασκαλίας ἰδέα δι᾽ ἐναντιώσεων ἀφορίζεσθαι τὸν λόγον, ἐπὶ τὴν χρησιμωτέραν ἐναντίω-
σιν ἀφικόμενος ὑπερεῖδε τῆς ἀχρηστοτέρας. χρησι-
109K μωτέρα δέ ἐστιν ἡ κατὰ τὰς διαθέσεις τῆς κατὰ τὰς
τρίψεις ἑνὶ μὲν καὶ πρώτῳ λόγῳ τῷ κατὰ τὸ τέλος τῆς
τέχνης· ἐστοχασμέναι γάρ εἰσιν αἱ τρίψεις τῆς τοῦ
σώματος διαθέσεως ὡς τέλους· ἀεὶ γὰρ τὸ τέλος τοῦ
πρὸ αὐτοῦ κυριώτερον, ὅσῳ καὶ τοῦ διὰ τί γινομένου
τὸ δι᾽ ὃ γίνεται· δευτέρῳ δὲ λόγῳ ⟨τῷ⟩[10] τῆς σαφη-
νείας· ἐκ μὲν γὰρ τοῦ μαθεῖν ἡμᾶς τὰ τῆς πολλῆς καὶ
τὰ τῆς μετρίας ἀνατρίψεως ἔργα ῥᾷστόν ἐστιν ἐπινοῆσαι καὶ τὰ τῆς ὀλίγης· οὐκέθ᾽ ὁμοίως δ᾽ εὐσύν-
οπτος ἡ τῆς συμμέτρου τρίψεως δύναμις.

οὐ μὴν ἀλλὰ καὶ τρίτον τῷδε τῷ λόγῳ κάλλιστα
ἂν ὁ Ἱπποκράτης εὑρίσκοιτο διδάσκων περὶ τρίψεων.
ἡγεῖσθαι μὲν γὰρ χρὴ τὴν τῶν ἐναργῶς ἀποτελου-
μένων διδασκαλίαν, ἔπεσθαι δὲ τὴν τῶν ἀμυδρῶς,
εἴτε γράφοι τις περὶ αὐτῶν ῥητῶς εἴτε τοῖς ἀναγινώ-
σκουσιν ἀπολείποι. ἔστι δ᾽ ἐναργὲς ἐπινοεῖν[11] μὲν τὸ
τῆς μετρίας τρίψεως ἔργον, ἡ σάρκωσις τοῦ σώματος,
οὐκ ἐναργὲς δὲ τὸ τῆς ὀλίγης· οὔτε γὰρ σαρκοῦν οὔτ᾽
ἰσχναίνειν οὔθ᾽ ὅλως οὐδὲν ἐναργὲς φαίνεται ποιεῖν,
110K ὅτι μὴ θερμαίνοι μόνον ἐπὶ βραχύ. κοινὸν δ᾽ ἦν ἀπά-

whereas neither does satisfactorily or sufficiently from a small amount of massage. Therefore, since the oppositions of the massages do not correspond to the same degree to the oppositions of the conditions, but the form of the briefly expressed teaching compelled him to define the argument through oppositions, focusing on the more useful opposition, he overlooked the more useless. More useful is what pertains to the conditions than what pertains to 109K the massages. One, and the primary, reason relates to the goal of the craft, for massages are aimed at the condition of the body as a goal, and the goal is always more important than what precedes it by as much as what occurs is to the means by which it occurs. The second reason is for the sake of clarity in the discussion. Thus, from our learning the effects of much and moderate massage, it is very easy to think also of those of little massage, whereas the potency of moderate massage is not similarly seen at once.

But Hippocrates may also be found teaching a third point about massage very well in this discussion. For the teaching of those things that are clearly accomplished must take the lead, while that of those things that are obscure must follow, whether someone were to write about them clearly or leave it to his readers to find clarity. The action of moderate massage is clearly conceived—an enfleshing of the body. What is not clear is the action of slight massage, for it seems neither to enflesh nor reduce, nor to do anything else at all that is obvious, apart from heating a little. Heating is, however, common to all mas- 110K

---

[10] *post* λόγῳ: ⟨τῷ⟩ Ko; ἕνεκεν Ku
[11] ἐπινοεῖν *om.* Ku

σης τρίψεως τὸ θερμαίνειν. ὥστ᾽, ἐπεὶ μήτ᾽ ἐναργὲς
ἀποτελεῖ μηδέν, καὶ ὅ τι φαίνεται ποιεῖν, οὐδὲ τοῦτο
ἴδιον αὐτῆς ἐστιν, ἀλλὰ τὸ κοινὸν ἁπάσης τρίψεως,
εὐλόγως παρελείφθη. τὸ μὲν οὖν, ὅτι κοινὸν ἁπάσης
τρίψεως ἔργον ἐστὶ τὸ θερμαίνειν, οὐκ ἄξιον Ἱπποκράτους συγγράμματος· ὁποῖον δέ τι πέφυκεν ἑκάστη
τρῖψις ἴδιον ἀποτελεῖν, Ἱπποκράτει τε διδάσκειν
ἀναγκαῖον ἦν ἡμῖν τε μανθάνειν χρήσιμον. ἀναγα-
γὼν γὰρ ὥσπερ εἰς στοιχεῖά τινα τὰς ἁπλᾶς διαφορὰς ἅπαντα τὸν περὶ τῆς τρίψεως λόγον ἐδίδαξεν,
ὅπως ἀναθρέψεις ἢ καθαιρήσεις ἢ μαλάξεις ἢ σφίγ-
ξεις τὸ σῶμα. τούτοις δ᾽ εὐθέως συνεμφαίνεται τά τε
μέσα τῶν ἔργων καὶ τὰ κατὰ συζυγίαν ἀποτελούμενα·
μέσα μέν, ὅταν ἤτοι μήτε δῆσαι τὸ σῶμα μήτε λῦσαι
μήτε σαρκῶσαι μήτε μινυθῆσαι προελώμεθα, κατὰ
συζυγίαν δέ, ὅταν, εἰ οὕτως ἔτυχεν, ἅμα δῆσαι καὶ
σαρκῶσαι. τίς γὰρ οὐκ ἂν ἐπινοήσειεν, ὡς, ἐπειδὰν
111K σκληρᾷ σαρκὶ σαρκῶσαι σῶμα προαιρώμεθα, τὴν
σκληρὰν ἡμῖν τρῖψιν ἅμα ποσότητι συμμέτρῳ παρα-
ληπτέον ἐστίν, ὥσπερ, κἀπειδὰν μαλακῇ, τὴν μαλακήν τε ἅμα καὶ σύμμετρον ἐν τῷ ποσῷ, καὶ κατὰ τὰς
ἄλλας συζυγίας ἀνάλογον;

   ἃς ἔνιοι τῶν γυμναστῶν ὡς ἴδια γράφοντες εὑρή-
ματα μετὰ καὶ τοῦ προσεγκαλεῖν Ἱπποκράτει μεγάλως πλημμελοῦσιν, ὅτι τε τὸν πρῶτον ὑπὲρ αὐτῶν
διδάξαντα δικαίων ἐπαίνων ἀποστεροῦσι καί, τὸ τούτου χεῖρον, ὅτι τὴν ἐκείνου γνώμην εἰς ἑαυτοὺς μετα-

sage. Consequently, since it does not clearly accomplish anything, and what it seems to do is not something specific to it, but is common to all massage, it was reasonably omitted. Therefore, the fact that heating is an action common to all massage was not worthy of a book by Hippocrates. What was necessary for Hippocrates to teach, and useful for us to learn, is what kind of specific effect each massage naturally brings about. For having reduced everything to simple differences, as if to elements, he taught the whole theory of massage—how you will restore, reduce, soften or compress the body. With these, he jointly revealed the means of their actions and their effects when used in conjunction—"means" are whenever we choose beforehand neither to bind nor loosen the body, neither to enflesh nor reduce, and by conjunction, whenever, as may happen, we choose beforehand to bind and enflesh at the same time. Who, then, would not realize that, when we choose to enflesh the body with hard flesh, hard massage which is at   111K the same time of moderate amount is what we must apply, just as, when we choose soft flesh, we must apply soft massage in moderate amount, and analogously in respect of the other conjunctions?

Some gymnastic trainers, when they write about these things as their own discoveries, combine rebuking Hippocrates with being greatly in error themselves in that they deprive the first man to have taught about those things of

φέροντες ἔτι καὶ διαβάλλειν αὐτὸν ἐπιχειροῦσιν, οὐδ᾽ οὖν οὐδὲ καλῶς ἐπαλλάττοντες ἀλλήλαις τὰς ἁπλᾶς διαφοράς. ἐχρῆν γάρ, οἶμαι, τόν γε κατὰ τρόπον ἐπὶ τὰς ἐν μέρει συζυγίας ἔρχεσθαι βουλόμενον οὔτε τὰς συμπάσας ἐξ ποιεῖν, ὥσπερ ὁ Θέων, οὔτε παραλιπεῖν τι κατὰ ταύτας ἔργον ἢ ποσότητος ἢ ποιότητος ἴδιον, ὥσπερ ἐπὶ μὲν τῆς μαλακῆς τε ἅμα καὶ πολλῆς τὸ τῆς μαλακῆς, ἐπὶ δὲ τῆς σκληρᾶς τε ἅμα καὶ πολλῆς

112K τὸ τῆς πολλῆς ἴδιον ἐδείκνυτο παραλελοιπώς, ἀλλὰ καὶ ταῦτα προστιθέναι, καὶ τὰς κατὰ μέρος συζυγίας ἐννέα ποιεῖν. αἱ γὰρ τρεῖς διαφοραὶ τῶν κατὰ ποιότητα τρίψεων ταῖς τρισὶ διαφοραῖς τῶν κατὰ ποσότητα τρίψεων ἐπαλλαττόμεναι[12] συζυγίας ἀποτελοῦσιν ἐννέα, ἐξ μὲν τὰς ὑπὸ Θέωνος εἰρημένας ἐν ᾗ παρεθέμην ὀλίγον ἔμπροσθεν ῥήσει, τρεῖς δ᾽ ἄλλας, ἃς ἐκεῖνος παρέλιπε, τὴν μέσην σκληρᾶς τε καὶ μαλακῆς ὑπερβάς· καίτοι γ᾽ οὐδ᾽ ἐπινοῆσαι δυνατόν ἐστιν οὔτε σκληρὰν οὔτε μαλακὴν τρίψιν ἄνευ τοῦ προσεπινοῆσαι τὴν σύμμετρον· ᾗ τάς γε τρεῖς διαφορὰς τῆς ποσότητος εἴπερ ἔζευξεν, ἐννέα τὰς πάσας ἂν οὕτως ἀπειργάσατο συζυγίας τρίψεων, οὐχ ἕξ. ἐκθήσομαι δὲ αὐτὰς ἐπὶ διαγράμματος, ἐν ᾧ τὸν μὲν πρότερον στοῖχον ἄνωθεν κάτω ποιοτήτων χρὴ νοεῖν, τὸν δὲ δεύτερον ποσοτήτων.

12 ἐπαλλαττόμεναι add. Ko

his rightful praise and, what is worse, they do so by trans-
ferring his opinion to themselves. And still they try to
slander him, not even properly comparing the simple differences with each other. I think that someone who wishes
to go through the individual conjunctions correctly, must
neither make them six in all, as Theon does, nor leave out
some specific action in relation to these, either of quantity
or quality, as he was shown to have left out the specific
action of soft in the combination of soft and much, and of
much in the combination of hard and much, but must add   112K
these too, making nine conjunctions individually. The
three *differentiae* of massage according to quality, when
combined with the three *differentiae* according to quantity, make nine conjunctions. There are six mentioned by
Theon in the statement I set down a little earlier, and three
others which he left out, overlooking the moderately hard
and the moderately soft. And indeed, it is not possible to
think of either hard or soft massage without thinking besides of the moderate. If he had joined the three *differentiae* of quantity with this he would have made, in this way,
all the nine conjunctions of massage and not six. I shall set
these out in a diagram in which it is necessary to think of
qualities in the first column from above down, and quanti-
ties in the second.

	ποιότητες	ποσότητες
113K	σκληρά	1. ὀλίγη
		2. πολλή
		3. σύμμετρος
	μαλακή	1. ὀλίγη
		2. πολλή
		3. σύμμετρος
	σύμμετρος	1. ὀλίγη
		2. πολλή
		3. σύμμετρος

τῶν ἐν τούτῳ τῷ διαγράμματι γεγραμμένων ἐννέα συζυγιῶν ἐξ τὰς πρώτας εἰπὼν ὁ Θέων, οὐκέτ᾽ ἐμνημόνευσεν τῶν ὑπολοίπων τριῶν, αὐτὸς ἑαυτῷ φανερῶς μαχόμενος. εἴπερ γάρ ἐστι τῆς ὀλίγης καὶ πολλῆς μέση τρίψις, ἣν μετρίαν τε καὶ σύμμετρον ὀνομάζομεν, εἴη ἂν δηλονότι καὶ σκληρᾶς καὶ μαλακῆς ἑτέρα τις μέση, σύμμετρός τε καὶ μετρία προσαγορευομένη. μεμνῆσθαι δὲ ἡμᾶς χρὴ παρὰ πάντα τὸν λόγον, ὡς
114K ἅπαντα ταῦτα κατὰ τὸ πρός τι λέγεται. καὶ γὰρ ἡ σκληρὰ τῷδέ τινι μαλακὴ γένοιτ᾽ ἂν ἑτέρῳ τινὶ καὶ ἡ σύμμετρος ἀσύμμετρος ἤ τ᾽ ὀλίγη πολλὴ καὶ ἡ πολλὴ τῷδέ τινι τοῖς ἄλλως πως διακειμένοις ὀλίγη. τοῦτο μὲν δὴ καὶ ὁ Θέων βούλεται, καὶ οὐκ ἔστιν ὥς τις ὑπερβὰς τὴν σύμμετρον ἐν ποιότητι τρίψιν οὐ μακρῶς ἁμαρτάνει. καί μοι δοκεῖ περιπεσεῖν ὁ Θέων αὐτῷ, διότι παραλέλειπται κατὰ τὴν Ἱπποκράτους ῥῆσιν. ὥστε κἀκ τούτου κατάφωρον γίνεσθαι τὸν ἄνδρα

Qualities	Quantities	113K
Hard/Firm	1. Little	
	2. Much	
	3. Moderate	
Soft/Gentle	1. Little	
	2. Much	
	3. Moderate	
Moderate	1. Little	
	2. Much	
	3. Moderate	

Of the nine conjunctions set out in this diagram, Theon spoke about the first six but never mentioned the remaining three, clearly being at odds with himself. For if there is an intermediate massage between little and much, which we call medium and moderate, clearly there would also be some other intermediate between hard and soft, called medium and moderate. However, we must remember throughout the whole discussion that all these are stated relatively. And [what is] hard to one person might be soft to another, what is moderate, immoderate, or what is little, much, and what is much to one person might be little to others constituted somehow differently. Theon also certainly means this, and it is not possible for someone who overlooks massage that is moderate in quality not to be very much mistaken. And it seems to me that Theon was caught in his own snare because it was omitted in the statement of Hippocrates, so that it became clear from this that the man discovered nothing on his own about massage

GALEN

μηδὲν μὲν ἴδιον ὑπὲρ τρίψεως ἐξευρηκότα, τὰ δ᾿ Ἱπποκράτους οὐκ ὀρθῶς μεταχειριζόμενον. οὐ γὰρ ἀνέγνω
τὰ συγγράμματα τοῦ παλαιοῦ παρὰ διδασκάλοις εὐθὺς ἐκ παίδων ὁρμώμενος.

ὁμολογεῖ γοῦν αὐτὸς ἀθλητὴς γενέσθαι τὰ πρῶτα,
καταλύσας δὲ τὴν ἄσκησιν ἐπὶ τὴν γυμναστικὴν ἀφικέσθαι τέχνην. καὶ ταῦτα μὰ τοὺς θεοὺς οὐχ ὑπὲρ
τοῦ ψέξαι τὸν ἄνδρα προεθέμην εἰπεῖν ἀποδέχομαι
γὰρ αὐτὸν οὐδενὸς ἧττον ἑτέρων ἀρίστων γυμναστῶν,
ἀλλ᾿ ὑπὲρ τοῦ τοῖς ἀναγνωσομένοις τήνδε τὴν πραγματείαν ἐνδείξασθαι τὸ μὴ ῥᾴδιον εἶναι παρακο-
λουθεῖν βιβλίοις παλαιοῖς ἄνευ τῶν ἐπιμελῶς ἐξ-
ηγουμένων αὐτά. διότι μὲν γὰρ ἔπρεπε βραχυλογίᾳ
115K παλαιᾷ τὴν μέσην σκληρᾶς τε καὶ μαλακῆς ὑπερβῆναι τρίψιν ἔν γε τῇ λέξει τῆς ἑρμηνείας, ἔμπροσθεν
εἴρηταί μοι· διότι δ᾿ ἡμᾶς οὐ χρὴ παρορᾶν τά γε
τοιαῦτα, καὶ πρόσθεν μὲν ἐπέδειξα καὶ νῦν δ᾿ οὐδὲν
ἧττον ἐπιδείξω. εἴπερ γὰρ ἡ μαλακὴ καὶ ἡ σκληρὰ
τρῖψις ἐν τῷ πρός τι τὴν ὕπαρξιν καὶ τὴν νόησιν
ἔχουσιν, ἀνάγκη συνίστασθαι καὶ τὴν σύμμετρον ἐν
τῷ πρός τι.

τίθεσο δή μοι σῶμα τοιοῦτον, οἷον καὶ τὸ τοῦ παιδὸς ὑπεθέμεθα τοῦ προκειμένου κατὰ τόνδε τὸν λόγον,
ἀκριβῶς ὑγιαῖνόν τε καὶ σύμμετρον ἤδη πάντη, ὡς
μήτε μαλακώτερον ἐθέλειν ἡμᾶς αὐτὸ μήτε σκληρότερον ἐργάσασθαι μήτε προσθεῖναί τι τῆς σαρκώσεως
μήτ᾿ ἀφελεῖν. ἆρ᾿ οὖν ἐπὶ τοῦ τοιούτου σώματος ἢ τὴν
σκληρὰν τρῖψιν ἢ τὴν μαλακὴν προσάξομεν ἢ τὴν

166

but took over Hippocrates' statement incorrectly, not being moved to read the writings of the ancient [doctor] with his teachers right from childhood.

Anyway, he himself acknowledges that he became an athlete at first, and then, setting aside the practice, came to the art of gymnastic training. And by the gods, I did not take it upon myself to say these things to censure the man, for I accept him as not inferior to any of the other excellent gymnastic trainers, but in order to demonstrate to those who are going to read about this particular matter, that it is not easy to follow the old books without [the guidance of] those who interpret them carefully. For it was in keeping with ancient brevity of expression to pass over the intermediate massage between hard and soft in the interpretation of the statement, as I said previously. But that we must not disregard such things, I both demonstrated before and shall also demonstrate no less now. For if soft and hard massage have their existence and meaning in relation to something, the moderate must of necessity also exist in relation to something.

Assume for me now such a body, like that of the boy we supposed set before us in this discussion—a body perfectly healthy and already moderate in every respect, such that we do not wish to make it softer or harder, nor to add flesh or remove it. Shall we, then, in the case of such a

115K

πολλὴν ἢ τὴν μετρίαν; ἐγὼ μὲν οὐδαμῶς ἡγοῦμαι
συμφέρειν. ὑπὸ μὲν γὰρ τῆς σκληρᾶς σκληρότερον,
ὑπὸ δὲ τῆς μαλακῆς μαλακώτερον, ὥσπερ γε καὶ ὑπὸ
μὲν τῆς πολλῆς ἰσχνότερον, ὑπὸ δὲ τῆς μετρίας
παχύτερον ἀπεργασθήσεται τὸ τοιοῦτον σῶμα. χρὴ
116K δ᾽ οὐδὲν τούτων, ἀλλὰ τὴν ἀρχαίαν ἀκριβῶς αὐτῷ
φυλάττεσθαι συμμετρίαν. ὥστε οὔτε σκληρῶς οὔτε
μαλακῶς αὐτὸ τρίψομεν οὔτε πολλαῖς οὔτε ὀλίγαις,
ἀλλὰ μετρίαις ἀνατρίψεσιν αὐτὸ τρίψομεν, οὐδὲν
πλέον ἐργαζόμενοι τοῦ παρασκευάζειν τε πρὸς τὰ
γυμνάσια καὶ αὖθις ἀποθεραπεύειν, ἐπειδὰν ἱκανῶς
γυμνάσηται. καλείσθω γὰρ οὖν δὴ καὶ ἡμῖν, ὥσπερ
γε καὶ τοῖς νεωτέροις γυμνασταῖς, 'ἀποθεραπεία' τὸ
μετὰ τὰ γυμνάσια μέρος τῆς τρίψεως. ὁ δέ γε Θέων
οὐδεμίαν ὧν εἶπε κατὰ τὰς τρίψεις συζυγιῶν ἐφαρμόσαι δύναται τῇ τοιαύτῃ φύσει τοῦ σώματος. ἓξ μὲν
οὖν εἶπε τὰς πάσας, τρεῖς μὲν τὰς πρώτας τῆς μαλακῆς, ἑτέρας δὲ τρεῖς τὰς δευτέρας[13] τῆς σκληρᾶς. οὔτε
δὲ τῶν μαλακῶν οὐδεμιᾶς τὸ τοιοῦτον σῶμα φαίνεται
δεόμενον οὔτε τῶν σκληρῶν, ἀλλὰ τῆς τούτων ἀμφοτέρων μέσης, ἣν σύμμετρον ὀνομάζειν χρὴ κατὰ ποσότητα[14] καὶ ποιότητα. καὶ δῆλον ἤδη γέγονεν, ἡλίκον
ὁ Θέων ἔσφαλται τὴν μέσην σκληρᾶς τε καὶ μαλακῆς
ὑπερβὰς τρίψιν.

ἐπὶ γοῦν τῆς ἀρίστης κατασκευῆς τοῦ σώματος
οὐδεμίαν ὧν εἶπον συζυγιῶν ἐφαρμόσαι δυνατόν, οὔθ᾽

---

[13] τὰς δευτέρας add. Ko
[14] ποσότητα add. Ko

body, apply hard massage or soft, or much or moderate? I
do not think it is in any way expedient. For by hard massage, we shall make such a body harder, and by soft massage, softer, just as we shall also make it thinner by much
massage and thicker by moderate massage. None of these   116K
are necessary; rather we should preserve the original balance in it precisely. As a result, we shall not massage it
either firmly or gently, or much or little. We shall massage
it with moderate rubbing, doing nothing more than to
prepare it for the exercises, and restore it again when
it has exercised sufficiently. Let us, then, name it as the
younger generation of gymnastic trainers do, "apotherapy,"[10] which is the part of the massage after the exercises.
But Theon is not able to adapt any of the conjunctions he
spoke of to such a nature of the body. He mentioned six
in all—the first three soft (gentle) and the other and second three hard (firm). However, such a body is obviously
in need of neither soft massage nor hard, but what is intermediate between both, which we must call moderate
in respect of quantity and quality. And it has already become clear how greatly Theon erred when he overlooked
the massage between hard and soft.

Anyway, in respect of the best constitution of the body,
it is not possible to apply any one of the conjunctions he

[10] The term ἀποθεράπεια has two distinct meanings listed in
LSJ: "regular worship" and "restorative treatment after fatigue."
Clearly, the latter is intended here, although Galen's definition
above adds "after exercise" specifically. The term is not listed in
a standard modern medical dictionary (S), but it is included in
the 1933 OED. The definition, attributed to Galen, is: "being
rubbed and anointed after exercise." Reference is also made to
the use by Rabelais, who was of course a doctor and student of
Galenic medicine.

117K ὑγιαίνοντος ἀκριβῶς οὔτ' ἐπανορθώσεως δεομένου
τοῦ τοιούτου σώματος. εἰ μὲν γὰρ ἀκριβῶς ὑγιαίνει,
μόνης τῆς παρασκευαστικῆς δεῖται τρίψεως, ἣν ὀλίγην τε ἅμα καὶ μέσην σκληρᾶς καὶ μαλακῆς ἐδείξαμεν ὑπάρχειν· εἰ δ' ἤτοι τοῦ δέοντος ἰσχνότερον ἢ
παχύτερόν ποτε γίνοιτο, μὴ μέντοι κατὰ ποιότητα τῆς
σαρκὸς ὑπαλλαγείη μηδέν, ἀλλ' εἰς τὸ μέσον ἀκριβῶς φυλάττοιτο μαλακοῦ τε καὶ σκληροῦ, τηνικαῦτα
τὴν πολλήν τε ἅμα καὶ σύμμετρον κατὰ ποιότητα
προσάξομεν ἰσχναίνειν βουλόμενοι, σαρκοῦν δὲ τὴν
ὀλίγην τε ἅμα καὶ σύμμετρον κατὰ ποιότητα, τὴν
σύμμετρον δὲ κατ' ἀμφότερα, τό τε ποιὸν λέγω καὶ τὸ
ποσόν, ἀνατρέφειν προαιρούμενοι.[15] ταύτας τὰς τρεῖς
συζυγίας ὑστάτας ἐπὶ ταῖς ἐξ ὀλίγον ἔμπροσθεν ἐξε-
θέμην ἐν τῷ διαγράμματι, δεικνὺς ὡς ἁπάσας αὐτὰς
ὁ Θέων παρέλιπεν.

ἐπειδὴ τοίνυν οὐ μόνον ὅτι παρέλιπεν, ἀλλὰ καὶ ὅτι
χρησιμωτάτας ὑπαρχούσας, ὁ λόγος ἀπέδειξεν, ἑξῆς
ἂν εἴη καιρὸς ἐπί τι τῶν ὀλίγον ἔμπροσθεν ἀναβληθέντων ἰέναι καὶ πρῶτόν γε εἰπεῖν, ὡς ἡ κατὰ τὰ στοιχεῖα τῶν πραγμάτων διδασκαλία χρησιμωτέρα τῶν
ἄλλων ἐστίν· εὐσύνοπτόν τε γὰρ ἐργάζεται τὸ πᾶν
πρᾶγμα καὶ τῇ μνήμῃ παρατιθέμενον εἰς ἀνάμνησίν
118K τε ῥαδίως ἐρχόμενον ἁπάντων τε τῶν κατὰ μέρος ἐπι-
δέξιον χρῆσιν ἑτοίμως δεχόμενον, ὡς ἂν εἰς ὀλίγα καὶ
ὡρισμένα στοιχεῖα τῆς ἀναφορᾶς γινομέμης αὐτῶν.
τίς γὰρ οὐκ ἂν ἐξεύροι ῥαδίως πάσας τὰς κατὰ μέρος
ἐν ταῖς τρίψεσι διαφοράς τε καὶ χρείας καὶ δυνάμεις,

spoke of, either when such a body is perfectly healthy, or
when it needs to be restored. For if it is perfectly healthy,
it only needs preparatory massage which I showed to be
small in amount and midway between hard and soft. If,
however, at any time it becomes thinner or thicker than it
needs to be, and doesn't change in any way in relation to
the quality of the flesh, but is to be maintained exactly at
the midpoint between soft and hard, then we shall apply
under these circumstances much massage along with what
is moderate in quality, if we wish to reduce, and little massage along with what is moderate in quality if we wish to
enflesh, choosing to nurture with moderation in both respects —I speak of quality and quantity. I put these three
conjunctions last after the six in the diagram a little earlier
to show that Theon overlooked all of them.

Accordingly, since the discussion showed that not only
did he omit them, but also that they are the most useful,
it would be timely to go next to some of those things we
put off a little earlier. The first thing to say is that the
teaching pertaining to the elements of the matters is more
useful than the others in that it makes the whole matter
easily taken in at a glance, and establishes it in the memory. Thus, it comes easily to recollection, readily demon-
strating the use of the things that individually come to
hand, as these are referred to a few defined elementary
principles. For who would not easily discover all the differences in the massages individually, and uses and potencies, if only he learns once and for all the opinion of the

---

[15] τὴν σύμμετρον δὲ κατ᾽ ἀμφότερα, τό τε ποιὸν λέγω καὶ
τὸ ποσόν, ἀνατρέφειν προαιρούμενοι *not included in* Ku

εἰ μόνον ἅπαξ ἐκμαθὼν τὴν γνώμην τοῦ παλαιοῦ
μετὰ ταῦτα πρόχειρον ἔχει τῇ μνήμῃ τὴν ῥῆσιν
αὐτοῦ, δι' ἧς ἡμᾶς ἐδίδαξεν, ὡς δῆσαι μὲν ἡ σκληρά,
λῦσαι δὲ ἡ μαλακή, μινυθῆσαι δὲ ἡ πολλή, σαρκῶσαι δ' ἡ μετρία δύναται τρίψις; ἐκ γὰρ τοῦ ταῦτά γε
νοῆσαί τε καὶ μνημονεῦσαι πρώτας μὲν τὰς ἐν τῇ
ῥήσει παραλειφθείσας δύο διαφορὰς εὑρήσομεν, εἶθ'
ἑξῆς ἐπιπλέκοντες ἀλλήλαις ἁπάσας ἐννέα συζυγίας
ἀπεργασόμεθα, τὰς ὀλίγον ἔμπροσθεν ἐπὶ τοῦ διαγράμματος ἐγκειμένας, ἃς οὔθ' εὑρεῖν οὔτε μνημονεῦσαι δυνατὸν ἄνευ τοῦ προηγήσασθαι τὴν στοιχειώδη
διδασκαλίαν, ἣν Ἱπποκράτης ἐποιήσατο, σύμπαντα
τὸν περὶ τῆς τρίψεως λόγον εἰς τὰς πρώτας ἀρχὰς
ἀναγαγών, ἐξ ὧν ἀκριβῶς εὑρημένων οὐ μόνον τὰ νῦν
εἰρημένα περιγίνοιτο ἂν ἡμῖν, ἀλλὰ καὶ τὸ κρίνειν
119K ἁπάσας τὰς μοχθηρὰς διδασκαλίας.

ἴδιον γὰρ μάλιστα μεθόδου τοῦτο, τὸ διὰ βραχείας
ἀρχῆς στοιχειώδους ἐπὶ τὰ πάντα δύνασθαι τὰ κατὰ
μέρος ἰέναι καὶ κρίνειν ἅπαντα τὰ μοχθηρῶς εἰρημένα καθάπερ κανόνι τινὶ τοῖς ἐπιστημονικοῖς θεω-
ρήμασι τὰς οὐκ ὀρθὰς δόξας δοκιμάζοντας. περὶ μὲν
δὴ τοῦ μήτε ἄλλον τινὰ περὶ τρίψεων ὀρθῶς ἐγνωκέναι μήτε τὸν γυμναστὴν Θέωνα, καίτοι τῶν ἄλλων
κάλλιον ὑπὲρ αὐτῶν ἀποφηνάμενον, ἀλλ' Ἱπποκράτην
τε καὶ ὅσοι τούτῳ παρηκολούθησαν ἱκανὰ καὶ ταῦτα.

5. Λείποιτο δ' ἄν τι τῶν τέως ἀναβληθέντων ὑπὲρ
τῶν ὀνομάτων εἰπεῖν, ἵνα μή τις ὑπὸ τοῦ πλήθους
αὐτῶν ἐξαπατώμενος ἰσαρίθμους οἰηθῇ τὰς διαθέσεις

ancient [doctor Hippocrates], and after this, has ready to
hand in the memory his statement through which he
taught us that firm massage is able to bind, soft massage
to loosen, much massage to diminish, and moderate massage to enflesh? For from observing and remembering
these things, we shall discover first the two differences
omitted in the statement, and then, next, combining them
all with each other, we shall complete the nine conjunctions, contained in the diagram a little earlier. It is not
possible to discover and remember these without being
guided by the elementary teaching which Hippocrates
created by referring the whole discussion about massage
to first principles. It is from these principles having been
accurately discovered that not only the things now said
would remain for us, but also the judging of all the bad   119K
teachings.

This is particularly characteristic of a method which is
able to proceed, by way of brief elementary principles, to
all the individual matters, and to judge all the things said
badly, examining wrong opinions by scientific theories, as
if by some canon. Certainly, no one else knew rightly about
massage—not even the gymnastic trainer Theon, although
he gave a better account of this than others—except Hippocrates and those who followed him closely. But this is
enough on these matters.

5. Of those things we put off for a time, what still remains is to speak about terms so that no one, deceived by
their great number, might think the conditions are equal

ὑπάρχειν ταῖς προσηγορίαις. τὸ μὲν γὰρ σκληρὸν
ὄνομα κυρίως ἐπιφέρεται κατά τινος μιᾶς τοῦ σώματος διαθέσεως, ἣν οὐδ᾽ ἐξηγεῖσθαι διὰ πολλῶν ὁποία
τίς ἐστι δεόμεθα, πάντων ἀνθρώπων ὑπὸ τῆς φωνῆς
ὁδηγουμένων[16] ἐπὶ τὸ πρᾶγμα. κατὰ δὲ τὸν αὐτὸν τρόπον καὶ τὸ μαλακόν. τὸ δ᾽ ἀραιὸν καὶ τὸ πυκνὸν

120K οὐκέθ᾽ ὁμοίως ἐναργῶς διασημαίνει τὰς διαθέσεις τοῦ
σώματος, ὅτι διττή τις ἡ τῶν ὀνομάτων χρῆσις ἐγένετο, κυρίως μὲν ὀνομαζόντων ἑτέρα, καταχρωμένων
δ᾽ ἑτέρα. τὸ μὲν οὖν κυρίως ἀραιόν ἐστι τὸ μεγάλοις
διαλαμβανόμενον πόροις, ὥσπερ γε καὶ πυκνὸν τὸ
μικροῖς· τὸ δ᾽ ἐκ μεταφορᾶς ἢ καταχρήσεως ἢ ὅπως
ἄν τις ὀνομάζειν ἐθέλῃ καὶ κατὰ τοῦ κεχυμένου τε καὶ
πεπιλημένου λέγεται. κατὰ τοῦτο γοῦν ἔστιν ὅτε τὸν
μὲν ἀέρα καὶ τὸ πῦρ ἀραιά, τὸ δ᾽ ὕδωρ καὶ τὴν γῆν
πυκνὰ λέγομεν, ἐπ᾽ αὐτὰ τὰ στοιχεῖα τὰς εἰρημένας
προσηγορίας ἐπιφέροντες, ἡνωμένα τε καὶ ὁμοιομερῆ
τὴν φύσιν ὑπάρχοντα καὶ μηδενὶ διαλαμβανόμενα
πόρῳ. πολὺ δὲ δὴ τούτων ἔτι μᾶλλον ἀποκεχώρηκε
τοῦ κυρίως ὀνομάζεσθαι τό τε ἐσφιγμένον καὶ τὸ
δεδεμένον, ἐκ μεταφορᾶς ἄμφω λεγόμενα, τὸ μὲν
ἐσφιγμένον ἐπὶ τοῦ πυκνοῦ καὶ τοῦ σκληροῦ, ποτὲ
μὲν ἑκατέρου καταμόνας ὑπάρχοντος, ἔστιν ὅτε δ᾽ εἰς

---

[16] ποδηγουμένων Ku

---

[11] The four terms considered in this short section are, for
obvious reasons, given in the transliterated form. The range of
meaning of the two adjectives is indicated by the following partial
list given in LSJ: ἀραιὸς: thin, narrow, slender, loose-textured,

in number to the terms. Thus the term *skleros* ("hard") is properly applied to one single condition of the body, and we do not need to expound at length what kind of condition this is, since all people are led to the matter by the sound. And the same also applies to *malakos* ("soft"). However, *araios* and *puknos*[11] no longer clearly signify conditions of the body in a similar way because a twofold use of the terms has arisen; one by those who use the terms correctly and the other by those who use them catachrestically.[12] *Araios* is properly applied to distinguish something with large pores, just as *puknos* is used in the case of something with small pores. On the other hand, the metaphorical or catachrestical use, or whatever you might wish to call it, is said of "flowing" and "condensed." Anyway, it is in this respect we sometimes say air and fire are *araios* while water and earth are *puknos*, applying the terms mentioned to the elements themselves, being unified and *homoiomerous* in nature, and not distinguished by any pore. Even further removed from the proper application of terms are *to esphigmenon* ("compressed") and *to dedemenon* ("bound up"), both being used metaphorically. The former is used in reference to *puknos* and *skleros*, which are sometimes alone and sometimes come to-

120K

empty, scanty, intermittent. πυκνὸς: close, compact, firm, narrow, constricted, close-packed, frequent, thick. *Esphigmenon* is the middle perfect participle of σφίγγω: to bind, tie, fetter, hinder, harden, brace up. *Dedemenon* is the middle perfect participle of δέω: to bind, hold together, tighten up, tie up, shut close.

[12] Catachresis is defined by a modern English dictionary (*The Chambers Dictionary* [11th ed., 2008]) as, "misapplication or incorrect use of a word." The corresponding definition for metaphor is, "a figure of speech by which a thing is spoken of as being that which it only resembles." These seem to be Galen's uses here.

ταὐτὸν ἀφιγμένων, τὸ δ' αὖ δεδεμένον ἐπὶ τῶν αὐτῶν
μέν, ἀλλ' οὐκ ἐκ τῆς αὐτῆς μεταφορᾶς· ἐπειδὴ γὰρ
ἅπαντα τὰ δεδεμένα δυσκίνητά ἐστιν, οὕτως ὀνομά-
121K ζουσι καὶ τὰ διὰ ξηρότητά τινα ἢ ψύξιν ἢ φλεγμονὴν
ἢ σκίρρον ἢ τάσιν ἢ πλήρωσιν ἢ βάρος ἐν δυσκινη-
σίᾳ καθεστῶτα. διὰ δὲ τὴν αὐτὴν αἰτίαν καὶ τοῖς
ἐναντίοις χρῶνται τῶν ὀνομάτων ἐπὶ τῶν ἐναντίων
διαθέσεων ἀνεῖσθαι λέγοντες ἢ ἐκλελύσθαι ἢ κεχαλά-
σθαι.

χρὴ δὲ μὴ τῷ πλήθει τῶν ὀνομάτων προσέχειν,
ἀλλ' ἡγεῖσθαι διττὰ γένη διαθέσεων ὑπάρχειν τὰ
πάντα, τὸ μὲν ἐν αὐτοῖς τοῖς ὁμοιομερέσι σώμασιν,
ὅπερ ἤτοι σκληρὸν ἢ μαλακόν ἐστι, τὸ δ' ἐν ταῖς τῶν
ὀργάνων ποροποιΐαις συνιστάμενον, ὅπερ ἤτοι πυκνὸν
ἢ ἀραιόν. αὗται γὰρ ἴδιαι τῶν σωμάτων αὐτῶν εἰσιν
αἱ διαθέσεις, ἐπίκτητοι δὲ καὶ ὡς ἂν εἴποι τις πρόσ-
καιροι ποτὲ μὲν ἐμπεπλησμένων τῶν πόρων ὑγρότητος περιττῆς, ἔστι δ' ὅτε καθαρῶν ὑπαρχόντων, καὶ
ποτὲ μὲν ἀναπεπταμένων, ἔστι δ' ὅτε μεμυκότων.

6. Ἀλλὰ περὶ μὲν τῶν διαθέσεων τούτων ἐν τοῖς
ἑξῆς ὑπομνήμασιν ὁ λόγος ἔσται· νυνὶ δὲ ἐπὶ τὸ προκείμενον ἐπανιέναι χρὴ καὶ πρῶτον μὲν διορίσασθαι
σαφέστερον ἔτι περὶ τρίψεως. ὡς ἐνίοτε μὲν αὐτὴ καθ'
122K ἑαυτὴν ἐργάζεταί τι περὶ τοῖς σώμασιν ἡμῶν χρηστόν, ἐνίοτε δὲ τοῖς ἐργαζομένοις ὑπηρετεῖ, καθάπερ
Ἱπποκράτης ἔλεγε περὶ ἐπιδέσεως· "ἐπίδεσις τὸ μὲν
αὐτὴ ἰῆται, τὸ δὲ τοῖς ἰωμένοις ὑπηρετεῖ." αὕτη μὲν
οὖν ἡ τρίψις ἐργάζεται τά τ' ἀραιὰ πυκνοῦσα καὶ τὰ

gether, whereas the latter is used in reference to the same things but not by the same metaphor. Since all those things that are "bound up" are hard to move, people apply the term in the same way to those things placed in the cate- 121K gory of hard to move due to dryness, cold, inflammation, induration, tension, fullness or weight. For the same reason too, they use the opposite terms to refer to the opposite conditions, speaking of loosened, released or relaxed.

However, we must not direct our attention to the great number of terms but realize that there are two classes of conditions in all: the one in the actual *homoiomerous* bodies, which is either hard or soft, and the one existing in the state of the pores of the organs, which is condensed (*puknos*) or rarefied (*araios*). These are specific conditions of the bodies themselves—acquired and as we might say temporary—sometimes when the pores are filled with superfluous moisture, sometimes when they are clear, sometimes when they are widely patent and sometimes when they are occluded.

6. But the discussion in the subsequent books will be about these conditions. Now, however, we must return to the matter before us, and first lay down definitions of massage more clearly still—that sometimes it brings about something useful for our bodies by itself and sometimes 122K assists the things bringing this about, as Hippocrates said about binding: "In part, binding itself cures; in part, it assists curative agents."[13] Thus, massage itself acts by condensing what is rarefied and hardening what is soft; such

[13] Hippocrates, *In the Surgery* 8, *Hippocrates* III, LCL 149, 64–65.

μαλακὰ σκληρύνουσα[17] σκληρὰ δ' ἂν εἴη πάντως ἡ
τοιάδε τρίψις καὶ μὲν δὴ καὶ τὰ σκληρὰ μαλάττουσα
καὶ τὰ πυκνὰ διευρύνουσα μαλακὴ δέ ἐστιν ἡ τοιαύτη
τρίψις. οὕτω δὲ καὶ ἡ τοῦ σαρκῶσαι χάριν καὶ ἡ τοῦ
λεπτῦναι παραλαμβανομένη καθ' ἑαυτὰς ἐργάζονταί
τι χρηστὸν ἐν τοῖς σώμασιν. ἡ μέντοι παρασκευάζουσα πρὸς τὰ γυμνάσια καὶ ἡ μετὰ ταῦτα παραλαμβανομένη τοῖς γυμνασίοις ὑπηρετοῦσιν, ἡ μὲν ἐκ τοῦ
θερμῆναι μετρίως τούς τε πόρους ἀναστομοῦσα τοῦ
σώματος καὶ τὰ κατὰ τὴν σάρκα περιττώματα χέουσα
καὶ τὰ στερεὰ μαλάττουσα, καλεῖται δὲ παρασκευαστικὴ τρίψις ἡ τοιαύτη. ἡ δὲ ἑτέρα προσαγορεύεται
μὲν ἀποθεραπευτική, γινομένη δὲ μετ' ἐλαίου πλέονος
ἐπιτέγγει τε ἅμα τῷ λίπει καὶ μαλάττει τὰ στερεὰ καὶ
τι καὶ διαφορεῖ τῶν ἐν τοῖς πόροις περιεχομένων·
123K ἀλλὰ περὶ μὲν τῆσδε καὶ αὖθις εἴποιμεν ἂν ἐφεξῆς
τοῖς γυμνασίοις.

ἡ μέντοι παρασκευαστικὴ τρίψις ἐπὶ τῆς ἀρίστης
φύσεως ἕνεκα τοῦ διαθερμῆναι τὰ σώματα παραλαμβανομένη καθ' ὃν εἴρηται τρόπον ὀλίγον ἔμπροσθεν
γινέσθω, μαλακὴ μὲν τὰ πρῶτα, προσαγόντων δὲ ἤδη
τοῖς πόνοις σκληρά. οὕτω γὰρ ἂν μάλιστα τό τε μαλάττειν ἔχοι καὶ τὸ πρὸς τὰς ἐνεργείας ἐπεγείρειν, καὶ
τὸ διαφυλάττειν ὁποίαν παρέλαβε τὴν φύσιν τοῦ
σώματος. εἰ μέντοι διαμαρτάνοιτο κατά τι, πρὸς τὸ
σκληρὸν ἐκτρεπέσθω μᾶλλον. αἱ γὰρ ἐπ' ὀλίγον
ὑπερβολαὶ τῆς συμμετρίας ἐν τῷ δέρματι καταπαύον-
ται τῶν ἐντὸς οὐδὲν ἀλλοιοῦσαι, βλάπτοιτο δ' ἂν ἦτ-

massage would be altogether firm. Furthermore, it acts by softening what is hard and rarefying what is condensed; such massage is gentle. In this way too, massage undertaken for the sake of enfleshing or thinning brings about something useful in bodies by itself. However, massage which prepares for exercises and that undertaken after exercises assists the exercises. By heating moderately it opens up the pores of the body, liquefies the superfluities in the flesh and softens solid bodies. Such massage is termed preparatory in one case and apotherapeutic in the other." When this occurs with more oil, it moistens as well as anoints, and both softens the solid bodies and disperses those things contained in the pores. But let me speak   123K
about this massage again subsequently in relation to the exercises.

Certainly the preparatory massage, undertaken in the case of the best nature for the purpose of heating the body, should be according to the manner described a little earlier—that is, gentle at first but firm when applied to those already exercising. For in this way it would be most softening, would stimulate the functions and would preserve the kind of nature of the body it received. If, however, it were to fail completely in some respect, it should turn more toward the hard. Slight excesses over the moderate end in the skin and change nothing that is within, and the skin

---

17 πυκνοῦσα καὶ τὰ μαλακὰ σκληρύνουσα Κο; πυκνὰ ποιοῦσα, καὶ τὰ μαλακὰ πυκνοῦσα Κυ

τον τὸ δέρμα πρὸς τὸ σκληρόν τε καὶ πυκνὸν ἐκτρε-
πόμενον. οὕτω γὰρ ἂν εἴη δυσπαθέστερον. ὡς, εἴ γε
καὶ διαπνεῖσθαι καλῶς ἠδύνατο τὸ τοιοῦτον, σκληρότατον ⟨ἂν⟩ αὐτὸ καὶ πυκνότατον ἀπειργαζόμεθα.

   νυνὶ δέ, ἐπειδὴ πρὸς ἄμφω χρὴ παρασκευασθῆναι
καλῶς αὐτό, καὶ πρὸς τὴν τῶν ἔνδοθεν περιττωμάτων
διαπνοὴν καὶ τὴν τῶν ἔξωθεν ὁμιλούντων βίαν, ἄρι-
124K στον ἂν εἴη τὸ μέσον ἑκατέρων τῶν ὑπερβολῶν. εἰ δ᾽
ἄρα ποτὲ μὴ φυλάττοιτο τοῦτο, βέλτιον τὸ σκληρότερόν τε καὶ πυκνότερον τοῦ μαλακωτέρου τε καὶ
ἀραιοτέρου. τὸ μὲν γὰρ τῆς διαπνοῆς ἐλλιπὲς ἐπανορθώσασθαι γυμνασίοις οἷόν τε· τὸ δὲ τῆς ἑτέρας διαθέσεως εὐεπηρέαστον ὑπὸ τῶν ἔξωθεν αἰτίων οὔτ᾽
ἐπανόρθωσίν τινα ἑτέραν ἑτοίμην ἔχει, καὶ πρόσεστιν
αὐτῷ βλάβη τις οὐ σμικρὰ διαφερομένων πολλάκις
οὐ τῶν περιττωμάτων μόνον, ἀλλὰ καὶ αὐτῆς τῆς τροφῆς. ἐν μὲν δὴ ταῖς κατὰ ποιότητα διαφοραῖς ἐπὶ τὸ
σκληρότερον ἁμαρτάνειν χρὴ μᾶλλον ἤπερ ἐπὶ τὸ
μαλακώτερον, ἐν μέντοι ταῖς κατὰ ποσότητα πρὸς τὸ
ἔλαττον ἐπὶ τῆς προκειμένης δηλονότι φύσεώς τε καὶ
ἡλικίας. ἀεὶ γὰρ χρὴ τούτου μεμνῆσθαι κατὰ τὸν
ἐνεστῶτα λόγον. αὐξάνεσθαί τε γὰρ ἔτι βουλόμεθα τὸ
τοιοῦτον σῶμα καὶ ἥκιστα ξηραίνεσθαι. τίνας μέντοι
χρὴ φύσεις σωμάτων καὶ τίνας διαθέσεις ἐπὶ πλέον
τρίβειν, αὖθις εἰρήσεται.

   7. Νυνὶ δὲ ἐπὶ τὴν ἐνεστῶσαν ὑπόθεσιν ἐπανελ-
125K θόντες ὑπὲρ τοῦ μέτρου τῆς τρίψεως ἐπισκεψώμεθα
τὸν αὐτὸν τρόπον, ὅνπερ ἀρτίως ἐπεσκέμμεθα περὶ

would be harmed less, if turned toward hard and condensed. In this way it would be less easily subject to affection (more *dyspathic*). So if, in fact, such skin were able
to transpire well, we would make it very hard and very
dense.

For the present, however, since it is necessary for it to
be well prepared regarding both—that is, for the transpiration of the superfluities within and the force of those
things contacting it from without—the best would be what
is intermediate between both the extremes. If, then, at any   124K
time this were not being preserved, harder and more condensed is better than softer and more rarefied, for then it
is possible for the deficiency of transpiration to be corrected by exercises. But the exposure to harm of the other
condition by external causes has no other ready correction, and injury befalls it to no small extent, since repeatedly not only the superfluities are carried through, but also
food itself. Therefore, in the differences pertaining to
quality, you must err more toward the harder than toward
the softer. But in the differences pertaining to quantity, it
is quite clearly better to err toward the less in the case of
the nature and age before us. It is always necessary to bear
this in mind in regard to the present discussion, for we still
wish to build up such a body, and least of all to dry it. I
shall speak again later about what natures of bodies and
what conditions it is necessary to massage more.

7. Returning now to the present subject, let us consider   125K
the amount of massage in the same way as we just considered the quality. Here too it seems reasonable to call to

181

τῆς ποιότητος. ἔοικα δὲ κἀνταῦθα, πρὶν ὁρίζειν τὸ μέτρον, ἑτέρου τινὸς ἐπιμνησθήσεσθαι σκέμματος, οὗ χωρὶς οὐδὲ τὸ μέτρον ὀρθῶς οὔτε ὁρισθῆναι δυνατὸν οὔτε γνωρισθῆναι. τὸ δὲ δὴ σκέμμα τοῦ περιέχοντος ἀέρος τὸν τριβόμενον ἡ κρᾶσίς ἐστιν, ἣν δεῖξαι μὲν ἐπὶ τῶν ἔργων ἐγχωρεῖ, διελθεῖν δὲ τῷ λόγῳ σαφῶς οὕτως, ὡς μηδὲν ἀπολείπεσθαι τὴν ἑρμηνείαν ἐναργοῦς ἐνδείξεως, ἀμήχανόν ἐστι καὶ ἀδύνατον παντάπασιν. ἀλλ᾽ εἰ μὴ πολὺ λείποιτο δείξεως ἐναργοῦς ὁ λόγος, ἱκανὸν καὶ τοῦτο. προηγεῖται δὲ δὴ καὶ τοῦδε πάλιν ἑτέρα τις ὑπόθεσις, ἐν τίνι μὲν ὥρᾳ τοῦ ἔτους, ἐν τίνι δὲ χώρᾳ τῆς οἰκουμένης ὁ γυμνασόμενος ἔσται. μόνα γὰρ ἐπ᾽ αὐτοῦ δύο διώρισται, τό τε τῆς κατασκευῆς τοῦ σώματος καὶ τὸ τῆς ἡλικίας· οὔτε δὲ ἐν ᾧτινι τέθραπται μέρει τῆς γῆς, οὔτε ἐν ᾧ μέλλει γυμνάσασθαι νῦν, οὐδὲ καθ᾽ ἥντινα τοῦ ἔτους ὥραν ἢ καὶ τῆς ἡμέρας, προσδιώρισται, καίτοι παρὰ ταῦτα πάντα τὸ μέτρον τῆς τρίψεως ὑπαλλάττεται. προσδιοριστέον οὖν αὐτὰ πάλιν, εἰρημένα μὲν ἤδη δυνάμει κατὰ τὴν ἀρχὴν τοῦδε τοῦ γράμματος, οὐ μὴν τῇ γ᾽ ἑρμηνείᾳ σαφῶς δεδηλωμένα.

126K

λέγοντες γὰρ οἷον κανόνα τινὰ πρῶτον ἐκτίθεσθαι πάντων τῶν ἐφεξῆς εἰρησομένων τὸν ἄμεμπτον τῇ κατασκευῇ τοῦ σώματος ἄνθρωπον, εὐθέως ἐν τούτῳ καὶ τὴν χώραν αὐτοῦ διορισόμεθα δυνάμει. οὔτε γὰρ εὐκρατότατον οὔτε ἀμεμπτότατον τῇ κατασκευῇ σῶμα γενέσθαι δυνατὸν ἐν τοῖς ἀμέτρως κεκραμένοις χωρίοις, ὡς ὅ τε λόγος ὑπαγορεύει καὶ ἡ πεῖρα δείκνυσι.

mind another issue before defining the due amount, without which this amount cannot be either correctly defined or become known. The issue is the *krasis* of the air surrounding the person being massaged, which it is possible to show in the case of the actions, but difficult and even impossible to go over clearly in the discussion, so as no explanation is left without clarification. But as long as the discussion doesn't leave out much in respect of those things visibly displayed, this is enough. However, another subject also precedes this in turn—in what season of the year and in what part of the inhabited world the person who will be exercising will be. Thus far only two things have been determined—the constitution of the body and the age. What have not yet been determined are in what part of the world he was brought up, in what part he is now going to exercise, and in what season of the year and part of the day, and yet the due measure of massage changes 126K depending on all these things. Therefore, we must in turn make a prior determination of these things, having already spoken about them at the beginning of this book in regard to potency, without having clarified them completely in the explanation.

For stating first as a kind of standard to be set up for all the things that will be spoken about in succession, there is the person faultless in the constitution of the body, and immediately in this person we shall also determine his place in terms of potency, for it is not possible for a body to become entirely *eukratic* or entirely faultless in constitution in places that are immoderately mixed, as the argument suggests and experience shows. Thus, people be-

ξηροὶ μὲν γὰρ καὶ ἰσχνοὶ καὶ οἷον ἐσκελετευμένοι
γίνονται κατὰ τὰς θερμὰς χώρας οἱ ἄνθρωποι, ἀνώμα-
λοι δὲ ταῖς κράσεσιν, ὡς τὰ μὲν ἔξω ψυχρά, τὰ δὲ
ἔνδον τε καὶ κατὰ τὰ σπλάγχνα θερμὰ περαιτέρω τοῦ
προσήκοντος ἔχειν, οἱ τῶν ψυχρῶν χωρίων οἰκήτορες.
τὸ δ' ἄριστον σῶμα, περὶ οὗ νῦν ὁ λόγος, ὥσπερ ὁ
Πολυκλείτου κανών ἐστιν, ᾧ κατὰ μὲν τὴν ἡμετέραν
χώραν, ὡς ἂν εὔκρατον ὑπάρχουσαν, ὦπται πολλῶν
127K πολλὰ παραπλήσια σώματα, παρὰ δὲ Κελτοῖς ἢ Σκύ-
θαις ἢ Αἰγυπτίοις ἢ Ἄραψιν οὐδ' ὄναρ ἔστιν ἰδεῖν
τοιοῦτον σῶμα.

   καὶ αὐτῆς δὲ τῆς ἡμετέρας χώρας ἱκανὸν ἐχούσης
πλάτος, εὐκρατότατόν ἐστι τὸ μεσαίτατον, οἷόνπερ
ὑπάρχει τὸ κατὰ τὴν Ἱπποκράτους πατρίδα· καὶ γὰρ
χειμῶνος αὕτη καὶ θέρους ἐστὶν εὔκρατος, ἔτι δὲ δὴ
μᾶλλον ἦρός τε καὶ φθινοπώρου. τοιαύτην οὖν τινα
χώραν ὑποθέμενοι τῷ προκειμένῳ σώματι, τὴν ὥραν
τοῦ ἔτους αὐτῷ προσυποθώμεθα τὸ μεσαίτατον τοῦ
ἦρος. ἔστω δὲ καὶ τῆς ἡμέρας ἐκείνης, ἐν ᾗ μέλλει
πρὸς ἡμῶν γυμνάζεσθαι τὸ πρῶτον, ὡς οἷόν τε τὸ
μεσαίτατον, ἵνα κατὰ μηδένα τρόπον ὑπὸ τοῦ περιέχοντος ἐξαλλαχθῇ πως ἡ φυσικὴ δύναμις τῆς κράσεως αὐτοῦ. διὰ δὲ τὸν αὐτὸν λογισμὸν οὐδὲ τὸν
οἶκον, ἐν ᾧ γυμνάζεσθαι μέλλει, θερμότερον ἢ ψυχρότερον εἶναι προσήκει κατά γε τὴν ἡμέραν ἐκείνην
τοῦ κοινοῦ τῆς πόλεως ὅλης ἀέρος. ἐν μέντοι χειμῶνι
καὶ θέρει τοσοῦτον αὐτῷ προσθετέον, ἐν μὲν χειμῶνι
θερμότητα, ἐν δὲ θέρει ψυχρότητα, ὡς εὔκρατον ἀκρι-

come dry and lean, and, as it were, wasted away in hot places, whereas the inhabitants of cold places are irregular in terms of *krasias,* so as to have what is external cold, but what is internal, including the internal organs, hot beyond what is appropriate. The best body, which is what the discussion is now about, is like the standard of Polyclitus.[14] In our own country, as it is *eukratic,* many similar bodies are seen among the populace, whereas among Celts, Scythians, Egyptians and Arabs, one would not dream of seeing such a body.   127K

And since this region of ours is relatively wide, the most central part is the most *eukratic,* as for example Hippocrates' fatherland is, for in winter and summer it is *eukratic,* and still more so in spring and autumn. Therefore, when we assume such a region for the proposed body, let us assume for it also the season of the year to be the very middle of spring. Let that day on which he is first going to be exercised by us be as far as possible at the very middle, so that in no way the physical capacity of his *krasis* may be changed to any degree by the ambient air. By the same argument, let the house in which he is going to be exercised be neither hotter nor colder on that day than the general air of the whole city. However, in winter and summer, we must apply this much to him: in winter, heat, and

14 Polyclitus was an Argive sculptor active during the middle to late third century BC. His most famous work was the Doryphorus (or Spearbearer). He is said to have written a book (his Canon) detailing the principles of his art as exemplified by that statue. Galen also refers to this work in several other places, such as *Ars M.,* I.343K; *Mixt.,* I.566K; and *Opt. Const.,* IV.744K. The key feature was the proportion of the parts of the sculpted body.

βῶς ἀποτελεσθῆναι τὸ σῶμα κατὰ τὸν τῆς τρίψεως
128K καιρόν. εἰ γὰρ ἤτοι θερμότερος ἢ ψυχρότερος εἴη περαιτέρω τοῦ προσήκοντος, ἐν μὲν τῷ θερμοτέρῳ φθά-
σειεν ἂν ἱδρῶσαι, πρὶν αὐτάρκως μαλαχθῆναι, κατὰ
δὲ τὸν ψυχρότερον οὐδ' ἂν ἐκθερμανθείη τὴν ἀρχὴν
οὐδὲ μαλαχθείη ποτὲ καλῶς οὐδ' ἐπανθήσειεν ἔρευθος
εὐανθὲς οὐδ' εἰς ὄγκον ἀρθείη τὸ σῶμα. ταῦτα γὰρ
δὴ τὰ γνωρίσματα συμμέτρου τρίψεώς ἐστιν ἐν ἀέρι
συμμέτρῳ περὶ τὴν εὔκρατον ἕξιν τοῦ σώματος, ἔρευ-
θός τε καὶ ὄγκος. ὥσπερ γὰρ ἐν τῷ καταχεῖν ὕδωρ
θερμὸν εἰς ὄγκον μὲν τὸ πρῶτον ἐξαίρεται τὰ σώματα,
πλεοναζόντων δὲ καθίσταται, καὶ διὰ τοῦθ' Ἱπποκράτης ἐπ' αὐτῶν εἶπε "τὸ μὲν γὰρ πρῶτον ἀείρεται,
ἔπειτα δ' ἰσχναίνεται," οὕτω καὶ ἡ τρίψις ἐξαίρει μὲν
τὸ πρῶτον, αὖθις δὲ συστέλλει τε καὶ καθαιρεῖ τὸ
σῶμα.

τοὺς μὲν δὴ σαρκώσεως ἕνεκα τριβομένους τηνικαῦτα παύεσθαι προσῆκεν, ὅταν εἰς ὄγκον ἀρθὲν τὸ
σῶμα πλησίον ἥκῃ τοῦ καθίστασθαι· τοὺς δ' εἰς τὰ
γυμνάσια παρασκευαζομένους οὐ χρὴ[18] τοῦτον ἀναμέ-
νειν τὸν καιρόν, ἀλλὰ πολὺ πρόσθεν παύεσθαι, καὶ
μάλιστα ὅταν ἄριστοι τὴν κρᾶσιν ὦσι καὶ παῖδες ἔτι
129K τὴν ἡλικίαν. ὑγρὰ γὰρ δὴ τούτων τὰ σώματα καὶ μαλακὰ καὶ βραχείαις τρίψεσι μαλαττόμενα, σκοπὸς δέ
ἐστι καὶ τέλος τῆς παρασκευαστικῆς τρίψεως τὸ μαλαχθῆναι μὲν τὰ στερεά, λυθῆναι δὲ τὰ ὑγρά, τοὺς
πόρους δ' εὐρυνθῆναι. ὁπόσον δ' ἐστὶ τὸ πλῆθος τῶν

in summer, cold, so the body is made precisely *eukratic* at
the time of the massage. If it were to be either warmer or   128K
colder beyond what is appropriate, he would sweat first
in the warmer house before being sufficiently softened,
whereas in the colder house, he would not be warmed at
the beginning, nor would he be softened properly at any
time, nor would a florid redness appear on the surface, nor
would the body be raised to a swelling. For these are the
signs of moderate massage in balanced air surrounding the
body in a *eukratic* state—redness and swelling. Just as, by
pouring warm water over bodies, they are first raised to a
swelling, so too, if this goes to excess, it stops. Because of
this, Hippocrates said, regarding these things: "First it is
raised and then it is reduced."[15] In the same way too, massage first produces swelling of the body and then contracts
and reduces it.

Now in the case of those being massaged for the purpose of enfleshing, it is appropriate to cease when the body
being raised to a swelling comes close to that state. However, in the case of those being prepared for exercises, one
should not wait for this time but stop much earlier, and
particularly when they are best in terms of *krasis* and are
still children in terms of age. For their bodies are moist   129K
and soft, and are softened by brief massaging, while the
objective and end of the preparatory massage is for solid
parts to be softened, the moisture released and the pores
dilated. It is impossible to make clear in a discussion how

---

[15] Hippocrates, *In the Surgery* 13, *Hippocrates* IV, LCL 149,
72–73, and Galen, *Hipp. Off. Med.*, XVIIIB.872K.

---

[18] οὐ χρὴ Ko; οὐχὶ Ku

ἀνατρίψεων, οὐχ οἷόν τε λόγῳ δηλῶσαι, ἀλλὰ χρὴ
τὸν ἐπιστατοῦντα, τρίβωνα τῶν τοιούτων ὑπάρχοντα,
κατὰ μὲν τὴν πρώτην ἡμέραν οὐκ ἀκριβεῖ στοχασμῷ
χρήσασθαι, κατὰ δὲ τὰς ἑξῆς ἐμπειρίαν ἤδη τινὰ τῆς
τοῦ σώματος ἐκείνου φύσεως ἔχοντα τὸν στοχασμὸν
ἀεὶ καὶ μᾶλλον ἐξακριβοῦν. καὶ μὲν δὴ καὶ κατὰ τὰ
γυμνάσια τῇ μὲν πρώτῃ τῶν ἡμερῶν οὐ δυνατὸν ἀκρι-
βῶσαι τὸ μέτρον, ἐν δὲ ταῖς μετὰ τήνδε καὶ πάνυ
δυνατόν. ἔστω δὴ κἀνταῦθα γνωρίσματα κατὰ μὲν
τὴν πρώτην ἡμέραν, ὅταν ἐγκελευομένῳ καὶ παρορ-
μῶντι γυμνάσασθαι μηκέθ᾽ ὁμοίως ὑπακούῃ, ἀλλὰ
βραδυτέρας τε καὶ ἀραιοτέρας καὶ ἀσχημονεστέρας[19]
καὶ τὸ σύμπαν εἰπεῖν ἀτονωτέρας ποιῆται τάς τε λα-
βὰς τῶν προσπαλαιόντων καὶ τὰς κινήσεις.

130K   εἶναι δὲ δηλονότι χρὴ τὸν γυμναζόμενον οὔτε θυ-
μικόν, ὡς ἔτι προθυμεῖσθαι γυμνάζεσθαι καμνούσης
ἤδη τῆς δυνάμεως, οὔτ᾽ ἄθυμον, ὡς ἀπαγορεύειν ἔτι
πονεῖν δυνάμενον. ἔστι δὲ δήπου τοιοῦτος οὐ τὸ σῶμα
μόνον, ἀλλὰ καὶ τὴν ψυχὴν ὁ νῦν ἡμῖν ὑποκείμενος
ἄνθρωπος. ὡς ὅσοι γε ψυχροὺς ἔχουσι χυμοὺς ἐν τῇ
γαστρὶ περιεχομένους ἢ κατὰ τὴν ὅλην ἕξιν ἠθροισμέ-
νους, ἀργότεροι πρὸς τὰς κινήσεις εἰσίν. ὡσαύτως δὲ
καὶ οἱ πληθωρικῶς διακείμενοι καὶ οἱ προσφάτῳ
κρύει καταπονηθέντες ἄθυμοί τέ εἰσι καὶ ὀκνηροὶ κινεῖσθαι, καὶ τούτων ἔτι μᾶλλον, ὅσοι φύσει ψυχρότε-
ροι, καὶ πολὺ δὲ δὴ μᾶλλον, εἰ τῇ ψυχρότητι καὶ
ὑγρότης τις προσείη. οὗτοι μὲν γὰρ νοθεύουσι τὰ τοῦ
μέτρου τῶν γυμνασίων γνωρίσματα, καθάπερ γε καὶ

much the amount of massaging is; rather, it is necessary
for the person in charge, being the masseur of such people, not to use a precise estimation on the first day, but on
the following days, having now some experience of the
nature of that body, to always make the estimation more
precise. And furthermore, in relation to the exercises, it is
not possible to make the measure precise on the first day,
whereas on the days after this is entirely possible. Even
here take note of the signs on the first day, when the person exercising is no longer attending in the same way to
the person exhorting and urging on the exercises, but
makes the holds and movements more slowly, loosely and
feebly, and to speak generally, more weakly than those
with whom he is wrestling.

Clearly it is necessary for the one exercising not to be
high-spirited, such that he is still eager to exercise when   130K
his power has already waned. Nor should he be without
spirit, such that he gives up while still able to exert himself.
The person who is our subject is, I presume, of such a kind
not only in his body but also in his soul. Thus, those who
have cold humors contained in the stomach or collected
together in the whole system, are more sluggish in their
movements. In like manner too, those who are in a pletho-
ric state and those brought down by recent cold are spiritless and hesitant in their movements. Still more does this
apply to those who are colder by nature, and certainly
much more, if some moisture is added to the cold. These
people depart from the norm in the signs of moderation
of the exercises, just as, in fact, do those who are hotter in

---

19 ἀσχημονεστέρας Ko; ἀσθενεστέρας Ku

οἱ θερμότεροι τὴν κρᾶσιν ἢ διὰ τὴν οἰκείαν φύσιν ἢ
δι' ἐπίκτητόν τινα διάθεσιν εὔθυμοί τέ εἰσι καὶ φιλό-
νεικοι καὶ πρὸς τὰς ἐνεργείας ἕτοιμοι περαιτέρω τοῦ
δέοντος. ὅσοι δὲ μήτε τὸ σῶμα θερμότερον ἢ ψυχρότερον ἔχουσι μήτ' ἄθυμοι τὴν ψυχήν εἰσι τὸ πάμπαν ἢ φιλόνεικοί τε καὶ φιλότιμοι καὶ θυμικοί, τούτοις
131K ἀκριβῆ διαφυλάττεται τοῦ μέτρου τῶν γυμνασίων τὰ
γνωρίσματα, καὶ κατὰ τὴν πρώτην μὲν ἡμέραν εὐθύς,
ἀτὰρ οὖν ἔτι δὴ καὶ μᾶλλον ἐπὶ προήκοντι τῷ χρόνῳ.
εἰ γάρ τι καὶ παρέλαθεν κατὰ τὴν ἀρχήν, ἀκριβωθή-
σεται τοῦτο τῇ πείρᾳ διδαχθέν.

οὕτως οὖν εἰς ἀκριβῆ στοχασμὸν ἔρχεται καὶ τὸ
τῶν σιτίων μέτρον οὐδενὶ τρόπῳ κατ' ἀρχὰς γνωρισθῆναι δυνάμενον· ἀλλ' ἡ καθ' ἑκάστην ἡμέραν πεῖρα
καὶ μνήμη τοῦ ποσοῦ τῶν σιτίων καὶ τῶν γυμνασίων,
οὐδὲν ἀργῶς ὁρῶντος τοῦ προεστῶτος, ἀλλ' ἀεὶ με-
μνημένου, ὅπως ἔπεψεν ἐπὶ τοσοῖσδε τοῖς γυμνασίοις
τὰ τοσάδε σιτία, πλησίον ἀκριβοῦς ἐπιστήμης ἄγει
τὴν διάγνωσιν ἐν τῷ χρόνῳ. ταυτὶ μὲν οὖν ἤδη πως
ἅπαντα κοινὰ καὶ ταῖς μοχθηροτέραις ὑπάρχει κατασκευαῖς τῶν σωμάτων, ὑπὲρ ὧν οὐδὲν νῦν πρόκειται
λέγειν. ἡ δ' ἀρίστη φύσις, ἡ νῦν ἡμῖν προκειμένη, τὰ
μέτρα πάντων εὔδηλα κέκτηται, μήτε τοῖς τῆς ψυχῆς
ἤθεσι μήτε ταῖς τοῦ σώματος δυσκρασίαις ἐπιθο-
λοῦσά τε καὶ νοθεύουσα τῶν εἰρημένων μηδέν, ἀλλὰ
καὶ τρίψεων καὶ γυμνασίων καὶ σιτίων καὶ ὕπνων
132K ἐναργῶς ἐνδεικνυμένη τὰ μέτρα, καὶ τηνικαῦτα πρῶτον ἀπαγορεύουσα πρὸς ἕκαστον, ὅταν μηκέτι δέηται,

krasis, either due to their specific nature or due to some added condition, and are spirited, contentious and ready for actions beyond what is required. However, in those who are neither hotter nor colder in the body, nor altogether spiritless in respect of the soul, nor contentious, ambitious or high-spirited, the signs of measure of the   131K
exercise are maintained precisely for them, and even immediately on the first day, but still more with the progression of time, for if something did escape notice at the beginning, this will be made exact when taught by experience.

In the same way, then, the measure of food also comes to an exact estimate, although there is no way it can be known at the beginning. But the experience of each day and the recollection of the amount of food and exercise when the one in charge is not remiss in his observations, but always remembers how much food was digested after how much exercise, leads to the recognition of a more exact knowledge over time. The same things are already in some way all common and exist in those with more abnormal constitutions of their bodies, although I do not propose to say anything about these now. However, the best nature, which is what now lies before us, has acquired the easily recognized measure of all things, not being disturbed by either the characteristics of the soul or the *dyskrasias* of the body, nor departing from normal in any of the things mentioned, but displaying clearly the measures of massage, exercise, food and sleep. And under these   132K
circumstances, first desisting from each of these things

ὥστε εἶναι καὶ τοῦτο τῷ προεστῶτι τοιούτου σώματος οὐ μικρὸν γνώρισμα μέτρου. λέγω δὲ τὴν πρὸς ἕκαστον ὧν ἂν πράττῃ προθυμίαν, αὐτῆς τῆς φύσεως ἑαυτῇ τὸ μέτρον εὑρισκούσης ἐν ταῖς ἀρίσταις κατασκευαῖς.

καὶ γὰρ ἀνατριβόμενοι τηνικαῦτα πρῶτον ἐξίασιν ἐπὶ τὸ γυμνάσιον οἱ ἄριστα πεφυκότες, ὅταν ἱκανῶς ἔχωσι μαλακότητός τε ἅμα καὶ θερμότητος ἅπαντι τῷ σώματι· καὶ γυμναζόμενοι τότε πρῶτον ὀκνήσουσιν, ὅταν αὐτάρκως γυμνασθῶσι· καὶ δὴ καὶ σιτίων τε καὶ πομάτων ἀποστήσονται τότε πρῶτον, ὅταν ἱκανῶς ἐμπλησθῶσιν, ὥστ᾽ οὐδεὶς φόβος οὔτε ὑπερπονῆσαι τὸν τοιοῦτον ἄνθρωπον οὔθ᾽ ὑπερεμπλησθῆναι, ταῖς ὁρμαῖς τῆς φύσεως οἰακιζόμενον. οὔκουν οὐδὲ τοῦ προεστῶτος εἰς ἄκρον ἥκοντος ἐπιστήμης ὁ τοιοῦτος ἄνθρωπος δεῖται, καθάπερ οἱ μετὰ ταῦτα λεχθησόμενοι πάντες, οἱ μοχθηρῶς κατεσκευασμένοι. αὐτὸ γὰρ ἑαυτῷ πάντ᾽ ἐξευρίσκει τὸ ὑγιεινὸν σῶμα, ταῖς τῆς
133K φύσεως ὁρμαῖς ἐπιτροπευόμενον, καὶ μάλιστ᾽ εἰ καλῶς εἴη τὰ τῆς ψυχῆς αὐτῷ πεπαιδευμένα. πολλοὶ γὰρ ἔθεσι μοχθηροῖς ἐντραφέντες ἀκολαστότερον ἢ ἀργότερον διαιτώμενοι διαφθείρουσι φύσεις χρηστάς, ὥσπερ αὖ πάλιν ἔνιοι μοχθηρῶς φύντες τὸ σῶμα βίῳ σώφρονι καὶ ἔργῳ[20] καὶ γυμνασίοις εὐκαίροις ἐπανωρθώσαντο τὰ πολλὰ τῶν ἐλαττωμάτων.

ἀλλὰ[21] τούτοις μὲν ὁ μετὰ ταῦτα λόγος ἅπας σύγκειται· περὶ δὲ τῶν ἄριστα κατεσκευασμένων τὸ σῶμα

whenever it is no longer required, is also no small sign of measure to the one in charge of such a body. I mean the eagerness for each of the things it does, since the actual nature discovers for itself the measure in the best constitutions.

Also, at the time of being massaged, those who are naturally the best first go out to exercise when they have sufficient softness and warmth in the whole body. And when they are exercising, they will, at that time, be the first to stop when they are sufficiently exercised. Furthermore, they will first desist from food and drink when they are sufficiently filled, so there is no fear of such a person either working too hard or overfilling himself, being directed by the impulses of nature. Therefore, such a person doesn't need a trainer who has come to the peak of knowledge, as do all those of defective constitution who will be spoken of after this. The healthy body will discover all this for itself, being governed by the impulses of nature, and particularly if there is proper training of the soul in it. For many who are brought up with defective customs, living a life that is too undisciplined or idle, destroy their good natures, just as some in turn, defectively nurtured in respect of the body, by a well-considered life and work, and by timely exercises, corrected the majority of the deficiencies.

133K

But the following discussion is, in its entirety, made up of these exercises and actions.[16] What I propose to go

16 The Kühn text is followed here.

---

20 καὶ ἔργῳ add. Ko   21 post ἀλλά: τούτοις μὲν Ko; τούτων μὲν γυμνασίων τε καὶ ἔργων Ku

καὶ τὸν νοῦν προσεχόντων ἐπιστάταις ὑγιεινοῖς ἐν
τῷδε τῷ γράμματι πρόκειται διελθεῖν. ὁ μὲν οὖν ἔμπροσθεν λόγος ὑπὲρ αὐτῶν εἰς τὸ μέτρον ἐτελεύτα
τῶν γυμνασίων· ἐφεξῆς δ᾽ ἂν εἴη καιρὸς ὑπὲρ τῶν
εἰδῶν διελθεῖν.

8. Εἴδη δὲ γυμνασίων ὀνομάζω πάλην καὶ παγκράτιον καὶ πυγμὴν καὶ δρόμον ὅσα τ᾽ ἄλλα τοιαῦτα,
τινὰ μὲν οὖν αὐτὸ τοῦτο, γυμνάσια μόνον, ὑπάρχοντα, τινὰ δὲ οὐ γυμνάσια μόνον, ἀλλὰ καὶ ἔργα·
γυμνάσια μὲν αὐτά γε δὴ ταῦτα τὰ εἰρημένα καὶ
προσέτι τὸ πιτυλίζειν, τὸ ἐκπλεθρίζειν, τὸ σκιομα-
134K χεῖν, τὸ ἀκροχειρίζεσθαι, τὸ ἄλλεσθαι, τὸ δίσκον
βάλλειν καὶ ἀποτομάδα²² καὶ διὰ κωρύκου καὶ διὰ
σφαίρας, ἢ μικρᾶς ἢ μεγάλης, καὶ δι᾽ ἁλτήρων ἐκπο-
νῆσαι τὸ σῶμα, γυμνάσια δ᾽ ἅμα καὶ ἔργα σκάπτειν,
ἐρέττειν, ἀροῦν, κλᾶν ἀμπέλους, ἀχθοφορεῖν, ἀμᾶν,
ἱππεύειν, ὁπλομαχεῖν, ὁδοιπορεῖν, κυνηγετεῖν, ἁλιεύ-
ειν, ὅσα τ᾽ ἄλλα κατὰ μέρος πράττουσιν ἄνθρωποι
τεχνῖταί τε καὶ ἄτεχνοι τῶν κατὰ τὸν βίον ἕνεκα
χρειῶν, ἢ οἰκοδομοῦντες ἢ χαλκεύοντες ἢ ναυπηγοῦν-
τες ἢ ἀροτρεύοντες ἤ τι τοιοῦτον ἕτερον πολέμιον ἢ
εἰρηνικὸν ἐργαζόμενοι. τοῖς πλείστοις δὲ τῶν τοιούτων

---

²² καὶ ἀποτομάδα add. Ko

17 This term and the following three that are simply transliterated pose difficulties as regards finding a precise English equivalent. Some comments on the terms are as follows: (1) παγκράτιον
seems to be a kind of all-in fighting—a mixture of boxing and

through in this book are matters concerning those best constituted in the body who devote themselves to expert hygienists. Thus, the previous discussion brought to a conclusion the measure of exercises themselves; in what follows, it would be timely to go over the kinds of exercises.

8. I term the kinds of exercise wrestling, *pankration*,[17] boxing and running, and other such things. Some are actual exercises alone while some are not only exercises but also activities. The actual exercises are those mentioned, and in addition, *pitulism, ekplethrism,* shadow fighting, *acrocheirism,* leaping, quoit throwing, javelins and working out the body with a heavy bag and balls, either small or large, and with weights, while exercises and activities together are digging, rowing, plowing, pruning vines, bearing burdens, reaping, horse riding, practicing the use of weapons, walking, hunting, fishing and other such individual things that men, both skilled and unskilled, do for a living—house building, metalworking, ship building, plowing, or making some other such thing pertaining to war or peace. The majority of these things can at times be

134K

wrestling, perhaps somewhat akin to the "cage-fighting" now popular; (2) pitulism is derived from the verb πιτυλίζω. The definition in LSJ is "to practice regular swinging of the arms, as with dumb-bells" with reference to this passage and 144K. It can, however, also mean "to dart about"; (3) ekplethrism is derived from the verb ἐκπλεθρίζω, used by Galen in chapter 8 (VI.133) and defined in LSJ as to "run round and round in a course that narrows every time." As defined above it seems somewhat different; (4) ἀκροχειρίζω has the basic meaning of "to take hold of." According to LSJ, in the middle voice, it means "to struggle at arm's length"—see Philostratus, *De gymnastica* 31.

ἔνεστί ποτε καὶ ὡς γυμνασίοις μόνον χρήσασθαι. τριττὴ γὰρ αὐτῶν τις ἡ σύμπασα χρεία, ποτὲ μὲν ὡς ἔργων μόνον αὐτὸ δὴ τοῦτο παραλαμβανομένων, ἐνίοτε δὲ ὡς ἀσκημάτων ἕνεκα τῆς τῶν μελλόντων ἔργων χρείας, ἔστιν ὅτε δὲ καὶ ὡς γυμνασίων. κατ᾽ ἀγρὸν γοῦν ποθ᾽ ἡμεῖς ληφθέντες ἐν χειμῶνι ξύλα τε σχίζειν ἠναγκάσθημεν ἕνεκα τοῦ γυμνάζεσθαι καὶ κριθὰς ἐμβάλλοντες ἐν ὅλμῳ κόπτειν τε καὶ ἐκλεπίζειν, ἅπερ ἑκάστης ἡμέρας οἱ κατ᾽ ἀγρὸν ἔπραττον ὡς 135K ἔργα. περὶ μὲν δὴ τῆς ὡς ἔργων αὐτῶν χρήσεως ἐν τῷ μετὰ ταῦτα λόγῳ διαιρήσομεν· ἐν δὲ τῷ παρόντι περὶ τῆς ὡς γυμνασίων ἐροῦμεν.

ἅπαντα γὰρ τὰ τοιαῦτα γυμνάσια γίνεται μήτε δι᾽ ὅλης ἡμέρας αὐτὰ πραττόντων μήτ᾽ ἐν ἄλλῳ καιρῷ τοῦ πρὸ τῶν σιτίων. φυλάττεσθαι δὲ δήπου προσήκει καὶ τὸ μέτρον ἐπ᾽ αὐτοῖς τῶν γυμνασίων καὶ τὴν καλουμένην ἀποθεραπείαν. εἰ δὲ καὶ προανατριψάμενός τις αὐτάρκως ἐπ᾽ αὐτὰ παραγίνοιτο, νῦν μὲν ἂν ἅπαντας ἀκριβῶς ἔχοι τοὺς ἀριθμοὺς τῶν γυμνασίων. ὑπέρ τε οὖν τούτων ἁπάντων, ἃ δὴ καὶ γυμνασίων εἴδη καλοῦμεν, ἐπίστασθαι χρὴ τὸν τὴν ὑγιεινὴν τέχνην μετιόντα, καλεῖν δ᾽ ἔξεστιν αὐτόν, ὡς ἔμπροσθεν εἶπον, ἢ ὑγιεινὸν ἢ γυμναστὴν ἢ ἰατρόν, κυριωτάτης μὲν ἐσομένης τῆς πρώτης προσηγορίας, ἐκ καταχρήσεως δὲ τῶν ἄλλων. εἴπερ γὰρ ἅπαντες οἱ τεχνῖται παρωνύμως ὀνομάζονται τῶν τεχνῶν, ἃς μεταχειρίζονται, πρόδηλον, ὡς ὁ τὴν ὑγιεινὴν τέχνην μετερχόμενος ὑγιεινὸς ἂν εὐλόγως προσαγορεύοιτο,

196

used as exercises alone. Their use overall is threefold; sometimes the actual thing is undertaken as an activity alone, sometimes as practicing for use in activities that are going to be done, and sometimes also as exercises. Anyway, on one occasion, when I was caught in the countryside in winter, I was forced to split wood for exercise, and throwing the barley into a mortar to cut and thresh, which those in the country were doing every day as work. I shall certainly define the use of these activities themselves in the discussion following this. In the present discussion, I shall speak about their use as exercises.

135K

All such exercises are not done throughout the whole day by those who do them, nor at any other time than before food. It is doubtless also appropriate to maintain in these cases a due measure of exercises and the so-called apotherapy. And if anyone, when previously massaged sufficiently, were to come to them, he would now have the whole number of exercises in their entirety. It is therefore necessary for someone who practices the art of hygiene to know about all those things that we also call kinds of exercises, and, as I said before, it is possible to call him a hygienist, gymnastic trainer or doctor, although the first will be the most appropriate term, the others arising from catachresis. Thus, if all those who are skilled in something are named derivatively from the skills they practice, it is clear that the one who practices the art of hygiene should logically be called a hygienist, just as one who is concerned

καθάπερ καὶ ὁ τὴν περὶ τὰ γυμνάσια μόνον γυμνα-
136K στὴς καὶ ὁ περὶ τὰς ἰάσεις ἰατρός. εἰ δέ τις ἢ γυμναστὴν ἢ ἰατρὸν ὀνομάζοι τὸν ὑγιεινὸν δὴ τοῦτον, ἀπὸ
μέρους τε προσαγορεύσει τὸ σύμπαν καὶ οὐ κυρίως,
ἀλλ' ἐκ καταχρήσεως ἢ ἐπὶ διαστάσεως, ἢ ὅπως ἄν
τις ἐθέλῃ καλεῖν, οὕτω ποιήσεται τὴν προσηγορίαν.
αἴτιον δὲ τούτου τό, μιᾶς οὔσης τῆς περὶ τὸ σῶμα
τέχνης, ἐφ' ὅλης αὐτῆς ὄνομα μηδὲν τετάχθαι κύριον,
ὑπὲρ ὧν ἐπὶ πλεῖον ἐν ἑνὶ βιβλίῳ τὸν λόγον ἐποιη-
σάμην, ὃ Θρασύβουλος ἐπιγράφεται.

εἴδη μὲν δὴ γυμνασίων τὰ εἰρημένα. ποιότης δὲ ἢ
διαφορά (καὶ γὰρ οὖν καὶ ταύτην ἔξεστιν ἑκατέρως
ὀνομάζειν) ὀξύτης τε κινήσεώς ἐστι καὶ βραδύτης εὐ-
τονία τε καί, ὡς ἂν εἴποι τις, ἀτονία καὶ πρὸς τούτοις
ἔτι σφοδρότης τε καὶ ἀμυδρότης. τρόποι δὲ τῆς χρήσεως ἁπάντων τῶν εἰρημένων εἰδῶν ἅμα ταῖς οἰκείαις
διαφοραῖς τοιοίδε εἰσίν· ἤτοι συνεχὴς ἡ κίνησίς ἐστιν
ἢ διαλείπουσα· καὶ εἰ μὲν συνεχής, ἤτοι ὁμαλὴ ἢ
ἀνώμαλος, εἰ δὲ διαλείπουσα, ἤτοι τεταγμένη ἢ
ἄτακτος. οὗτοι μὲν οὖν οἱ κατ' αὐτὸ τὸ πρᾶγμα τρόποι
τῆς χρήσεως· οἱ δ' ἀπὸ τῶν ἔξωθεν αὐτῷ προσιόντες
137K τοιοίδε· ἢ ἐν ὑπαίθρῳ χωρίῳ γίνεται τὸ γυμνάσιον ἢ
ἐν καταστέγῳ ἢ ἐν ὑποσυμμιγεῖ σκιᾷ. κατὰ δὲ τὸν
αὐτὸν τρόπον ἢ θερμόν ἐστι τὸ χωρίον ἢ ψυχρὸν ἢ
εὔκρατον καὶ ἤτοι ξηρὸν ἀκριβῶς ἢ ὑγρὸν ἢ μέτριον.
οὕτω δὲ καὶ τὸ μετὰ κόνεως ἤτοι πλείονος ἢ ἐλάττονος
ἐλαίου τε κατὰ τὸν αὐτὸν λόγον ἢ πλείονος ἢ ὀλίγου[23]

---

[23] post ἢ πλείονος: ἢ ὀλίγου Ko; ἢ ἐλάττονος Ku

with exercises alone should be called a gymnastic trainer
(exercise therapist), and one who is concerned with treatment, a doctor. If, however, someone were to name a par-   136K
ticular hygienist either a gymnastic trainer or a doctor, he
would be naming the whole from the part, and not properly, but would be making the name in this way catachrestically or ambiguously, or however one might wish to term
it. The reason for this is that, although there is one art
concerning the body, no proper name has been established for the whole art itself. I gave a more detailed
account of these matters in one book, inscribed *Thrasybulus*.[18]

The kinds of exercises are those stated. The quality or
difference—for it is possible to use either of these terms—
is in the swiftness and slowness of movement, the vigor
and, as one might say, the lack of vigor, and in addition to
these, the violence or gentleness. The ways of use of all
the kinds mentioned, along with the specific differences,
are as follows: the movement is either continuous or intermittent; if it is continuous it is either even or uneven; if it
is intermittent it is either regular or irregular. These, then,
are the modes of use in relation to the matter itself. Those
arising from what is external to the person are as follows:
whether the exercise occurs in an open place, in a covered   137K
place, or in one partly mixed with shadow, and in the same
way, whether the place is hot, cold or *eukratic*, or perfectly
dry or moist, or moderate. In this way too, if it is with
powder, either more or less, and oil on the same basis

---

[18] This (*Thras.*, V.806–98K) is Galen's other major work on
hygiene, and is included in LCL 536.

ἢ καὶ χωρὶς ἑκατέρου τρόπος ἐστὶ χρήσεως γυμνασίου.

9. Ἐπεὶ τοίνυν διώρισται πάνθ᾽ ἡμῖν, ὧν ἔμπειρον
εἶναι χρὴ τὸν ὑγιεινόν, ἐπὶ τὰ τῶν γυμνασίων εἴδη
καιρὸς μετιέναι, καὶ πρῶτον μὲν διελέσθαι, τί τε κοινὸν ἅπασιν ὑπάρχει καὶ τί καθ᾽ ἕκαστον ἴδιον, ἐφεξῆς
δὲ τοὺς καιροὺς τῆς χρήσεως²⁴ ἀφορίσασθαι. τὸ μὲν
δὴ κοινὸν ἁπάντων γυμνασίων ἐστὶ θερμότητος αὔξησιν ἐξ αὐτῶν τοῖς ζῴοις ἐργάζεσθαι. θερμαίνεται
γὰρ ἡμῶν τὰ σώματα κατά τε τὰ βαλανεῖα καὶ τὰ
τῶν θερμῶν ὑδάτων λουτρὰ καὶ τὰς θερμὰς ὥρας τοῦ
ἔτους, ἡλιοθερούντων τε καὶ παρὰ πυρὶ θαλπομένων
καὶ θερμοῖς φαρμάκοις ἀνατριβομένων. ἀλλ᾽ ἔξωθεν
αἱ τοιαῦται πᾶσαι θερμότητες, οὐκ ἔνδοθεν, οὐδ᾽ ἐκ
τῆς οἰκείας ἀρχῆς ἀνάπτονται καὶ αὐξάνονται.

κατὰ δὲ τὰ γυμνάσια τῆς ἐμφύτου τοῖς ζῴοις θερ-
μασίας²⁵ αὔξησίς ἐστιν, ἐξ αὐτῶν τῶν ἰδίων σωμάτων
γινομένη. καὶ τοῦτο κοινὸν μὲν ἁπάντων τῶν γυμνασίων, οὐ μήν γε ἴδιόν ἐστιν, εἴ γε δὴ καὶ τοῖς θυ-
μωθεῖσι καὶ τοῖς ἀγωνιάσασι καὶ τοῖς αἰδεσθεῖσιν
αὔξησις τῆς ἐμφύτου γίνεται θερμότητος. ὁ μέν γε
θυμὸς οὐδ᾽ ἁπλῶς αὔξησις, ἀλλ᾽ οἷον ζέσις τίς ἐστι
τοῦ²⁶ κατὰ τὴν καρδίαν θερμοῦ· διὸ καὶ τὴν οὐσίαν
αὐτοῦ τῶν φιλοσόφων οἱ δοκιμώτατοι τοιαύτην εἶναί
φασι· συμβεβηκὸς γάρ τι καὶ οὐκ οὐσία τοῦ θυμοῦ
ἐστιν ἡ τῆς ἀντιτιμωρήσεως ὄρεξις. αὐξάνεται δὲ καὶ
τοῖς αἰδεσθεῖσιν ἡ ἔμφυτος θερμότης, εἴσω μὲν τὰ
πρῶτα συνδραμόντος ἅπαντος τοῦ θερμοῦ, μετὰ

either more or less, or without either, is a mode of use of
an exercise.

9. Accordingly, since we have already determined all
the things the hygienist must be experienced in, it is time
to proceed to the kinds of exercises, and first to go over
what is common to all of them, and what is specific to each
one, and next to determine the times of use. Common,
certainly, to all exercises is to produce an increase in heat
in animals from them. Our bodies are heated in bath
houses, baths of warm waters, the hot seasons of the year,
lying in the sun, being warmed beside a fire, and by being
massaged with warm medicaments. But all such things   138K
heat externally, not internally, nor are they kindled or increased from one's own source.

In exercises, there is an increase in the innate heat in
animals, arising from our own bodies themselves. And this
is common to all exercises and is not specific [to exercise],
if in fact an increase of the innate heat occurs in those who
have become angry, have contended or are ashamed. Anger is not simply an increase but a kind of seething of heat
in relation to the heart, on which account also the most
notable of philosophers say the essence of it is of this kind,
for the desire for vengeance is something contingent and
not the essential component of anger. Also, the innate heat
is increased in those who are ashamed, all the heat at first
running together inward, and after this having collected

---

24 τῆς χρήσεως Ko; τῆς τρίψεως Ku

25 post ἐμφύτου: τοῖς ζῴοις θερμασίας Ko; θερμότητος τοῖς
ζῴοις Ku

26 τίς ἐστι τοῦ add. Ko

201

ταῦτα δ᾽ ἀθροισθέντος ἐν τῷ βάθει, κἄπειτα αὐξηθέν-
τος καὶ διὰ τὴν ἄθροισιν μὲν τὴν ἐνταῦθα καὶ διὰ τὴν
κίνησιν δὲ τὴν συνεχῆ. οὐ γὰρ ἡσυχάζει τὸ πνεῦμα
τῶν αἰδουμένων, ἀλλ᾽ ἔνδον τε καὶ περὶ αὐτὸ μετὰ τοῦ
139K   σύμπαντος αἵματος κυκᾶται πολυειδῶς, ὥσπερ γε καὶ
τὸ τῶν ἀγωνιώντων. εἰρήσεται δ᾽ ἐπὶ πλέον ὑπὲρ τῶν
τοιούτων ἁπάντων παθῶν τῆς ψυχῆς ἐπὶ προήκοντι
τῷ λόγῳ. νυνὶ μὲν γὰρ διὰ τὸ κοινὸν ἕπεσθαι σύμπτωμα τούτοις τε τοῖς πάθεσι καὶ τοῖς γυμνασίοις
ἠναγκάσθην αὐτῶν μνημονεῦσαι δεικνύς, ὡς ἁπάντων κοινὸν γυμνασίων ἐστὶν ἡ τῆς ἐμφύτου θερμότητος αὔξησις ἔνδοθέν τε καὶ ἐξ αὑτῆς, οὐ μὴν ὅτι γε
μόνοις ὑπάρχει τοῦτο τοῖς γυμνασίοις, ἀλλὰ καὶ τοῖς
εἰρημένοις ἄρτι πάθεσιν.

ἤδη δ᾽ ἐπὶ τὰ καθ᾽ ἕκαστον τῶν γυμνασίων ἴδια
τὸν λόγον ἄγειν καιρός, ἐπισημηνάμενόν γε πρότερον, ὡς καὶ κατὰ ταῦτα πλείους εἰσὶν αἱ διαφοραί. τὰ
μὲν γὰρ ἄλλοτε ἄλλο τι μέρος ἕτερον ἑτέρου γυμνάζει
μᾶλλον, καὶ τὰ μὲν ἐλινυόντων γίνεται, τὰ δὲ ὀξύτατα
κινουμένων, καὶ τὰ μὲν εὐτόνως, τὰ δὲ ἀτόνως, καὶ
πρὸς τούτοις ἔτι τὰ μὲν σφοδρῶς, τὰ δ᾽ ἀμυδρῶς.
εὔτονον μὲν οὖν γυμνάσιον ὀνομάζω τὸ βιαίως ἄνευ
τάχους διαπονεῖν, σφοδρὸν δὲ τὸ βιαίως τε καὶ σὺν
140K   τάχει· βιαίως δὲ ἢ ῥωμαλέως λέγειν οὐ διοίσει. τὸ μὲν
οὖν σκάπτειν εὔτονόν τε καὶ ῥωμαλέον ἐστίν, οὕτω δὲ

---

19 The issue here is the difference (if any) in the use of
εὔτονος: when applied to men's bodies and limbs, it is taken to

together in the depths, is increased, due to both the collection there and the movement that is continuous. For the *pneuma* of those who are ashamed is not at rest but within and around it is stirred up in various ways along with all the blood, just as it also is in those who are contending. I shall, however, say more about all such affections of the soul as the discussion advances. For the present, because a common symptom follows both these affections and exercises, I was compelled to make mention of them, showing that an increase of innate heat within is common to all exercises, intrinsically and from this, and that this is not in exercises alone, but also in the affections spoken of just now.

Now is an appropriate time to lead the discussion to the specifics in relation to each of the exercises, indicating first that there are also many differences in these. There are those that, at one time or another, exercise one part more than another, those that are restful, those that involve very quick movements, those done vigorously, those done in a relaxed fashion, and in addition to these, those done violently, and those done weakly. I term an exercise "vigorous" (*eutonos*) when a man works out violently but without speed, and "violent" (*sphodros*) when he works out violently with speed. Whether we say "violently" or "strongly" makes no difference.[19] Thus digging is vigorous and strong, and in the same way, controlling four horses

139K

140K

mean "well-strung and vigorous" (Hippocrates, *Aphorisms* 3.17); when used more generally, it is taken to mean "active and energetic." Among the meanings listed in LSJ for σφόδρος are "vehement," "violent," "active," and "excessive." In the present passage εὔτονος is rendered "vigorous."

καὶ τὸ τέτταρας ἵππους ἅμα κατέχειν ἡνίαις εὔτονον
μὲν ἱκανῶς γυμνάσιον, οὐ μὴν ὠκύ γε. κατὰ δὲ τὸν
αὐτὸν τρόπον εἴ τις ἀράμενος ὁτιοῦν μέγιστον φορτίον ἢ μένοι κατὰ χώραν ἢ προβαίνοι σμικρὰ καὶ οἱ
ἀνάντεις περίπατοι τούτου τοῦ γένους εἰσίν. ἀναφέρε-
ται γοῦν καὶ ἀναβαστάζεται κατ' αὐτοὺς ὑπὸ τῶν
πρώτων κινουμένων ὀργάνων ἅπαντα τὰ λοιπὰ μόρια
τοῦ σώματος, ὥσπερ τι φορτίον. οὕτω δὲ καὶ ὅστις
ἀναρριχᾶται διὰ σχοινίου, καθάπερ ἐν παλαίστρᾳ
γυμνάζουσι τοὺς παῖδας εἰς εὐτονίαν παρασκευάζον-
τες. ὡσαύτως δὲ καὶ ὅστις ἢ σχοινίου λαβόμενος ἤ
τινος ὑψηλοῦ ξύλου μέχρι πλείστου κατέχει κρεμάμε-
νος ἐξ αὐτοῦ, ῥωμαλέον μέν τι καὶ ἰσχυρὸν γυμνάζεται γυμνάσιον, οὐ μὴν ὠκύ γε, καὶ ὅστις προτείνας ἢ
ἀνατείνας τὼ χεῖρε πὺξ ἔχων ἀτρεμίζει μέχρι πλείστου. εἰ δὲ καὶ παραστήσας τινὰ κελεύει καθέλκειν
κάτω τὼ χεῖρε, μὴ ἐνδιδοὺς αὐτός, ἔτι δὴ μᾶλλον
οὗτος εἰς εὐρωστίαν παρασκευάζει τούς τε μῦς καὶ τὰ
141K νεῦρα· τούτων γὰρ ἴδια τὰ τοιαῦτα μάλιστα σύμπαντα γυμνάσια· πολὺ δὲ δὴ μᾶλλον, εἴ τι βάρος ταῖς
χερσὶν ἄκραις περιλαβὼν ἑκατέραις καταμόνας, οἷοίπερ οἱ κατὰ παλαίστραν εἰσὶν ἁλτῆρες, ἀτρέμας ἔχει
προτείνας ἢ ἀνατείνας αὐτάς. εἰ δὲ δὴ καὶ κελεύσειε
τινι καθέλκειν τε καὶ κάμπτειν βιαίως, ἑαυτὸν ἀκίνη-
τόν τε καὶ ἄκαμπτον οὐ ταῖς χερσὶ μόνον, ἀλλὰ καὶ
τοῖς σκέλεσι καὶ τῇ ῥάχει διαφυλάττων, οὐ σμικρὸν
γυμνάσεται γυμνάσιον εἰς εὐτονίαν ὀργάνων.

with reins is a very vigorous exercise, although not in fact rapid. By the same token, if someone, having lifted a great weight, were either to remain on the spot, or step forward a little, and then walk uphill, [the movements] would be of this class. Anyway, all the remaining parts of the body are carried and lifted up during the latter by the first-moving organs, like some weight. This also applies to someone who climbs up a rope with hands and feet, like they train children to do in the wrestling school, preparing them for vigorous activity. And similarly, when someone takes hold of a rope or some high piece of wood, and remains hanging from it as long as possible, he is practicing a robust and strong exercise, but not in fact one that is rapid. The same applies to someone who stretches out or holds up both hands as clenched fists, and stays still for as long as possible. Also, if having made somebody stand next to him, he orders this person to drag down both his hands while he himself doesn't yield, this prepares the muscles and sinews for strength even more. For all such exercises   141K
are particularly specific, and much more so if, taking hold of some weight with each of the outstretched hands apart, such as the weights that are in the wrestling school, he holds the hands motionless, having extended them forward or raised them up. If he then were to direct someone to draw them down or bend them forcibly, while keeping himself immobile and rigid, not with his hands only but also with his legs and spine, he will be carrying out a substantial exercise for the vigor of the organs.

οὕτω τοι λόγος ἔχει κἀκεῖνον τὸν Μίλωνα γυμνάζειν ἑαυτόν, ἐνίοτε μὲν ἀποσαλεῦσαί τε καὶ μετακι-
νῆσαι τῆς ἕδρας ἐπιτρέποντα τῷ βουλομένῳ (ἀλλὰ
τοῦτο μὲν σκελῶν ἂν εἴη μάλιστα γυμνάσιον), ἐνίοτε
δέ, εἰ τὰς χεῖρας γυμνάζειν βούλοιτο, τὴν πυγμὴν
διαλύειν κελεύοντα· αὖθις δ᾽ ἄν, ὥς φασιν, ἐν ταῖν
χεροῖν ἔχων ἢ ῥοιὰν ἢ ἕτερόν τι τοιοῦτον ἀφαιρεῖσθαι
τῷ βουλομένῳ παρεῖχε. ταυτὶ μὲν οὖν τὰ γυμνάσια
μεγίστης ἰσχύος ἐπίδειξίν τε ἅμα καὶ ἄσκησιν ἔχει,
τόνον δὲ μορίων γυμνάζει τε καὶ ῥώννυσι, κἀπειδὰν
142K ἤτοι διαλαβὼν ἕτερόν τινα μέσον ἢ διαληφθεὶς αὐτὸς
ἐπηλλαγμένων πρὸς ἀλλήλας τῶν χειρῶν τε καὶ τῶν
δακτύλων ἤτοι τῷ κρατουμένῳ προστάξῃ διαλύειν ἢ
αὐτὸς λύῃ τοῦ κρατοῦντος· οὕτω δὲ κἀπειδάν, ἑτέρου
προνεύσαντος ἐκ πλαγίων αὐτῷ προσελθὼν ἐν κύκλῳ
τοῖς λαγόσι περιβαλὼν τὰς χεῖρας, ὥσπερ γέ τι φορτίον ἀράμενος ἀναφέρῃ τε ἅμα καὶ περιφέρῃ τὸν ἀρ-
θέντα, καὶ μᾶλλον εἰ ἐπινεύοι τε καὶ ἀνανεύοι βαστά-
ζων· ὧδε γὰρ ἂν ἀκριβῶς τις ἅπασαν τὴν ῥάχιν εἰς
ῥώμην παρασκευάσειεν. οὕτω δὲ καὶ ὅσοι τὰ στέρνα
πρὸς ἀλλήλους ἀπερεισάμενοι βιαίως ὠθοῦσιν εἰς
τοὐπίσω καὶ ὅσοι τῶν αὐχένων ἐκκρεμάμενοι κατα-
σπῶσιν, εἰς εὐτονίαν παρασκευάζουσιν.

---

[20] Milo, a noted strongman, is mentioned earlier (see Book 1,
n. 14 above). The feat for which he was most renowned was
carrying a dead bull on his shoulders around the arena. Galen's
comment on this is: "But what about the story of Milo of Kroton?
He once did a lap of the stadium with a sacrificed bull on his

There is a story, let me tell you, that the famous Milo[20] exercised himself in this way, sometimes presenting himself to anyone wishing to loosen and remove him from his seat (but this would be particularly an exercise of the legs), and sometimes, if he should wish to exercise his hands, directing another to unclench his clenched fists, or again, so they say, having a pomegranate or some other such thing in his two hands, he would offer it to anyone who wished to take it away. These exercises are, then, a demonstration of the greatest strength while at the same time being training. And they exercise and strengthen the vigor of the parts, and whenever, either seizing some other person about the middle or being seized thus oneself, and interlocking the hands and the fingers with each other, he himself enjoins the one being held to release himself, or he releases himself from the one holding him. In the same way too, when someone approaches another from the side, when the other has bent forward, throwing his arms around the flanks to encircle them, lifts him up like some burden and holds him lifted while at the same time carrying him forward, and particularly if the one carrying bends forward and backward while carrying him. In this way someone might perfectly prepare the whole spine for strength. In the same way too, those who place their chests against one another, forcibly thrusting in opposition, and those who, hanging by the neck, pull downward, prepare themselves for vigor.

142K

shoulders. What incredible stupidity that was! Not to realise that just a little earlier, while it was alive, the animal's body was lifted up by a soul which drove it and made it run with much less effort than Milo's" (*Protr.*, I.34K [trans. after Singer, *Galen: Selected Works*, 50]).

ἀλλὰ τὰ μὲν τοιαῦτα καὶ χωρὶς παλαίστρας ἢ
βαθείας κόνεως δύναται γίνεσθαι καθ' ὁτιοῦν χωρίον
ἐπίκροτον ὀρθῶν ἑστώτων· ὅσα δὲ παλαίοντες εἰς ἀλλήλους δρῶσιν ἀσκοῦντες τὸν τόνον, ἤτοι κόνεως βαθείας ἢ παλαίστρας δεῖται. ἔστι δὲ τὰ ⟨τοιαῦτα⟩²⁷
τοιάδε· περιπλέξαντες τοῖς ἑαυτοῦ δύο σκέλεσι τὸ ἕτερον σκέλος τοῦ προσπαλαίοντος, ἔπειθ' ἅψαντες πρὸς
143K ἀλλήλας τὼ χεῖρε, τὴν μὲν ἐπὶ τὸν αὐχένα βιαίως
ἐρείδειν, ἥτις ἂν ᾖ κατ' εὐθὺ τοῦ κατειλημμένου σκέ-
λους, τὴν δ' ἑτέραν ἐπὶ τὸν βραχίονα. δύναιτο δ' ἂν
καὶ περὶ τὴν κεφαλὴν ἄκραν τὸ ἅμμα περιθεὶς ἀνα-
κλᾶν εἰς τοὐπίσω βιαζόμενος. τὰ τοιαῦτα γὰρ παλαί-
σματα πρὸς εὐτονίαν ἑκάτερον τῶν παλαιόντων ἀσκεῖ,
καθάπερ γε καὶ ὅσα ζώσαντος τοῖς σκέλεσι θατέρου
τὸ ἕτερον ἢ κατ' ἀμφοῖν ἄμφω καθέντος γίνεται· καὶ
γὰρ καὶ ταῦτ' ἀμφότερα εἰς ῥώμην παρασκευάζει.

μυρία δὲ ἕτερα τοιαῦτα κατὰ παλαίστραν ἐστὶν
εὔτονα γυμνάσια, περὶ ὧν ἁπάντων τὴν ἐμπειρίαν τε
ἅμα καὶ τριβὴν ὁ παιδοτρίβης ἔχει, ἕτερος δέ τις ὢν
ὅδε τοῦ γυμναστοῦ, καθάπερ ὁ μάγειρος τοῦ ἰατροῦ.
καί πως ἔοικεν αὖ καὶ τοῦθ' ἡμῖν ἥκειν εἰς σκέμμα·
περὶ οὗ λέλεκται μὲν ἤδη καὶ κατ' ἐκεῖνο τὸ βιβλίον,
ὃ Θρασύβουλον ἐπιγράφομεν, εἰρήσεται δὲ καὶ νῦν
ὅσον αὔταρκες εἰς τὰ παρόντα, πρότερόν γε διελθόν-
των ἡμῶν τὰς τῶν γυμνασίων διαφοράς. ὅσα μὲν οὖν
εὔτονα, καὶ δὴ λέλεκται.

144K   10. Μεταβαίνειν δὲ ἤδη καιρὸς ἐπὶ τὰ ταχέα χωρὶς
εὐτονίας καὶ βίας. δρόμοι δ' εἰσὶ ταῦτα καὶ σκιομα-

But such things can also be done away from the wrestling school or deep sand, while standing upright in any place that is trampled down. All the things they do to each other when wrestling and exercising their strength need either deep sand or the wrestling school. These exercises are as follows: gripping with their own two legs one leg of the wrestling partner and then, clasping the two hands together, forcibly to press on the neck, using whichever   143K hand is on the side of the gripped leg and putting the other hand on the arm. And he would also be able, placing the arm around the top of the head, to forcibly bend it backward. Such feats of strength train each of the wrestlers for vigor, just as those which happen when the one has girded the other's leg with his legs; both these maneuvers also prepare for strength.

There are countless other such vigorous exercises in the wrestling school, and the gymnastic trainer is experienced as well as practiced in all of them, although he himself is as different from the gymnast as the cook is from the doctor. And it seems to me that in some way this too has come to us for consideration, which I have already spoken about in that book I entitled *Thrasybulus*.[21] I shall also say now as much as is sufficient for my present purpose, having gone over the different exercises. Those that are vigorous, I have already described.

10. It is time now to pass on to the exercises that are   144K rapid without being vigorous and violent. These are run-

[21] See note 18 above.

---

[27] τοιαῦτα *add.* Ku

χίαι καὶ ἀκροχειρισμοὶ καὶ τὸ διὰ τοῦ κωρύκου τε καὶ
τῆς σμικρᾶς σφαίρας γυμνάσιον, ὅταν ἐκ διεστώτων
τε καὶ διαθεόντων γίνηται. τοιοῦτον δέ τι καὶ τὸ ἐκπλεθρίζειν ἐστὶ καὶ τὸ πιτυλίζειν. τὸ μὲν ἐκπλεθρίζειν
ἐστίν, ἐπειδάν τις ἐν πλέθρῳ πρόσω τε ἅμα καὶ ὀπίσω
διαθέων ἐν μέρει πολλάκις ἐφ᾽ ἑκάτερα χωρὶς καμπῆς
ἀφαιρῇ τοῦ μήκους ἑκάστοτε βραχὺ καὶ τελευτῶν εἰς
ἓν καταστῇ βῆμα· τὸ δὲ πιτυλίζειν, ἐπειδὰν ἐπ᾽ ἄκρων
τῶν ποδῶν βεβηκὼς ἀνατείνας τὼ χεῖρε κινῇ τάχιστα, τὴν μὲν ὀπίσω φέρων, τὴν δὲ πρόσω. μάλιστα
δὲ τοίχῳ προσιστάμενοι γυμνάζονται τοῦτο τὸ γυμνάσιον, ἵν᾽, εἰ καί ποτε σφάλλοιντο, προσαψάμενοι
τοῦ τοίχου ῥᾳδίως ὀρθῶνται· καὶ οὕτω δὴ γυμναζομένων λανθάνει τε τὰ σφάλματα καὶ ἀσφαλέστερον[28]
γίνεται τὸ γυμνάσιον. ὠκεῖαι δὲ κινήσεις εἰσίν, οὐ
μὴν βίαιοί γε, καὶ ὅσαι κατὰ παλαίστραν ἐπιτελοῦν-
145K   ται κυλινδουμένων ὀξέως μεθ᾽ ἑτέρων τε καὶ καταμόνας.

ἐγχωρεῖ δὲ καὶ ὀρθοὺς ἐνειλουμένους τε ἅμα καὶ
μεταβαλόντας ἐν τάχει τὸν πέλας ὀξὺ γυμνάσασθαι
γυμνάσιον. ἐγχωρεῖ δὲ καὶ διὰ τῶν σκελῶν μόνων
ὀρθὸν ἐφ᾽ ἑνὸς χωρίου γυμνάσασθαι γυμνάσιον ὀξύ,
πολλάκις μὲν εἰς τοὐπίσω μόνον ἐφαλλόμενον, ἔστιν
ὅτε δὲ καὶ εἰς τοὔμπροσθεν ἀναφέροντα τῶν σκελῶν
ἑκάτερον ἐν μέρει. καὶ μὲν δὴ καὶ διὰ τῶν χειρῶν
ἔστιν ὀξὺ γυμνάσιον ὁμοιόρροπον γυμνάσασθαι χωρὶς τοῦ κατέχειν ἁλτῆρας, ἐπισπεύδοντα τὰς κινήσεις
αὐτῶν εἰς πυκνότητά τε ἅμα καὶ τάχος, εἴτε πὺξ

ning, shadow fighting, *acrocheirism*,[22] and the exercise with the punching bag and small ball, done both standing at distance and running. *Ekplethrism* is such an exercise, as is *pitulism*. *Ekplethrism* is when someone runs forward and back repeatedly over a plethron (approx. one hundred ft.) course without deviating to either side, but takes away a short length each time until finally coming to a stop at one pace. *Pitulism* is when someone, standing on tiptoe, stretches his arms up and, moving very quickly, carries them backward and forward. For the most part people practice this exercise against a wall, so if at any time they should start to fall, they could easily right themselves by touching the wall. But when they exercise in this way concealing their mistakes, the exercise also becomes weaker. The movements are rapid but not violent, and when they carry them out in the wrestling school, they roll around swiftly, either with others or alone.   145K

It is possible to exercise rapidly while standing upright, engaging and at the same time changing position with those adjacent. It is also possible to practice a rapid exercise with the legs alone, upright in one place, repeatedly springing backward only, and sometimes also advancing each of the legs forward in turn. And furthermore, it is possible to practice a similar swift exercise with the hands without holding jumping weights, urging on their movements in frequency and rapidity, either with the fist closed,

---

[22] For these and the following transliterated terms, see note 17 above.

[28] ἀσφαλέστερον Ko; ἀσθενέστερον Ku; imbecillius L

ἐθέλοι τις εἴτε καὶ χωρὶς πυγμῆς ἀνασείειν[29] ἁπλῶς.
τοιοῦτον μὲν δή τι καὶ τὸ ταχὺ γυμνάσιόν ἐστιν ἐν
οἷς εἴπομεν εἴδεσιν ἀφωρισμένον.

ἐπὶ δὲ τὸ σφοδρὸν ἰέναι καιρός. ἔστι δ᾽, ὡς εἴρηται,
τοῦτο σύνθετον ἐξ εὐτόνου τε καὶ ταχέος. ὅσα γὰρ
εὔτονα τῶν γυμνασίων εἴρηται, τούτοις ἅπασιν ὡς
σφοδροῖς ἄν τις χρῷτο, ταχείας κινήσεις προστιθείς.
οὐχ ἥκιστα δὲ καὶ τὰ τοιάδε γυμνάσια σφοδρά, σκά-
ψαι καὶ δισκεῦσαι καὶ κινῆσαι καὶ πηδῆσαι συνεχῶς
146K ἄνευ τοῦ διαναπαύεσθαι. οὕτω δὲ καὶ τὸ ἀκοντίζειν
ὁτιοῦν τῶν βαρέων βελῶν συνείροντα τὴν ἐνέργειαν
ἢ βαρέσιν ὅπλοις ἐσκεπασμένον ἐνεργεῖν ὀξέως. ἀμέ-
λει καὶ οἱ γυμναζόμενοι διά τινος τῶν τοιούτων ἀνα-
παύονται κατὰ βραχύ. καί σοι καὶ ἡ κατὰ τὸ συνεχές
τε καὶ διαλεῖπον γυμνάσιον ἤδη πως γινωσκέσθω
διαφορά. τὰ γὰρ εἰρημένα νῦν δὴ πάντα διαλείποντες
μᾶλλον εἰς χρείαν ἄγουσι, καὶ μάλισθ᾽ ὅσα πόνοι
τινές εἰσι καὶ ἔργα, μὴ μόνον γυμνάσια, καθάπερ τὸ
ἐρέσσειν τε καὶ σκάπτειν. ὅσα δ᾽ ἀσθενέστερα τῶν
γυμνασίων ἐστίν, ἄνευ τοῦ διαναπαύεσθαι γίνεται
μᾶλλον, ὥσπερ ὁ δόλιχός τε καὶ αἱ ὁδοιπορίαι.

11. Ταῦτ᾽ οὖν ἅπαντα γυμνασίων ἐστὶν εἴδη, τάς
γε νῦν εἰρημένας ἔχοντα διαφορὰς καὶ πρὸς τούτοις
ἔτι τὸ τὰ μὲν ὀσφὺν μᾶλλον ἢ χεῖρας ἢ σκέλη διαπονεῖν, τὰ δὲ τὴν ῥάχιν ὅλην ἢ τὸν θώρακα μόνον ἢ
τὸν πνεύμονα. βάδισις μέν γε καὶ δρόμος ἴδια σκελῶν
γυμνάσια, ἀκροχειρισμοὶ δὲ καὶ σκιομαχίαι ἴδια χειρῶν, ὀσφύος δὲ τὸ ἐπικύπτειν τε καὶ ἀνακύπτειν συν-

if one wishes, or simply swinging to and fro without making a fist. This, then, is the rapid exercise I spoke of, divided into kinds.

It is time now to go on to violent exercise. This, as I said, is a combination of vigor and rapidity. Someone might use as violent all those exercises described as vigorous, if swift movements are added. Not least also the following exercises are violent: digging, throwing a quoit, and moving and jumping continuously without a rest. The 146K same also applies to the act of hurling any one of the heavy weapons, while joining together the activity, or acting rapidly while clad in heavy armor. Of course, those who exercise with one of these things stop after a short time. And you already know to some degree the difference between continuous and intermittent exercise. All those exercises mentioned now are particularly brought into use intermittently, and especially those that are hard work and activities, and not only exercises, like rowing and digging. Those exercises that are gentler mostly occur without an interval—examples are long-distance running and walking.

11. These, then, are all the kinds of exercises, some having the differences now mentioned. In addition to these, there are those that work out the loins more, or the arms, or the legs, and in other cases the whole spine, or the chest alone, or the lungs. In fact, walking and running are exercises specific for the legs, *acrocheirism* and shadow fighting are specific for the arms, bending forward and backward continuously for the loins, or lifting a weight 147K

---

29 ἀνασείειν Ko; ἀναχθήσειεν Ku

213

147K ἐχὼς ἢ αἴροντά τι βάρος ἀπὸ τῆς γῆς ἢ ἐν ταῖν
χεροῖν βαστάζοντά τι διαπαντός. ἔνιοι μὲν γὰρ
καταθέντες ἁλτῆρας ἐν τῷ πρόσθεν διεστῶτας ἀλλήλων ὀργυιάν, εἶτ᾽ ἐν τῷ μέσῳ στάντες αὐτῶν ἀναι-
ροῦνται προκύπτοντες, τῇ μὲν δεξιᾷ χειρὶ τὸν ἐν τοῖς
ἀριστεροῖς, τῇ δὲ ἀριστερᾷ τὸν ἐν τοῖς δεξιοῖς, καὶ
αὖθις ἑκάτερον εἰς τὴν οἰκείαν κατατίθενται χώραν
καὶ τοῦτο δρῶσιν ἐφεξῆς πολλάκις ἀτρεμίζοντες τῇ
βάσει. τὰ δὲ πλάγια μέρη τῆς ῥάχεως ἡ κίνησις ἥδε
διαπονεῖ μᾶλλον, ὥσπερ ἡ προειρημένη τὰ κατ᾽ εὐθύ.
θώρακος δὲ καὶ πνεύμονος αἱ μέγισται τῶν ἀναπνοῶν
οἰκεῖα γυμνάσια, καθάπερ γε καὶ αἱ μέγισται φωναὶ
πρὸς τοῖς εἰρημένοις ἁπάντων τῶν φωνητικῶν ὀργάνων. εὔρηται δ᾽ ὁ κατάλογος αὐτῶν ἐν τοῖς Περὶ φωνῆς
ὑπομνήμασι.

ἐπεὶ δ᾽ ἐνταῦθα τοῦ λόγου γεγόναμεν, οὐ χεῖρον
ὑπὲρ ἁπάντων διελθεῖν τῶν τοῦ ζῴου μορίων, ὅσα τε
κινήσεις ἐναργεῖς ἔχει καὶ ὅσα βραχείας τε καὶ ἀμαυ-
ράς, καὶ τίνα μὲν ἐξ ἑαυτῶν κινεῖται, τίνα δὲ ὑφ᾽
ἑτέρων· ἡ γὰρ τοιαύτη διαίρεσις οὐ σμικρὰν εὐπορίαν
148K παρέξει τῷ γυμναστῇ πρὸς τὸ κινεῖν ἅπαντα τοῦ
ζῴου τὰ μόρια, ποτὲ μὲν ἐξ ἑαυτῶν τε καὶ κατὰ τὰς
οἰκείας δυνάμεις, ἔστιν ὅτε δ᾽ ὑφ᾽ ἑτέρων τε καὶ δι᾽
ἑτέρων. αἱ μέν γε κατὰ προαίρεσιν ἐνέργειαι πᾶσαι
μυῶν τε καὶ νεύρων καὶ τενόντων ἴδιαι κινήσεις εἰσίν·
εἰ δὲ καὶ σφοδρότεραι γίνοιντο, τὰ μὲν εἰρημένα μόρια πρῶτά τε καὶ μάλιστα γυμνάζουσι, κατὰ δέ τι
συμβεβηκὸς καὶ τὰς ἀρτηρίας. ὀστᾶ δὲ καὶ φλέβας

from the ground or supporting something in a sustained fashion in both arms. Some people place jumping weights in front of them that are six feet apart, then standing in between them and bending forward, raise them up, the one on the left with the right hand and the one on the right with the left hand, and replace each again in its proper place; and they do this in sequence repeatedly, keeping their base still. This movement particularly works out the lateral parts of the spine, just as the previously mentioned movement does those parts that are central (vertical). Very deep breaths are specific exercises for the chest and lungs, just as very loud sounds, in addition to the aforementioned, are for all the organs of speech. The list of these is set out in the treatises *On the Voice*.[23]

Since we have come to this point of the discussion, it would not be bad to go over all the parts of the animal, both those which have visible movements and those which have small and indistinct movements; and which of them are moved by themselves and which of them by other parts. Such a division will provide no little advantage to the gymnastic trainer regarding moving all the parts of the 148K animal, sometimes from themselves and in accord with their intrinsic powers, and sometimes by other parts and through other parts. All the voluntary functions of muscles, sinews and tendons are intrinsic movements. If, however, they were to become more violent, they would exercise the aforementioned parts primarily and particularly, while incidentally also exercising the arteries. They move

[23] The work *De voce* was in four books and dedicated to Boethius. The original has been lost, although some fragments and an Arabic summary remain. See Boudon, *Galien,* 419n3.

καὶ σάρκας καὶ συνδέσμους καὶ τἆλλα σύμπαντα τοῦ
ζῴου μόρια συγκινοῦσιν ἴσως τοῖς προειρημένοις.

αἱ δ' ἄλλαι κινήσεις ἐν τοῖς τῶν ζῴων σώμασιν,
ὅσαι μὴ κατὰ κοινὴν προαίρεσιν, ἀλλ' ἤτοι φυσικαί
τινες ἢ κατὰ τὰ τῆς ψυχῆς γίνονται πάθη, διτταὶ μέν
εἰσι κατὰ γένος· αἱ μὲν ἕτεραι καρδίας τε καὶ ἀρτηριῶν, αἱ δὲ ἕτεραι φλεβῶν τε καὶ ἥπατος ἐνέργειαι,
πρῶτον μὲν τούτων καὶ μάλιστα, δεύτερον δὲ τῶν ἄλλων ἁπάντων μορίων, ἐπειδὴ πάντα ταῖς τέτταρσι
φυσικαῖς διοικεῖται δυνάμεσιν. ἡ μὲν δὴ τῶν ἀρτηριῶν τε καὶ τῆς καρδίας κίνησις οὐκ ἔστιν ὅτ' ἐκλεί-
πει τὸ πάμπαν, ἐπιτείνεται μέντοι[30] δὲ καὶ ἀνίεται
κατὰ μέγεθος καὶ σμικρότητα καὶ τάχος καὶ βραδυ-
149K τῆτα καὶ σφοδρότητα καὶ ἀμυδρότητα παρὰ πολλὰς
αἰτίας, ἃς συμπάσας μὲν ἐν τῇ Περὶ τῶν ἐν τοῖς
σφυγμοῖς αἰτιῶν πραγματείᾳ διῆλθον, ἀρκεῖ δ' ἐν τῷ
παρόντι τὰ κεφάλαια μόνον αὐτῶν εἰπεῖν τὰ πρῶτα.
μία μὲν οὖν αἰτία τῆς τῶν σφυγμῶν μεταβολῆς ἐστιν
αὔξησίς τε καὶ μείωσις τῆς ἐμφύτου θερμασίας, ἑτέρα
δὲ ἡ κατὰ τὴν ποσότητα τοῦ ψυχικοῦ πνεύματος
ἀνάλωσις,[31] τρίτη δὲ ἡ κατὰ τὴν τῆς δυνάμεως εὐρωστίαν τε καὶ ἀρρωστίαν, τετάρτη δὲ ἡ κατὰ τὰ τῶν
ὀργάνων πάθη.

---

[30] μέντοι add. Ku
[31] ἀνάλωσις Ko; ἀλλοίωσις Ku (here and subsequently)

---

[24] See particularly Book 1 of Galen's *Nat. Fac.*, II.1–73K
(English trans., Brock, *On the Natural Faculties*).

bones, veins, flesh and ligaments, and all other parts of the animal together equally with those previously mentioned.

The other movements in the bodies of animals—those not in relation to ordinary volition—which are either physical or occur in relation to the affections of the soul, are twofold in terms of class. Some are functions of the heart and arteries, and others of the veins and liver, and are first and foremost of these and secondarily of all the other parts, since all are governed by the four natural capacities (physical powers).[24] Now it is not possible for the movement of the heart and arteries to fail altogether. It does, however, increase and decrease in largeness and small- 149K ness, quickness and slowness, and strength and weakness due to many causes, all of which I went over in the work, *The Causes of the Pulses*.[25] It is sufficient for the present to state only the chief and primary of these. Thus, one cause of the change of the pulses is an increase or decrease of the innate heat; another is the change of the quantity of psychic *pneuma*;[26] a third relates to strength or weakness of the capacity; and a fourth relates to the affections of the organs.

[25] This is one of Galen's four major treatises on the pulses— *Caus. Puls.*, IX.1–204K. The other three are on classification (*Diff. Puls.*, VIII.493–765K); diagnosis (*Diagn. Puls.*, 766–961K); and prognosis (*Praesag. Puls.*, IX.205–430K), respectively. Also extant are three shorter treatises on the pulses: *Puls. ad Tir.*, VIII.451–492K; *UPuls.*, V.149–180K; and *Syn. Puls.*, IX.431–549K.

[26] Of the textual variations, the Kühn version is followed: both here and below "consumption" would seem to be the appropriate rendering if Koch were followed.

ἀλλὰ περὶ μὲν ταύτης οὐ νῦν διδάσκειν καιρός· ἤδη
γὰρ νοσεῖν ἀνάγκη τὸ ζῷον ἐν τοῖς τοιούτοις πάθεσιν·
αἱ δὲ ἄλλαι πᾶσαι τῶν σφυγμῶν ἀλλοιώσεις καὶ ἐν
τοῖς ὑγιαίνουσι γίνονται, παρὰ μὲν τὴν αὔξησίν τε
καὶ μείωσιν τῆς ἐμφύτου θερμότητος αἵ τε παρὰ τὰς
ὥρας καὶ τροφὰς καὶ πόματα καὶ λουτρὰ καὶ τρίψεις
καὶ ὕπνους καὶ ἐγρηγόρσεις ἔτι τε τὰ ψυχικὰ πάθη
καὶ τὰς κατὰ προαίρεσιν ἐνεργείας, ἡ δὲ παρὰ τὴν
τοῦ ψυχικοῦ πνεύματος ἀνάλωσιν ἐν ταῖς τῶν καθ'
ὁρμὴν κινήσεων διαφοραῖς, ἡ δὲ παρὰ τὴν τῆς δυνά-
150K   μεως εὐρωστίαν τε καὶ ἀρρωστίαν ἐν ταῖς εὐκρασίαις
τε καὶ δυσκρασίαις αὐτοῦ τοῦ σώματος τῆς τε καρδίας καὶ τῶν ἀρτηριῶν. ἡ δὲ τῶν φλεβῶν ἐνέργεια
σὺν καὶ τοῖς ἄλλοις ἅπασιν ὀργάνοις, ὅσα περὶ τὴν
τῆς τροφῆς οἰκονομίαν ὑπὸ τῆς φύσεως ἐγένετο, κατὰ
τὰς ἐδωδάς τε καὶ πόσεις, ἀναδόσεις τε καὶ πέψεις καὶ
θρέψεις ἀλλοιοῦται· μέρος δέ τι ταύτης ἐστὶν κἀν ταῖς
ἀρτηρίαις.

ὅσαι δὲ τῶν κινήσεων οὔκ εἰσιν ἐνέργειαι, τριττὴ
τούτων ἡ διαφορά· τινὲς μὲν γὰρ ἐξ ἑαυτῶν τοῖς ζῴοις
ἐγγίνονται, τινὲς δὲ ἔξωθεν προσέρχονται, τινὲς δὲ
ὑπὸ φαρμάκων καταναγκάζονται. ἐξ ἑαυτῶν μέν, ἃς
ἔμπροσθεν εἶπον· ἔξωθεν δὲ κατά τε τοὺς πλοῦς, τὰς
ἱππασίας, τὰς αἰωρήσεις, ὅσαι τε δι' ὀχημάτων γίνονται καὶ ὅσαι διὰ σκιμπόδων κρεμαμένων ἢ λίκνων
σειομένων ἢ ἐν ταῖς τῶν τροφῶν ἀγκάλαις τοῖς βρέφεσιν. ἐκ δὲ τῶν ἔξωθεν κινήσεων εἴη ἂν καὶ ἡ ἀνάτριψις (εἴθ' οὕτω τις αὐτὴν ὀνομάζειν ὁμοίως τοῖς

But now is not the time to teach about this, for the animal is, of necessity, already diseased in such affections, whereas all the other changes of the pulses occur also in those who are healthy—those on account of an increase or decrease of the innate heat, those on account of the seasons, nutriments, drinks, baths, massages, sleep and wakefulness, and further, the psychical affections and the voluntary functions. That due to the change of the psychic *pneuma* is in the differences of the movements initiated voluntarily; that due to the strength or weakness of the capacity is in the *eukrasias* or *dyskrasias* of the actual body and of the heart and arteries. The function of the veins, along with all the other organs created by Nature for the management of nutrition, changes in relation to foods and drinks, and their distribution, concoction and nourishing. Some part of this is even in the arteries.

150K

There is a threefold difference in those movements that are not functions: some spring up in animals of themselves, some come upon them from without, and some are forced upon them by medications. Those from themselves I spoke of previously. Those from without relate to sailing, horse riding, and the passive (oscillatory) movements, such as occur from chariots, from being suspended in a traveling hammock,[27] or infants being shaken in a cradle or in the arms of nurses. Also, among the extrinsic movements is rubbing (whether someone wishes to so name it

[27] LSJ lists two meanings for σκίμπους: (1) a small pallet or couch; (2) "a kind of hammock used by invalids travelling" with reference to this passage, although there is no specific mention of invalids here.

παλαιοῖς εἴτε καὶ χωρὶς τῆς "ἀνά" προθέσεως ὁμοίως
τοῖς νεωτέροις βούλοιτο· διαφέρει γὰρ οὐδὲν εἰς τὰ
151K παρόντα). ἔνιαι μέντοι κινήσεις εἰσὶ μικταί, καθάπερ
καὶ ἡ ἱππασία· οὐ γὰρ ὥσπερ ἐν τοῖς ὀχήμασι καὶ
μάλιστα ἐν οἷς κατακλινάμενοι ἀτρεμίζομεν, οὕτω
κἂν ταῖς τῶν ἵππων ὀχήσεσι συμπίπτει σείεσθαι μόνον ὑπὸ τοῦ φέροντος ἐνεργοῦντα μηδέν, ἀλλὰ τήν τε
ῥάχιν ὄρθιον ἀπευθύνειν χρὴ καὶ τοῖς μηροῖς ἀμφοτέροις ἀκριβῶς ἔχεσθαι τῶν πλευρῶν τοῦ ἵππου καὶ
τετάσθαι τὰ σκέλη καὶ προορᾶσθαι τὰ πρόσθεν· ἐν
τούτῳ δὲ καὶ ἡ ὄψις γυμνάζεται καὶ ὁ τράχηλος πο-
νεῖ. μάλιστα δ' ἐν τῷ τοιούτῳ γυμνασίῳ σείεται τὰ
σπλάγχνα. σείεταί γε μὴν οὐδὲν ἧττον τὰ σπλάγχνα
καὶ τοῖς ἀλλομένοις, ἐν μέντοι ταῖς ἐπὶ τῶν ὀχημάτων
αἰωρήσεσιν ἧττον. ὥστε εἴ τις ἐθέλοι τὰ κάτω τῶν
φρενῶν σπλάγχνα κινῆσαι βιαιότερον, ἐπί τε τοὺς
εἰρημένους ἡκέτω πόνους καὶ πρὸς τούτοις ὅσα διὰ
τῆς τῶν ἀμμάτων περιθέσεως τρίβουσι. χρὴ δ' ὄπισθεν εἶναι τὸν τρίβοντα, περιφέροντα τὼ χεῖρε ποτὲ
μὲν ἐπ' ἀριστερά, ποτὲ δ' ἐπὶ δεξιά, συνεπικλινομένου
πρὸς ταῦτα καὶ τοῦ τριβομένου. συγκινοῦσι μέν πως
152K τὰ κάτω τῶν φρενῶν σπλάγχνα καὶ αἱ μέγισται τῶν
ἀναπνοῶν τε καὶ φωνῶν, ὥσπερ γε καὶ αἱ ἐκφυσήσεις
αἵ τε καταμόνας γινόμεναι καὶ αἱ μετὰ καταλήψεως
ἐν αὐλήσεσί τε καὶ φωναῖς, καὶ αὐτὴ δ' ἡ κατάληψις
τοῦ πνεύματος γυμνάσιόν ἐστιν οὐχ ἧττον τῶν κατ'
ἐπιγάστριον ἢ τῶν κατὰ θώρακα μυῶν. ἀλλὰ περὶ μὲν
ταύτης αὖθις εἰρήσεται· πρὸς γὰρ ταῖς εἰρημέναις

like the ancients or prefers to omit the prefix ἀνά in the manner of those of later times, makes no difference to present matters).[28] Some movements are, however, mixed, as horse riding is. For it is not the case that, just as in carriages, and particularly those in which we lie quiet and reclining, and even in the carriages drawn by horses, where it happens that one is shaken only by the bearer and does nothing oneself, in horse riding one must keep the spine straight, hold the sides of the horse firmly with both thighs, stretch the legs out, and look forward. In the last, the vision is also exercised and the neck works. Particularly in such an exercise, the internal organs are shaken. And they are shaken no less in jumping, although of course less in the case of those passively exercising in carriages. As a result, if someone should wish to move the internal organs below the diaphragm more vigorously, he should come to the aforementioned exercises and, in addition to these, those that massage through the application of bindings. It is necessary for the one doing the massaging to be behind, applying both hands at one time to the left and at another to the right, the person being massaged inclining at the same time toward these. The largest of the movements of inspiration and phonation in some way move the internal organs below the diaphragm, as do the expirations, both those that occur alone and those that accompany a holding of breath in flute playing and phonation. And the holding of breath itself is an exercise no less of the muscles of the epigastrium or those of the thorax. But I shall speak of this again, for in addition to the movements mentioned just

151K

152K

---

[28] This is a distinction made by Galen earlier—see note 5 above.

ἄρτι καὶ ἄλλην οὐ σμικρὰν ἔχει χρείαν, ἧς ἕνεκεν ἐπὶ
τελευτῇ τῶν γυμνασίων αὐτὴν παραλαμβάνομεν.

ἐπὶ δὲ τὸ προκείμενον ἐπανιτέον, ὡς πολλαὶ τῶν
ἐγγινομένων κινήσεων τοῖς τοῦ ζῴου μορίοις οὔτ'
ἐνέργειαι τῶν μορίων αὐτῶν εἰσιν οὔτ' ἐνεργείαις ἀκολουθοῦσιν, ἀλλ' ὑφ' ἑτέρων τε καὶ δι' ἑτέρων ἀποτελοῦνται, ὡς ἐπί τε τῶν ὀχουμένων γίνονται καὶ πλεόν-
των καὶ τριβομένων καὶ καθαιρομένων ὑπὸ φαρμάκων
ἐμετηρίων τε καὶ ὑπηλάτων. ἀλλ' ἡ τοιαύτη κίνησις
οὔκ ἐστι τῆς ὑγιεινῆς πραγματείας, αἱ δ' ἄλλαι πᾶσαι, καὶ μάλισθ' αἱ διὰ τρίψεως ἀναγκαιόταται γινώσκεσθαι τοῖς ὑγιεινοῖς. τῆς γὰρ τούτων τέχνης ἐστὶν
153K ἁπασῶν τῶν κινήσεων ἐπίστασθαι τὰς δυνάμεις,
ὥσπερ, οἶμαι, τῶν τεχνιτῶν αὐτῶν ἁπάσας τὰς κατὰ
μέρος· ἐκεῖνοι μὲν γὰρ τῆς κατὰ τὴν ὕλην ποικιλίας,
ὁ γυμναστὴς δὲ τῆς δυνάμεως αὐτῶν ἔχει τὴν γνῶσιν.

εἰ γοῦν τίς μοι κελεύσειεν ὁπλομαχικὰς κινήσεις ἢ
ἕτερον διδάσκειν ἢ αὐτὸν εὐρύθμως κινεῖσθαι, οὐκ ἂν
δυναίμην καλῶς ἐνεργῆσαι μίαν ἐξ αὐτῶν τὴν ἐπι-
τυχοῦσαν· εἰ μέντοι παρείην τινὶ τῶν ὁπλομαχικῶν
ἐνεργοῦντι, καὶ τίνα δύναμιν ἑκάστη τῶν ἐνεργειῶν
ἔχει καὶ τί μάλιστα μόριον ἐκπονεῖ, πάντων ἂν ἐκείνων ἀκριβέστερον εἰδείην. μᾶλλον δ', εἰ χρὴ τἀληθὲς
εἰπεῖν, ὁ μὲν ὁπλομαχικὸς οὐδὲν ἂν εἴποι τῆς δυνάμεως αὐτῶν, ὁ δὲ τὴν περὶ τὰ γυμνάσια τέχνην ἐπιστάμενος ἀκριβῶς ἁπάσας διαγνώσεται πρὸς ὡρι-
σμένους σκοποὺς ἀναφέρων· ἢ γὰρ βίαιοί τινές εἰσι
καὶ βαρεῖαι καὶ εὔτονοι ἢ κοῦφαι καὶ ταχεῖαι καὶ σύν-

now, massage has another not insignificant use, for the sake of which we are taking it up at the end of the exercises.

I must return to what is before us since many of the movements arising in the parts of the animal are not functions of the parts themselves, nor do they follow functions, but are brought about by and through other things, such as those that occur in riding in carriages, sailing, massage and purification by emetic and purging medications. But such a movement is not a matter of hygiene, whereas all the others, and particularly those through massage, are very necessary for hygienists to know, for to know the 153K powers of all these movements is part of the art, just as, I think, for craftsmen themselves to know all the powers of the individual arts, for the latter have the knowledge of the variations of the material while the gymnastic trainer has the knowledge of their power.

Anyway, if someone were to order me to teach the movements of fighting with heavy arms, or something else, or to teach him to move in a coordinated manner, I would not be able to carry out properly whichever one of these it happened to be. If, however, I were to be present with someone carrying out fighting with heavy arms, I would know quite accurately in all cases what power each of the actions has, and what part especially it works out hard of all those. Much more, however, if one must speak the truth, the fighter with heavy arms might say nothing of the power of these, whereas the one who knows the art pertaining to exercise will recognize them all precisely, referring to defined objectives. Thus, some are violent, strong and vigorous, some are light, swift and intense, or violent

τόνοι ἢ βίαιοί τε ἅμα καὶ ὀξεῖαι. ταῦτα τε οὖν γνωρίσεις ῥᾳδίως ἐν αὐτῷ θεασάμενος γινόμενα καὶ πρὸς
τούτοις ἔτι, τίνες μὲν ἐνέργειαι σκέλη μᾶλλον ἢ χεῖ-
154K ρας ἢ θώρακα, τίνες δὲ ὀσφὺν ἢ κεφαλὴν ἢ ῥάχιν ἢ
γαστέρα, τίνες δ' ὁτιοῦν ἄλλο μέρος ὑπὲρ τἆλλα δια-
πονοῦσιν.

ὁ μὲν γὰρ ὁπλομαχικὸς εὐρύθμως μὲν ἐνεργήσει
κινήσεις ταχείας ἤ, εἰ οὕτως ἔτυχεν, εὐτόνους τε ἅμα
καὶ βαρείας, οὐ μὴν ὅτι γε πυκνοῦσιν καὶ ἰσχναίνου-
σιν αἱ τοιαῦται κινήσεις οἶδεν, ὥσπερ οὐδ' ὅτι σαρ-
κοῦσί τε καὶ ἀραιοῦσιν αἱ βραδύτεραι. κατὰ δὲ τὸν
αὐτὸν τρόπον ἐνεργήσει μέν ποτε κινήσεις εὐτόνους
καὶ βαρείας καὶ βραδείας, οὐ μὴν ὅτι γε ῥώμην αὗται
καὶ βάρος σώματος κατασκευάζουσιν οἶδεν. οὕτω δὲ
καὶ ὁ ἡνιοχικὸς ἐνεργήσει μὲν ἁπάσας τὰς κατὰ
μέρος ἐνεργείας εὐρυθμότατα, ἅμα δὲ καὶ τῇ χρείᾳ
συμφορώτατα, ποῖαι δ' αὐτῶν ἤτοι λεπτύνουσιν ἢ
σαρκοῦσιν ἢ ῥώμην ἢ συντονίαν ἀποτελοῦσιν, ἢ μαλακὸν ἢ σκληρὸν ἢ πυκνὸν ἢ ἀραιὸν ἐργάζονται τὸ
σῶμα, παντάπασιν ἀγνοεῖ. κατὰ δὲ τὸν αὐτὸν τρόπον
ὁ σφαιριστικὸς ἁπάσας μὲν ἐπίσταται τὰς τῆς σφαίρας βολάς τε καὶ λήψεις, οὐ μὴν ἥντινά γε διάθεσιν
ἑκάστη τῷ σώματι περιποιεῖ. οὕτω δὲ καὶ ὁ παιδοτρίβης ἁπασῶν μὲν τῶν κατὰ παλαίστραν ἐνεργειῶν
155K ἐπιστήμων ἐστίν, ὅ τι δὲ ἑκάστη πέφυκε δρᾶν ἀγνοεῖ.
καὶ συλλήβδην εἰπεῖν ἅπαντες ἄνθρωποι τεχνῖταί τε
καὶ ἄτεχνοι, διὰ τῶν σωμάτων ἐνεργοῦντες, ἀγνοοῦσι
τῶν ἐνεργειῶν τὰς δυνάμεις, ὀρχησταὶ ναυτίλοι τέκτο-

224

and rapid at the same time. You will recognize these easily
by actually seeing them occurring in him, and in addition
to these things, which actions particularly work the legs   154K
out hard, or the arms, or the chest, and which work out
the loins, head, spine, or abdomen, and which any part
more than others.

Thus the fighter with heavy arms carries out rapid
movements in a coordinated manner and, should it so happen, [movements that are] vigorous and strong. But he
doesn't know that such movements condense and reduce,
just as he doesn't know that the slower movements enflesh
and rarefy. In the same way, he will on occasion perform
movements that are vigorous, strong and slow without
knowing that these in fact make for strength and weight
in a body. In this way too, the charioteer will perform all
the actions individually in very coordinated fashion that
are at the same time also most suitable for use, but is altogether ignorant as to which kinds of these either thin,
enflesh, bring about strength or tension, or make the body
soft, hard, condensed or rarefied. In the same way, the
ballplayer knows all the throws and catches of the ball, but
not what condition each produces in the body. And in like
manner, the physical trainer is knowledgeable about all
the actions pertaining to the wrestling school, but doesn't   155K
know what each one naturally does. In summary, all men,
trained and untrained, when acting with their bodies, do
not know the powers of the actions, and this applies generally to all those who do anything whatsoever, whether they

νες ἁλιεῖς γεωργοὶ χαλκεῖς οἰκοδόμοι σκυτοτόμοι
πάντες ἁπλῶς οἱ ὁτιοῦν πράττοντες.

ἀλλ᾽ ὁ γυμναστικός, ἀφ᾽ ὧν εἶπον ὀλίγον ἔμπροσθεν ὁρμώμενος, εἰ καὶ νῦν πρῶτον εἴη θεώμενος
ἡντινοῦν ἐνέργειαν, οὐκ ἀγνοήσει τὴν δύναμιν αὐτῆς.
οἷον αὐτίκα τῶν ὀρχηστῶν αἱ σύντονοι κινήσεις, ἐν
αἷς ἅλλονταί τε μέγιστα καὶ περιδινοῦνται στρεφό-
μενοι τάχιστα καὶ ὀκλάσαντες ἐξανίστανται καὶ
προσσύρουσι καὶ διασύρουσι[32] καὶ διασχίζουσιν ἐπὶ
πλεῖστον τὰ σκέλη καὶ ἁπλῶς εἰπεῖν ἐν αἷς ὀξύτατα
κινοῦνται, λεπτὸν καὶ μυῶδες καὶ σκληρὸν καὶ πυκνὸν
ἔτι τε σύντονον ἀποτελοῦσι τὸ σῶμα. κατὰ δὲ τὰς
ἐκλύτους τε καὶ βραδείας καὶ μαλακὰς κινήσεις οὐ
μόνον οὐκ ἂν γένοιτο τὸ σῶμα τοιοῦτον, οἷον εἴρηται
νῦν, ἀλλ᾽ εἰ καὶ φύσει μυῶδές τε καὶ σύντονον ὑπάρχοι, τὴν ἐναντίαν ἀμείψει διάθεσιν. ὅπερ οὖν ὀλίγον
156K ἔμπροσθεν ἔλεγον, ὡς ὁ παιδοτρίβης ὑπηρέτης ἐστὶ
τοῦ γυμναστοῦ τοιοῦτος, οἷόσπερ ὁ μάγειρος τοῦ ἰατροῦ, τοῦτο καὶ νῦν ἐπιδέδεικται. σκευάζει γὰρ ὁ μάγειρος ἢ τεῦτλον ἢ φακῆν ἢ πτισάνην ἄλλοτε ἀλ-
λοίως, οὔτε δὲ τὸ σκευαζόμενον ὁποῖόν τι τὴν δύναμίν
ἐστιν ἐπιστάμενος οὔθ᾽ ἥτις τῶν σκευασιῶν ἡ βελτί-
στη· ὁ δ᾽ ἰατρὸς οὐδὲν μὲν τούτων ὁμοίως τῷ μαγείρῳ
παρασκευάσαι δυνατός ἐστιν, παντὸς δὲ τοῦ παρα-
σκευασθέντος ἐπίσταται τὴν δύναμιν.

12. Ὁ τοίνυν γυμναστὴς τοῦ προκειμένου νῦν ἐν τῷ
λόγῳ μειρακίου, τοῦ τὴν ἀρίστην ἔχοντος κατασκευήν, ἐπίσταται μὲν τῶν γυμνασίων ἁπάντων τὰς

be dancers, sailors, carpenters, fishermen, farmers, smiths, builders or cobblers.

But the gymnastic trainer, among those whom I began to speak about a little earlier, if now he is first aware of any action whatsoever, will not be ignorant of its power. For example, the vigorous movements of dancers, in which they jump very high and whirl around, turn very quickly and crouch down, rise up, drag along, disperse and separate their legs to a great degree, and to speak generally, move very quickly in their actions, make the body thin, muscular, hard, condensed and vigorous. However, in the relaxed, slow and gentle movements, not only would the body not become as I described just now, but even if it were naturally muscular and taut, it would change to the opposite condition. Therefore, as I was saying a little earlier, the physical trainer is the same sort of servant to the gymnast as the cook is to the doctor; and this I have now demonstrated. Thus the cook prepares beets, lentils and barley sometimes in one way and sometimes in another, not knowing what kind of preparation it is in terms of potency or what is the best of the preparations. The doctor, however, while he is not able to prepare these things like the cook, knows the potency of every preparation.

156K

12. Therefore, the gymnastic trainer of the young lad now proposed in the discussion—a boy having the best constitution—knows the powers of all exercises and

---

32 καὶ διασύρουσι *add.* Ko

δυνάμεις, ἐκλέγεται δὲ καθ᾽ ἕκαστον εἶδος τὰ σύμμετρά τε καὶ μέσα τῶν ἀμετριῶν ἑκατέρων. οὔτε γὰρ
ὀξέος οὔτε βραδέος ἡ ἀρίστη κατασκευὴ τοῦ σώματος, ἀλλὰ τοῦ συμμέτρου τε καὶ μέσου δεῖται γυμνασίου, κατά τε τὸν αὐτὸν λόγον οὔτε βιαίου καὶ σφο-
δροῦ οὔτ᾽ ἐκλύτου καὶ ἀμυδροῦ, ἀλλὰ κἀνταῦθα τὸ
σύμμετρον ἄριστον. οὐ γὰρ ὑπαλλάττειν προσήκει
τὴν ἀρίστην κατασκευὴν τοῦ σώματος, ἀλλὰ φυλάττειν.

157K εἶτ᾽ οὖν ἐν ὅπλοις ἐθέλοι γυμνάζεσθαι τὸ τοιοῦτον
μειράκιον ὁ γυμναστὴς αὐτοῦ, τὸν ἐμπειρότατον τῆς
ὕλης τῶν ὁπλομαχικῶν ἐνεργειῶν παραλαβὼν ἁπάσας αὐτῷ δειχθῆναι κελεύσει, κἄπειτ᾽ αὐτὸς ἐκλέξεται
καὶ διακρινεῖ καὶ προστάξει, κατὰ ποίας μὲν αὐτῶν
ἐπὶ πλέον χρὴ γυμνάζεσθαι, κατὰ ποίας δ᾽ ὀλιγάκις
ἢ συμμέτρως ἢ οὐδ᾽ ὅλως ἢ διαπαντός. οὐ γὰρ δύναται λαθεῖν αὐτὸν οὔθ᾽ ὅ τι μόριον ἑκάστη διαπονεῖ
μᾶλλον οὔθ᾽ ἥτις αὐτῆς ἐστιν ἡ ποιότης τε καὶ ἡ
δύναμις. εἰ δὲ διὰ σφαίρας ἐπιθυμήσειε γυμνάζεσθαι,
κἀνταῦθα πάλιν ἐξευρήσει τό τε εἶδος τῶν ἐνεργειῶν
καὶ τὴν ποιότητα καὶ τὸ μέτρον, ὑπηρέτην λαβὼν τὸν
σφαιριστικὸν ὑπὲρ τοῦ τὴν ὕλην ἅπασαν θεάσασθαι
τῶν ἐνεργειῶν. αὐτῆς μὲν γὰρ τῆς κατὰ μέρος ὕλης
ἐν ἑκάστῃ τῶν τεχνῶν ἄπειρός ἐστιν ὁ γυμναστής, ᾗ
γυμναστής ἐστιν· εἰ δὲ ἅπαξ αὐτὴν θεάσαιτο, τήν τε
ποιότητα καὶ τὴν δύναμιν αὐτίκα γνωρίσει. μυρίους
γοῦν ἡμεῖς ἀσθενέστερά τινα μέρη τοῦ σώματος
158K ἔχοντας, ὡς συνεχέστατα τοῖς κατ᾽ αὐτὰ νοσήμασιν

chooses from each kind, those that are moderate and midway between the excesses of each. For the best constitution of the body does not require exercise that is fast or
slow, but that which is moderate and intermediate. By the
same token, it requires neither strong and violent nor relaxed and light exercise, but here too moderate exercise is
best because it is not appropriate to change the best constitution of the body but to preserve it.

If, then, such a young lad should wish to exercise in   157K
arms, his gymnastic trainer, taking the one most practiced
in the range[29] of actions of fighting, will order all of them
to be demonstrated to him, then he himself will choose,
distinguish and assign according to their kinds, which of
them he must exercise with more and the kinds he must
exercise with rarely, moderately, not at all, or continually,
for it cannot escape him what part each works out more,
nor what the quality and potency of this is. If he is keen to
use exercise with a ball, even here again he will discover
the kind of actions, and the quality and measure, taking as
an assistant a ballplayer for the purpose of seeing the
whole range of the actions. This is because the gymnastic
trainer himself is ignorant of the range in each of the arts
individually. But if he were to look at the whole range all
at once, he would immediately recognize both the quality
and the potency. Anyway, I have restored strength in a
countless number of people who have certain parts of the
body weaker so as to very frequently be attacked by dis-   158K

[29] The translation of ὕλη is "range" (Soranus 1.46, 2.15) rather
than the more usual "matter," "material," or "stuff."

ἁλίσκεσθαι, διὰ μόνης γυμναστικῆς ἀνερρώσαμεν
οὐκ ἀπαγαγόντες ἀπὸ τῶν οἰκείων γυμνασίων, ἀλλ᾽,
εἴτ᾽ ὀρχηστικὸς ἦν ὁ ἄνθρωπος εἴθ᾽ ὁπλομαχικὸς εἴτε
παγκρατιαστικὸς εἴτε παλαιστρικὸς εἴθ᾽ ὁτιοῦν ἄλλο,
τὰς ἐν ἐκείνῃ τῇ τέχνῃ κινήσεις ἁπάσας αὐτὸν ἐπιτά-
ξαντες ἡμῶν παρόντων κινηθῆναι κἀξ αὐτῶν ἐκλεξά-
μενοι τὰς ἐπιτηδειοτάτας ἅμα καιρῷ τε καὶ μέτρῳ
προσετάττομεν ταύταις χρῆσθαι·[33] περὶ μὲν δὴ τῶν
τοιούτων ἐπανορθώσεων ἐπὶ προήκοντι τῷ λόγῳ δια-
λέξομαι κατ᾽ ἐκεῖνο τὸ μέρος τῆς πραγματείας, ἐν ᾧ
περὶ τῶν μοχθηρῶν κατασκευῶν τοῦ σώματος δι-
έξειμι.

νυνὶ δὲ τὸ μὲν ἄριστον σῶμα ἐπειδὴ[34] πρόκειται
φυλάττειν ἄριστον, ἐξ ἁπάντων οὖν ἐκλεκτέον αὐτῷ
τὸ σύμμετρον, ἐκ τρίψεων, ἐκ γυμνασίων, ἐκ λουτρῶν,
ἐκ τροφῶν, ἐξ ὕπνων, μήτε μαλακωτέραν αὐτοῦ τὴν
ἕξιν τοῦ σώματος ἐργαζομένους μήτε σκληροτέραν ἡ
μὲν γὰρ εὐνίκητος ὑπὸ τῶν ἔξωθεν αἰτίων, ἡ δὲ τὴν
αὔξησιν κωλύει μήτε πυκνοτέραν, ὡς ἴσχεσθαί τι
τῶν κατὰ τὴν σάρκα περιττωμάτων, μήτ᾽ ἀραιοτέραν,
ὡς ἀπορρεῖν τι καὶ τοῦ χρηστοῦ. κατὰ δὲ τὸν αὐτὸν
159K τρόπον οὐδ᾽ ἰσχνοτέραν ἑαυτῆς ποιητέον οὐδὲ παχυ-
τέραν, εἴπερ ἄριστα διέκειτο, γινώσκοντας, ὡς τὸ μὲν
ἰσχνότερον εὐεπηρέαστον ὑπὸ τῶν ἔξωθεν αἰτίων, τὸ
δὲ παχύτερον ὑπὸ τῶν ἔνδοθεν κἀξ αὐτοῦ τοῦ σώματος ὁρμωμένων. τί δεῖ λέγειν, ὡς οὐδὲ θερμότερον
αὐτὸν ἢ ψυχρότερον ἢ ξηρότερον ἢ ὑγρότερον ἀπο-
δεικτέον, εἴπερ ἀμέμπτως ἐκέκρατο; εἷς οὖν ἐπὶ τοῦ

eases in these parts, through exercise alone, without taking them away from their own exercises. But whether the man were to be a dancer, or a fighter with heavy weapons, or an all-in fighter (*pancratist*), or a wrestler, or whatever else, I would enjoin him to practice all the movements in that art, to be carried out in my presence and then I would choose from these the most suitable for use in time and measure. I shall discourse about such restorations in the discussion to come, in that part of the treatise in which I shall go over in detail the defective constitutions of the body.

For the present, since what lies before us is to maintain the best body as the best, one must choose for it moderation in all things—[this applies to] massage, exercises, baths, nutriments and sleep—making the state of the body itself neither softer nor harder, for the one is easily overcome by external causes while the other prevents growth, nor more dense, as this retains some of the superfluities in the flesh, nor more loose textured, as also some of what is useful flows away. And in the same way, we must not make it thinner or thicker (fatter) than it was, if it was in the best state, recognizing that what is thinner is exposed to harm from external causes while what is thicker (fatter) from those arising within from the body itself. What need is there to say that we must not make it warmer, colder, drier or moister, if it had been mixed faultlessly? There-

159K

---

33 προσετάττομεν ταύταις χρῆσθαι add. Ko
34 ἐπειδὴ add. Ko

GALEN

τοιούτου σώματος ὁ σκοπὸς ἐν ἁπάσαις ταῖς ὑγιει-
ναῖς ὕλαις τὸ σύμμετρόν τε καὶ μέτριον, ὅπερ ἀκριβῶς ἐστι μέσον ἑκατέρων τῶν ἀμετριῶν. ὡς δ' ἄν τις
μάλιστα τοῦ σκοποῦ τυγχάνοι, λέλεκται μὲν ἤδη καὶ
πρόσθεν, ἀλλ' οὐδὲν χεῖρον ὑπὲρ τῶν ἀναγκαιοτάτων
ἀναμιμνήσκειν πολλάκις.

ἡ μὲν γὰρ πρώτη τῶν ἡμερῶν πλατὺν ἔχει τὸν
στοχασμόν, ἡ δευτέρα δὲ καὶ ἡ τρίτη καὶ ἡ τετάρτη
καὶ τῶν λοιπῶν ἑκάστη κατὰ τὸ ἑξῆς ἀκριβέστερον.
ἐν μὲν γὰρ τῇ πρώτῃ διὰ τῶν εἰρημένων σκοπῶν
ἅπαντά σοι πραττέσθω. ἀποδυέσθω μὲν ἐπὶ πεπεμμέ-
νοις ἀκριβῶς τοῖς οὔροις, ὡς Αἰγίμιος ἐκέλευσεν.
160K ἐφεξῆς δὲ τῆς μὲν ἀνατρίψεως ὁ σκοπός, ὡς μαλαχθῆναι τὰ μόρια· δηλώσει δὲ τό τε ἐπιτρέχον ἄνθος
αὐτοῖς καὶ τὸ ῥᾳδίως ἐκμαλάττεσθαι τὰ κῶλα καὶ τὸ
πρὸς τὰς κινήσεις ἁπάσας ἑτοίμως ἔχειν. μετὰ ταῦτα
δὲ ἤδη γυμναζέσθω, μέχρις ἂν εἰς ὄγκον αἴρηται τὸ
σῶμα καὶ εὐανθὲς ὑπάρχῃ καὶ αἱ κινήσεις ἕτοιμοί τε
καὶ ὁμαλαὶ καὶ εὔρυθμοι γίνωνται. ἐν τούτῳ δὲ καὶ
ἱδρῶτα θεάσῃ θερμὸν ἀτμῷ συμμιγῆ. παύεσθαι δὲ
τηνικαῦτα πρῶτον, ἐπειδὰν ἔν τι τῶν εἰρημένων ἀλ-
λοιωθῇ. καὶ γὰρ εἰ φανείη συστελλόμενος ὁ τοῦ σώματος ὄγκος, αὐτίκα παύειν τὸ μειράκιον· εἰ γὰρ ἐπὶ
πλέον γυμνάζοις, ἐκκενώσεις τι καὶ τῶν χρηστῶν,
ὥστ' ἰσχνότερον ἀποδείξεις τὸ σῶμα καὶ ξηρότερον
καὶ ἀναυξέστερον.

ὡσαύτως δὲ καὶ εἰ τὸ τῆς χρόας εὐανθὲς μαραί-

232

fore, in the case of such a body, the one objective in all the healthy materials is moderation and balance, which is precisely in the middle of each of the extremes. How someone might best attain the objective has already been stated previously, but it is no bad thing to call to mind frequently those things that are most essential.

The first day provides a broad estimate, while the second, third, fourth, and each of the remaining days to follow allow greater precision. Thus, on the first day, you should do everything by way of the previously mentioned objectives. Let the person strip off when the urine is entirely concocted, as Aegimius[30] directed. Next, the aim of   160K the rubbing is that the parts be softened. The blush running over them will show this, as will the limbs being easily relaxed and ready for all movements. After this, let the person now be exercised up to the point where the body is raised to a swelling and a blush exists, and the movements become ready, even and well-coordinated. In this you will also see warm sweat mixed with vapor. Under these circumstances, first cease whenever there is a change in one of the aforementioned things. And if the swelling of the body seems to you to be reducing, you should stop the young lad immediately, for if you exercise him more, you will also empty out some of what is useful, so that you will make the body thinner, drier and less likely to grow.

In like manner too, if the bloom of color dies away,

[30] Aegimius (Aigimios of Elea, fl. 325–300 BC) was a Greek doctor credited by Galen with a work on the pulses (*Diff. Puls.*, VIII.498 and 751–52K). He is said to have attributed disease to foods and superfluities (*Londonensis Medicus* 13.2–14.3). See also Manetti, EANS, 47–48.

νοιτο, παύεσθαι· καὶ γὰρ καταψύξεις τὸ σῶμα καὶ
διαφορήσεις ἐπὶ πλέον, εἰ γυμνάζοις ἔτι. καὶ μὲν δὴ
καὶ τὸ τῶν κινήσεων ἕτοιμον ἢ εὔρυθμον ἢ ὁμαλὲς
ἐπειδὰν ἐνδιδόναι που φαίνηται καὶ ὀκλάζειν κατά τι,
161K παύειν αὐτίκα, καὶ εἰ περὶ τὸν ἱδρῶτα γίνοιτό τις ἢ
κατὰ τὸ πλῆθος ἢ κατὰ τὴν ποιότητα μεταβολή.
πλείονα γὰρ αὐτὸν ἀεὶ καὶ μᾶλλον χρὴ γίνεσθαι καὶ
θερμότερον, ἐς ὅσον ἂν αἱ κινήσεις ἀνάγωνται πρὸς
τὸ σφοδρότερον. ὅταν οὖν ἢ ἐλάττων ἢ ψυχρότερος
γένηται, διαφορεῖται ἤδη τὸ σῶμα καὶ ψύχεται καὶ
ἐκλύεται³⁵ καὶ ξηραίνεται περαιτέρω τοῦ προσήκοντος. ἀκριβῶς οὖν προσέχειν τὸν νοῦν τῷ γυμναζομένῳ σώματι καὶ διαπαύειν εὐθέως, ἐπειδὰν προ-
φαίνηταί τι τῶν εἰρημένων σημείων, οὐ μὴν αὐτίκα
γε ἀπολύειν λουσόμενον, ἀλλὰ τῆς μὲν ἀκμῆς τῶν
γυμνασίων ἐπισχεῖν καὶ στῆναι κελεῦσαι, καὶ εἰ βου-
ληθείης μετὰ κατοχῆς πνεύματος πληρῶσαι τὸν λα-
γόνα, περιχέοντα τοὔλαιον ἀποθεραπεύειν τοὐντεῦθεν.
οἷον δέ τι πρᾶγμά ἐστιν ἡ ἀποθεραπεία καὶ τίνες ἐν
αὐτῇ σκοποὶ καὶ τί τὸ μέτρον, ἐξ ὧν συμπληροῦνται
κινήσεών τε καὶ τρίψεων, ἐν τῷ μετὰ ταῦτα λόγῳ δηλωθήσεται.

νυνὶ γάρ μοι δοκῶ καταπαύειν ἤδη τὸν ἐνεστῶτα
λόγον³⁶ αὔταρκες ἔχοντα μέτρον, ἐκεῖνο μόνον ἔτι
προσθείς, ὡς καὶ τὸ λουτρὸν τὸ ἀκριβῶς εὔκρατον ἐπὶ
162K τῆς προκειμένης ἡλικίας τε καὶ φύσεως ἐκλέγεσθαι
προσήκει. προύκειτο δ᾽, εἴ τι μεμνήμεθα,³⁷ τρίτη τις
ἑβδομὰς ἐτῶν ἀπὸ τῆς γενετῆς, τουτέστιν ἡ μετὰ ⟨τὸ⟩

stop, for you will cool the body and cause more sweating, if you continue to exercise [him]. Furthermore, whenever the readiness, coordination or evenness of the movements appears to give way or abate somewhat, stop immediately, and also if some change occurs in either the amount or quality of the sweats. For this latter must always become more abundant and warmer to the degree that the movements are carried toward a greater vigor. Therefore, whenever it becomes less or colder, the body is already dissipated and is cooled, loosened and dried more than is appropriate. So direct your attention completely to the exercising body and immediately stop when one of the aforementioned signs appears. Do not, however, release him it to bathe at once, but prevent the completion of the exercises and direct him to stop, and if you should wish to fill the flanks with retention of *pneuma,* after pouring on oil, apply apotherapy thereafter. What kind of thing apotherapy is, what objectives are in it, what its measure is, and by what movements and massage they are accomplished will be shown in the book following this one.

161K

Now it seems to me I should stop the present book which has reached a sufficient length, although I shall add this one further thing only: the bath should also be exactly *eukratic,* chosen appropriately for the age and nature under consideration. What is before us, if I remember rightly, is someone in the third of the seven-year periods from

162K

---

35 καὶ ἐκλύεται *add.* Ko

36 λόγον *add.* Ko

37 si recte meminimus L

τεσσαρεσκαιδέκατον ἔτος ἡλικία μέχρι ⟨τοῦ⟩ πρώτου
καὶ εἰκοστοῦ, καθ' ἣν οὐδέπω κελεύω ψυχρολουτεῖν τὸ
μειράκιον, ἵν' ἐπὶ πλεῖστον αὔξοιτο. τελειωθέντος δ'
αὐτοῦ κατὰ τὸ μέγεθος, ἐπισκεψώμεθά τι καὶ περὶ τῆς
ψυχρολουσίας. ὡσαύτως δὲ καὶ περὶ τῶν ἐπὶ κόνει
γυμνασίων ἀκριβέστερον ἐπισκεψώμεθα κατὰ τὸν
ἑξῆς λόγον. ἐν δέ γε τῷ παρόντι τοῦτο εἰπεῖν ἀρκέσει,
τὸ μηδὲ κόνεως χρήζειν τὸ μειράκιον, εἰ τὸν οἶκον, ἐν
ᾧ γυμνάζεται, καθ' ὃν ἐν ἀρχῇ τρόπον ἐκέλευσα
παρεσκευασμένον ἔχοι. εἰ δέ γε θερμότερος εἴη ποτέ,
καὶ κόνει χρηστέον. εἰ μὲν δὴ κονίσαιτο, πάντως λου-
στέον· εἰ δὲ μὴ κόνει χρήσαιτο, δυνατὸν καὶ μὴ λε-
λοῦσθαι, καὶ μάλιστα χειμῶνος.

ἐφεξῆς δὲ ὅσα περὶ τροφὴν ἢ ποτὸν ἢ ὕπνον ἢ
περίπατον ἐστοχάσθαι χρὴ τὸν ἐπιστατοῦντα τοῦ
μειρακίου, λεχθήσεται μὲν ἐν τοῖς ἑξῆς· ἀδύνατον δ'
163K ἀκριβῶς τυχεῖν αὐτῶν ἐν τῇ πρώτῃ τῶν ἡμερῶν· ἀλλὰ
κατὰ τὴν δευτέραν, ἐπὶ πόσοις γυμνασίοις ὅπως διαι-
τηθῇ γινώσκων, εἰ μὲν ἀκριβῶς φαίνοιτο διαφυλάττων τὴν ἑαυτοῦ φύσιν, ἐν τοῖς αὐτοῖς μέτροις διαιτή-
σει τε καὶ γυμνάσει, μὴ φυλάττοντα δέ, καθότι ἂν
ἐξίστηται τῶν ἀρχαίων, ἐπανάγειν πειράσεται μετα-
βάλλων τὰ μέτρα. καὶ τοῦτ' οὐ παύσεται ποιῶν, ἄχριπερ ἂν ἐφ' ἑκάστῳ τῶν πραττομένων ἀκριβῶς ὁρίσῃ
τὸ μέτρον. ὁπόσαι δ' εἰσὶ καὶ τίνες αἱ εἰς τὸ παρὰ
φύσιν ἐκτροπαὶ καὶ πῶς ἑκάστην χρὴ διαγινώσκειν
τε καὶ θεραπεύειν, ὁ ἐφεξῆς λόγος ἐξηγήσεται.

236

birth—that is to say, the age between fourteen and twenty-one—in which I do not yet direct the young lad to take a cold bath, so that he may grow to the maximum. If, however, he has grown to full size, let us also consider cold baths. In similar manner too, let us give more precise consideration to exercises with powder in the next book. In the present book it will be enough to say this: the young lad has no need of powder if the house in which he is exercising is prepared in the manner I directed it to be prepared at the beginning. If, however, it is too warm at some time, we must use powder also. If he is powdered, we must by all means bathe him. If he doesn't use powder, it is also possible not to bathe, and especially in winter.

Next in order, I shall speak in what follows about those things regarding nutriment, drink, sleep and ambulation with which the guardian of the young lad must concern himself. It is impossible to hit the mark accurately regarding these on the first day, but on the second day, recognizing after how much exercise he has been fed, if he is obviously preserving his own nature exactly, he will diet and carry out hard exercise within these limits. However, if he is not preserving his previous nature, depending on how he may deviate from the original circumstances, he will attempt restoration by changing the measures. And he will not stop doing this until he determines the measure accurately in each of the things he is doing. The next book will expound on how many and what the deviations contrary to nature are, and how we must diagnose and treat each one.

163K

Γ

1. Τῶν γραψάντων ὑγιεινὰς πραγματείας ἰατρῶν τε
καὶ γυμναστῶν ἔνιοι μὲν ἅπασιν ἀνθρώποις κοινάς
τινας ὑποθήκας ἐποιήσαντο, μηδὲν νοήσαντες ὅλως
ὑπὲρ τῆς κατ᾽ εἶδος ἐν τοῖς σώμασιν ἡμῶν διαφορᾶς·
ἔνιοι δ᾽, ὅτι μὲν οὐ μικρῷ τινι διαφέρομεν ἀλλήλων,
ἐδήλωσαν, ὡς δ᾽ ἀδυνάτου ὄντος ἁπάσας ἐπελθεῖν τὰς
διαφορὰς ἑκόντες παρέλιπον· ὀλίγοι δέ τινες εἴδεσί τε
καὶ γένεσιν ἐπιχειρήσαντες ἀφορίσασθαι πλέονα δι-
ήμαρτον ὧν κατώρθωσαν. ἡμεῖς δὲ κατὰ τὸ πρῶτον
εὐθέως βιβλίον ἐδηλώσαμεν, ὁπόσα τὰ πάντα ἐστὶν
εἴδη τῶν ἀνθρωπίνων σωμάτων, ἐπηγγειλάμεθά τε
καθ᾽ ἕκαστον ἰδίᾳ γράφειν ὑποθήκας ὑγιεινάς.

ἠρξάμεθα δ᾽ ἀπὸ τοῦ τὴν κατασκευὴν ἄμεμπτον
ἔχοντος. ἐπεὶ δὲ καὶ ὁ τοιοῦτος ἄνθρωπος ἐν περιστάσεσι πραγμάτων ἐνίοτε γινόμενος ἢ ἑκὼν ἢ ἄκων
ἐμποδίζεται κατὰ τὰ προστάγματα τῆς ὑγιεινῆς τέχνης διαιτᾶσθαι, κάλλιον ἔδοξεν ὑποθέσθαι πρῶτον
αὐτὸν ἐλεύθερον ἀκριβῶς, αὐτῇ μόνῃ τῇ τοῦ σώματος ὑγείᾳ σχολάζοντα. ὅπως μὲν οὖν ἀνατρέφεσθαι
χρὴ τὸν τοιοῦτον ἄνθρωπον, ὁ πρῶτος λόγος ἐδίδαξε[1]
μέχρι τῆς τεσσαρεσκαιδεκαέτιδος ἡλικίας ἐκτείνας

BOOK III

1. Of the doctors and gymnastic trainers who have written 164K
on matters of hygiene, some have devised certain instructions common to all men, not giving any consideration at
all to the differences in kinds of our bodies, whereas some
have declared that we differ from each other to no small
degree. However, as it is impossible to go over all the differences, I have deliberately left them out. Those few who
have attempted to make distinctions in kinds and classes
have erred more than they have been correct. I showed
right from the first book, how many kinds of human bodies 165K
there are in all and promised I would write instructions on
hygiene for each one individually.

I began from the person who has a faultless constitution. And although such a man is sometimes involved in
states of affairs, whether voluntarily or involuntarily, and
is prevented from following a regimen in accordance with
the dictates of the art of hygiene, it seemed better to assume first that he was completely free to devote his time
solely to the actual health of the body. The first book
taught how such a man must be reared, extending his care
up to the fourteenth year of age. How he should be

---

1 ἐδίδαξε Ko; ἔδειξε, Ku

αὐτοῦ τὴν ἐπιμέλειαν· ὅπως δ' ἀνδροῦσθαι, διὰ τοῦ
δευτέρου γράμματος ἐδηλώσαμεν. ἐμηκύνθη δ' ὁ λόγος εἰς κοινὰ κεφάλαια τῆς ὑγιεινῆς πραγματείας
ἀφικόμενος, ὧν χωρὶς οὐχ οἷόν τ' ἦν οὐδὲ περὶ τῆς
ὑποκειμένης ἡλικίας τε καὶ φύσεως ἀκριβῶς διελθεῖν.
ἁπάσας γὰρ ἐξηριθμησάμεθα τάς τε τῶν τρίψεων καὶ
τῶν γυμνασίων διαφοράς, οὐ μὰ Δία τὰς κατὰ μέρος,
ὅτι μὴ πάρεργον ἕνεκα παραδείγματος εἰς χρείαν ἐλ-
166K θούσας, ἀλλὰ τάς γε ἐν εἴδεσί τε καὶ γένεσιν ἀφωρι-
σμένας, ἐν μὲν ταῖς τρίψεσι τὴν σκληρὰν καὶ τὴν
μαλακὴν καὶ πρὸ τούτων γε τὴν σύμμετρον ἥντινα
δύναμιν ἔχουσιν ἐξηγησάμενοι, προσθέντες δ' αὐταῖς
τὰς κατὰ τὸ ποσὸν διαφοράς, τρεῖς οὔσας καὶ αὐτάς,
εἶτα κατὰ συζυγίαν ἐννέα τὰς πάσας ἐπιδείξαντες,
ἑκάστης τε τὴν δύναμιν εἰπόντες.

ἐν δὲ τοῖς γυμνασίοις τίνα μὲν ὀξέα τε καὶ ταχέα
προσαγορεύομεν, τίνα δ' ἀμβλέα τε καὶ βραδέα, καὶ
τίνα τούτων μέσα τε καὶ πρῶτα κατά γε τὴν φύσιν
ὑπάρχοντα καὶ περὶ τῶν εὐτόνων καὶ μαλακῶν καὶ
βαρέων δὴ καὶ κούφων ὁμοίως, ὅσα τε τούτων ἐστὶ
τὰ μέσα διελθόντες ἐδείξαμεν, ὡς χρὴ τὴν ἀρίστην
κατασκευὴν ἐν ἅπασι τοῖς μετρίοις τε καὶ συμμέτροις,
ἃ δὴ καὶ μέσα τῶν ἀμέτρων ἐστί, διαιτᾶσθαι κατά τε
τρίψεις καὶ γυμνάσια καὶ λουτρὰ καὶ τροφὰς ὅσα τ'
ἄλλα συμπληροῖ τὴν ὑγιεινὴν δίαιταν. ὑπεσχόμεθα
δέ, καθάπερ ἐπὶ τρίψεών τε καὶ γυμνασίων ἐποιήσα-
μεν, εἰς εἴδη τινὰ κοινὰ τὴν θεωρίαν ἀναγαγόντες,
167K ὥστε εὐμνημόνευτόν τε ἅμα καὶ μεθοδικὴν εἶναι τὴν

brought to manhood, I showed in the second book. This book was extended because it had to come to the general headings of the matter of hygiene, without which it was not possible to go over the assumed age and nature precisely. I enumerated all the differences of massages and exercises, although not, by Zeus, individually, in that they did not come to be useful in a subordinate manner for the sake of exemplification, but for the differentiation into kinds and classes. In the massages I explained the hard and soft and preferable to these, the moderate, and what potency they have. In addition to these, I explained the differences in terms of quantity, there being three of these. I then showed that, in respect of conjunction, there are nine combinations in all, and discussed the potency of each.

166K

In exercises, I termed some sharp and quick, some gentle and slow, some the means of both of these, having primary features in accord with nature, and similarly for the vigorous, gentle, heavy and light. Going through the means of these, I showed that in respect of the best constitution, in all cases the mean and moderate forms, which are the means of the extremes, must be administered in massages, exercises, baths and nutriments, and those other things that complete the hygienic regimen. And I undertook, just as I did in the cases of massages and exercises, to refer the theory to certain common kinds, so that the teaching is easy to remember and at the same time

167K

διδασκαλίαν, οὕτω κἀπὶ τῶν ἄλλων ἁπάντων ποι-
ῆσαι· καὶ πρῶτόν γε περὶ τῆς καλουμένης ἀποθεραπείας, ἐπειδὴ τοῖς εἰρημένοις[2] ἔμπροσθεν ἐφεξῆς
ἐτέτακτο· δηλώσαντες γάρ, εἰς ὅσον ἐπιτείνειν τε καὶ
παραύξειν χρὴ τὰ γυμνάσια πρὸς τὸ τέλειον, ὡς ἐν
ὑγείας λόγῳ, ἀκολουθεῖν ἔφαμεν αὐτοῖς τὴν καλουμένην ἀποθεραπείαν, ὑπὲρ ἧς ἤδη λέγομεν.

2. Ἓν μὲν καὶ πρῶτον, ὡς διττὴ κατὰ γένος ἐστί·
ἡ μέν τις ὡς μέρος, ἡ δ' ὡς εἶδος γυμνασίου. περὶ μὲν
δὴ τῆς ὡς εἶδος ἐξῆς ἐροῦμεν, περὶ δὲ τῆς ὡς μέρος
ἤδη λέγομεν. ἅπαντος γυμνασίου καλῶς ἐπιτελουμέ-
νου τὸ τελευταῖον μέρος ἀποθεραπεία καλεῖται· δύο δ'
αὐτῆς οἱ σκοποί, κενῶσαί τε τὰ περιττώματα καὶ ἄκοπον φυλάξαι τὸ σῶμα. κοινὸς μὲν οὖν ὁ πρότερος[3]
ὅλῳ τῷ γυμνασίῳ.[4] καὶ γὰρ κἀκείνου δύο τοὺς πάντας
168K ἐλέγομεν εἶναι σκοπούς, ἐπιρρῶσαί τε τὰ στερεὰ μόρια τοῦ ζῴου καὶ κενῶσαι τὰ περιττώματα.

ὁ δ' ἴδιος τῆς ἀποθεραπείας σκοπὸς ἐνστῆναί τε
καὶ διακωλῦσαι τοὺς εἰωθότας ἐπιγίνεσθαι τοῖς ἀμε-
τροτέροις γυμνασίοις κόπους. ἐπὶ μὲν οὖν τῶν ἀθλητῶν[5] καὶ τῶν ὁτιοῦν ἔργον ἀναγκαῖον ἐν τῷ βίῳ διαπραττομένων, οἷον ἤτοι σκαπτόντων ἢ ὁδοιπορούντων

---

[2] τοῖς εἰρημένοις add. Ko

[3] post πρότερος; σκοπὸς (Ku) om.   [4] post τῷ γυμνασίῳ: τῶν ἀθλητῶν τε καὶ ὁτιοῦν ἔργον ἀναγκαῖον ἐν τῷ βίῳ
διαπραττομένων, οἷον ἤτοι σκαπτόντων Ku   [5] post τῶν
ἀθλητῶν: καὶ τῶν ὁτιοῦν ἔργον ἀναγκαῖον ἐν τῷ βίῳ διαπραττομένων om. Ku—see note 4 above.

methodical, and to do likewise in all the other things. First, regarding what is termed "apotherapy,"[1] since order had been brought to what was previously said, and having shown how much one must extend and increase the exercises toward the goal, as in the discussion on hygiene, I said so-called apotherapy follows these, so let me speak about this now.

2. One thing is primary; apotherapy is twofold in terms of class—it is both a part of exercise and a kind of exercise. About apotherapy as a kind of exercise, I shall speak in due course: about apotherapy as a part of exercise, let me speak now. Apotherapy is called the final part of all exercise properly completed. Its two aims are to evacuate the superfluities and to keep the body free of fatigue. The first aim is common to all exercise, both for athletes and for those who perform any necessary work whatever in their lives, as for example diggers.[2] I also said there are two objectives of exercise in all: to strengthen the solid parts 168K of the organism and to evacuate the superfluities.

The specific aim of apotherapy is to resist and prevent the customary fatigues supervening in the more immoderate exercises. Thus, in the case of athletes and those who do any kind of work necessary in their lives, like those who

[1] Essentially, this term, no longer in medical use, means "restorative therapy" in a medical context, although the primary meaning is given as "regular worship of the gods."

[2] The Kühn text is followed here. The additional material is present in Linacre's Latin translation.

ἢ ἐρεσσόντων ἤ τι τοιοῦτον διαπραττόντων, ἑτοιμότε-
ρον οἱ κόποι συνίστανται, πλὴν εἴ τις ἀποθεραπείᾳ
χρῷτο· κατὰ δὲ τὸ προκείμενον ἐν τῷ νῦν λόγῳ σῶμα,
τὸ κάλλιστά τε κατεσκευασμένον ἀπηλλαγμένον τε
δουλείας ἁπάσης, ὡς μόνῃ σχολάζειν ὑγείᾳ, σπάνιος
ἡ τοῦ κόπου γένεσις. ὥσπερ γὰρ οὐδ' οἱ πλεῖστα πο-
νοῦντες ἀθληταὶ κατ' ἄλλο τι γυμνάσιον ἐφεδρεύοντα
κόπον ἔχουσι, πλὴν τὸ καλούμενον ὑπ' αὐτῶν τέλειον,
οὕτως οὐδ' οἱ βίον ἐλευθέριον ζῶντες ὑγείας μόνης
ἕνεκα γυμναζόμενοι κοπωθήσονταί ποτε διὰ τὸ μηδ'
εἰς ἀνάγκην ἀφικνεῖσθαι τοῦ τοιούτου γυμνασίου.
τοῖς μὲν γὰρ ἀθληταῖς ἀναγκαῖόν ἐστιν, ὡς ἂν παρα-
σκευάζωσι τὰ σώματα πρὸς τοὺς ἐν τοῖς ἄθλοις
πόνους ἀμέτρους ἔσθ' ὅτε καὶ δι' ὅλης ἡμέρας γινο-
169K μένους, γυμνάζεσθαί ποτε τὸ τελεώτατον ἐκεῖνο γυμνάσιον, ὃ δὴ καὶ κατασκευὴν ὀνομάζουσι. τοῖς δ'
ὑγείας μόνης ἕνεκα γυμναζομένοις οὔτ' ἀναγκαῖον
οὔτε χρήσιμον ὅλως ἐστὶν εἰς ὑπερβάλλοντας ἄγε-
σθαι πόνους, ὥστ' οὐδεὶς φόβος ἁλῶναι κόποις. ἀλλ'
ὅμως ἀποθεραπεύειν αὐτῶν χρὴ τὰ σώματα, κἂν εἰ μὴ
διὰ κόπου προσδοκίαν, ἀλλά τοι τοῦ κενῶσαι ἕνεκα
τὰ περιττώματα. προσγίνεται δ' ἐξ ἐπιμέτρου τῷδε
καὶ ἡ πρὸς τὸν κόπον ἀσφάλεια.

καὶ γὰρ εἰ καὶ ὅτι μάλιστα τὸ μέτρον αὐτοῖς τῶν
γυμνασίων ἄκοπόν ἐστιν καὶ ὁ τῆς ἐνεργείας τρόπος
ἀβίαστος, ἐνδέχεταί ποτε λαθεῖν τὸν γυμναστὴν ἐν
ἑκατέρῳ τι σμικρόν, ὃ παροφθὲν ἐργάσεταί τινα, κἂν
μὴ μέγαν, ἀλλὰ βραχύν γε τῷ γυμναζομένῳ τὸν κό-

dig, walk or row, or do some other such thing, the fatigues more readily arise unless the person uses apotherapy. However, in the body proposed in the present discussion—perfectly constituted and removed from all servitude so as to have time for health alone—the creation of fatigue is rare. For just as not even athletes who work the most have any other kind of exercise which has fatigue lying in wait, except what is called by them the goal, so those living a free life and exercising for the sake of health alone will never at any time be fatigued because they do not come to the point of necessity of such exercise. It is essential for athletes, so their bodies might be prepared for the excessive exertions in competitions that sometimes last a whole day, to exercise at times to the most complete level, which they also call preparation. However, for those exercising for the sake of health alone, it is neither necessary nor at all useful to lead themselves into excessive exertions, so there is no fear of being overcome by fatigues. Nonetheless, apotherapy is necessary for their bodies, even if there is no expectation of fatigue; it is for the purpose of evacuating superfluities. To this is added, for extra measure, safety with respect to fatigue.

169K

And even if the measure of the exercises is for them particularly free from fatigue and the manner of the activity without violence, it is sometimes possible for the gymnastic trainer to overlook some small matter in each respect, and for what is neglected to have some effect, even if it is a small rather than large fatigue in the one exercis-

πον. οὐ μὴν προσήκει τὸν ἑαυτῷ ζῶντα καὶ μόνῃ
σχολάζοντα τῇ τοῦ σώματος ὑγείᾳ βλάπτεσθαί ποτε
οὐδὲ τὸ σμικρότατον. ἀσφαλέστατον οὖν ἀποθεραπείᾳ χρῆσθαι διαπαντός. ὁποίαν δέ τινα ποιητέον αὐτήν, ἡ τῶν σκοπῶν φύσις ἐνδείξεται.

170K   ἐπειδὴ γὰρ πρόκειται τῶν ἐν τοῖς στερεοῖς τοῦ ζῴου
μέρεσι περιττωμάτων, ὅσα θερμανθέντα καὶ λεπτυν-
θέντα πρὸς τῶν γυμνασίων ἔτι μένει κατὰ τὸ σῶμα,
κένωσιν ἀκριβῆ ποιήσασθαι, χρὴ δήπου τάς τε δι'
ἑτέρων ἀνατρίψεις παραλαμβάνεσθαι μετὰ τοῦ συν-
τείνειν τὰ τριβόμενα μόρια καὶ πρὸς τούτοις ἔτι τὴν
καλουμένην τοῦ πνεύματος κατάληψιν. ἐπεὶ δ' οὔτε
τῶν τρίψεων ἓν ἁπασῶν ἐστιν εἶδος οὔτε τῆς τοῦ
πνεύματος καταλήψεως, ἄμεινον ἐκλέξασθαι τὸ χρησιμώτατον ἐξ ἑκατέρου. τῶν μὲν δὴ τρίψεων αἱ σκλη-
ραὶ συνδεῖν ἐδείκνυντο, τουτέστι πυκνότερόν τε ἅμα
καὶ σκληρότερον ἀποφαίνειν τὸ σῶμα· ὥστ' οὐκ ἂν
ἁρμόττοιεν αἱ τοιαῦται τοῖς παροῦσιν, εἴ γε δὴ τὸ μὲν
πυκνούμενον ἐντὸς ἑαυτοῦ στέγει, τὸ δ' ἀραιούμενον
ἐπιτρέπει διαρρεῖν τοῖς περιττοῖς. οὕτω δὲ καὶ τὸ μὲν
σκληρύνεσθαι τοῖς συντεταμένοις ἐναντιώτατον (αὐ-
ξάνει γὰρ αὐτῶν τὴν διάθεσιν), τὸ δὲ μαλάττεσθαι
χρησιμώτατον. εἴπερ οὖν ἅμα τε διαφορεῖν χρὴ τὰ
περιττώματα καὶ μαλάττειν τὰ συντεταμένα, τὰς
σκληρὰς τῶν τρίψεων φευκτέον. οὐδὲν δ' ἧττον, οἶμαι,

---

[3] The main account of this maneuver and its application is

ing. It is not appropriate for the person living for himself and having leisure time for the health of the body alone to be harmed to the slightest degree at any time. It is therefore safest to use apotherapy throughout. The nature of the objectives will show what kind of apotherapy must be carried out.

Since what is proposed is to effect complete evacuation   170K
of the superfluities in the solid parts of the organism, such as those which are heated and thinned by the exercises and still remain in the body, it is obviously necessary that massages by others should be undertaken, along with stretching of the massaged parts, and in addition to these things, the so-called suppression of the breath (*pneuma*).[3] Since the kind is not the same in all the massages, nor in the suppression of the breath (*pneuma*), it is better to choose the most useful of each. The hard massages were already shown to bind—that is to say, to render the body more condensed (with small pores) and at the same time harder—so that such massages would not be suitable for present purposes, if in fact that which is made dense retains within itself what is superfluous, whereas what is loose textured allows the superfluities to flow through. In this way too, to be hardened is most inimical to what is drawn tight (for it increases the condition of these), whereas to be softened is most useful. Therefore, if it is necessary at the same time to dissipate the superfluities and soften those things that are drawn tight, one must avoid the hard massages. No less, I think, must one avoid

here, although it is also described in *UPart.* (III.582K; May, *On the Usefulness,* 1.358–59). Its use in clearing superfluities is mentioned in *Diffic. Resp.,* VII.941K.

171K   φευκτέον ἐστὶ καὶ τὰς βραδείας. ἐπειδὴ γὰρ οὐκέτ᾽ ἐξ
ἑαυτοῦ κινεῖται τὸ σῶμα, κίνδυνος αὐτῷ ψυχθῆναί τε
καὶ πυκνωθῆναι, μηδὲν ἐπικούρημα θερμαῖνον ἔξωθεν
προσλαβόντι.

διὰ ταῦτ᾽ οὖν οὐ μόνον εἰς τάχος χρὴ τρίβειν, ἀλλὰ
καὶ πολλαῖς χερσίν, ἵν᾽ ὡς οἷόν τ᾽ ἐστὶ μάλιστα μηδὲν
ᾖ μέρος τοῦ τριβομένου γυμνόν. ἀλλ᾽ εἴπερ μήτε βρα-
δεῖαν εἶναι προσήκει τὴν τρίψιν μήτε σκληράν, ἔλαιον
δαψιλὲς χρὴ περικεχύσθαι τῷ τριβομένῳ σώματι· καὶ
γὰρ εἰς τάχος τοῦτο καὶ εἰς μαλακότητα τῇ τρίψει
συντελεῖ, καὶ πρόσεστιν αὐτῷ τι καὶ ἄλλο μέγιστον
ἀγαθόν· ἐκλύει γὰρ τὰς τάσεις καὶ μαλάττει τὰ πεπονηκότα κατὰ τὰς σφοδροτέρας ἐνεργείας. διὰ ταῦτα
μὲν δὴ φευκτέον ἐστὶ τὴν σκληρὰν τρίψιν, δι᾽ ἕτερα
δὲ τὴν μαλακήν· οὔτε γὰρ ἐξικνεῖται πρὸς τὸ βάθος
ἡ τοιαύτη τρίψις, ἀλλ᾽ αὐτόθι που κατὰ τὸ δέρμα καὶ
τὰς πλησίον αὐτοῦ σάρκας ἐκλύεται, οὔτ᾽ ἐκθλίβει τὰ
περιεχόμενα κατὰ τοὺς στενοὺς τῶν πόρων περιττώματα. δι᾽ ὃ δὴ τήν τε συνέντασιν[6] τῶν τριβομένων
παραλαμβάνομεν καὶ τὴν τοῦ πνεύματος κατάληψιν.
ἡ μέση τοίνυν μαλακῆς καὶ σκληρᾶς, ἥπερ δὴ καὶ
172K   σύμμετρός ἐστιν, ἐκπεφευγέναι φαίνεται τό τε τῆς
μαλακῆς ἄπρακτον καὶ τὸ τῆς σκληρᾶς βίαιον καὶ
βλαβερόν.

ἐνεργηθήσεται δὲ τῶν μὲν τοῦ τρίβοντος χειρῶν
ἐρρωμένως ἐπιβαλλομένων, ὡς ἐγγύς που τὴν ἀπ᾽ αὐτῶν θλῖψιν εἶναι τῇ σκληρᾷ τρίψει, διὰ δὲ τὸ πλῆθος
τοῦ λίπους καὶ τὸ τάχος τῆς φορᾶς ἐκλυομένων τοσ-

248

the slow massages. Since the body is no longer moving of   171K
itself, it is dangerous for it to be cooled and condensed,
receiving no heating remedy from without.

Because of these things, then, it is necessary to massage not only quickly but also with many hands, so that, as
far as possible, no part of the person being massaged is
bare. But if it is fitting for the massage to be neither slow
nor hard, it is necessary to pour oil in abundance over the
body being massaged, for this contributes to quickness
and softness in massage, and in addition to this, has another major benefit because it relaxes tensions and softens
those things that have suffered in the stronger actions.
Because of these things, what one must certainly avoid is
hard massage; because of other things, one must avoid soft
massage, for such massage does not reach the depths but
relaxes the skin at the site and the flesh near it, and doesn't
squeeze out the superfluities contained in the constricted
pores. This is why I undertake the stretching of the parts
being massaged and suppression of the breath (*pneuma*).
Moreover, the mean between soft and hard, which is in
fact also moderate, appears to escape the ineffectiveness   172K
of soft massage and the violence and harm of hard massage.

[Moderate massage] will be done with the hands of the
masseur firmly applied so that the pressure from them is
close to the hard massage, but due to the amount of fat
and the speed of the motion, with the hands being reduced

---

6 συνέντασιν Ko; σύντασιν Ku

οὗτον, ὡς ἀκριβῶς γίνεσθαι σύμμετρον. τό τε γὰρ λίπος οὐ σμικρὸν ἀλεξητήριόν ἐστι βιαίας ἐπιβολῆς, τό τε βραχυχρόνιον τῆς ὁμιλίας τοσοῦτον ἀφαιρεῖ τῆς βίας, ὅσον καὶ τοῦ χρόνου. τείνειν δ' ἀξιοῦμεν ἐν τούτῳ τὰ τριβόμενα μόρια χάριν τοῦ πᾶν ὅσον ἐστὶ μεταξὺ τοῦ δέρματος καὶ τῆς ὑποκειμένης σαρκὸς περίττωμα διὰ τοῦ δέρματος ἐκκενοῦσθαι. χαλαρῶν γὰρ ἀμφοτέρων ὑπαρχόντων οὐδὲν μᾶλλον εἴσω φέρεσθαι τοῖς περιττώμασιν ἢ ἔξω συμβήσεται· ταθέντων δὲ τῶν ὑποκειμένων τῷ δέρματι πάντ' ἐκτὸς ἐκκρίνεται, καθάπερ ὑπὸ δυοῖν πιεζόμενα χεροῖν, μιᾶς μὲν αὐτῆς τῆς ἔξωθεν ἐπιβεβλημένης τοῦ τρίβοντος, ἑτέρας δὲ τῶν τεταμένων ἔνδον μερῶν.

173K   δι' αὐτὰ δὲ ταῦτα καὶ ἡ τοῦ πνεύματος κατοχή τε καὶ κατάληψις οὐ σμικρὸν μόριον ἀποθεραπείας ἐστίν· ὀνομάζεται δὲ οὕτως, ἐπειδὰν ἅπαντας ἐντείναντές τε καὶ προσστείλαντες τοὺς μῦς τοῦ θώρακος, οἳ κατὰ τὰς πλευράς εἰσιν, ἐπέχωμεν τὴν ἐκπνοήν. συμβαίνει γὰρ τηνικαῦτα τὸ θλιβόμενον ὑπὸ τῶν πλευρῶν πνεῦμα κεκωλυμένον ἐκπνεῖσθαι διὰ τὸ κεκλεῖσθαι τὸν λάρυγγα πᾶν ὠθεῖσθαι κάτω πρὸς τὸ διάφραγμα· τούτῳ δὲ ὑποκειμένων ἥπατός τε καὶ σπληνὸς καὶ γαστρὸς καί τινων ἑτέρων, συνεξαίρεται ταῦτα σύμπαντα τῷ διαφράγματι. χρὴ δ' ἐν τούτῳ συνεντείνειν ἀτρέμα τοὺς κατ' ἐπιγάστριον μῦς, ἵν' ὑπὸ τούτων τε ἅμα καὶ τοῦ διαφράγματος, ὥσπερ ὑπὸ δυοῖν θλιβόμενα χεροῖν, ὅσα μεταξὺ κεῖται μόρια τὸ περιεχόμενον ἐν αὐτοῖς περίττωμα πρὸς τοὺς εἴκοντας

to such an extent as to become precisely moderate. For the fat is able to protect against the forceful application to no small extent, while the brevity of contact takes away the force as much as the time does. I think it worthwhile in this to stretch the parts being massaged for the purpose of evacuating through the skin all superfluities between the skin and the underlying flesh. For when both the skin and flesh are relaxed, what will happen to the superfluities is that they are no more carried inward than they are carried outward.[4] But when the parts underlying the skin are stretched, all the superfluities are evacuated outward, as if squeezed by two "hands"; one of these hands is what the masseur applies externally and the other is of the stretched parts within.

Due to these same things too, retention and suppression of the breath (*pneuma*) is no small part of apotherapy. 173K It is so termed whenever we restrain the exhalation when we stretch and draw tight all the muscles of the chest which are related to the ribs. What happens under these circumstances is that the breath (*pneuma*), compressed by the ribs, is prevented from being exhaled due to the larynx having been closed, so it is all thrust downward toward the diaphragm. By this, the underlying structures—liver, spleen, stomach and certain others—are all lifted up by the diaphragm. In this, there must be gentle straining of the muscles in the epigastrium so that, by these and the diaphragm squeezing like two hands, those parts that lie in between urge the superfluity contained in them toward

[4] Inward and outward are reversed in order in the Kühn text.

τόπους⁷ ὠθῇ. τὰ μὲν οὖν μεταξὺ κείμενα τό τε ἧπάρ
ἐστι καὶ ὁ σπλὴν καὶ ἡ γαστὴρ καὶ κώλων τε καὶ
λεπτῶν ἐντέρων τὰ προὔχοντα· χῶραι δ᾽ εἰς ὑποδοχὴν
ἕτοιμοι τοῖς ἐκθλιβομένοις περιττώμασιν ἥ τε τῆς γαστρὸς εὐρυχωρία πᾶσα καὶ τῶν ἐνταῦθα κειμένων
ἐντέρων αἱ κοιλότητες. εἰ δ᾽ ἀργοὺς ἀκριβῶς ἐάσαις
τοὺς κατ᾽ ἐπιγάστριον μῦς, οὐδενὸς μὲν τῶν εἰρη-
174K   μένων ἐκκενώσεις τὰ περιττώματα, τὰ δ᾽ ἐν θώρακί τε
καὶ πνεύμονι μεταστήσεις κάτω. προσήκει δ᾽ ἐνταῦθα
μεθίστασθαι μᾶλλον αὐτοῖς τοῦ μένειν ἐν ἐκείνοις,
ὅσῳ καὶ ἡ κένωσις ἑτοιμοτέρα τῶν ἐν τῇ κοιλίᾳ περιεχομένων ἤπερ ἡ τῶν ἐν πνεύμονί τε καὶ θώρακι. τὰ
μὲν γὰρ ἐμεῖταί τε καὶ ἀποπατεῖται ῥᾳδίως, τὰ δὲ
μετὰ συντονίας τε καὶ βίας ὑπὸ βηχὸς ἐκβάλλεται.

εἰ μέντοι τις ὁμοίως ταῖς φρεσὶν ἐντείνει τοὺς κατ᾽
ἐπιγάστριον μῦς ἐν ταῖς τοῦ πνεύματος καταλήψεσιν,
ἀκριβέστερον μὲν ἐκκαθαρθήσεται τὰ κάτω τῶν φρενῶν σπλάγχνα, μεταστήσεται δὲ οὐδὲν ἐκ τῶν τοῦ
πνεύματος ὀργάνων εἰς τὰ τῆς τροφῆς, ἀλλ᾽ ἅπαν ἐν
θώρακί τε καὶ πνεύμονι μενεῖ τὸ περίττωμα. διὸ τὴν
τοιαύτην κατάληψιν τοῦ πνεύματος οὐκ ἐπαινῶ,⁸ ἔτι δὲ
μᾶλλον, ὅταν ἤτοι μηδ᾽ ὅλως ἐντείνας τις τὰς φρένας
ἰσχυρῶς καὶ βιαίως ἢ ἐπ᾽ ὀλίγον⁹ προσστείλῃ τοὺς
κατ᾽ ἐπιγάστριον μῦς. ἀνάγκη γὰρ ἐν τῷδε ἐμπίπλασθαι μὲν αἵματός τε καὶ πνεύματος ἅπαντα τὰ κατὰ
τὸν τράχηλον ἀγγεῖα καὶ μόρια, φέρεσθαι δ᾽ ἄνω τε

───────

⁷ τοὺς εἴκοντας τόπους Ko; τοὐκτὸς Ku

the places that yield. The parts lying between are the liver, spleen, stomach and the ventral (forward lying) parts of the colon and small intestines. The places prepared for the reception of the compressed superfluities are the whole open space of the stomach and the cavities of the intestines lying there. If, however, you allow the muscles in the epigastrium to be entirely inactive, you will not evacuate the superfluities of any of the organs mentioned, but you will transfer downward those in the chest and lungs. It is more appropriate for them to be transferred there than to remain in those structures by virtue of the fact the evacuation of those superfluities contained in the belly occurs more readily than that of the superfluities contained in the lungs and thorax. For the former are easily evacuated by vomiting and defecation whereas the latter are expelled with exertion and force by coughing.

174K

If, however, someone stretches the muscles in the epigastrium in like manner to those in the diaphragm in the suppressions of the breath (*pneuma*), the organs below the diaphragm will be more completely evacuated, while nothing from the organs of respiration will be transferred to those of nutrition, but all the superfluity will remain in the thorax and lungs. This is why I do not recommend such a suppression of breath (*pneuma*)—and still more so, whenever someone, having not completely strained the diaphragm strongly and forcibly or for a little while, compresses the muscles in the epigastrium. For inevitably in this all the vessels and parts in the neck are filled quite full of blood and *pneuma*, and the superfluities are carried

---

8 post ἐπαινῶ; νῦν (Ku) om.
9 ἢ ἐπ᾽ ὀλίγον add. Ko

καὶ πρὸς τὴν κεφαλήν, οὐκ ἐπὶ τὴν γαστέρα τε καὶ
175K κάτω τὰ περιττώματα. θεάσασθαι δ᾽ ἔστιν αὐτὸ κἀπὶ
τῶν αὐλούντων ἢ μέγιστον ἢ ὀξύτατα φωνούντων·
εὐρύνεται γὰρ αὐτῶν ἅπας ὁ τράχηλος, οἰδίσκεται δὲ
τὸ πρόσωπον, ἥ τε κεφαλὴ πληροῦται σφοδρῶς, ὅτι
καὶ κατὰ τοῦτο τὸ ἔργον οἱ κατ᾽ ἐπιγάστριον ἐντείνον-
ται μύες εἴκοντος αὐτοῖς τοῦ διαφράγματος. ἔστι γὰρ
δὴ τὸ τοιοῦτον ἔργον ἅπαν, ὡς ἐν τοῖς Περὶ φωνῆς
ἀποδέδεικται, μικτὸν καὶ σύνθετον ἔκ τε μεγίστης ἐκπνοῆς καὶ πνεύματος καταλήψεως, ἐν μὲν ταῖς ἀθρόαις
ἐκφυσήσεσι μεγίστης ἐκπνοῆς γινομένης ἐπὶ σφοδροτάταις ἐντάσεσι τῶν κατὰ τὰς πλευρὰς μυῶν ἅμα
τοῖς κατ᾽ ἐπιγάστριον, ἐν δὲ ταῖς τοῦ πνεύματος κατα-
λήψεσι τῆς μὲν αὐτῆς ἐντάσεως γινομένης ἑκατέρων
τῶν μυῶν, ἐκπνεομένου δ᾽ οὐδενός, ἐπὶ δὲ τῶν αὐλούντων τε καὶ φωνούντων ὀξὺ τῶν μὲν μυῶν ὡσαύτως
τεινομένων, οὔτε δ᾽ ἐπεχομένης ἀκριβῶς τῆς ἐκπνοῆς
οὔτε ἀθρόως ἐπιτελουμένης, ἀλλὰ μέσην ἐχούσης
κατάστασιν, ὥστε ταῖς τρισὶν ἐνεργείαις κοινὴν μὲν
176K εἶναι τὴν τάσιν τῶν μυῶν, ἰδίαν δὲ κατὰ μὲν τὰς
ἀθρόας ἐκφυσήσεις τὴν ταχεῖαν ἔξω φορὰν τοῦ πνεύματος, κατὰ δὲ τὰς καταλήψεις τὴν ἐπίσχεσιν, ἐν δὲ
ταῖς αὐλήσεσί τε καὶ φωναῖς σύμμετρον κένωσιν.
αἰτία δὲ τῆς διαφορᾶς τῶν τριῶν ἐνεργειῶν ἡ φάρυγξ,
ἀνοιγνυμένη μὲν ἐπὶ πλεῖστον ἐν ταῖς ἀθρόαις ἐκφυ-
σήσεσιν, ἀκριβῶς δὲ κλειομένη καταλαμβανόντων τὸ
πνεῦμα, μέσην δ᾽ ἔχουσα κατάστασιν αὐλούντων τε

upward and toward the head and not downward to the
stomach. It is possible to see this same thing in the case of   175K
flute players, when they are playing either very loudly or
shrilly. For the whole neck broadens, the face becomes
swollen and the head is filled excessively. This is because,
in this action, the muscles in the epigastrium are strained
and those of the diaphragm yield to them. Such an action
in its entirety is a mixture and combination of maximal
exhalation and stoppage of breath (*pneuma*), as was shown
in my work *On the Voice*,[5] in the concentrated emissions
of breath, when maximal exhalation occurs immediately
following the most violent strainings of the muscles in
relation to the ribs along with those in the epigastrium. On
the other hand, in the stoppages of the breath (*pneuma*),
when the same tension occurs in each of the groups of
muscles, no air is expelled, whereas in the case of flute
players and those crying out shrilly, when the muscles are
strained in a similar way, exhalation is neither completely
held back nor accomplished in a concentrated fashion, but
is in an intermediate state. Consequently, in the three actions, the tension of the muscles is a common factor; what
is specific in the concentrated exhalations is the swift out-   176K
ward passage of the breath, the retention of breath in
the stoppages, and moderate expulsion in flute players
and those phonating (crying out). The cause of the difference of the three actions is the pharynx, which opens to
the greatest extent in the concentrated exhalations, closes
completely in holding the breath, and has an intermediate

[5] The work *De voce* was in four books and dedicated to Boe-
thus. The original has been lost, although some fragments and an
Arabic summary remain. See Boudon, *Galien*, 419n3.

καὶ φωνούντων ὀξύ τε ἅμα καὶ μέγα. περὶ μὲν δὴ
τούτων ἀνάγκη ποτὲ καὶ αὖθις εἰπεῖν ἐν τῷ περὶ τῆς
ἀναφωνήσεως λόγῳ.

ἡ δ᾽ εἰς τὴν ἀποθεραπείαν ἐπιτήδειος ἐξαίρει τὴν
γαστέρα, τεινομένων μὲν ἁπάντων τῶν τοῦ θώρακος
μυῶν, ἀνιεμένων δὲ τῶν κατ᾽ ἐπιγάστριόν τε καὶ τὰς
φρένας· οὕτω γὰρ ἐνεχθήσεται κάτω τὰ περιττώματα.
δευτέραν δὲ ἔχει τάξιν ἡ μετρίως ἐντείνουσα τοὺς κατ᾽
ἐπιγάστριον μῦς ὑπὲρ τοῦ τὰ κάτω τῶν φρενῶν ἀπο-
θεραπεῦσαι σπλάγχνα. τῶν δ᾽ αὐτῶν τούτων ἕνεκα
καὶ αἱ τῶν ἀμμάτων περιφοραὶ γινέσθωσαν, ἃς ἐξ-
όπισθεν τῶν τριβομένων οἱ τρίβοντες ἱστάμενοι ποι-
177K   οῦνται περὶ τὴν γαστέρα σύμπασαν· ἄλλας δ᾽ ἐκ τῶν
πρόσθεν ἱστάμενοι τῶν μεταφρένων περιβολὰς ἀμμάτων ποιοῦνται τὼ χεῖρε περιάγοντες· ἄλλας δὲ ταῖς
πλευραῖς τε καὶ τῇ ῥάχει καὶ τῷ στέρνῳ συνεπιστρεφομένου πως αὐτοῖς τοῦ τριβομένου. κατὰ τῆς ὀσφύος
δὲ γινέσθωσαν ἀμμάτων ὅμοιαι περιβολαί τε καὶ
περιφοραί, συνεντεινομένου μὲν ἁπάσαις αὐταῖς τοῦ
τριβομένου, συνεπιστρεφομένου δὲ οὐχ ἁπάσαις·
οὐδὲ γὰρ συνεχεῖς ἔτ᾽ αὐτὸν ἀπὸ τῆς θεραπείας[10] χρὴ
ποιεῖσθαι τὰς κινήσεις, ὥσπερ οὐδὲ σφοδράς, ἀλλὰ
ποιεῖσθαι μέν τινας, ἐκ διαλειμμάτων δ᾽ ἐχόντων τρίψεις.

αἱ μὲν γὰρ συνεχεῖς τε καὶ σφοδραὶ τρίψεις ἴδιαι
τῶν κατασκευαστικῶν γυμνασίων εἰσίν, αἱ δὲ μήτε
συνεχεῖς μήτε σφοδραὶ τῆς ἀποθεραπείας οἰκεῖαι.

state in flute players and those crying out shrilly and loudly. It is necessary to speak again about these things in the discussions on vocal exercises.[6]

The usefulness (of retention of breath) to apotherapy is that it lifts up the stomach, strains all the muscles of the thorax and relaxes those in the epigastrium and diaphragm, for in this way, the superfluities will be carried downward. The moderate straining of the muscles in the epigastrium has a secondary importance in the apotherapy for the organs below the diaphragm. Also serving these same purposes are the encircling bindings which masseurs, standing behind the person being massaged, make around the whole abdomen. When they stand in front, 177K they make other bindings encircling the broad parts of the back, twisting them around with their hands, and others around the ribs, spine and sternum, somehow turning the person being massaged around with them. Similar encirclings and windings around of bindings should be applied to the loins, putting the person being massaged on the stretch together with all these, but not turned around at the same time with them all. It is necessary not to make his movements from the apotherapy continuous, just as you should not make them violent, but make some movements with massage occurring in the intervals.

Continuous and vigorous massages are specific to the preparatory exercises, whereas neither continuous nor vigorous massages are fitting for apotherapy. Conse-

6 On ἀναφωνήσις as "vocal exercises," see Soranus 1.49, and Aretaeus, *Treatment of Chronic Disease*, 2.7.13.

---

10 θεραπείας Ko; ἀποθεραπείας Ku

ὥστε πολλάκις μὲν ἐνανειλείσθω, πολλάκις δὲ καὶ
διανωθείσθω, πολλάκις δὲ καὶ μεταβαλλέτω τὸν προ-
γυμναζόμενον ἐν τούτῳ τῷ καιρῷ. πολλάκις δὲ καὶ
κατὰ νώτου γινόμενος αὐτός, ἑκάτερον ἐν μέρει τῶν
σκελῶν περιπλέκων τῷ προγυμναστῇ μετ᾽ ἐντάσεώς
τινος οὐκ ἠπειγμένης, ὑφ᾽ ἑτέρων εὐκαίρως ἐπαφω-
178K μένων τριβέσθω· οὕτω γὰρ ἂν μάλιστα διαφυλάττοι
τε τὴν ἐν τοῖς γυμνασίοις ηὐξημένην θερμότητα καὶ
συνεκκρίνοιτο ταῖς ἰδίαις ἐντάσεσί τε καὶ κινήσεσι τὰ
περιττώματα· πρὸς ὃ δὴ καὶ ἡ τοῦ πνεύματος κατάληψις οὐκ ὀλίγον μοι ἔοικεν ἐπιβοηθεῖν· ὠθούμενον γὰρ
τοῦτο πανταχόθεν εἰς τοὺς λεπτοὺς πόρους ἀναγκάζε-
ται καταδύεσθαι, καὶ ἢν ἐπὶ πλέον θλίβηταί τε καὶ
προωθῆται, πάντας αὐτοὺς διεξέρχεται, συναποφερό-
μενον ἑαυτῷ τι καὶ τῶν λελεπτυσμένων περιττωμάτων.
οὕτω γέ τοι καὶ τρήματα πολλάκις ὀργάνων λεπτὰ
τοὺς δημιουργοὺς ἔστιν ἰδεῖν ἐκκαθαίροντας ἐμφυσή-
σει σφοδροτέρου πνεύματος.

εἰς ὅσον γὰρ τοῦτο φέρεται πρόσω βιαίως ἐπαναγ-
καζόμενον, εἰς τοσοῦτον τὰ μὲν ὠθεῖται πρὸς αὐτοῦ,
τὰ δὲ παρασύρεται διεξελθεῖν ἐφιέμενα τὴν ὁδὸν ἅπα-
σαν· ὠθεῖται μὲν τὰ πρόσω, παρασύρεται δὲ τὰ πλάγια, τῇ ῥύμῃ τῆς φορᾶς ἄμφω βιαζόμενα. καὶ τοίνυν
καὶ τῶν γυμνασίων αὐτῶν μεταξὺ παραλαμβάνουσιν
οἱ ἄριστοι γυμνασταὶ κατάληψιν πνεύματος, ὥσπερ
179K γε καὶ τὴν προειρημένην ἀποθεραπευτικὴν τρίψιν,
ἅμα μὲν ἀναπαύοντες, ὅταν ἄρχωνται κάμνειν, ἅμα δὲ
κατὰ βραχὺ διακαθαίροντες τοὺς πόρους, ἵν᾽ εὔπνουν

quently, let someone prior to exercise be rolled back often, thrust away often and turned quickly often in this time. Often also the one who is at his back, when each of the legs in turn are intertwined with those of the one prior to exercise, with some tension but not excessive, let him be massaged by others touching him lightly at opportune times. For in this way particularly, he would preserve the 178K heat increased in the exercises which, with the specific stretchings and movements, would help in clearing out the superfluities. Toward this also the stoppage of the breath seems to me to help to no small extent, for this, being pushed on all sides, is compelled to go down into the fine channels, and if it is compressed still more and thrust forward, passes through all these and helps to carry off with itself some of the superfluities that have been thinned. In this way too, it is often possible for practitioners to see the fine orifices of organs when purging them with inflations of more violent breath.

For to the extent that this is carried forward, being strongly compelled by force, so to this extent also are some superfluities thrust on before it, while some are swept along, desiring to complete their whole journey. Those in front are thrust forward while those to the sides are swept along, both being compelled by the strength of the movement. And accordingly, the best gymnastic trainers undertake suppression of breath between the actual exercises, just as they also use the previously mentioned apotherapeutic massage. At the same time, they make [their subjects] rest when they begin to tire, at the same time thoroughly purging the pores briefly, so the body is breathing 179K

τε ἅμα καὶ καθαρὸν ᾖ τὸ σῶμα πρὸς τοὺς ἑξῆς πόνους, ὡς κίνδυνός γε μηδενὸς τοιούτου προνοηθέντα
τὸν γυμναστὴν ἐμφράξαι μᾶλλον ἢ καθῆραι τοὺς
πόρους.

αἱ γάρ τοι σφοδρόταται φοραὶ τῶν ὑλῶν τἀναντία
πεφύκασιν ἐργάζεσθαι, κατὰ διαφέροντας καιροὺς
καὶ τρόπους ἐνεργούμεναι· ἐμφράξεις μέν, ὅταν
ἀθρόον τε ἅμα καὶ πολὺ καὶ παχὺ τὸ φερόμενον ᾖ,
καθάρσεις δέ, ἐπειδὰν ὀλίγον τε καὶ λεπτομερὲς ὑπάρχον μὴ πάνυ κατεπείγηται καὶ καταναγκάζηται
πᾶν ἀθρόως ἐκκενοῦσθαι. φαίνεται γὰρ οὕτω ταῦτα
γινόμενα κατά τι τῶν ἐκτὸς ἁπάντων ὀργάνων τε καὶ
πλοκάμων. ἀποπλύνεται γάρ τοι καὶ ἀπορρύπτεται τὸ
περιττὸν ἅπαν ἐξ αὐτῶν, οὐχ ὅταν, ὑπεξιόντων ἔτι τῶν
προτέρων, ἕτερα βιαίως ἐπιφέρηται (κίνδυνος γὰρ ἐν
τῷδε, σφηνωθέντα καὶ διερεισθέντα πρὸς ἄλληλα τὰ
διεξερχόμενα ἐμφράξαι τὴν ὁδόν),[11] ἀλλ' ὅταν τῶν
προτέρων ἤδη κεκενωμένων αὖθις ἕτερα κενωθῇ. καὶ
180K   τί δεῖ περὶ τῶν μικροτέρων[12] θαυμάζειν, ὅπου γε καὶ
τῶν θεάτρων ἀθρόως ἐξιόντες πολλοὶ κατὰ τὰς διεξ-
όδους ἴσχονται; διὰ ταῦτα μὲν δὴ καὶ τοὺς ἐν μέσοις
τοῖς πόνοις ἀποθεραπείᾳ χρωμένους ἐπαινῶ, καὶ
μάλιστα ἐπὶ τῶν τοὺς βαρεῖς καλουμένους ἄθλους
ἀσκούντων. ἀλλὰ περὶ μὲν τούτων αὖθις, ὁ δὲ νῦν
ἡμῖν ὑποκείμενος ἄνθρωπος οὐκ ἀθλητικὴν εὐεξίαν,

---

[11] διαιρεθέντα . . . διερχόμενα φράξαι . . . Ku
[12] post μικροτέρων; πόρων (Ku) om.

freely and is pure for the subsequent exertions, as there is
in fact a danger, if no provision is made for such a thing,
that the gymnastic trainer will obstruct rather than clear
the pores.

Mark you, the most violent movements of the materials
naturally create effects that are opposite, operating at different times and in different ways. Obstructions occur
when what is carried is concentrated, large in amount and
thick, while purifications occur whenever it is small in
amount and fine-particled, and does not greatly urge on
and compel the whole to be evacuated in a concentrated
fashion. It is obvious that these things occur according
to something outside all organs and plexuses [of veins].[7]
For certainly, all the superfluity is washed away and
cleansed from them, not when, while the former are still
going out beneath, other things are forcibly carried against
them—for there is a danger in this of those things passing
through being plugged up and thrust against each other,
blocking the path—but when the first things have already
been evacuated, others in turn are evacuated. And what
about the smaller pores, I wonder! Is this like a large num-   180K
ber of people leaving the theater all at once and blocking
up the exits? For these reasons, I also commend those who
use apotherapy in the middle of exertions, and particularly
in the case of those practicing the so-called strong con-
tests. But I shall speak again about these. The matter before us now is a man whose objective is simply health and

---

[7] I am uncertain as to what is being described here, and particularly about the meaning of πλοκάμων. The Kühn Latin has
calathorum plexibus. See *UPart.*, 9.4.

ἀλλ᾽ ἁπλῶς ὑγείαν ἔχει τὸν σκοπόν. οὔτ᾽ οὖν πολλῶν
αὐτῷ χρεία γυμνασίων ἐστὶν οὔτε πρὸς ἀνάγκην ἐδωδῆς, ἀλλ᾽ οὐδὲ πλήθους κρεῶν χοιρείων οὐδ᾽ ἄρτων
τοιούτων, οἵους ἐσθίουσιν οἱ βαρεῖς ἀθληταί. διὰ
ταῦτα γοῦν ἅπαντα τῷ μὲν οὐδεὶς κίνδυνος ἐμφραχθῆναι τοὺς πόρους, εἰ καλῶς προπαρασκευάσαιτο,
τουτέστι ἐπιτηδείᾳ τε τρίψει καὶ πόνοις ἐξ ὀλίγου τε
ἅμα καὶ κατὰ βραχὺ προϊοῦσιν, ἀθλητῇ δὲ βαρεῖ κίνδυνός ἐστι διά τε τὴν ποιότητα καὶ τὸ πλῆθος τῶν
ἐδεσμάτων, εἰ μὴ πάντα γίνοιτο καλῶς, ἐμφραχθῆναι
μᾶλλον ἐν τοῖς γυμνασίοις ἢ καθαρθῆναι τοὺς πόρους.

181K   3. Ἅλις μὲν ἤδη μοι τῶν περὶ τῆς ἀποθεραπείας
λόγων. ἴωμεν δ᾽ ἐξῆς ἐπὶ τὰ λουτρά, τοσοῦτον ἔτι
μόνον εἰπόντες ὑπὲρ τῶν προκειμένων, ὡς, ὅστις ἂν
ἐλάττοσιν ἔπεσιν ἑρμηνεύσῃ ταῦτα, μακρολογίαν
ἡμῖν ἐγκαλείτω. εἰ δὲ τῶν ἀναγκαιοτάτων τι θεωρη-
μάτων ἢ τῶν ταῦτα πιστουμένων ἀποδείξεων ὑπερβάς
τινας βραχὺν ἡγοῖτο πεποιηκέναι τὸν λόγον, οὐκ
ἀγάλλεσθαι προσῆκεν, ἀλλ᾽ αἰσχύνεσθαι μᾶλλον
αὐτῷ ταῖς τοιαύταις βραχυλογίαις. ἐγὼ δέ, καίτοι βιβλίον ὅλον ὑπὲρ τῆς καλουμένης ἀποθεραπείας γράψαι δυνάμενος, οὐκ ἐδικαίωσα ποιεῖν οὕτως, συν-
τέμνειν ὅτι μάλιστα τὸ μῆκος τῆσδε τῆς πραγματείας
προῃρημένος. εἰ γὰρ ἐπὶ τὸ διελέγχειν ὅσα κακῶς
εἴρηται τοῖς πλείστοις ἐτραπόμην, οὐ μικρὰν οὐδὲ
φαύλην ὕλην εἰς μῆκος λόγων ἔσχον ἄν· ἀλλ᾽ ἐξ ὧν

not the high condition of an athlete. For him, then, there is no need of many exercises, nor of food beyond what is necessary; nor of an abundance of pig's flesh or such breads as the heavy athletes eat. Anyway, for all these reasons, there is no danger to him of the pores being obstructed, if he prepares properly beforehand—that is to say, with beneficial massage and exertions, proceeding from the small and brief. The danger to the heavy athlete is through the quality and amount of foods; if everything does not occur properly, the pores are more likely to be obstructed in the exercises than cleaned out.

3. This is already enough from me on the discussion of 181K
apotherapy. Let me come next to baths, saying only as much about these matters that, if someone should explain them in fewer words, let him accuse me of prolixity. On the other hand, if someone passes over any of the most essential concepts or any of the confirmatory demonstrations of these, and thinks he has made the discussion of them brief, it is inappropriate to exalt him; rather, he should be particularly ashamed of such brief descriptions. And yet I, though I could have written a whole book about the so-called apotherapy, did not think it fit to do so, choosing as far as possible to cut short the length of this work. For if I had turned to the detailed refutation of those things stated badly by the majority, I would have had no small amount of material, nor would it have been trivial, for the length of the discussion. But from the things I

ἀπέδειξα, νομίζω τοῖς ἔχουσι νοῦν ἁπάσας τῆς ἀντιλογίας τὰς ἀφορμὰς παρεσχῆσθαι.

λέγοντος γοῦν Ἀσκληπιάδου τὴν κατάληψιν τοῦ πνεύματος ἐμπιπλάναι τὴν κεφαλήν, ἐξ ὧν ἐγὼ διωρισάμην ὀλίγον ἔμπροσθεν ἐπιδεικνὺς αὐτῆς τὰς διαφοράς, ἔνεστι τῷ βουλομένῳ τὴν πρὸς αὐτὸν ἀντιλογίαν ποιεῖσθαι. οὕτω δὲ καὶ κατὰ τὸ δεύτερον βιβλίον ἐξῆν δήπου κἀμοὶ τὰ κατὰ μέρος ἅπαντα γυμνάσια διηγουμένῳ μηκῦναι τὸν λόγον, ὥσπερ ἄλλοι τέ τινες ἐποίησαν ὅ τε κάλλιστα μεταχειρισάμενος ὅλην τὴν πραγματείαν Θέων ὁ Ἀλεξανδρεύς· τέτταρα γὰρ οὗτος ἔγραψε βιβλία περὶ τῶν κατὰ μέρος γυμνασίων, ἃ πάντα κἀμοὶ λέγειν ἐξῆν ἄμεινόν γε ἑρμηνεύειν ἐκείνου δυναμένῳ καὶ προσέτι καὶ ἄλλων πολλῶν μνημονεύειν γυμνασίων ἔργοις κοινῶν. ἐκεῖνος μὲν γὰρ ὡς ἂν ἀθλητὰς μάλιστα γυμνάζειν ἔργον πεποιημένος ἐν τοῖς ἐπ᾽ ἐκείνων ἐπλεόνασε γυμνασίοις, ἔξεστι δὲ τῷ βουλομένῳ περὶ πάντων τῶν κατὰ πάσας τὰς τέχνας διεξέρχεσθαι. ταῦτα μὲν οὖν μοι λελέχθω πρὸς τοὺς ἀγανακτήσοντας τῷ μήκει τῆς πραγματείας.

4. Καιρὸς δ᾽ ἤδη περὶ λουτρῶν διέρχεσθαι, πρῶτον μέν, ὁπόσα γλυκέων ὑδάτων ἐστὶ θερμαινομένων, ἐπειδὴ τούτων ἡ χρεία πλείων, ἑξῆς δὲ τῶν ψυχρῶν, εἶθ᾽ οὕτω τῶν αὐτοφυῶν ὀνομαζομένων, καὶ αὐτῶν δή που τὰ μὲν εὔκρατα, τὰ δὲ ζέοντα, τὰ δὲ χλιαρά, τὰ δὲ ψυχρὰ παντάπασιν ὑπάρχει. ἔστι δὲ ἡ τῶν γλυ-

did show, I think for those who have the intelligence, I have provided all the means for the refutation.

Anyway, when Asclepiades says that suppression of the breath (*pneuma*) fills the head,[8] it is possible for anyone who so wishes to fashion the refutation against him from those things I distinguished a little earlier, when I demon- 182K strated the differences of this. In the same way too, in the second book, I could, of course, have lengthened the discussion by describing in full all the exercises individually, as certain others did, and especially Theon the Alexandrian,[9] who handled the whole matter best, for he wrote four books about the individual exercises, all of which I might say too, being better able to explain them than that man, and besides, to mention many other exercises common to actions. For that man, since he created a work chiefly for training athletes, wrote at length in these on the exercises for those men. However, it is possible for someone who wishes to do so to go through all the exercises pertaining to all the arts. Let this, then, be my response to those who are vexed by the length of the work.

4. It is now time to go over baths: first, those of sweet waters that are being heated, since the use of these is 183K considerable, and next of cold waters, and then, in like manner, those of the so-called natural waters; of the last, there are of course the *eukratic,* the bubbling, the lukewarm and those that are altogether cold. The potency of

[8] For Asclepiades' views on respiration, see J. Vallance, *The Lost Theory of Asclepiades of Bithynia* (1990), 67–74 and 82–85.

[9] On Theon, frequently referred to in the present work, see EANS, 795, and the General Introduction to the present work (pp. xix–xx).

κέων ὑδάτων θερμῶν δύναμις εὐκράτων μὲν ὄντων
ὑγρὰ καὶ θερμή, χλιαρωτέρων δὲ γενομένων ὑγρὰ καὶ
ψυχρά, θερμοτέρων δὲ τοῦ δέοντος ἀποτελεσθέντων
θερμὴ μέν, οὐκέτι δ' ὁμοίως ὑγρά. φρίττειν γὰρ ἀναγ-
κάζει τὰ σώματα καὶ πυκνοῦσθαι τοὺς πόρους αὐτῶν,
ὡς μήτ' ἀπολαύειν ἔτι τῆς ἔξωθεν ὑγρότητος μήτ' ἐκ-
κενοῦσθαί τι τῶν ἔνδον περιττωμάτων. ἀλλὰ γὰρ ἀπὸ
τῶν εὐκράτων ἀρκτέον, ἃ διὰ παντὸς ὑγραίνει μὲν καὶ
θερμαίνει καθ' ἑαυτά. συμβαίνει δ' αὐτοῖς ἐνίοτε κατά
τι συμβεβηκὸς ἤτοι διαφορεῖν τὰς ὑγρότητας ἢ πληροῦν ῥεύματος περιττοῦ τὰ μόρια τοῦ σώματος ἢ
μαλάττειν ἢ πέττειν ἢ ῥωννύναι τὴν δύναμιν ἢ κατα-
λύειν. εἰς ταῦτα μέν γε καὶ ἡ ποσότης αὐτῶν οὐκ
ὀλίγα συντελεῖ. πολλὰ δὲ καὶ ἄλλα τῶν τοῖς εἰρημένοις ἑπομένων ἔνεστι καταλέγειν ἔργα τῶν εὐκράτων
184K   ποτίμων λουτρῶν, ὧν τὰ πρέποντα τῇ νῦν ἡμῖν ἐν-
εστώσῃ πραγματείᾳ λεχθήσεται καθ' ὅσον ἐγχωρεῖ
διὰ βραχυτάτων, ἀναβεβλημένης ἐν τῷ παρόντι τῆς
εἰς τὰ νοσήματα χρείας αὐτῶν.

εἰς δὲ τὴν ὑγιεινὴν χρῆσιν ὁ προκείμενος ἐν τῷ
λόγῳ νεανίσκος ἡκέτω γεγυμνασμένος ὡς εἴρηται
πρόσθεν. ἐπὶ τούτου τοίνυν ὀλίγη μὲν ἡ ἐξ αὐτοῦ
ὠφέλεια· πάντα γὰρ ἔχει φθάνων ἐκεῖνος ἔκ τε τῶν
συμμέτρων γυμνασίων καὶ τῆς εἰρημένης ἀποθεραπείας. ὅμως δ' οὖν, εἰ καλῶς παραλαμβάνοιτο τὰ λουτρά, μέρος ἄν τι γένοιτο καὶ αὐτὰ τῆς ἀποθεραπείας,
εἴ γε δὴ μαλάττει μὲν τὰ σκληρὰ καὶ τὰ τεταμένα
μόρια, διαφορεῖ δ', εἴ τι περίττωμα καὶ σύντηγμα

the sweet, warm waters, if they are *eukratic,* is moist and
hot; of those that are more lukewarm, it is moist and cold;
and of those that are made hotter than they need be, it is
hot but no longer similarly moist. For the last compels
bodies to shiver and condenses their pores, so they do not
still enjoy the benefit of the external moisture, nor is there
any evacuation of the superfluities within. But I must begin with the *eukratic,* which are throughout moistening
and heating in themselves. What happens with them is
that sometimes, *per accidens,* they either disperse the fluids, or fill the parts of the body with excessive flow, or
soften, or concoct, or strengthen the capacity, or break it
down. The quantity of these baths contributes to these
effects to no small extent. And it is possible to add to those
mentioned many other actions of those that follow *eukratic* baths of potable waters. I shall speak now, as briefly   184K
as possible, about those most clearly relevant to our present matter, postponing for the moment the use of these
baths for those who are sick.

Let the young man proposed in the discussion come to
the hygienic use [of a bath], having exercised, as was said
earlier. In this case, benefit from the bath is slight, for he
has gained everything beforehand from his moderate exercises and the apotherapy mentioned. Nevertheless, if he
should undertake the baths properly, these would also
become part of the apotherapy, if in fact they soften the
hard and tense parts and disperse whatever superfluity

πρὸς τοῦ δέρματος ἐντὸς ἴσχοιτο. ἀλλὰ τούτων γε
οὐδετέρου χρῄζειν ἔοικεν ὁ ὑποκείμενος ἄνθρωπος ἐν
τῷ λόγῳ· οὔτε γὰρ σύντηγμα σαρκὸς ἀπαλῆς ἢ πιμελῆς εἰκὸς αὐτῷ γεγονέναι τι κατὰ τὰ γυμνάσια.
ταῖς γὰρ ἀμέτροις τε καὶ σφοδραῖς κινήσεσιν εἴπετο
τὰ τοιαῦτα, κεκένωται δὲ πάντ' αὐτῷ τὰ περιττώματα
185K καὶ μεμάλακται τὰ στερεὰ μόρια κατὰ τὸν τῆς ἀποθεραπείας καιρόν, ὥστ' ἀποπλύνασθαι τὸν ἱδρῶτα
καὶ τὴν κόνιν, εἰ καὶ ταύτῃ ποτὲ χρήσαιτο, δεῖται
μᾶλλον ἢ θερμανθῆναι κατὰ τὸ βαλανεῖον. διαβαδί-
σαι τοιγαροῦν χρῄζει μόνον ἄχρι τῆς δεξαμενῆς, οὐκ
ἐνδιατρῖψαι τῷ βαλανείῳ καθάπερ οἱ χωρὶς τοῦ γυμνάσασθαι καθέψοντες ἑαυτούς. οὐ μὴν οὐδ' ἐγχρονί-
ζειν ἐν τῇ κολυμβήθρᾳ δεῖται, περιπλυνάμενος δ', ὡς
εἴρηται, πρὸς τὸ ψυχρὸν ὕδωρ ἐπειγέσθω. σύμμετρον
δ' ἔστω καὶ τοῦτο τῇ συμμέτρῳ φύσει τοῦ σώματος.
ἐφ' ὅσον γὰρ ἤτοι σφόδρα ψυχροῦ χρεία τοῦ ὕδατος
ἢ χλιαροῦ[13] τε καὶ οἷον εἰληθεροῦς, αὖθις εἰρήσεται.
τὸ δ' ἄριστα πεφυκὸς σῶμα, μέχρι μὲν αὐξάνεται,
λέλεκταί που καὶ πρόσθεν, ὡς οὐ χρὴ ψυχρῷ λούειν,
ἵνα μή τι τῆς αὐξήσεως αὐτοῦ κωλύσωμεν· ηὐξημένου
δ' ἱκανῶς, ἐθίζειν ἤδη καὶ τῷδε· κρατύνει τε γὰρ ἅπαν
τὸ σῶμα καὶ τὸ δέρμα πυκνὸν καὶ σκληρὸν ἀποτελεῖ·
κράτιστον δὲ τοῦτο πρὸς τὴν ἀπὸ τῶν ἔξωθεν βλάβην.

ὡς δ' ἄν τις ὑπάρξαιτο λούσασθαι ψυχρῷ, μηδενὸς
ἀπὸ τῆς ἐξαιφνιδίας μεταβολῆς ἀπολαύσας βλαβε-
186K ροῦ, παντὸς μᾶλλον ἐπίστασθαι χρή. πολλοὶ γὰρ κα-

and waste material is retained within by the skin. But the person proposed in the discussion seems to need neither of these things, as it is unlikely that waste products of soft flesh or fat have arisen in him during the exercises. Such things follow the excessive and violent movements, whereas in him all the superfluities have been evacuated and the solid parts softened during the time of the apo-   185K
therapy, so he needs to wash away the sweat and the powder, if he should also use the latter at any time, more than he needs to be heated by the bath. For that very reason, he needs only to walk across as far as the receiving tank and not to spend time in the bath, like those who refresh themselves apart from exercising. There is no need to delay in the swimming bath—having washed, as I said, he should hurry on to the cold water. However, this should be moderate too, like the moderate nature of the body. I shall speak again on how much there is a strong need for cold water, or lukewarm, or, as it were, water warmed by the sun. The naturally best body, until it grows up, as I have said somewhere else before, we must not wash in cold water, so that we do not hinder its growth in any way. However, when it has grown sufficiently, it should now be made accustomed to this, for it strengthens the whole body and makes the skin thick and hard. This is best against damage from external factors.

It is necessary to know above all that, should someone begin to bathe in cold water, he would experience nothing harmful from the sudden change. Many, having started off   186K

---

13 ἢ χλιαροῦ Ko; ἢ μαλακοῦ Ku

GALEN

κῶς ἀρξάμενοι διεβλήθησαν οὕτω πρὸς ὅλον τὸ ἐπι
τήδευμα τῆς ψυχρολουσίας, ὥστε μηδὲ τοῖς ἀσφαλῶς
αὐτὸ μεταχειριζομένοις ὑπομένειν ἑαυτούς ποτε παρα
σχεῖν. ἔστω τοίνυν ὁ μὲν τοῦ ἔτους καιρὸς ἀρχόμενον
θέρος, ἵνα πρὸ τοῦ χειμῶνος ἐν ἅπαντι τῷ μεταξὺ
χρόνῳ γένηταί τις ἐθισμὸς ἀξιόλογος. ἔστω δὲ δή
που καὶ ἡ ἡμέρα, καθ᾽ ἣν ἀρχόμεθα, νήνεμος ὡς ἔνι
μάλιστα καὶ εἰς ὅσον οἷόν τε κατ᾽ ἐκεῖνον τὸν καιρὸν
θερμοτάτη. δῆλον δ᾽, ὡς καὶ τῆς ἡμέρας αὐτῆς ἐκλέ
γεσθαι χρὴ τὸ θερμότατον ὥσπερ γε καὶ τὸ γυμνα
στήριον εὐκρατότατον.

ἡ μὲν οὖν ἔξωθεν αὕτη παρασκευή. τὸ σῶμα δ᾽
αὐτὸ τὸ μέλλον χρῆσθαι τῷ ψυχρῷ παρεσκευάσθω
κατὰ τάδε. τῇ μὲν ἡλικίᾳ ἔστω κατὰ τὴν τετάρτην
ἑβδομάδα μεσοῦσαν μάλιστα, μηδὲν ὑπὸ μηδενὸς
ἠλλοιωμένος αἰτίου προσφάτου κατ᾽ ἐκείνην τὴν ἡμέ
ραν ἢ τὴν πρὸ αὐτῆς νύκτα, τὴν δ᾽ ὑγιεινὴν κατάστα
σιν, ἣν εἶχεν ἔμπροσθεν, ἀκριβῶς διαφυλάττων. ἔστω
δὲ καὶ κατὰ τὴν ψυχὴν εὔθυμός τε καὶ φαιδρὸς ὁ μέλ
λων χρήσασθαι τῷ ψυχρῷ νεανίας, εἴπερ ποτὲ ἄλ
187K λοτε, καὶ τότε μάλιστα. πρῶτον μὲν οὖν ἀνατριβέσθω
σινδόσιν ἐπιπλέον ἢ πρόσθεν·[14] ἔστωσαν δὲ καὶ σφο
δρότεραι νῦν μᾶλλον ἢ πρόσθεν αἱ τρίψεις καὶ διὰ
σκληροτέρων ὀθονίων· εἰ δὲ καὶ χειρίδας ῥαπτὰς
περιθέμενοι ταῖς χερσὶν οἱ προγυμνασταὶ τρίβοιεν,
ὡς ὁμαλωτέραν γενέσθαι τὴν ἐνέργειαν, οὐδὲν ἂν εἴη
χεῖρον. ἐφεξῆς δὲ δι᾽ ἐλαίου τριβέσθω, καθότι σύν
ηθες ἦν αὐτῷ· κἄπειτα γυμναζέσθω τῷ πλήθει μὲν

270

badly, were filled with doubt in this way toward the whole usefulness of the cold bath so as not to permit those who administered such a bath safely, to provide one for themselves at any time. Accordingly, let the time of year be the beginning of summer, so that, in the whole time intervening before the winter, significant habituation occurs. Let the day on which we begin be without wind as far as possible, and be as warm as possible for that time. It is clear also that it is necessary to choose the hottest part of the day itself, just as the gymnastic school should also be most *eukratic*.

The external preparation is this. Prepare the actual body that is going to use the cold bath as follows: in age, let it be particularly in the middle of the fourth seven-year period (i.e., approximately twenty-four), not changed by any fresh cause on that day or during the night before it, and maintaining perfectly the healthy state it had before. Also, let the young man who is going to use the cold bath be in good spirits and cheerful generally at other times and particularly at this time. First, let him be rubbed down 187K more than before with muslin. Also, let the massages now be more vigorous than before and through harder linen. And if the gymnastic trainers massage with stitched gloves on their hands so as to make the action more even, it would be no bad thing. Next, let the skin be rubbed with oil in the manner customary for him and then let him be exercised with exercises equal in amount but now quicker than

---

14 ἢ πρόσθεν *add.* Ko

ἴσα γυμνάσια, θάττονα δὲ νῦν ἢ πρόσθεν· ἔπειθ᾽ οὕτως εἰς τὸ ψυχρὸν ὕδωρ καταβαινέτω μὴ βλακεύων,
ἀλλ᾽ ἐπωκύνων τὴν ἐνέργειαν ἢ ἀθρόως ἐναλλέσθω,
σκοπὸν ἐν ἀμφοῖν ἔχων, ὡς μάλιστα καθ᾽ ἕνα χρόνον
ἅπασι τοῖς σώματος μέλεσι περιχυθῆναι τὸ ὕδωρ· τὸ
γὰρ κατὰ βραχὺ πλησιάζειν αὐτῷ φρίκης ἐστὶ ποιη-
τικόν. ἔστω δὲ μήτε χλιαρὸν τὸ ὕδωρ μήτε ἀτέραμνόν
τε καὶ παγετῶδες· τὸ μὲν γὰρ οὐ ποιεῖ θερμασίας
ἐπανάκλησιν, τὸ δὲ ψύχει καὶ καταπλήττει[15] τοὺς
ἀήθεις. ὡς ἔν γε τῷ χρόνῳ προϊόντι καὶ τοιούτῳ ποτὲ
χρήσαιτ᾽ ἂν ἡμῖν ὁ νεανίσκος, ἀνάγκης καταλαβού-
σης· ἀλλὰ κατά γε τὴν πρώτην ἡμέραν ἀκριβῶς χρὴ
188K   φυλάττεσθαι τὸ λίαν ψυχρόν. ἐξελθόντα δὲ τοῦ ὕδατος ἐπιπλέον[16] ἀνατρίβεσθαι προσήκει δι᾽ ἐλαίου, μέχρις ἂν ἐκθερμανθῇ τὸ δέρμα· καὶ μετὰ ταῦτα σιτία
μὲν πλείω τῶν εἰωθότων, ἔλαττον δὲ προσφερέσθω τὸ
πόμα.

    ταῦτα δ᾽, εἰ καὶ σὺ μὴ κελεύσειας, αὐτὸς ἂν οὕτω
ποιήσειεν ἁπάντων ὀρθῶς γενομένων. καὶ γὰρ ὀρέγον-
ται πλεόνων ἐπὶ ταῖς ψυχρολουσίαις καὶ πέττουσιν
ἄμεινον καὶ διψῶσιν ἧττον. ἀφικνοῦνται δὲ καὶ κατὰ
τὴν ὑστεραίαν ἐπὶ τὰ γυμνάσια σαφῶς εὐεκτικώτεροι,
τὸν μὲν τοῦ σώματος ὄγκον ἴσον ἔχοντες τῷ πρόσθεν,
ἐσφιγμένον δὲ καὶ μυωδέστερον καὶ συντονώτερον
καὶ τὸ δέρμα σκληρότερόν τε καὶ πυκνότερον. ὁμοίως
οὖν ἐπ᾽ αὐτοῦ κατὰ τὴν δευτέραν ἡμέραν πρακτέον
ἅπαντα καὶ κατὰ τὴν τρίτην τε καὶ τετάρτην. εἶθ᾽
οὕτως ἐπὶ προήκοντι τῷ χρόνῳ κελεύσομεν αὐτὸν ἐμ-

before. Then, in the same way, let him go down into the cold water, and not in a leisurely manner, but quickening the action, or leaping in all of a sudden, having as the aim in both, that water be especially poured over all parts of the body at the same time; a gradual approach will produce shivering in him. The water should be neither lukewarm nor unsoftened and ice-cold, for the former does not produce the reaction of heating, while the latter chills and surprises those unaccustomed to it. It may be that at some future time our young man could use such water, being compelled by necessity, but on the first day it is absolutely necessary to guard against extreme cold. When he comes   188K
out of the water, it is appropriate for him to be rubbed with oil further until the skin is heated. After these things, let him be given more food than is customary but less drink.

He himself would do these things in this way, even if you didn't order him to, when everything occurs properly. For people desire more food after cold baths and digest better, and they are less thirsty. During the following day too, they clearly come to exercises in a more healthy state, having a body mass the same as before but tightened up, more muscular and more toned, and with skin that is harder and thicker. One must do everything for him similarly during the second day, and during the third and fourth days. Then, in like manner, with the advance of

---

15 τὸ δὲ ψύχει καὶ καταπλήττει Ko; τὸ δὲ πλήττει καὶ καταψύχει Ku

16 ἐπιπλέον Ko; ὑπὸ πλείονων Ku

βαίνειν τὸ δεύτερον τῷ ψυχρῷ μετὰ τὴν ἐπὶ τῷ προ-
τέρῳ τρύψιν, ὡς εἴρηται πρόσθεν. τὸ δὲ καὶ τὸ τρίτον
ἔτι τοῦτο ποιεῖν, ὡς ἔνιοί τινες ἐκέλευσαν, οὐκ ἐπαινῶ·
καὶ γὰρ καὶ τὸ δεύτερον αὔταρκες εἶναί μοι δοκεῖ δυ-
189K   ναμένων γε ἡμῶν, εἰς ὅσον ἂν ἐθελήσωμεν, ἐν αὐτῷ
κελεῦσαι διατρίβειν. ὁ δὲ σκοπὸς κἀνταῦθα τοῦ χρόνου τῆς διατριβῆς ἐκ τῆς καθ' ἑκάστην ἡμέραν λαμ-
βανέσθω πείρας. εἰ μὲν γὰρ ἀνελθὼν ἐκ τοῦ ὕδατος
ἐπὶ ταῖς ἀνατρίψεσιν εὔχρους ἐν τάχει γίνοιτο, μετρίως ἐν αὐτῷ διέτριψεν· εἰ δὲ δυσεκθέρμαντός τε καὶ
ἄχρους διαμένοι μέχρι πλείονος, ἀμετρότερον ἐχρήσατο τῷ ψυχρῷ. τοῦ χρωτὸς τοίνυν προσέχων τοῖς
γνωρίσμασιν ἐξευρήσεις ῥᾳδίως, εἴτε τὸν ἴσον χρόνον αὖθις ἐν τῷ ψυχρῷ διατρίβειν προσήκει εἴτε καὶ
μετακινῆσαί τι πρὸς τὸ ἔλαττον ἢ τὸ πλέον. καὶ περὶ
μὲν ψυχρολουσίας ὡς πρὸς τὴν ἀρίστην φύσιν ἱκανὰ
καὶ ταῦτα.

5. Τὰς δὲ τῶν παρεμπιπτόντων ἁμαρτημάτων ἐπανορθώσεις ὅπως ἄν τις κάλλιστα ποιοῖτο, διελθεῖν
ἤδη καιρός· εἰ γὰρ καὶ ὅτι μάλιστα τὴν κατασκευὴν
τοῦ σώματος ἄμεμπτον ἔχοι τις ἀπηλλαγμένος τε εἴη
τῶν κατὰ τὸν βίον ἁπάντων πραγμάτων καὶ ἑαυτῷ
μόνῳ ζῶν, ἀλλὰ τό γε μηδέποθ' ἁμαρτάνειν μηδὲν ἢ
αὐτὸν ἢ τὸν ἐπιστάτην αὐτοῦ παντάπασιν ἀδύνατον.
190K   εἰκὸς τοίνυν ἐστὶ πρῶτόν τε καὶ μάλιστα καὶ συνεχέστατα περιπίπτειν ἁμαρτήματι νεανίσκον γυμναστι-
κόν, οἷον ὁ κόπος ἐστίν· ὑπὲρ οὗ πολλάκις μὲν ἤδη
πολλοῖς οὐκ ἰατροῖς μόνον ἢ γυμνασταῖς, ἀλλὰ καὶ

time, we shall direct him to step into a cold bath a second time, after the massage following the previous one, as I said before. I do not, however, recommend doing this again a third time, as some have directed, for truly, the second seems to me to be sufficient, if we are able to direct 189K him to spend as much time in it as we should wish. The aim here is to take the time to be spent from the experience of each day. For if, when he comes out of the water, he were to become of good complexion from the massages quickly, he spent a moderate time in it, but if he should be hard to warm and stays a poor color for a longer time, he used the cold water too immoderately. Accordingly, if you direct your attention to the signs of the body surface, you will easily discover whether it is appropriate to spend an equal time again in the cold water or to change toward less or more. These things are sufficient about cold baths with regard to the best nature.

5. It is now time to go over how someone might effect in the best way the corrections of the faults that befall [people]. For even if someone were to have the most fault-less constitution of the body possible and be freed from all the matters pertaining to life, and to live for himself alone, it would still be altogether impossible for either he himself or his supervisor [of training][10] never to err. Moreover, first and foremost, it is likely that a young man exercising 190K will very frequently fall into faults such as fatigue. This has already been spoken about often by many men—and not only by doctors or gymnastic trainers but also by philoso-

---

[10] On the term ἐπιστάτης in this sense, see Plato, *Republic,* 412a.

φιλοσόφοις εἴρηται, ὥσπερ καὶ Θεοφράστῳ βιβλίον
ὅλον ὑπὲρ αὐτοῦ γέγραπται. ἐγὼ δὲ κἀνταῦθα μήκους
φειδόμενος, ὅσα μὲν εἴρηται κακῶς ὑπὲρ αὐτοῦ τισιν,
ὑπερβῆναι διέγνωκα, τὰ δ' ἀναγκαιότατα τοῖς ὑγιεινοῖς ἅμα ταῖς οἰκείαις ἀποδείξεσι διελθεῖν. καὶ πρῶτόν
γε περὶ τῆς ἐννοίας αὐτοῦ· δίκαιον γὰρ ἀπὸ ταύτης
ἀρξάμενον οὕτως ἐπὶ τὴν οὐσίαν μεταβῆναι.

ἔννοιαν δ' ἔχουσι κόπου τινὲς μὲν ἐν τῷ ἐμπίπρασθαι τε καὶ τετάσθαι δοκεῖν ἤτοι πάντα τὰ μέλη τοῦ
σώματος ἢ τὰ πονήσαντα μόνον· τινὲς δ' ἐν τῷ δυσ-
χερῆ τινα καὶ ἀνιαρὰν αἴσθησιν ἑαυτῶν κατὰ τὰς
κινήσεις λαμβάνειν, ἥν τινες μὲν ἄρρητον ὑπάρχειν[17]
ἔφασαν, ἔνιοι δὲ ἑλκώδη προσηγόρευσαν· ἄλλοι δέ
τινες, ὅτι ὡς τεθλασμένων τε καὶ φλεγμαινόντων
αἰσθανόμεθα τῶν μελῶν. εἰσὶν μὲν οἳ καὶ μιγνύουσιν
191K ἀλλήλαις τὰς ἁπλᾶς ταύτας διαθέσεις, τήν τε τῆς
τάσεως καὶ τὴν ἑλκώδη καὶ τὴν φλεγμονώδη· τινὲς δὲ
δύο μιγνύουσιν ἐξ αὐτῶν, ἤτοι τὴν ἑλκώδη τῇ μετὰ
τάσεως ἢ τὴν φλεγμονώδη μεθ' ἑκατέρας αὐτῶν ἀνὰ
μέρος. ὥσθ' ἑπτὰ τὰς πάσας γίνεσθαι δόξας περὶ τῆς
κατὰ τὸν κόπον ἐννοίας, ἐκ μέρους μέν τινος ἁπάσας
ἀληθεῖς, τὸ σύμπαν δὲ οὐχ ἁπλῶς. ἐάν τε γὰρ ἑλκώδης αἴσθησις γίνηται κινουμένοις, ἐάν τε φλεγμονώδης, ἐάν τ' ἐμπίπρασθαι καὶ τείνεσθαι δοκῶσιν, ἐάν
τε κατὰ συζυγίαν τινὰ τούτων, ἄν θ' ὁμοῦ συνέλθῃ τὰ

17 ὑπάρχειν add. Ko

phers; the whole book written by Theophrastus about this is an example.[11] Even here, refraining from prolixity, I have decided to pass over those things said badly about this by some, and go through those things most essential for hygienists, together with suitable demonstrations. And first, I shall go over the concept of fatigue, for I think it right and proper to begin from this, and then pass on to the essence.

Some have a concept of fatigue in which either all the parts of the body or only those that have been working seem to be burning up and under tension. Others take it to be in a vexatious and distressing sensation of themselves, in relation to movements, while others are accustomed to saying it is ineffable and others again term it "wound-like." Certain others liken it to when we perceive the limbs as bruised and inflamed. There are also those who mix these simple conditions with each other—that of the tensive, and the wound-like, and the inflammation-like. Some mix two of these—either the wound-like with the tensive, or the inflammation-like with each of the former two in turn. Consequently, there are in all seven ideas on the concept of fatigue, all in some part true, but none wholly and absolutely so. For if a wound-like sensation occurs to those moving, and if an inflammation-like sensation occurs, and if they think they are burning up and stretched, and if there is some conjunction of these, or if

191K

[11] Theophratus of Eresos (ca. 340–286 BC) wrote a short work *On Fatigue*. There is an English translation of this by W. W. Fortenbaugh, R. W. Sharple, and Michael G. Sollenberger (2003). For a summary of his theory on fatigues, see the EANS, 800, and the General Introduction to the present work (p. xlii).

πάντα, κόπος ὀνομάζεται τῶν εἰρημένων διαθέσεων
ἑκάστη. ὥστ᾽ εἶναι τὰς συμπάσας διαφορὰς τῶν κόπων ἑπτά· ἁπλᾶς μὲν τρεῖς, συνθέτους δὲ τέτταρας.
ἐπανορθώσεις δὲ τῶν οὕτως ἐχόντων σωμάτων αἱ μέν
τινες ἴδιαι καθ᾽ ἑκάστην εἰσὶ διάθεσιν, αἱ δέ τινες
ἁπασῶν κοιναί. λεχθήσεται δ᾽ ἑξῆς ὑπὲρ αὐτῶν, ἐὰν
πρότερον ὑπὲρ τῆς παρακειμένης τοῖς κόποις διαθέσεως εἴπωμεν, ἣν ἐξαπατώμενοί τινες ὀνομάζουσι
κόπον. ὁποία δ᾽ ἐστὶν αὕτη καὶ τίνα κέκτηται γνωρίσματα, δηλωθήσεται σαφέστερον, ἂν τὴν οὐσίαν
192K πρότερον[18] ἑκάστης τῶν κοπωδῶν διαθέσεων εὕρωμεν.

ἄχρι μὲν γὰρ τοῦδε τὰ συμπτώματα μόνον εἴρηται·
καὶ γὰρ τὸ πίμπρασθαί τε καὶ τείνεσθαι δοκεῖν τὰ
μόρια καὶ τὸ τοῖς κινουμένοις ἤτοι φλεγμονώδη τινὰ
ἢ ἑλκώδη γίνεσθαι τὴν αἴσθησιν, οὐ διαθέσεις εἰσίν,
ἀλλὰ συμπτώματα. διαθέσεις δέ γε τῶν σωμάτων αὐτῶν, ἐφ᾽ αἷς εἴωθε τὰ τοιαῦτα συμπίπτειν, ἁπλαῖ μὲν
τρεῖς, σύνθετοι δὲ τέσσαρες. ἡ μὲν οὖν ἑλκώδης διάθεσις ἐπὶ πλήθει γίνεται περιττωμάτων λεπτῶν τε
ἅμα καὶ δριμέων, ἅπερ ἐν τῷ γυμνάζεσθαι γεννᾶται
κατὰ διττὴν αἰτίαν, ἤτοι τῶν παχυτέρων περιττωμάτων χυθέντων τε καὶ λεπτυνθέντων, οὐχ ἁπάντων
δὲ ἐκκριθέντων, ἢ τακείσης τινὸς πιμελῆς ἢ σαρκὸς
ἁπαλῆς. ἀνάγκη γὰρ ὑπὸ τῶν τοιούτων ὑγρῶν, λεπτῶν καὶ δριμέων ὑπαρχόντων, κεντᾶσθαί τε καὶ οἷον
τιτρώσκεσθαι τὸ δέρμα καὶ τὰς σάρκας, ὥστε καὶ
φρίκην ἐνίοτε γίνεσθαι καί τι καὶ ῥίγους, ὅταν ἰσχυ-

they all come together in the same place, each of the conditions mentioned is called a fatigue. As a result, there are seven *differentiae* of fatigues in all—three simple and four compound. In regard to the restorations of the bodies so affected, some are specific to each condition, while others are common to all. I shall speak about these in due course, if I first speak about the predisposing condition for the fatigues, which some being misled term a fatigue. What this is and what signs it has acquired will be revealed more clearly, if we first discover the essence of each of the fatigue conditions.

192K

For up to this point, I have spoken only of the symptoms; the thought that the parts are burning up and stretched, and the sensation of some inflammation or wounding in those who are moving are not conditions but symptoms. However, conditions of the bodies themselves, in which such things are wont to occur, are three simple and four compound ones. Thus, the wound-like condition arises due to an abundance of thin and, at the same time, acrid superfluities, which in exercise are generated due to a twofold cause—either when the thicker superfluities have been poured out and thinned but not all evacuated, or when there is some dissolution of fat or soft flesh. For of necessity, there is, by such fluids, which are thin and pungent, a stinging and wounding, as it were, of the skin and flesh, so that sometimes shivering also occurs, and even rigors, when it is strongly acrid and, at the same

---

18 πρότερον *add.* Ko

ρῶς ἢ δριμέα τε ἅμα καὶ πολλά. τοιοῦτος μὲν δή τις
ὁ οἷον ἑλκώδης κόπος.

ἐν ᾧ δὲ τείνεσθαι δοκεῖ τὰ μόρια μόνον, ἑλκώδης
δ᾽ οὔκ ἐστιν αἴσθησις, ἐν τῷδε τῷ κόπῳ περίττωμα
193K μὲν οὐδέν, ὅ τι καὶ ἄξιον λόγου, περιέχεται τοῖς σώμασι, κατὰ δὲ τοὺς μῦς καὶ τὰ νεῦρα διάθεσίς τις ἐπὶ
ταῖς σφοδροτέραις ἐντάσεσιν, ἃς ἐποιήσαντο κατὰ τὰ
γυμνάσια, συνίσταται τοῦ ποιήσαντος αἰτίου τὴν δύναμιν ἐνδεικνυμένη. συμβαίνει γὰρ ἐν ταῖς σφοδροτέραις ἐντάσεσιν ἁπάσας μὲν τῶν μυῶν τείνεσθαι τὰς
ἶνας, οὐχ ὁμοίως δὲ ἁπάσας κάμνειν, ἀλλ᾽ ὅσαι μάλιστα κατὰ τὴν εὐθύτητα τῆς τάσεώς εἰσιν· ὡς ὅσαι γε
λοξότεραί πως ὑπάρχουσιν, ἧττον εὐθύνονται τεινό-
μεναι. ὥστε ταύταις μὲν οὐδεὶς ἐφεδρεύει κίνδυνος, ἐν
δὲ ταῖς ἐπὶ πλέον ἐκτεινομέναις, ὡς ἐγγὺς ἥκειν τοῦ
διασπασθῆναι, καταλείπεταί τις διάθεσις ὁμοία τῇ
κατὰ τὰς ἐνεργείας ἐγγινομένῃ· τείνεσθαι γὰρ ἔτι δο-
κοῦσι, κἂν μηκέτι τείνωνται.

ἡ δὲ δὴ τρίτη τοῦ κόπου διαφορά, καθ᾽ ἣν ὥσπερ
τεθλασμένων ἢ φλεγμαινόντων αἰσθανόμεθα τῶν μορίων, τηνικαῦτα μάλιστα συμπίπτειν ἔωθεν, ὅταν ἐκ-
θερμανθέντες ἱκανῶς οἱ μύες ἐπισπάσωνταί τι τῶν
περικεχυμένων ἑαυτοῖς περιττωμάτων. εἰ δὲ καὶ περὶ
τοὺς τένοντας ἢ τὰ νεῦρα τὴν αὐτὴν γενέσθαι διάθε-
194K σιν συμβαίνει, ὀστοκόπον ὀνομάζουσι τὸ πάθημα, τῷ

---

12 ὀστοκόπος (also ὀστεοκόπος, ὀστακόπος) is defined in
LSJ as, "an inflammatory attack which makes one feel as if one's

time, abundant. Such, certainly, is the kind of wound-like fatigue.

In that fatigue in which the parts seem only to be stretched to the uttermost, and there is not a wound-like sensation, there is, in this particular fatigue, no superfluity worth speaking of retained in the bodies, but in the muscles and sinews, there is a condition following the more violent exertions which they made during the exercises, indicative of the power associated with the effecting cause. What happens in the more violent strainings is that all the fibers of the muscles are stretched, but are not all similarly distressed. Rather, it is particularly those that are in a straight line with the tension; those that are in fact somewhat more oblique are kept straight less when being stretched. As a result, no danger lies in wait for these, whereas in those that are stretched still more, so as to come near to being ruptured, a certain condition remains like that which may develop during the actions, for they seem to be still being stretched, even if they are no longer stretched.

193K

The third *differentia* of fatigue, during which we have a sensation of the parts as bruised and inflamed, is especially wont to happen under those circumstances whenever the muscles are heated enough to attract some of the surrounding superfluities to themselves. And if the same condition happens to arise involving the tendons and sinews, they call the affection *ostokopos* (bone fatigue)[12] due

194K

bones are giving way." See also Hippocrates, *Acute (Sp.),* 1, Theophrastus, *On Fatigue,* and particularly Galen's *Sympt. Caus.* 2, VII.178–79K; Johnston, *Galen: On Diseases and Symptoms*, 253.

βάθει τῆς αἰσθήσεως ἐπὶ τὰ διὰ βάθους κείμενα
μόρια τοὔνομα φέροντες. ἐπιπολῆς μὲν γὰρ τὸ δέρμα,
δευτέραν δὲ ἔχουσι θέσεως τάξιν οἱ μύες, ἐν κύκλῳ
τοῖς ὀστοῖς περικείμενοι, συμφυεῖς δὲ οἱ τένοντες
ὑπάρχουσι τοῖς ὀστοῖς, ὥστ᾽ εὐλόγως, ὅταν οὗτοί τι
τῶν εἰρημένων πάσχωσιν, ἐν τῷ βάθει τε καὶ περὶ
τοῖς ὀστοῖς αὐτοῖς ἡ διάθεσις εἶναι δοκεῖ.

αὗται μὲν[19] δὴ τρεῖς ἁπλαῖ τῶν κόπων εἰσὶ διαφο-
ραί· σύνθετοι δ᾽ ἐξ αὐτῶν, ὡς ἔμπροσθεν εἴρηται, τέτταρες, ὑπὲρ ὧν ἑξῆς ἐροῦμεν, ἐὰν πρότερον τὸν περὶ
τῶν ἁπλῶν διέλθωμεν λόγον. ἔστι γὰρ δή τις καὶ
ἄλλη διάθεσις ἐξαπατῶσά τινας ὡς κόπος, ἧς ἡ μὲν
γένεσις ἐν τῷ ξηρανθῆναι τοὺς μῦς περαιτέρω τοῦ
προσήκοντος, ὥστε ἅπαν αὐχμηρὸν καὶ προσεσταλ-
μένον φαίνεσθαι τὸ σῶμα καὶ πρὸς τὰς κινήσεις
ὀκνεῖν ἀτρέμα, ἄλλο δ᾽ οὐδὲν ὑπάρχειν αὐτῷ τῶν
ἔμπροσθεν εἰρημένων, οὔτε τὴν οἷον ἕλκους αἴσθησιν
ἢ τάσεως, οὔτε δὲ πολὺ μᾶλλον τὴν οἷον φλεγμονῆς.
ἐναντιωτάτη γὰρ ἡ ὄψις τοῦ γε τοιούτου καὶ τῆς νῦν
195K λεγομένης διαθέσεώς ἐστιν. αὕτη μὲν γὰρ αὐχμώδεις
καὶ προσεσταλμένους ἀπεργάζεται τοὺς μῦς,[20] ὁ δὲ
φλεγμονώδης κόπος ἐν ὄγκῳ μείζονι καὶ αὐτοῦ τοῦ
κατὰ φύσιν. ὥστ᾽ εἶναι τὰς πάσας τέτταρας ἁπλᾶς
διαθέσεις, ἰδίας ἑκάστην ἐπανορθώσεως δεομένην.

6. Ἀρκτέον οὖν ἀπὸ τοῦ τὴν ἑλκώδη φέροντος
αἴσθησιν, ὃν καὶ διὰ δριμύτητα περιττωμάτων ἐλέγομεν συνίστασθαι. οὗτος ὁ κόπος συμπίπτει μὲν μάλι-

to the depth of the sensation in the case of the parts lying in the depths bearing the name (i.e., bones). For the skin is on the surface, while the muscles have the second order of position, lying around the bones in a circular fashion, while the tendons naturally grow with the bones. As a consequence, it is reasonable that whenever any of the things mentioned are affected, the condition seems to be in the depths and involving the bones themselves.

These are the three simple *differentiae* of the fatigues. Those compounded from them, as I said before, are four in number and I shall speak about them next, after I first complete the discussion about the simple fatigues. For there is, in fact, also another condition which deceives some people into taking it as a fatigue. Its origin is in the muscles being dried out beyond what is appropriate, so that the whole body appears parched and drawn tight and hesitates motionless with regard to movements. But there is nothing in it of those things previously mentioned—that is, no perception like wounding or tension, nor much more importantly, like inflammation. For the appearance is the very opposite of such a thing and of the condition described just now. This condition makes the muscles   195K
parched and drawn tight, whereas the inflammatory fatigue makes it greater in mass than normal. So there are four simple conditions in all, each requiring specific correction.

6. We must begin, then, from the fatigue producing the wound-like sensation which we said is established by acridity of superfluities. This fatigue happens particularly

---

19 *post* μὲν: δὴ Ko; αἱ μόναι Ku
20 τοὺς μῦς Ko; τὸ σῶμα Ku

στα τοῖς κακοχύμοις τε καὶ περιττωματικοῖς σώμασιν. ἐπιγίνεται δὲ καὶ ταῖς ὑπογυίοις ἀπεψίαις, ὅταν
ἤτοι γυμνάσωνται προπετέστερον ἢ ἐν ἡλίῳ διατρί-
ψωσιν. οὐ μὴν ἀδύνατόν γε αὐτὸν συστῆναί ποτε χωρὶς ἀπεψίας ἐν εὐχύμῳ σώματι δι' ὑπερβολὴν ἀμέτρων
γυμνασίων. εἰώθασι δ' αὐτὸν ὀξεῖαί τε καὶ πολλαὶ
φέρειν κινήσεις. πυκνὸν δὲ καὶ φρικῶδες φαίνεται τῶν
ἐν τούτῳ τῷ κόπῳ τὸ δέρμα, καὶ ὁμολογοῦσιν ἐν τῷ
κινεῖσθαι καθάπερ ἕλκος ἀλγεῖν, οἱ μὲν τὸ δέρμα μόνον, οἱ δὲ καὶ τὰς ὑπ' αὐτῷ σάρκας.

ἡ δ' ἴασις ἐξ ὑπεναντίου τῇ διαθέσει· διαφορῆσαι
196K γὰρ χρὴ τὰ περιττώματα, καὶ πέπαυται τὸ πάθημα.
διαφορηθήσεται δὲ τρίψει πολλῇ καὶ μαλακῇ σὺν
ἐλαίῳ μηδεμίαν ἔχοντι στύψιν, ὁποῖον μάλιστ' ἐστὶ
τὸ Σαβῖνον. ἐναντιώτατον δὲ τῇ διαθέσει τό τ' ἐκ τῆς
Ἱσπανίας καὶ τὸ ἐκ τῆς Ἰβηρίας, ὅπερ Ἱσπανὸν ὀνομάζουσι, τό τε καλούμενον ὀμφάκινον ἢ ὠμοτριβές·
ἑνὶ δὲ λόγῳ τὸ αὐστηρὸν ἅπαν οὐκ ἐπιτήδειον, ὡς
ἔνεστί σοι γευομένῳ διαγινώσκειν τὴν δύναμιν αὐτοῦ,
κἂν μήπω πρότερον ᾖς πεπειραμένος. οὕτω γοῦν καὶ
ἡμεῖς ἐν Μακεδονίᾳ ποτὲ γευσάμενοι τοῦ κατὰ τὸν
Αὐλῶνα τὸν περὶ τῷ Στρυμόνι γεννωμένου[21] τῆς αὐτῆς
εἶναι δυνάμεως ἐγνωρίσαμεν αὐτὸ τῷ καλουμένῳ
Ἱσπανῷ. καὶ μὲν δὴ καὶ τῶν ἄλλων ἐλαίων ἁπάντων
οὕτω γνωρίσεις τὴν δύναμιν, ὅσα τε καταχρηστικῶς

---

[21] γεννωμένου Ko; γινομένου ἐλαίου Ku

in *kakochymous* and excrementitious bodies. It also supervenes in the sudden *apepsias* (failures of concoction) when people either exercise too rashly or spend time in the sun. It is not impossible for it to exist sometimes apart from *apepsia* in a *euchymous* body due to an excess of immoderate exercises. Movements that are rapid and many are wont to produce it. The skin of those with this fatigue appears condensed and rough, and [the sufferers] agree that in moving they feel pain like a wound, some in the skin alone and some in the underlying flesh.

The cure is from what is opposite to the condition: it is necessary to disperse the superfluities and the affection ceases. They will be dispersed by much gentle massage with oil having no astringency; a particular example is the Sabine. The most opposite to the condition is the oil from Spain and that from Iberia, which they call Hispanic, and the so-called *omphakinos* or *omotribes* (oil from unripe olives);[13] in a word, all astringent oil is not beneficial, as it is possible for you to recognize by tasting its the potency, even if you have not experienced it before. At all events, in this way, once in Macedonia, when I tasted the oil arising from Aulis around Strymon,[14] I recognized its potency to be the same as that of the so-called Hispanic. Furthermore, you will also recognize the potency of all the other oils in this way—those that are named catachrestically and

196k

[13] Both are taken to refer to the oil from unripe olives—see Dioscorides 1.29.

[14] Aulis was a small Greek town near Tanagra. It was where Iphigenia was sacrificed to ensure a safe journey for the Achaean fleet.

ὀνομάζεται καὶ ὅσα σκευάζεται δι᾽ ἀνθῶν[22] ἢ ῥιζῶν ἢ
βοτανῶν ἢ φύλλων ἢ βλαστῶν ἢ καρπῶν.

εἴρηται δ᾽ αὐτάρκως μὲν ὑπὲρ αὐτῶν ἁπάντων ἐν
τῇ περὶ φαρμάκων πραγματείᾳ· λεχθήσεται δὲ καὶ
κατὰ τὸν ἐνεστῶτα λόγον ἐν οἰκείῳ καιρῷ. νῦν δὲ
197K τοῦτ᾽ ἀρκεῖ μόνον εἰπεῖν, ὡς, ὅπερ ἂν ᾖ γλυκύτατον
ἔλαιον, ἐπιτηδειότατόν ἐστιν εἰς τὰ παρόντα. τούτῳ
τοίνυν χρῆσθαι δαψιλεῖ μετὰ τρίψεως πολλῆς, ἐν μὲν
τῇ πρώτῃ τῶν ἡμερῶν ὑπὲρ τοῦ μηδ᾽ ὅλως γενέσθαι
τὸν ὑποπτευόμενον ἔσεσθαι κόπον, ἐν δὲ τῇ δευτέρᾳ
χάριν τοῦ λῦσαι τὸν ἤδη γεγονότα. λύει δ᾽ αὐτὸν τὸ
καλούμενον ἀποθεραπευτικὸν γυμνάσιον, ἐν ᾧ καὶ κινήσεις ἔνεστι ποιεῖν συμμέτρους μὲν τῇ ποσότητι,
βραδυτέρας δὲ τῇ ποιότητι, μετὰ πολλῶν τῶν μεταξὺ
διαναπαύσεων, ἐν αἷς χρὴ τρίβειν τὸν ἄνθρωπον,
ἐφαπτομένων ὁμοῦ πλειόνων, ὅπως μήτε καταψύχοιτό
τι μέρος αὐτοῦ καὶ τάχιστα διαφοροῖτο τὰ περιττώματα. πλεονάζειν δὲ χρὴ ταῖς μὲν τρίψεσι κατὰ τὸν
ἐν τῷ δέρματί τε καὶ ὑπὸ τῷ δέρματι τὸ πλῆθος τῶν
περιττωμάτων ἔχοντα, ταῖς δ᾽ ἐξ ἑαυτοῦ κινήσεσι
κατὰ τὸν ἕτερον κόπον, ᾧ τὸ πλέον ἐν τοῖς μυσὶν
ἤθροισται. τὰ γὰρ ἐν τούτοις περιττώματα τρῖψις
μόνη διαφορεῖν οὐχ ἱκανή. δεῖται γὰρ οὐκ ἔξωθεν μόνον ἕλκεσθαι πρός τινος, ἀλλὰ καὶ συναπωθεῖσθαι
198K πρὸς ἑτέρου τινὸς ἔσωθεν. ἐπωθεῖ δ᾽ αὐτὰ τό τ᾽ ἀν-
απτόμενον θερμὸν ἐν ταῖς κινήσεσι καὶ τὸ συνεκκρι-

---

[22] ἀνθῶν Ko; ἁλῶν Ku

those that are prepared from flowers, roots, herbs, shoots, leaves or fruits.

Enough has been said about all these in the work on medications[15] and will be said in the present work at an appropriate time. For now it suffices to say this alone— 197K whichever oil is sweetest is most suitable for the present purposes. Accordingly, use this abundantly with much massage. On the first of the days, in order that the fatigue which is suspected of being imminent may not happen at all, and on the second day, for the sake of resolving that which has already come about. The so-called apothera-peutic exercise, in which it is also possible to make movements which are moderate in quantity and slower in quality, along with many rests in between, dispels this. During rests, it is necessary to massage the person, many together applying it, so no part of him is chilled and the superfluities are very quickly dispersed. It is, however, necessary to use the massages in excess in someone who has a large amount of superfluities in and under the skin, but to increase the movements of the person himself in another kind of fatigue in which the majority [of the superfluities] are collected in the muscles. Massage alone is not sufficient to disperse the superfluities in these, for they not only need to be drawn toward something externally but also to be pushed on by something else internally. The 198K heat which is kindled in the movements pushes them on,

[15] Galen's three major works on materia medica / pharmacology are listed in note 37, p. 106 above. See particularly *Simpl. Med.*, XI.471Kff.

νόμενον πνεῦμα καὶ ἡ αὐτῶν τῶν μυῶν ἔντασις, ἐξ
ἐπιμέτρου δὲ καὶ ἡ καθ᾽ ἕκαστον τῶν μορίων ἀποκρι-
τικὴ τῶν ἀλλοτρίων δύναμις.

ὁ δὲ ἕτερος κόπος, ἐφ᾽ οὗ συντάσεως αἰσθάνονται,
τὸν σκοπὸν τῆς ἰάσεως ἔχει τὴν πρὸς Ἱπποκράτους
ὀνομαζομένην χάλασιν· ἐναντίον γὰρ τοῦτο τῇ συν-
τάσει, καθάπερ τῇ σκληρότητι ἡ μάλαξις. ἔλεγεν
οὖν ὧδε· "δέρματος σκληροῦ μάλαξις, συντεταμένου
χάλασις," ὡς ἐναντίον ὑπάρχον τῷ μὲν σκληρῷ τὸ
μαλακόν, τῷ δὲ συντεταμένῳ τὸ χαλαρόν. χαλᾶται δὲ
τὸ συντεταμένον ἐν μὲν ταῖς ἄλλαις διαθέσεσιν, ἃς ἐν
τῷ πέμπτῳ Περὶ τῆς τῶν ἁπλῶν φαρμάκων δυνάμεως
εἴπομεν, ἑτέρως, ἐπὶ δὲ τῇ διὰ τὰ γυμνάσια τρίψει μὲν
ὀλίγῃ τε ἅμα καὶ μαλακῇ δι᾽ ἐλαίου γλυκέος εἰληθεροῦς, ἀναπαύσει τε ὅλως ἢ ἡσυχίᾳ ἢ καὶ λουτροῖς
εὐκράτοις καὶ διατριβῇ πλέονι κατὰ τὸ θερμὸν ὕδωρ,
ὥστε, εἰ καὶ δὶς αὐτὸν ἢ καὶ τρὶς λούσαις, ὀνήσεις
μειζόνως. οὗτοι καὶ μετὰ τὰ βαλανεῖα ἀλείφεσθαι
δέονται πρὶν ἀμφιέννυσθαι· καὶ εἰ δι᾽ ἱδρῶτά τινα τύχοιεν ἀπομάξαντες τὸ λίπος, αὖθις ἀλείφεσθαι χρή-
ζουσι. καὶ μέντοι καὶ κατὰ τὴν ἑξῆς ἡμέραν ἀναστάν-
τες ἐκ τῆς κοίτης ἀλειφθῆναι δέονται, μηδέποτ᾽ ἄκρως
ψυχρῷ τῷ ἐλαίῳ μηδὲ σκληρῶς ἀνατριβόμενοι. γίνεται δ᾽ ὁ τοιοῦτος κόπος εὐχύμοις ἀνδράσι πονήσασιν
εὔτονα μᾶλλον ἢ ὀξέα γυμνάσια, καὶ δεινῶς ὀκνηρούς
τε καὶ δυσκαμπεῖς ἐργάζεται τοὺς κοπωθέντας, οὐ
μὴν πυκνοί τε καὶ φρικώδεις οἱ τοιοῦτοι φαίνονται,
καθάπερ οἱ μικρὸν πρόσθεν εἰρημένοι. προσεσταλμέ-

as does the accompanying evacuated breath (*pneuma*) and the tension of the muscles themselves, and over and above these, there is the eliminative capacity for alien things in each of the parts.

The other fatigue in which people are aware of tension has, as the aim of the cure, what Hippocrates called relaxation; for this is opposite to tension, just as softening is to the hardness. What he said was this: "Softening of hard skin and relaxation of tension," as the soft is opposite to the hard and the relaxed is opposite to the tense.[16] Relaxation of what is tense in the other conditions, which I spoke about in the fifth book of *On the Nature and Powers of Simple Medications*,[17] is different from that due to exercises, which will stop completely after a small amount of soft massage with sweet oil warmed by the sun, or with rest, *eukratic* baths and spending a longer time in warm water, so that if you also bathe the person two or three times, you will help rather more. These people also need to be anointed with oil after the baths before they are dressed. And if, due to some sweat, they happen to have wiped off the oil, they need to be anointed again. And, of course, also on the following day, when they get out of bed, they need to be anointed, but never with extremely cold oil, nor should they be massaged firmly. Such a fatigue occurs in *euchymous* men who have exerted themselves in vigorous rather than rapid exercise, and it makes those who are fatigued terribly sluggish and inflexible, although such men do not appear condensed and shivering, like those I spoke of a little earlier. But these men are no less

199K

---

[16] Hippocrates, *On the Use of Liquids* 1, *Hippocrates* VIII, LCL 482, 320–21.   [17] *Simpl. Med.*, XI.741K.

νοι δ᾽ οὐδὲν ἧττον ἐκείνων οὗτοι καὶ αὐχμώδεις ὁρῶν-
ται, καὶ θερμότεροι τοῖς ἁπτομένοις εἶναι δοκοῦσιν οὐ
μόνον τῶν τὴν ἑλκώδη διάθεσιν ἐχόντων, ἀλλὰ καὶ
σφῶν αὐτῶν, ὅθ᾽ ὑγίαινον.

7. Ὁ δὲ τρίτος τῶν κόπων ἐπὶ σφοδροτάταις γίνεται
κινήσεσι καὶ μόνος ἐξαίρει τοὺς μῦς εἰς ὄγκον ὑπὲρ
τὸ κατὰ φύσιν, ὡς ἐοικέναι φλεγμονῇ τὴν διάθεσιν
αὐτῶν. ταῦτά τοι καὶ ψαυόντων ὀδυνῶνται καὶ θερμότεροι φαίνονται· ὀδυνῶνται δὲ καί, ἢν αὐτοὶ καθ᾽
ἑαυτοὺς ἐπιχειρήσωσι κινεῖσθαι. ἀήθεσι δὲ γυμνασίων ἀνθρώποις ὁ τοιοῦτος κόπος ὡς τὰ πολλὰ συμπίπτει, γυμνάζεσθαι δ᾽ εἰθισμένοις ὀλιγάκις ἐγένετο
κατὰ τὰς σφοδροτάτας τε ἅμα καὶ παμπόλλας κινήσεις. ἡ δὲ ἴασις αὐτοῦ τρεῖς ἔχει τοὺς σκοπούς,
οὕσπερ σχεδόν τι καὶ τὰ φλεγμαίνοντα σύμπαντα,
κένωσιν τοῦ περιττώματος καὶ ἀνάπαυσιν τοῦ συντεταμένου καὶ ἀνάψυξιν τοῦ φλογώδους. ἔλαιόν τε οὖν
πολὺ χλιαρὸν αἵ τε τρίψεις μαλακώτεραι καὶ ἡ ἐν τοῖς
εὐκράτοις ὕδασι διατριβὴ πολυχρονιωτάτη τοὺς τοιούτους ἰᾶται κόπους. εἰ δὲ καὶ βραχύ τι χλιαρώτερον
εἴη τὸ ὕδωρ, ὀνήσει μᾶλλον. οὕτω δὲ καὶ ἡσυχία
πολλὴ καὶ ἀλείμματα συνεχῆ καὶ πάνθ᾽ ὅσα τὸ μὲν
κεκμηκὸς ἀναπαύει τε ἅμα καὶ παρηγορεῖ, τὸ δὲ
περιττὸν διαφορεῖ.

τάχα δ᾽ ἄν τις οἰηθείη τὸν τοιοῦτον κόπον οὐχ
ἁπλοῦν οὐδὲ τρίτον ἐπὶ τοῖς εἰρημένοις ἔμπροσθεν
δύο, σύνθετον δ᾽ ὑπάρχειν ἐξ αὐτῶν, οὐδὲν ἔχοντα
πλέον τῆς τε τάσεως τῶν νευρωδῶν σωμάτων καὶ τῆς

drawn tight than those, and look dried out and seem hotter to those who touch them, not only than those with the wound-like condition, but also than they themselves were when healthy.

7. The third of the fatigues occurs after very violent movements and alone raises the muscles to a swelling beyond an accord with nature, so their condition seems like inflammation. They are certainly painful when touched and seem overly hot; and [those affected] suffer pain too, if they attempt to move by themselves. Such a fatigue for the most part befalls men unaccustomed to exercises; it has rarely occurred in those accustomed to exercises, even with movements that are very violent and prolonged. The cure of this has three aims which are those of almost all the inflammations—evacuation of the superfluity, putting an end to the tension, and cooling of what is inflamed. Thus, oil that is abundant and lukewarm, massages that are quite gentle, and a very prolonged time spent in *eukratic* waters cure such fatigues. If the water is slightly more lukewarm, it will help more. In this way too, a lot of rest, continuous use of unguents, and all those things that relieve and at the same time provide comfort for what is fatigued, and disperse the superfluity [will help].

Perhaps someone might think such a fatigue is neither simple nor a third form following the two previously spoken of, but is a compound of these, being nothing more than the stretching of the sinewy bodies and the wound-

200K

ἑλκώδους αἰσθήσεως· τὴν γὰρ θερμότητα τοῖς τοιούτοις κόποις ἄλλως μὲν ὑπάρχειν φύσει, καθάπερ καὶ
τῶν προειρημένων ἑκατέρῳ, οὐ μὴν συμπληρωτικήν
γε τῆς ἐννοίας ἢ τῆς οὐσίας εἶναι. ἀλλά τοι τό γε τοῦ
παρὰ φύσιν ὄγκου τούτῳ τῷ κόπῳ μόνῳ παρὰ τοὺς
ἄλλους ἐξαίρετον ὑπάρχει καὶ τὸ τῆς ἀλγεινῆς αἰσθήσεως οὐχ ὅμοιον ἔν γε τῷ τονώδει κόπῳ καὶ τῷδε.
τείνεσθαι μὲν γὰρ ἐν ἐκείνῳ, τεθλάσθαι δὲ τὰ νεῦρα
σύμπαντα μέχρι καὶ τῶν ὀστῶν οἱ τούτῳ τῷ κόπῳ
κατεχόμενοι νομίζουσιν· ὥστε κατά γε ταῦτα διαφοράν τινα ἐξαίρετον ἔχει παρὰ τοὺς ἄλλους δύο κόπους, οὐχὶ σύνθετός ἐστι μόνον.

αὗται μὲν δὴ τρεῖς εἰσιν, εἴτε καταστάσεις σώματος εἴτε διαθέσεις εἴθ' ὅ τι βούλεταί τις ὀνομάζειν.
ἄλλη δ' ἐπ' αὐταῖς τετάρτη, παραπλησία μὲν ὑπάρχουσα κόπῳ, κόπος δ' οὐκ οὖσα, τῷ μήτε τὴν ἑλκώδη
μήτε τὴν τονώδη μήτε τὴν φλεγμονώδη διάθεσιν
ἔχειν, ἀλλὰ μηδὲ φρίκην τινὰ μηδ' ἄλγημα μηδὲ τὸν
πρὸς τὰς κινήσεις ὄκνον ὅμοιον τοῖς κόποις ἐπιφέρειν,
ἰσχνότητα δὲ μόνην ἅμα ξηρότητι. γίνεται μὲν οὖν ἐν
εὐχύμοις τε ἅμα καὶ γυμναστικοῖς σώμασιν, ὅταν
ἀμετρότερον γυμνασθέντα μὴ καλῶς ἀποθεραπευθῇ.
διαφορεῖται γὰρ οὕτω τὰ περιττώματα καὶ χαλᾶται
τὰ τεταμένα καὶ οὐδὲν ἄλλο ὑπολείπεται κατὰ τὸ
σῶμα πλὴν ξηρότητος, ἣν ἐκ τῆς ἀμετροτέρας κινήσεως ἔσχον.

δεῖται δὲ κατὰ μὲν τὴν πρώτην ἡμέραν οὐδενὸς
ἐξηλλαγμένου παρὰ τὰ πρόσθεν, ὅτι μὴ θερμοτέρου

like sensation, for the heat in such fatigues is different in nature, as is that in each of those previously mentioned, and is not an essential part of either the concept or the essence. But certainly the presence of the swelling contrary to nature in this fatigue alone is remarkable compared to the others, and the presence of the painful sensation is not the same in the tensive fatigue and in this one. For there is tension in the former, whereas those possessed by this fatigue think all the sinews are bruised right to the bones. So, because of this feature, there is a certain notable difference compared to the other two fatigues—it is not just a compound. 201K

Surely, then, there are three [fatigues], and they are either states of the body or conditions or whatever someone wishes to call them. But there is another and fourth in addition to these, which is like a fatigue without actually being a fatigue in that it is neither the wound-like, nor the tensive, nor the inflammation-like condition. But neither is there shivering, pain or hesitancy of movement like that which accompanies the fatigues; there is only thinness along with dryness. It occurs in *euchymous* and trained bodies whenever they are exercised immoderately without proper apotherapy. For in this way, the superfluities are dispersed, those structures that are tense are relaxed, and nothing else remains in the body except dryness, which 202K comes from the more excessive movement.

During the first day nothing needs to change from the previous days, unless the water is hotter so as to draw to-

τοῦ ὕδατος, ὡς συναγαγεῖν ἀτρέμα καὶ θερμῆναι καὶ
τονῶσαι τὸ δέρμα, κατὰ δὲ τὴν δευτέραν ἀποθεραπευτικοῦ γυμνασίου βραχέος τε ἅμα καὶ μαλακοῦ καὶ
βραδέος ἔν τε κινήσεσι καὶ τρίψεσι καὶ τῆς δεξαμενῆς ὁμοίως θερμῆς. ἐκπηδάτωσαν δ᾽ εὐθέως εἰς τὴν
ψυχρὰν ὑπὲρ τοῦ μένειν αὐτοῖς τὸν ἐν τῷ δέρματι
τόνον ἅμα θερμότητι. καὶ γὰρ ἧττον ἐν τῷ μετὰ ταῦτα
χρόνῳ διαφοροῦνται καὶ ῥᾳδίως εἴς τε τὰς σάρκας
καὶ τὸ δέρμα τὴν τροφὴν ἀναλαμβάνουσιν, οὗ μεῖζον
ἀγαθὸν οὐδὲν ἂν ἐξεύροις αὐτοῖς, οὐδεμίαν γε διάθεσιν ἐξαίρετον ἔχουσι παρὰ τὴν τῆς σαρκὸς ἰσχνότητα
καὶ ξηρότητα. δεῖται δ᾽, οἶμαι, τό γε τοιοῦτον ἀνατρα-
φῆναί τε ἅμα καὶ ὑγρανθῆναι, καὶ ταῦτ᾽ ἄμφω κάλ-
λιστ᾽ αὐτῷ γίνεσθαι πέφυκεν ἐκ τῆς ὑγραινούσης
τροφῆς.

203K   8. Ἐπειδὴ δὲ ἅπαξ ἐμνημόνευσα τῆς ἐπὶ τοῖς λουτροῖς διαίτης, οὐ χεῖρον ἂν εἴη καὶ τὰ περὶ τῶν κοπωδῶν διαθέσεων ἐπεξελθεῖν. ὁ μὲν οὖν ἑλκώδης κόπος, εἰ μὲν ἱκανῶς ἀποθεραπευθείη, τῆς συνήθους
δεῖται τροφῆς ἤ τι βραχὺ μείονος, ἔτι δ᾽ ὑγροτέρας
τε ἅμα καὶ ἐλάττονος· εἰ δὲ κατὰ τὴν ἀποθεραπείαν
εἰς τὴν τετάρτην μεταπέσοι διάθεσιν (εἴωθε γὰρ οὕτω
γίνεσθαι τὰ πολλά), κατ᾽ ἐκείνην καὶ λουέσθω καὶ
τρεφέσθω. ὁ δὲ τονώδης ἔτι δὴ καὶ μᾶλλον ὀλιγω-
τέρας δεῖται τροφῆς· ὁ δὲ φλεγμονώδης ὑπὲρ ἅπαντας
ὑγροτάτης τε καὶ βραχυτάτης καί τι καὶ ψῦχον ἐχούσης. εὐχύμου δ᾽ ὁμοίως πάντες οἱ κεκοπωμένοι δέονται τροφῆς, ὁποίᾳ δηλονότι καὶ ὑγιαίνων ὁ ὑποκείμε-

gether gently, heat and brace the skin. However, during
the second day, there is need of apotherapeutic exercise
that is brief, soft, and slow in movements and massages;
similarly, there is need of the hot tank. Let those so affected jump out immediately into the cold tank, for the
sake of preserving in them the tone in the skin along with
the heat. For in the time after these things, they are dissipated less and easily take up nourishment to the flesh
and skin. You would find nothing better for them as they
have no notable condition apart from thinness and dryness
of the flesh. Such a body needs, I think, to be built up and
moistened, and both these things best come about in it
naturally from nutriment that moistens.

8. Since I have only once mentioned the diet after   203K
baths, it would be no bad thing to go over those matters
also concerning the fatigue conditions. Thus, the wound-
like fatigue, if it is to be treated adequately with apotherapy, requires the customary nourishment or slightly
less, and less of those nutrients that are more moist. If, in
relation to apotherapy, it should undergo a change to the
fourth condition (for this is wont to occur in many instances), in that condition, both bathe and nourish [the
person]. However, the tensive fatigue needs even more
reduction of nutriment, while the inflammation-like fatigue above all needs nutriment that is very moist and very
small in amount, and which also contains something cold.
All those who are fatigued have a similar need for *euchymous* nutriment of the kind, clearly, that the healthy young

νος ἐν τῷ λόγῳ νεανίσκος ἐχρῆτο. φυλάττεσθαι δὲ
προσήκει τὸ γλίσχρον ἐν αὐτῇ κατά τε τὸν ἑλκώδη
καὶ τὸν φλεγμονώδη κόπον, ὡς ἂν κωλῦον διαφορεῖσθαι τὰ περιττώματα. κατὰ μέντοι τὸν τονώδη καὶ τὰ
τοιαῦτα σιτία δοτέον, ἀφαιροῦντα τοῦ πλήθους αὐτῶν.

204K   οὐδὲν οὖν θαυμαστὸν ἐναντιολογίαν εἶναι πολλὴν
οὐκ ἰδιώταις μόνον ἐν ἀλλήλοις, ἀλλὰ καὶ τεχνίταις
πρὸς ἑαυτούς τε καὶ τοὺς ἰδιώτας, οὔτε περὶ γυμνασίων οὔτε περὶ τρίψεων οὔτε περὶ λουτρῶν οὔτε
περὶ διαιτημάτων ὁμολογοῦσιν ὑπὲρ τῶν κοπωθέντων. ἀκοῦσαι γοῦν ἔστι τῶν μὲν φασκόντων, ὡς κόπῳ
χρὴ λύειν τὸν κόπον, ἑτέρων δέ, ὡς ἀνάπαυσις ἰᾶται
τὸν κάματον, καὶ τῶν μέν, ὡς ἐνδεῶς χρὴ διαιτᾶσθαι
τοὺς κοπωθέντας, ἄλλων δέ, ὡς οὐ μόνον ⟨οὐκ⟩²³
ἀφαιρεῖν προσήκει τῶν εἰθισμένων, ἀλλὰ καὶ τοσούτῳ πλείω προσφέρειν, ὅσῳπέρ τις ἔτυχε πλείω
γυμνασάμενος ἀνὰ λόγον γὰρ χρῆναι τοῖς πόνοις
προσφέρεσθαι²⁴ τὰς τροφάς, ἄλλων δέ τινων, ὡς οὔτε
προστιθέναι χρὴ ταῖς τροφαῖς οὔτ᾽ ἀφαιρεῖν. οὕτω δὲ
καὶ λούουσιν οἱ μὲν εὐκράτοις ὕδασιν, οἱ δὲ θερμοτέροις, οἱ δὲ χλιαρωτέροις. ἤ τε γὰρ ἐμπειρία πρὸς
ἐκεῖνο μόνον ἕκαστον ἀπάγει, πρὸς ὅπερ ἔτυχε θεασάμενος πολλάκις, ὅ τε λόγος, ὡς ἂν μὴ τέλειος
ὑπάρχων ἑκάστῳ, καθάπερ ἐδείκνυτο πρόσθεν, ἀλλὰ
μίαν τινὰ διάθεσιν ἐκδιδάσκων κοπώδη, τὰς δ᾽ ἄλλας
205K   ὥσπερ οὐκ οὔσας ὑπερβαίνων, ἐκείνης μόνης ἐκδιδά-
σκει τὴν ἐπανόρθωσιν, ἧς ἔγνω μόνης.

man assumed in the discussion might use. It is appropriate to guard against what is viscous in the nutriment, both in the wound-like and inflammation-like fatigues, as it would prevent the superfluities being dispersed. However, in the tensive fatigue, one must give such foods while keeping away from an excess of them.

It is not surprising, then, that there is a great difference 204K of opinion, not only in laymen between one another, but also between experts among themselves and with respect to layman; since they agree neither about exercises, nor massages, nor baths, nor regimen for those who are fatigued. Anyway, it is possible to hear some claiming it is necessary to resolve fatigue with fatigue, while there are others who say that rest cures the weariness; and some, that it is necessary to feed those who are fatigued sparingly, and others who say it is not only inappropriate to take away those things that are customary nourishment, but also more should be provided to the extent that they happen to have been exercising more, for it is necessary to provide nutriments in proportion to their exertions. But there are others who say it is neither necessary to add to the nutriments nor to take them away. It is the same with baths. Some use *eukratic* waters, some hotter waters and some more lukewarm waters. For experience leads each person only to that which he has happened to see often and theory is not conclusive in each case, as was shown before. And when a person is taught one particular fatigue condition, he passes over the others as not existing and teaches the correction only of that one which alone he has 205K known.

---

23 οὐκ *add.* Ko   24 προσφέρεσθαι Ko; προσαίρεσθαι Ku

ἀληθὲς γοῦν ἐστι καὶ τὸ κόπῳ λύεσθαι τὸν κόπον,
ὅταν γε δὴ φαίνηταί ποτε δέον εἶναι τοῖς ἐν τῇ προ-
τεραίᾳ γυμνασίοις ἴσα κατὰ τὴν ὑστεραίαν γυμνάζεσθαι, καὶ τὸ τὴν ἴασιν τῶν κόπων ἡσυχίαν ὑπάρχειν.
τούτων γὰρ τὸ μὲν ἐν τοῖς περιττωματικοῖς κόποις,
καὶ μάλισθ᾽ ὅσοι κατὰ τοὺς μῦς ἔχουσι τὰ περιττώματα, τὸ δ᾽ ἐν τοῖς τονώδεσί τε καὶ φλεγμονώδεσι
φαίνεται συμφέρον. ἀληθὲς δὲ καὶ τὸ χρῆναι τοὺς
κοπωθέντας ἐνδεῶς διαιτᾶσθαι· τοῦτο γὰρ ἑώραται
τοὺς φλεγμονώδεις ὠφελοῦν. ἀληθὲς δὲ καὶ τὸ τὰ
συνήθη διδόναι· τοῦτο γὰρ ἐπὶ τῶν ἑλκωδῶν, ὅταν
ἀποθεραπευθῶσιν, ὀρθῶς τετήρηται γινόμενον, ὥσπερ
γε καὶ τὸ πλείω λαμβάνειν ἐπὶ τῆς ὁμοιουμένης τοῖς
κόποις διαθέσεως, εἰς ἣν κἀκ τῆς ἑλκώδους ἔνιοι
μεταπίπτουσιν.

οὕτω δὲ καὶ τὸ λούειν εὐκράτοις ὕδασιν ἀληθὲς ἐπὶ
τοῖς περιττωματικοῖς κόποις. ἀληθὲς δὲ καὶ τὸ μὴ
206K λούειν τοιούτοις ἐπί τε τῶν φλεγμονωδῶν χλιαρωτέρων γὰρ οὗτοι δέονται κἀπὶ τῆς ὁμοίας κόπῳ διαθέσεως οὗτοι γὰρ οὐ χρῄζουσι θερμοτέρων. οἴονται δ᾽
ἔνιοι καὶ τὴν τοιαύτην διάθεσιν εἶναι κόπον. ἕτεροι δέ
τινες ὁποία μέν τίς ἐστιν ἡ διάθεσις οὐκ εἶπον, ἐφ᾽ ὧν
δ᾽ ἂν ὑποψία τις ᾖ γενησομένου κόπου συμβουλεύ-
ουσι θερμοτέρῳ χρῆσθαι τῷ κατὰ τὸ λουτρὸν ὕδατι
καὶ προστιθέασι δὲ τὴν αἰτίαν, ὡς ἡμεῖς ἔμπροσθεν
ἔφαμεν, οἱ μὲν ἀναδόσει φάσκοντες, οἱ δὲ θρέψει συν-
τελεῖν τὸ τοιοῦτον λουτρόν, οὐ μὴν καθ᾽ ὅντινά γε
λόγον οἷά τε κωλύειν ἐστὶ κόπον ἀνάδοσίς τε καὶ θρέ-

At any rate, both are true: fatigue is resolved by fatigue, whenever there obviously seems to be a need, on a particular occasion, for the exercises of the previous day to be practiced to an equal extent on the day after, and rest is the cure of the fatigues. Of these [two approaches], the former applies in the excrementitious fatigues, and particularly those who have the superfluities in the muscles, while the latter seems beneficial in the tensive and inflammation-like fatigues. It is also true that you must feed those who are fatigued a reduced diet, for this is seen to benefit the inflammation-like fatigues. And it is also true that you must give the customary nutriments, when people are receiving apotherapy. In the case of the wound-like fatigues, this is properly protective, just as it also is to take more [food] in the case of the condition corresponding to the fatigues, into which some change from the wound-like fatigues.

In the same way too, it is true that you must bathe in *eukratic* waters for the excrementitious fatigues. It also true that you must not bathe in such waters for the inflammation-like fatigues, for these require more lukewarm 206K water; in the condition resembling fatigue, they do not need warmer water. Some also think that such a condition is a fatigue. Some others, however, do not say what kind of condition it is; among these are those who, when there is some suspicion it will become a fatigue, advise the use of warmer water in the bath, adding the reason, as I said before, that such a bath contributes to distribution and others that it contributes to nutrition. But they do not expound any reason as to why distribution and nutrition

ψις ἐκδιδάσκουσιν, ἀλλ᾽ οὐδὲ τὴν ἀρχὴν ὁποία τίς
ἐστιν ἡ τοῦ κόπου διάθεσις, οὐδὲ τοῦτο γράφουσιν. οἱ
γοῦν πλείους αὐτῶν οὐδὲν εἶπον, ἔνιοι δὲ τολμήσαντες
εἰπεῖν ἀπεφήναντο ξηρότητα. τὸ μὲν οὖν ὑπὸ τροφῆς
θεραπεύεσθαι ξηρότητα καὶ μάλιστα τῆς ὑγραινούσης, ὅπερ οὐδ᾽ αὐτὸ προστιθέασιν οἱ πλείους αὐτῶν,
ἀληθέστατον, οὐ μὴν οὔτ᾽ ἴαμα τῶν κόπων ἐστὶν οὔτε
προφυλακὴ τὸ τοιοῦτον. ἀλλ᾽ ὅταν, ὡς εἴρηται, ξηρότερα μὲν ἀπεργασθῇ τὰ μόρια, μήτε δὲ τάσις ᾖ
207K   κατ᾽ αὐτὰ μηδεμία μήτε περιττώματα λεπτὰ καὶ δριμέα μήτε φλεγμονώδης διάθεσις, ἀνατρέφειν δήπου
προσήκει τὰ διὰ τὴν κένωσιν ἐξηραμμένα τροφαῖς
ὑγραινούσαις, ὡς, εἴ γε μὴ ἀναθρέψειας αὐτάρκως,
ἰσχνότερόν τε καὶ ξηρότερον ὄψει κατὰ τὴν ἑξῆς ἡμέραν ἑαυτοῦ τὸ σῶμα.

    καὶ ταύτην τὴν διάθεσιν ἔνιοι κόπον ὑπολαβόντες
ὑπάρχειν οἴονται θερμοτέρῳ λουτρῷ καὶ τροφῇ δαψιλεῖ κεκωλυκέναι τὴν γένεσιν αὐτῆς. ἀλλ᾽ ἐκεῖνο θαυμάζειν ἐνίων ἄξιον, ὥσπερ καὶ Θέωνος, εἰ τὸν ἤδη
γεγενημένον κόπον οὐκ ἀξιοῦσιν ὁμοίως ἰᾶσθαι. εἴτε
γὰρ ἦν νῦν εἴρηκα διάθεσιν ὑπολαμβάνει τις εἶναι
κόπον, ἀναθρέψει χρὴ λύειν αὐτὸν καὶ κωλύειν γενέσθαι, εἴτε τῶν ἄλλων τινὰ τῶν ὄντως κόπων, ἐνδε-
έστερον ἐπὶ πάντων χρὴ τρέφεσθαι τὸν ἄνθρωπον, ἔτι
τε προσδοκωμένων καὶ ἤδη γεγονότων. οὐδὲ γὰρ
τοῦτ᾽ ἔστιν εἰπεῖν, ὡς ἡ τετάρτη[25] διάθεσις ἡ παρακει-

────────

[25] τετάρτη Ko; τοιαύτη Ku

are the kinds of things that prevent fatigue, nor do they
set down what sort of condition fatigue is in the first place.
Anyway, the majority of them said nothing, although
some, when they dared to speak, declared flatly that it was
dryness. What is most true—and the majority of them do
not add this—is that dryness is treated by nutriment and
particularly by moist nutriment, but this is not a cure of
the fatigues, nor is it prophylactic for such a thing. Rather
when, as I said, the parts are made too dry, there is not any
tension in them at all, nor are the superfluities thin and   207K
acrid. Nor is it an inflammation-like condition, although it
may of course be fitting to restore with moist nutriments
what has been taken away from these parts dried by the
evacuation, because, if you were not to feed sufficiently,
you would see on the following day the body thinner and
drier than it was.

There are some who, since they assume this particular
condition to be a fatigue, think they have prevented the
genesis of it with a hotter bath and abundant nutriment.
But what is deserving of surprise among these people, of
whom Theon is an example, is that if the fatigue has already come about, they do not think it worthwhile to treat
it in a similar fashion. For if someone assumes the condition I have spoken about just now is a fatigue, he must
resolve it or prevent it occurring by restoration,[18] and if it
is one of the other genuine fatigues, the greater need in
all cases must be to nourish the person, whether the fatigue is anticipated or has already occurred. For this cannot be said: that the fourth condition lying close to the

---

[18] On the use of ἀναθρέψεις here (given in LSJ as "renewal"
or "restoration"), see Hippocrates, *Aphorisms* 1.3.

μένη τοῖς κόποις, εἰ μὴ κατὰ τὴν πρώτην ἡμέραν
ἐπανορθωθείη δι᾽ εὐτροφίας, εἴς τινα τῶν τριῶν μετα-
πίπτειν εἴωθεν. εἰ μὲν γὰρ αὐτὸ τοῦτο μόνον εἴη κατ᾽
208K   αὐτὴν ἡ ξηρότης, ἰσχνότητος οὐδὲν ἀκολουθήσει
πλέον· εἰ δὲ καὶ θερμότης τις συνείη, πυρέξαι κίνδυνος. οὐ μὴν ταὐτόν γ᾽ ἐστὶ κόπος καὶ πυρετός, εἰ καὶ
ὅτι μάλιστα τῶν κοπωθέντων ἐπύρεξαν ἔνιοι.

θαυμάζειν οὖν ἐπέρχεταί μοι Θέωνος ἐν τῷ τετάρτῳ
τῶν κατὰ μέρος γυμνασίων, ἐν οἷς περὶ τοῦ τελείου
γυμνασίου διεξέρχεται, τάδε γράφοντος· "καὶ κόπου
τινὸς τοῖς οὕτω γυμνασθεῖσιν ὡς τὰ πολλὰ τῇ ἑξῆς
ἡμέρᾳ παρακολουθοῦντος, ἡ ζεστολουσία παραιτεῖται
τὴν πρὸς τὸν κόπον ἐπιτηδειότητα, πυροῦσα τὴν ἐπιφάνειαν, ἵνα αὕτη σικύας τρόπον τὴν λαμβανομένην
τροφὴν ἐπισπωμένη τοῖς κεκμηκόσι ἀντιδιέληται νεύροις." οὗτος μὲν γὰρ πρὸς τοῖς ἄλλοις ὥσπερ αἴνιγμά
τι τὸ "ἀντιδιέληται" ῥῆμα παρέλαβεν ἐν τῷ λόγῳ.
δύναται μὲν γάρ τις ἀκούειν, ἵνα ἡ ἐπιφάνεια σικύας
τρόπον ἐπισπωμένη τὴν τροφὴν μέρος ἐξ αὐτῆς τι καὶ
τοῖς νεύροις παρέχῃ· δύναται δὲ καὶ τοὐναντίον, ἵνα
ἡ ἐπιφάνεια σικύας τρόπον ἐφ᾽ ἑαυτὴν ἀντισπῶσα
τὴν ἐπὶ τὰ νεῦρα φερομένην τροφὴν μερίζηται. ὥστε
κατὰ μὲν τὸν πρότερον λόγον εἰς εὐτροφίαν τοῖς νεύ-
209K   ροις συναίρεσθαι[26] θερμανθὲν τὸ δέρμα (τοῦτο γὰρ

---

[26] συναίρεσθαι Ko; ἐπιτηδεύεσθαι Ku

[19] The term rendered "bathing in hot water" is ζεστολουσία,
for which LSJ has "wash in hot water," citing only the uses in this
chapter.

302

fatigues, if it is not corrected on the first day with good nutrition, is wont to change into one of the three (fatigues). If the dryness itself is the only factor, nothing more than thinness will follow; if, however, there is also some heat present with it, there is a danger of fever. Fatigue and fever are not the same things, even if it is particularly the case that some of those who are fatigued have become febrile.

208K

It comes, then, as a surprise to me when Theon, in his fourth book on the particulars of exercises, in which he goes over the completion of exercise, writes as follows: "Since fatigue on the following day is a close association in those who have undertaken gymnastic exercise in such a way, as a general rule bathing in hot water[19] provides the most suitable relief for fatigue, heating the surface so that this, in the manner of a cupping glass, draws in the nutriment that is being taken and distributes it to the fatigued sinews." For in addition to other things, he has made use of the word *antidielētai* enigmatically (like a riddle) in his treatise.[20] For someone is able to hear, "in order that the surface, like a cupping glass, drawing to itself the nourishment, may provide some part of it also to the sinews." But it also possible to hear the opposite, "that the surface, like a cupping glass, may draw the nutriment back to itself which is being carried to the sinews and may distribute it." Consequently, according to the prior argument, the skin being heated takes care of the good nourishment for the

209K

[20] This is the aorist subjunctive middle third-person singular of the verb ἀντιδιαιρέω, which basically means "to distinguish logically." A second use, labeled "medical" is listed in LSJ as "perhaps distribution," citing the present passage.

ἡγοῦμαι λέγειν αὐτὸν "ἐπιφάνειαν"), κατὰ δὲ τὸν δεύτερον εἰς ὀλιγοτροφίαν.

ὅσον μὲν οὖν ἐπ᾽ αὐτῇ τῇ ῥήσει, τὴν γνώμην τοῦ Θέωνος οὐκ ἄν τις ἐξεύροι· ἐξ ὧν δ᾽ ἐν ἄλλοις τε λέγει κἂν τῷ τρίτῳ[27] τῶν Γυμναστικῶν, ἐλάττονα τροφὴν δίδοσθαι βούλεται μετὰ τὸ τέλειον γυμνάσιον. καίτοι γ᾽ οὐδ᾽ αὐτὸ τοῦτο διεσάφησε, τίνι ποτὲ λογισμῷ συμβουλεύει. δίδοται μὲν γὰρ ἐλάττων τροφὴ τῆς συνήθους ἢ τῷ μὴ δεῖσθαι τῆς ἴσης ἢ τῷ μὴ δύνασθαι πέψαι. τὸ μὲν οὖν πρότερον οὐκ ἀληθὲς ἐπὶ τῶν πολλὰ γυμνασαμένων, τὸ δὲ δεύτερον ἔστιν ὅτε μὲν ἀληθές, ἔστιν ὅτε δὲ ψευδές. εἰ μὲν γὰρ ἀρρωστοτέραν ἔχοιεν τὴν πεπτικὴν δύναμιν, ἀληθές, εἰ δὲ μὴ ταύτην, ἀλλ᾽ ἑτέραν τινά, ψευδές. ἀκούειν δὲ χρὴ πεπτικὴν δύναμιν οὐ τὴν ἐν γαστρὶ μόνον ἢ φλεψὶν ἢ καθ᾽ ἧπαρ, ἀλλὰ καὶ τὴν καθ᾽ ἕκαστον μόριον, ὥσπερ ἐν τοῖς παροῦσι τὴν ἐν τοῖς μυσίν, ἐν οἷς δὴ καὶ μάλιστά ἐστιν ἡ διάθεσις, ὑπὲρ ἧς διαλεγόμεθα. διαφορη-
210K θέντες γὰρ ἐπὶ πλέον ἐν τοῖς γυμνασίοις ἰσχνότεροί τε καὶ ξηρότεροι γίνονται. εἰ μὲν οὖν ἐπ᾽ ὀλίγον αὐτοῖς ταῦτα συμβαίη, κατεργάζεσθαι δύνανται τροφήν, ὅσησπερ δέονται· εἰ δέ τι πολὺ τοῦ κατὰ φύσιν ἀποχωρήσειαν, ἀδυνατοῦσιν. ὧν ὁ Θέων οὐδὲν ὅλως οὔτ᾽ ἐνενόησεν οὔτε διωρίσατο.

τὴν ἀρχὴν γὰρ οὐδ᾽ ἐκ λόγου τινός, ἀλλ᾽ ἐξ ἐμπειρίας, ὡς καὶ αὐτὸς ὁμολογεῖ, τὸ θερμότερον ὕδωρ ἐπὶ τῷ τελείῳ γυμνασίῳ χρήσιμον ὑπάρχον ἐτήρησεν.

sinews (this is what I think he means by "surface"), whereas on the second interpretation, it would be for little nourishment.

Thus, as far as the particular statement is concerned, one would not discover the opinion of Theon. However, from what he says in other places, and also in the third book of his *Gymnastics,* he wishes less food to be given after the completion of exercise. However, he did not make clear what reason he might have for recommending this. Less nourishment than is customary is given either because there is no need of an equal amount, or because it cannot be concocted. The former is not, however, true in the case of those who have exercised a lot, while the latter is sometimes true and sometimes false. For if someone has a digestive capacity that is weaker, it is true, whereas if he does not have this but something else, it is false. It is necessary to understand "digestive capacity" not only in respect of the stomach, veins or liver, but also in each part, just as, in the present circumstances, that in the muscles. Certainly, the condition which I am discussing is in these particularly. For being more dissipated in the exercises, they become thinner and drier. If, then, these things happen to them to a slight extent, they are able to acquire as much nutrition as they need. If, however, there is a major departure from an accord with nature, they are unable to do so. On the whole, Theon neither understood nor distinguished any of these things. 210K

In the first place, not from any theory but from experience, as he himself acknowledges, he observed that warmer water was useful after completed exercise. Thus,

---

²⁷ τρίτῳ Ko; *sign for* ἐκκαίδεκα Ku

οὕτως οὖν ἐφεξῆς γράφει· "τοῦτο δέ, εἰ μὲν καὶ τὸν λόγον ἔχει παρακείμενον, εὐτυχήματος ἔργον, εἰ δὲ μή γε, τὸ πρὸς τῶν ἀποτελεσμάτων ἐπιμαρτυρούμενον οὐ παραδεκτέον, εἰ μὴ καὶ τὸν λόγον ἐξ εὐκαίρου ἔχοι συμπροσπίπτοντα." εἰ μὲν οὖν ἀκριβῶς ἐξευρὼν τὴν διάθεσιν, ἐφ' ἧς τὸ θερμὸν ὕδωρ ἐπαινεῖ, τὴν αἰτίαν ἀγνοεῖν ὡμολόγει, συγγνωστὸς ἂν ἦν εἰκότως· ἐπεὶ δ' ἁπλῶς εἶπεν ἐπὶ τῶν τελείων γυμνασίων ἁρμόττειν τὴν ζεστολουσίαν οὕτω γὰρ αὐτὴν καὶ ὀνομάζει, δύνανται δὲ πολλαὶ διαθέσεις ἀκουλουθῆσαι τῷ τοιούτῳ γυμνασίῳ, μέμψαιτ' ἄν τις αὐτῷ μὴ δι- 211K ορισαμένῳ περὶ πασῶν ἐφεξῆς. αὐτὸς γοῦν οἶδε καὶ τὸν φλεγμονώδη κόπον ἀκολουθοῦντα τῷ τοιούτῳ γυμνασίῳ, καθ' ὃν εἰς ὄγκον μείζονα τοῦ κατὰ φύσιν ἐξαίρεται τὰ πεπονηκότα, καὶ τὸν ἕτερον, ὃν ὡς ἕνα γράφει, τῷ κοινῷ συμπτώματι προσέχων τὸν νοῦν, ὅπερ ἀντίκειται τῷ παρὰ φύσιν ὄγκῳ.

λεπτότεροι μὲν γὰρ ἐν τοῖς ἄλλοις δύο κόποις ἀποτελοῦνται καὶ προσέτι γε τῇ τετάρτῃ[28] διαθέσει, περὶ ἧς ὁ λόγος συνέστηκεν. ἀλλ' οὐχ, ὥσπερ ἓν κοινὸν σύμπτωμα τῶν τριῶν διαθέσεών ἐστιν, οὕτω καὶ ἡ διάθεσις μία. κατὰ μὲν οὖν τὴν ἐπὶ τῇ τάσει τῶν νευρωδῶν σωμάτων οὐ χρὴ λούειν θερμοτέρῳ τοῦ συνήθους· οὕτω δὲ οὐδὲ κατὰ τὴν ἐπὶ τοῖς περιττώμασι· κατὰ δὲ τὴν ἄνευ τούτων ἰσχνότητα συμφέρει λούειν θερμοτέρῳ τοῦ συμμέτρου· γίνεται γὰρ

[28] τῇ τετάρτῃ Ko; τῇδε τῇ Ku

he writes as follows: "This, even if it does have an acccpt-
able rationale, is an action of good fortune. If not, it must
not be admitted as evidence for the resulting effects, if it
does not have a simultaneously occurring theory that is
appropriate." If, therefore, having accurately discovered
the condition for which he recommends the warm water,
he admits he doesn't know the cause, his confessing this
would be reasonable. However, when he simply said that,
in the case of completed exercises, *zestalousia*[21] (for that
is what he terms it) is suitable, and that many conditions
can follow such exercise, one might blame him for not
making a distinction concerning all of them in order. At all   211K
events, he himself also knew the inflammation-like fatigue
that follows such exercise, in which those parts affected
are raised into a swelling greater than accords with nature,
and the other one, which is the one he writes about, directing his attention to the common symptom, which stands
opposite to the swelling contrary to nature.

In the other two fatigues, people become thinner, and
also in the fourth condition, which the discussion is about.
But just as there is not one common symptom of the three
conditions, so too there is also not one condition. Thus, in
the condition due to tension of the sinew-like structures,
you must not bathe in water that is hotter than customary,
just as you must not in the condition due to superfluities.
In the thinness without these, it does help to bathe in
water hotter than moderate. For the condition itself arises

---

[21] See note 19 above. The term is attributed to Theon on the
basis of its use here and subsequently in this chapter.

ἡ διάθεσις αὕτη, διαφορηθέντων ἐπὶ πλέον ἐν τοῖς
γυμνασίοις τῶν σωμάτων, οὕτως ὡς ἐν ταῖς μακρο-
τέραις ἀσιτίαις εἴωθε συμβαίνειν. ὥστε καὶ ἡ ἐπανόρθωσις αὐτῶν ἐν προσθέσει τε καὶ ἀναπληρώσει τοῦ
κενωθέντος ἐστίν. οὐ δύναται δ᾽ αὕτη γενέσθαι τοῦ
212K δέρματος ἀραιοῦ μένοντος.

συναγαγεῖν οὖν αὐτὸ χρὴ καὶ πυκνῶσαι καὶ σφίγξαι πρότερον, εἰ μέλλει τι τῆς δαψιλοῦς τροφῆς ὄφε-
λος ἔσεσθαι. συνάγει δὲ καὶ στεγνοῖ τό τε ψυχρὸν
ὕδωρ καὶ τὸ ζέον. ἀλλ᾽ ὑπὸ μὲν τοῦ ψυχροῦ κίνδυνος
βλαβῆναι τὸν ἄνθρωπον, ἀραιόν τε ἅμα καὶ κενὸν ἐπὶ
τῷ πλήθει τῶν γυμνασίων γεγενημένον· ὑπὸ δὲ τοῦ
ζέοντος βλάβη μὲν οὐδεμία, πυκνότης δ᾽ ἀσφαλὴς
ἐγγίνεται τῷ δέρματι, συνεπιλαμβανούσης τι καὶ τῆς
ἐγκαταλειπομένης αὐτῷ θερμότητος. ὅθεν οὐδὲ χρονί-
ζειν ἐπὶ πλέον ἐν τῷ ψυχρῷ προσήκει τὸν οὕτω λου-
σάμενον, ἀλλ᾽, ὥσπερ καὶ αὐτὸς ὁ Θέων τοῦτό γε
παρετήρησεν ὀρθῶς, φυλακτέον ἐστὶ τὴν ἐν τῷ ψυχρῷ διατριβήν, ὡς τὴν ἐκ τῆς ζεστολουσίας ἀναλύου-
σαν ὠφέλειαν. ἡ δὲ αἰτία τοῦ σφαλῆναι τὸν Θέωνα,
τὸ μιᾷ διαθέσει συμφέρον ὡς πάσαις ἁρμόττον γρά-
φοντα, τῶν γυμναζομένων ὑπ᾽ αὐτοῦ σωμάτων ἡ ἕξις
ἐστίν· ἀθλητὰς γὰρ ἐγύμναζε τοὺς μετὰ τὸ τέλειον
γυμνάσιον εἰς μὲν τὴν τετάρτην διάθεσιν ἑτοίμως
ἐμπίπτοντας, εἰς δὲ τὴν τρίτην σπανιάκις. ὅπερ οὖν
213K ἐθεάσατο πολλάκις, ὡς διηνεκὲς ἔγραψεν. εἰ δέ γε κα-
κοχύμους ἢ καχέκτας ἢ ἀήθεις γυμνασίων ἢ ἀσθενεῖς
ἢ μὴ νεανίσκους ἐγύμναζε, σπανιάκις μὲν ἂν εἰς τὴν

when there is greater dissipation of the bodies in the exercises, as is wont to happen in prolonged fasts. Consequently, the correction of these is in the administration of food and the replenishment of what is evacuated, But this cannot occur if the skin remains thin. 212K

Therefore, it is necessary to contract the skin, and condense and compress it first, if some of the abundant nutriment is going to be of benefit. Both cold and seething water contract the skin and make it impervious, but with cold water, there is a danger of the person who has become thin and empty due to the amount of exercises being harmed. There is no injury with seething water as a safe thickness arises in the skin with the heat being present remaining in it. On this account, it is not appropriate for the one being bathed in this way to spend too long in the cold water but (and Theon himself correctly observed this at least) what must be avoided is spending time in the cold water, as this does away with the benefit from bathing in hot water. The reason for Theon's error in writing that what benefits one condition is suitable for all conditions is the state of the bodies trained by him. For he trained athletes who, after the completion of exercise, readily fell into the fourth condition but rarely into the third. Then, what he often saw, he wrote as invariable. However, if he 213K were also training those who were *kakochymous, kachektic* (cachectic), unaccustomed to exercise, weak or not young men, he would rarely have seen them fall into such

τοιαύτην διάθεσιν ἐμπίπτοντας ἐθεάσατο, μυριάκις δ᾽ εἰς τὰς ἄλλας. ἢ τοῦτο μὲν ἴσως καὶ λέγειν περιττόν· αὐτὸς γὰρ ὁμολογεῖ μετὰ τὰ κατασκευαστικὰ γυμνάσια χρῆσθαι τῇ ζεστολουσίᾳ· τὰ δὲ τοιαῦτα γυμνάσια μόνοις ἀθληταῖς ἐπιτηδεύεται. καίτοι φήσει τις, ὡς ἐκείνοις μὲν ἑκουσίως καὶ κατὰ περίοδον, ἄλλοις δὲ πολλοῖς ἢ δι᾽ ἀνάγκην ἢ φιλονεικίαν ἤ τι τοιοῦτον ἕτερον.

ἀλλ᾽ ἐπί γε τῶν τοιούτων ἐπὶ τοῖς ἀμέτροις γυμνασίοις ἀνάγκη πρότερον ἢ καὶ μίαν ἢ καὶ πλείους συστῆναι κόπων διαθέσεις. ὥστ᾽ ἐν μόνοις τοῖς εὐεκτικοῖς σώμασιν ἡ τετάρτη γίνεται διάθεσις, οἷάπέρ ἐστι τά τε τῶν καλῶς ἀγομένων ἀθλητῶν καὶ τοῦ νῦν ἡμῖν ἐν τῷ λόγῳ προκειμένου νεανίσκου. καὶ εἴ τις ἁπλῶς ἀποφαίνοιτο, μετὰ τὰ σφοδρότατα γυμνάσια τὴν τοῦ θερμοτέρου λουτροῦ χρῆσιν ἐπιτήδειον ὑπάρχειν, ἐπὶ μιᾶς μὲν ἀληθεύσει καταστάσεως, ἐπὶ τριῶν δὲ ψεύσεται. πολλὰ δὲ καὶ ἄλλα τοιαῦτα καθ᾽ ὅλην τὴν ὑγιεινὴν πραγματείαν ἰατροῖς τε καὶ γυμνασταῖς γέγραπται ψευδῆ. κεφάλαιον δ᾽ αὐτῶν ἐστι παρὰ τὸ τῆς ῥήσεως ἀδιόριστον, ὅταν, ὅπερ ἐπὶ μιᾶς ἕκαστος ἐθεάσατο διαθέσεως, ἐπὶ πολλῶν ἀναγράφῃ.

9. Καὶ μὲν δὴ καὶ πέμπτη[29] τίς ἐστι διάθεσις ἐγγύς τι ταῖς προειρημέναις, ἣν ὀνομάζουσι[30] στέγνωσιν, ὑπὲρ ἧς, ἐπειδὰν συμπεραίνωμεν τὸν ἐπὶ τοῖς κόποις λόγον, ἐφεξῆς ἐρῶ. αἱ μὲν γὰρ ἁπλαῖ διαφοραὶ τρεῖς

214K

[29] πέμπτη Ko; ἔτι Ku   [30] ὀνομάζουσι Ko; ὀνομάζω Ku

a condition, but countless times into the others. Perhaps it is superfluous to say this because he himself acknowledges he uses a hot bath after the preparatory exercises. Such exercises are only practiced by athletes. And indeed, someone will say those men [do the exercises] voluntarily and according to a prescribed schedule, whereas in many others [they are done] through necessity, love of fighting, or some other such thing.

But in the case of such people, there is a prior necessity that after the excessive exercises, they will sustain one or more of the fatigue conditions. Consequently, only in healthy bodies does the fourth condition occur—bodies such as those of well-trained athletes and of the young man we are now proposing in the discussion. And if someone were simply to declare that the use of the hotter bath is suitable after very violent exercises, he will be speaking truly in the case of one condition but falsely in the case of 214K three. Many other such things in relation to the whole matter of hygiene have been falsely written by doctors and gymnastic trainers. The chief of these is due to the indefiniteness of the statement whenever each person sees one condition and documents it as occurring in many instances.

9. Furthermore, there is also a fifth condition, closely related to those previously spoken of, which they term "stoppage of the pores."[22] I shall speak about this next after I have brought the discussion on the fatigues to completion. There are three simple *differentiae* of these which

---

[22] This specific use of the term στέγνωσις is listed in LSJ as attested by this passage and also Oribasius, *Synopsis* 5.16. Galen defines the term at 218K below.

αὐτῶν εἰσιν, ὑπὲρ ὧν ἤδη μοι λέλεκται· κατὰ σύνδυο δὲ λαμβανομένων ἄλλαι γίνονται τρεῖς· ἑβδόμη δ' ἐπὶ πάσαις ἐστὶν ἡ τῶν τριῶν ἅμα συνερχομένων. ἡ μὲν δὴ διάγνωσις αὐτῶν ἐστιν ἀπὸ τοῦ συνδυάζεσθαι τὰ γνωρίσματα, σκοπὸς δὲ τῆς ἐπανορθώσεως ὁ μὲν κοινὸς ἁπασῶν ἀποβλέποντα πρὸς τὸ ἐπικρατοῦν μηδὲ τοῦ λοιποῦ παντάπασιν ἀμελεῖν, ὁ δ' ἴδιος ἐπὶ τῷ κοινῷ κατὰ τὰς ἐν μέρει διαθέσεις λαμβάνεται.

πάσας μὲν οὖν ἐπέρχεσθαι τὰς συζυγίας μακρόν,
215Κ ἕνεκα δὲ σαφηνείας ἐπὶ μιᾶς ὡς παραδείγματος ὁ λόγος περανθήσεται. ἐὰν τοίνυν ὄγκος τε ἅμα περὶ τοὺς μῦς ὑπάρχῃ καὶ τεθλάσθαι δοκῶσιν αὐτοὶ καὶ ἢ ἑλκώδης αἴσθησις ἢ φλεγμονώδης κόπος ἅμα τῷ περιττωματικῷ κατειλήφῃ τὸν ἄνθρωπον, ἡ ἀποθεραπεία γενήσεται στοχαζομένη μὲν ἀμφοῖν, ἀλλὰ μᾶλλον τοῦ μείζονος. οὐχ ἁπλῆ δ' ἐν ἅπασι τοῖς οὖσιν ἡ τοῦ μείζονος φύσις, ἀλλ' ἡ μὲν κατὰ δύναμίν τε καὶ τὸ οἷον ἀξίωμα τοῦ πράγματος, ἡ δὲ κατὰ τὴν οἰκείαν οὐσίαν. ἀξιώματι μὲν οὖν καὶ δυνάμει μείζων ἐστὶν ὁ φλεγμονώδης κόπος τοῦ περιττωματικοῦ· κατὰ δὲ τὴν οἰκείαν οὐσίαν ἑκάτερος αὐτῶν οὕτω δύναται γενέσθαι μέγας τε καὶ μικρός, ὡς εἰ καὶ κατὰ μόνας ἑκάτερος ἦν. εἰ μὲν οὖν ἴσον ἐξειστήκει τοῦ κατὰ φύσιν ἑκάτερος, ὁ φλεγμονώδης ἐφ' ἑαυτὸν ἐπισπάσεται τὸ κῦρος τῆς θεραπείας, ἐπικρατῶν γε κατὰ δύναμιν· ἂν δ' ὀλίγιστον μὲν ὁ φλεγμονώδης, πλεῖστον δ' ὁ ἑλκώδης ἀποκεχωρήκῃ τοῦ κατὰ φύσιν, ἐπισκεπτέον, εἴτε τοσοῦτον ὑπερέχει κατὰ τὸ μέγεθος ὁ ἑλκώδης,

I have already described, and there are three others consisting of two of the simple ones combined; there is a
seventh, in addition to all these, which consists of a combination of the three at the same time. The diagnosis of
these is from combining the signs of the two, while the aim
of their correction is common to all, focusing on the prevailing one without altogether neglecting the remaining
one, the specific being taken in addition to the common
in relation to the conditions one by one.

To go over all the conjunctions would then be a long
job; for the sake of clarity the discussion will concentrate   215K
on one as an example. Accordingly, if at the same time a
swelling exists in the muscles and they seem to be bruised,
and a wound-like sensation, or an inflammation-like fatigue along with the excrementious has taken hold of the
person, apotherapy will be aimed at both, but more at the
greater. The nature of the greater in all these entities is
not simple but relates to the power and rank, as it were,
of the matter, or to the specific essence. Thus, in rank and
power, the inflammation-like fatigue is greater than the
excrementitious. In relation to the specific essence, each
of them is able to become great or small in such a way as
if each existed alone. If, then, each deviates equally from
an accord with nature, the inflammation-like will draw to
itself the prime importance in treatment, since it pre-
dominates in terms of power, but if the inflammation-like
departs a slight amount and the wound-like more from
an accord with nature, one must consider whether the
wound-like is higher in terms of magnitude to the degree

216K ὅσον ὁ φλεγμονώδης κατὰ δύναμιν, ἢ ἔλαττον ἢ μεῖζον, καὶ οὕτως ἐξευρίσκειν τὸν ἐπικρατοῦντα· κἂν
ἰσοσθενεῖς δέ ποτε φαίνωνται, πρὸς ἀμφοτέρους
ὁμοίως ἀποβλεπτέον. αὕτη μὲν οὖν ἡ μέθοδος ἔστω
σοι κοινὴ πασῶν τῶν ἐπιπεπλεγμένων διαθέσεων.

ὥσπερ δ᾽ οἱ κόποι τρεῖς ὄντες, εἶτ᾽ ἀλλήλοις ἐπι-
πλεκόμενοι τέτταρας ποιοῦσι τὰς συζυγίας, οὕτως, εἰ
καὶ τὴν τετάρτην αὐτοῖς ἐπιπλέξειέ τις διάθεσιν, αἱ
συζυγίαι πολὺ πλείους γενήσονται. μάθοις δ᾽ ἂν
ἐναργῶς, ὃ λέγομεν, ἐπὶ διαγράμματος. ἔστω δ᾽ ἐν
αὐτῷ πρώτη μὲν διάθεσις ἡ ἑλκώδης· δευτέρα δὲ ἡ
τονώδης· τρίτη δ᾽ ἡ φλεγμονώδης· τετάρτη δ᾽ ἡ τῆς
ἰσχνότητος. ἢ τοίνυν ἡ πρώτη μετὰ τῆς δευτέρας συ-
στήσεται διαθέσεως ἢ μετὰ τῆς τρίτης ἢ μετὰ τῆς
τετάρτης· ἢ πάλιν ἡ δευτέρα μετὰ τῆς τρίτης ἢ τῆς
τετάρτης· ἢ πάλιν ἡ τρίτη μετὰ τῆς τετάρτης. ὥστ᾽
εἶναι τὰς πάσας ἓξ συζυγίας, ἀνὰ σύνδυο λαμβανομένων τῶν διαθέσεων, ἄλλας δὲ τέτταρας, ὅταν ἅμα
τρεῖς ἐπιπλέκωνται διαθέσεις ἀλλήλαις. ἤτοι γὰρ ἡ
πρώτη μετὰ τῆς δευτέρας ⟨τε⟩ καὶ τρίτης ἢ μετὰ τῆς
δευτέρας τε καὶ τετάρτης ἢ μετὰ τῆς τρίτης τε καὶ
217K τετάρτης[31] συστήσεται· ἢ πάλιν ἡ δευτέρα μετὰ τῆς
τρίτης τε καὶ τετάρτης. ὑστάτη δὲ πασῶν ἐπιπεπλεγ-
μένη διάθεσις ἔσται τῶν τεττάρων ἅμα διαθέσεων
ἀλλήλαις μιγνυμένων. ὥστ᾽ εἶναι τὰς πάσας ἕνδεκα
τὸν ἀριθμόν. ἦσαν δέ γε καὶ αἱ τῶν ἁπλῶν τέτταρες.

γενήσονται τοίνυν αἱ σύμπασαι πεντεκαίδεκα· α
(ἑλκώδης), β (τονώδης), γ (φλεγμονώδης), δ (ἰσχνό-

that the inflammation-like is either less or more in respect   216K
of power; in this way you discover the prevailing one. And
if at any time, they appear equipollent, one must give
consideration to both equally. Let this, then, be for you
the method common to all the mixed conditions.

Just as there are three fatigues which, when combined
with each other, make four conjunctions, so too, if you
combine some fourth condition with them, the conjunctions will become much more in number. You would
clearly understand what I am saying by means of a diagram. Let the wound-like fatigue be the first condition in
this, the second, the tensive, the third, the inflammation-
like, and the fourth that of thinness. Accordingly, either
the first will be associated with the second condition, or
with the third, or with the fourth. Or again, the second
with the third or the fourth; or again, the third with the
fourth. As a result, there are six conjunctions in all when
you take the conditions in twos at the same time. But there
are four others, when three conditions are combined with
each other at the same time—either the first with the
second and third; or with the second and fourth; or with
the third and fourth; or again, there will be the second   217K
with the third and fourth. Last of all, there will be a combined condition when the four conditions are mixed with
each other at the same time. There are also the four simple
conditions.

Accordingly, there will be fifteen in all: the first, wound-
like (A); the second, tensive (B); the third, inflammation-
like (C); the fourth, thinness (D); and then AB, AC, AD,

---

31 post τετάρτης: συστήσεται· ἢ πάλιν ἡ δευτέρα μετὰ τῆς
τρίτης τε καὶ τετάρτης. Κο; πάλιν. Κυ

της), αβ, αγ, αδ, βγ, βδ, γδ, αβγ, αβδ, αγδ, βγδ,
αβγδ.[32] εἰ δὲ καὶ τὰς τῆς στεγνώσεως διαφορὰς ἐπι-
πλέκοις ἀλλήλαις τε καὶ ταῖς πεντεκαίδεκα, παμπλη-
θεῖς ἑτέρας ἐργάσῃ συζυγίας. εἰ δὲ καὶ τὰς τῆς ἀπεψίας αὐταῖς ἢ τὰς ἐπὶ τοῖς ἀφροδισίοις ἢ ταῖς
ἐγκαύσεσιν ἢ ταῖς ἀγρυπνίαις ἢ ταῖς λύπαις ἐπιπλέ-
ξειας καταστάσεις τοῦ σώματος, οὐδ' ἀριθμηθῆναι
ῥᾳδίως ἅπασαι δυνήσονται. καὶ οὔπω λέγω τὰς τῆς
πληθώρας ἢ κακοχυμίας ἢ ἐπισχέσεως γαστρὸς ἢ
διαρροίας ἢ ἐμέτων ἢ βάρους κεφαλῆς ἤ τινος ἄλλου
μέρους ἢ ὅλως ὅσαι κατά τι σύμπτωμα συνίστανται·
λεχθήσεται γὰρ αὖθις ὑπὲρ τῶν τοιούτων ἁπάντων.
ἀλλὰ νῦν γε τούτου χάριν ἐπεμνήσθην αὐτῶν ὑπὲρ
218K τοῦ δεῖξαι τὸ πλῆθος τῶν ἐπιπλοκῶν ὁπόσον ἐστί.
θαυμάσαι γὰρ οἶμαί[33] τινα τοὺς ἐπὶ τοῖς ἀθροίσμασιν
αὐτῶν, ἃς ὀνομάζουσι συνδρομάς, ἤτοι θεραπείαν ἢ
πρόγνωσιν τῶν ἀποβησομένων ἐπαγγελλομένους τε-
τηρηκέναι. καθ' ἕνα γὰρ τρόπον οἷόν τ' ἐστὶ καὶ προ-
γνῶναί τι καὶ θεραπεῦσαι δεόντως, ὡς Ἱπποκράτης
ἐδίδαξεν, ἑκάστου τῶν ἁπλῶν πραγμάτων ἀξιῶν ἐπίστασθαι τὴν δύναμιν, ὡς ἐγὼ νῦν ἐπέδειξα περὶ τῶν
τεττάρων διαθέσεων. εἰ μὲν γὰρ ἑκάστη καθ' ἑαυτὴν
συσταίη, τὴν ἐπανόρθωσιν ἁπλῆν ἐνδείξεται, μι-
χθεῖσα δ' ἑτέρᾳ, κατὰ τὴν ὀλίγον ἔμπροσθεν εἰρημένην μέθοδον ὑπὲρ ἁπασῶν τῶν ἐπιπεπλεγμένων
διαθέσεων.

10. Ὁπότ' οὖν τοῦτ' ἔχον οὕτω φαίνεται, ἰτέον αὖθις
ἐπὶ τὰς ἁπλᾶς διαθέσεις, ὧν ἐφεξῆς ταῖς εἰρημέναις

BC, BD, CD, ABC, ABD, ACD, BCD and ABCD.[23] If you also combine the *differentiae* of *stegnosis* (stoppage of the pores) with each other and with the fifteen, you will create very many other conjunctions. If you also combine the *differentiae* of the *apepsias* with them, or those after sexual intercourse, or heatstrokes, or insomnias, or griefs, all the combined states of the body will not even be able to be easily enumerated. And I haven't yet mentioned the *differentiae* of the plethora, *kakochymia,* stoppages of the stomach, diarrhea, vomiting, heaviness of the head or of some other part, or altogether those things that exist in relation to some symptom. I shall speak again about all such things. But for the present I make mention of them for this purpose—to show how great the number of combinations is. One may, I think, marvel at someone who, by gathering these together, calls them syndromes, or therapy or prognosis of things that are going to occur, proclaiming they have been observed. Only in one way is it possible to prognosticate anything and treat properly, as Hippocrates taught, when he claims to know the potency of each of the important simple matters, as I showed just now concerning the four conditions. For if each should exist by itself, it will indicate a simple correction, whereas if it is mixed with another, the restoration will involve the method I spoke about a little earlier in regard to all the combined conditions.

218K

10. Therefore, since this is obviously the case, we must move on in turn to the simple conditions of which, in suc-

---

[23] In the translation, *C* replaces the γ of the Greek.

---

[32] *The sequence here has been slightly reordered in the interest of clarity.*   [33] οἶμαί Ko; οἷόν τ᾽ ἐστὶ Ku

ἦσαν αἱ κατὰ τὴν στέγνωσιν· οὕτω δ' ὀνομάζω τὴν
βλάβην τῶν πόρων, ἐφ' ᾗ κωλύεται διαφορεῖσθαι τὰ
περιττώματα. γίνεται δ' αὕτη δι' ἔμφραξιν ἢ πύκνωσιν, ἣν δὴ καὶ μύσιν ὀνομάζουσι τῶν πόρων. ἔμφρα-
219K ξις μὲν οὖν ὑπὸ γλίσχρων ἢ παχέων γίνεται περιττωμάτων ἀθρόωτερον ὁρμησάντων ἐπὶ τὸ δέρμα,
πύκνωσις δὲ ὑπό τε τῶν στυφόντων καὶ ψυχόντων.
ἀλλ' ἐμφράξει μὲν ἁλῶναι τὸ προκείμενον ἐν τῷ λόγῳ
σῶμα, κατὰ τὴν εἰρημένην ἐπιμέλειαν ἀγόμενον, οὐκ
ἐγχωρεῖ, πυκνωθῆναι δὲ δύναταί ποτε διά τε κρύος
καρτερὸν καὶ λουτρὸν στυπτηριῶδες. ἐγχωρεῖ δέ ποτε
καὶ μετὰ βαλανεῖον ἢ ἱδρῶτα καὶ ἄλλως ἀραιοῦ τοῦ
δέρματος ἔκ τινος ἑτέρας αἰτίας γενομένου καταπνεύ-
σασαν αὔραν ἔμψυξίν τέ τινα καὶ πύκνωσιν ἀποτελέ-
σαι. διαγινώσκεται μὲν οὖν ἡ εἰρημένη διάθεσις
εὐθὺς μὲν ἀποδύντων ἀχροίᾳ τε λευκῇ καὶ σκληρότητι
καὶ πυκνώσει τοῦ δέρματος, κατὰ δὲ τὸ γυμνάζεσθαι
τῷ δυσεκθερμάντῳ. οὔτε γὰρ ἱδροῦσιν ὁμοίως ὡς
πρόσθεν οὔτ' εὐχροοῦσιν, ἀλλ' εἰ καὶ βιάσαιντο τῇ
εὐτονίᾳ τῶν γυμνασίων ἱδρῶτός τι προκαλέσασθαι,
καὶ μείων οὗτος γίνεται τοῦ συνήθους καὶ ψυχρότερος
καὶ ἧττον ἀτμώδης. ἡ δ' ἴασις τῆς τοιαύτης διαθέσεως θέρμανσίς ἐστιν· ἐναντίον γὰρ τοῦτο τῇ ψύξει.
220K   συντονωτέροις τε οὖν γυμνασίοις χρηστέον ἐστὶ
καὶ βαλανείοις θερμοτέροις. ἄμεινον δὲ καὶ καλινδεῖ-
σθαι κατὰ τὸν πρῶτον οἶκον ἐπὶ ἐλαίου λιπαρῶς.
ἔστω δὲ καὶ τὸ ἔλαιον τῶν χαλαστικῶν, οἱόνπέρ ἐστιν
ἐν Ἰταλίᾳ τὸ Σαβῖνον· ἄμεινον δ' ἐπὶ τῶν τοιούτων

cession to those spoken of, were those related to *stegnosis*.
I term in this way damage of the pores due to which the
superfluities are prevented from being dispersed. This
arises through blockage or constriction (condensation)[24]
which people also call occlusion of the pores. Blockage
arises from viscid or thick superfluities when they come to   219K
be overly collected together in the skin, while constriction
arises due to astringents and cooling agents. But it is not
possible for the body proposed in the discussion to be
seized by obstruction, if it is treated with the care mentioned, whereas it can be constricted on occasion due to
severe cold and bathing in alum-containing water. It is also
possible on occasion after a bath or sweating, and otherwise, if the skin becomes loose textured from any other
cause, as when a cooling breeze blows over [him] to produce some cooling and constriction. The condition mentioned will, then, be diagnosed, as soon as people undress,
from the lack of color, whiteness, hardness and condensation of the skin, and by being hard to warm thoroughly
during exercise. For they do not sweat in the same way as
before, nor do they have a good color, but even if someone
were to compel them, by the vigor of the exercises, to call
forth sweat, this too is less than usual, colder and less vaporous. The cure of such a condition is heating because
this is the opposite of cooling.

One must, then, use more vigorous exercises and   220K
warmer baths. It is also better to roll oneself in fatty oil in
the first chamber. The oil should be of the relaxing kind
as, for example, the Sabine is in Italy. In such conditions,

[24] Although not listed in LSJ, "constriction" is preferred to
"condensation" for πύκνωσις based on context.

διαθέσεων ἐτῶν εἶναι δύο ἢ τριῶν αὐτό· καὶ γὰρ λε-
πτομερέστερον τοῦτο καὶ θερμότερον. ἡ δὲ ἐν τῇ ψυχρᾷ κολυμβήθρᾳ διατριβὴ μὴ πολυχρόνιος γινέσθω,
μηδ᾽ αὐτὸ τὸ ὕδωρ ἄγαν ἔστω ψυχρόν. ἐνδύεσθαι δὲ
μέλλοντες ἀλειφέσθωσάν τινι τῶν μετρίως θαλπόν-
των, ἐλαίων μέν, ὅσα γε κατ᾽ Αἴγυπτόν εἰσι, κικίνῳ
καὶ ῥαφανίνῳ, κατὰ δὲ τὴν ἄλλην οἰκουμένην τῷ γλυκεῖ καὶ λεπτομερεῖ καὶ μετρίως παλαιῷ, μύροις δὲ
Σουσίνῳ τε καὶ γλευκίνῳ καὶ ἰρίνῳ καὶ ἀμαρακίνῳ
καὶ Κομμαγηνῷ. τὸ μὲν δὴ γλεύκινον ἄκοπον ἀκριβῶς ἐστιν καὶ χαλαστικόν, ὥστε καὶ τοῖς ἰσχυρῶς
κοπωθεῖσιν ἐπιτήδειον ὑπάρχει· βραχεῖ δὲ αὐτοῦ τὸ
Σούσινον θερμότερόν ἐστι καὶ μαλακτικώτερον·[34] ἴρι-
νον δὲ καὶ ἀμαράκινον καὶ Κομμαγηνὸν ἱκανώτερον
τούτων θερμῆναι, ὥστε καὶ ταῖς καλουμέναις ἰδίως
221K ψύξεσι χρήσιμα τετύχηκεν ὄντα. τὰς μέντοι πυκνώ-
σεις τοῦ δέρματος αὐτάρκως ἰᾶται καὶ τὸ ἀνήθινον
ἔλαιον, καὶ μάλιστ᾽ εἰ χλωρὸν εἴη τὸ ἄνηθον. ἁρμόττει δὲ ταῖς τοιαύταις διαθέσεσιν, ὥσπερ οὖν καὶ τοῖς
ἰσχυροῖς κόποις, τὸ διὰ τοῦ σπέρματος τῆς ἐλάτης
ἄκοπον· εἰρήσεται δὲ ἐν τοῖς ἐφεξῆς ὅπως χρὴ σκευάζειν αὐτό. νυνὶ μὲν γάρ μοι δοκῶ καὶ ταῦτα περαιτέρω τῆς ὑποθέσεως εἰρηκέναι. τῷ γὰρ ἄριστα κατεσκευασμένῳ τὸ σῶμα καὶ βίον ἐλεύθερον ἐπανηρημένῳ
καὶ μηδὲν αὐτῷ πλημμελοῦντι καὶ τὸν ἐπιστατοῦντα

---

[34] μαλακτικώτερον Κο; χαλαστικώτερον Κu

it is better for this to be two or three years old, for this is more fine-particled and hotter. The time spent in the cold swimming bath should not be long, nor should the water itself be excessively cold. When people are about to dress, they should be anointed with one of the moderate warming agents; of the oils which are in Egypt, these are that from the *kiki* tree and from radishes; from the rest of the inhabited world, oil that is sweet, fine-particled and moderately old, or with unguents such as that from Susa;[25] and with *gleukinos*,[26] iris, amaracus and from Commagenus.[27] *Gleukinos* is entirely refreshing and emollient, so it is suitable even for those who are strongly fatigued. The Susene is a little warmer and more relaxing than this, while the iris, amaracus and Commagene oils are more adequate than these for warming, so they happen to be suitable for the so-called chills specifically. Also dill (anis) oil cures the constriction of the skin adequately, and particularly if the dill is green. The oil from the seed of the silver fir is refreshing and is suitable for such conditions, just as it also is for the strong fatigues. I shall describe how this must be prepared in what follows. For the present, it seems to me I have spoken about these matters more than the subject requires. The body which is best in constitution and chooses a free life, which never indulges in excess, and

221K

[25] Susa was an ancient city in the early Iranian empire and one of the most important cities of the ancient Near East. It was located in the lower Zagros mountains about 150 miles east of the Tigris.   [26] According to LSJ this is oil made with sweet, new wine or grape juice as a vehicle—see Dioscorides 1.57, and Galen XIII.1039K.   [27] Commagenus was the northern province of Syria; the chief town was Samosata.

τῆς ὑγείας ἄριστον ἔχοντι τὰς νοσωδεστέρας διαθέσεις οὐ πάνυ τι συμπίπτειν εἰκός.

11. Ἐπανέλθωμεν οὖν αὖθις ἐπὶ τὴν ἐξ ἀρχῆς ὑπόθεσιν καὶ παραλιπόντες ἐκκαύσεις καὶ ψύξεις καὶ ἀπεψίας καὶ διαρροίας ὅσα τ' ἄλλα τοιαῦτα (βέλτιον γὰρ ἀναβαλέσθαι σύμπαντα ταῦτα εἰς ἕνα λόγον, τὸν περὶ τῶν νοσωδῶν συμπτωμάτων ἐπιγραφησόμενον) ἐν τῷ παρόντι περὶ τῶν ἐπ' ἀφροδισίοις γυμνασίων ἐπισκεψώμεθα. καὶ γὰρ διαπεφώνηταί πως ὑπὲρ αὐτῶν, ἐνίων μὲν οἰομένων οὕτω χρῆναι γυμνάζειν ἐπ' 222K αὐτοῖς ὡς κατὰ τὸ καλούμενον ἀποθεραπευτικόν, ἐνίων δὲ ὡς κατὰ τὸ παρασκευαστικόν. ἔστι δὲ δήπου τὸ παρασκευαστικὸν γυμνάσιον ἐν μὲν τῇ ποσότητι τῶν κινήσεων ἔλαττον τοῦ συμμέτρου, κατὰ δὲ τὴν ποιότητα συντονώτερόν τε καὶ ὀξύτερον. οἱ μὲν οὖν ἀποθεραπεύειν ἀξιοῦντες, ὥσπερ τοὺς ἀπὸ καμάτου, τήν τε κατάλυσιν τῆς δυνάμεως ὑφορῶνται καὶ τὴν ξηρότητα τοῦ σώματος· ἄμφω γὰρ ταῦτα πάσχομεν ἐπ' ἀφροδισίοις τε καὶ πλήθει τῶν γυμνασίων· οἱ δὲ τῷ παρασκευαστικῷ χρῆσθαι γυμνασίῳ, τὴν ἀραιότητά τε καὶ τὸ εὔιδρωτον, ἅπερ ἐπιτείνεσθαι μὲν ὑπὸ τῶν ἀποθεραπευτικῶν, ἐπανορθοῦσθαι δ' ὑπὸ τῶν παρασκευαστικῶν. ἐγὼ δ' ἑκατέρους ἐπαινέσας, ὡς ἑωρακότας ἐκ μέρους τἀληθές, ἐς ταὐτὸν συνθήσω τὰς δόξας αὐτῶν. ὅτι μὲν γὰρ ἀναρρῶσαί τε χρὴ τὴν δύναμιν καὶ σφίγξαι τὴν ἀραιότητα καὶ μὴ παραυξῆσαι τὴν ξηρότητα, συγχωρήσουσιν ἑκάτεροι. λείπει δ' αὖ εἷς διορισμὸς ἀμφοτέροις, ὥσπερ ἐν ταῖς ἐπιπε-

which has the best supervisor of health, is not very likely to fall into these more morbid conditions.

11. Let us return, then, once more to the original hypothesis, and leaving aside heatstrokes, chillings, *apepsias,* diarrheas and other such things (for it is better to put all these off to one book I shall write about disease symptoms),[28] and for the moment consider the exercises following sexual intercourse. There is disagreement in some respects about these: there are some who think a person ought to exercise after it, as with the so-called apotherapeutic exercise, whereas there are others who think of exercise as preparatory. Now, of course, preparatory exercise, in terms of the amount of movement, is less than moderate, but in terms of quality, it is quite vigorous and rapid. Those, then, who regard apotherapy as worthwhile, as in those fatigued from toil, are anticipating the dissipation of the capacity and dryness of the body. For we suffer both of these immediately after sexual intercourse and excessive exercise. However, those who think it right to use preparatory exercise are anticipating looseness of texture and easy sweating, which are intensified by the apotherapeutic but corrected by the preparatory exercises. I commend each, and seeing the truth on each side, will combine their opinions into one and the same. Both sides agree it is necessary to strengthen the capacity and condense the looseness but not increase the dryness. However, one distinction still remains in both, just as in all the

222K

[28] *Sympt. Caus.,* VII.85–272K—see for example VII.206 and 263K (English trans., Johnston, *Galen: On Diseases and Symptoms*). See also Book 6 in the present treatise.

πλεγμέναις ἁπάσαις διαθέσεσιν ὑφ' ἡμῶν ἔμπροσθεν
εἴρηται, ῥηθήσεται δὲ καὶ νῦν οὐδὲν ἧττον.

    ἐπειδὰν γὰρ ἐς ταὐτὸν συνέλθωσι[35] πλείους διαθέ-
223K σεις, εἰ μὲν ἕνα τρόπον ἐνδείκνυται θεραπείας, ἐπιτείνεσθαι χρὴ τὸν τρόπον μᾶλλον ἤπερ εἰ κατὰ μόνας
ἑκάστη τῶν διαθέσεων ἦν, εἰ δ' ἐναντιούμεναι, κατὰ
τὴν ἐπικρατοῦσαν διάθεσιν ἐπανορθωτέον ἐστὶ πρότερον, οὐδὲ τῶν ἄλλων ἀμελοῦντας τὸ σύμπαν. ὅσοι μὲν
οὖν ἤτοι δι' ἡλικίαν ἢ καὶ ἄλλως ἀσθενεῖς ὑπάρχον-
τες ἀφροδισίοις ἐχρήσαντο, τούτοις μὲν ἀναγκαῖόν
ἐστιν ἐπικρατεῖν τὴν ἀρρωστίαν τῆς δυνάμεως· ὅσοι
δ' ἰσχυροί τε καὶ νέοι, καθάπερ ὁ νῦν[36] ὑποκείμενος ἐν
τῷ λόγῳ νεανίσκος, ἐν τούτοις ἡ τοῦ σώματος ἕξις εἰς
ἀραιότητα πλέον ἤπερ ἡ δύναμις εἰς ἀρρωστίαν ἀλλοιοῦται. καὶ τοίνυν ἡ ἐπανόρθωσις οὐ διὰ τῶν ἀραι-
ούντων, οἷόν ἐστι τὸ ἀποθεραπευτικὸν γυμνάσιον,
ἀλλὰ διὰ τῶν συναγόντων τε καὶ σφιγγόντων, ὁποῖόν
ἐστι τὸ παρασκευαστικόν, ἐν τοῖς τοιούτοις γίνεται
σώμασιν.[37] εἰ δὲ δὴ καὶ ψύξις τις ἐπὶ τοῖς ἀφροδισίοις
ἐγγίνεται τοῖς σώμασι, καὶ κατὰ τοῦτ' ἂν εἴη τῷ
παρασκευαστικῷ γυμνασίῳ χρηστέον· ἐπεγείρει γὰρ
ἐκεῖνο τὴν θερμότητα τῷ τε τῶν κινήσεων ὀξεῖ καὶ
συντόνῳ καὶ τῷ συνάγειν καὶ σφίγγειν τὴν ἕξιν. ὅτι
224K δὲ τὴν ἀραιότητα τοῦτο μόνον τῶν γυμνασίων ἰᾶται,
πρὸς ἁπάντων ὡμολόγηται τῶν γυμναστῶν, ὑπὸ τῆς
πείρας δεδιδαγμένων. ὥστ' οὐδὲν ἂν εἴη βέλτιον εἰς
τὰ παρόντα τοῦ τοιούτου γυμνασίου.

combined conditions which I spoke about earlier and will
also speak no less about now.

Whenever we bring together a number of conditions
into one, if they indicate one kind of treatment, it is more   223K
necessary to increase the intensity of that kind than if each
of the conditions existed by itself. If, however, they are
opposing, what must first be restored pertains to the pre-
dominating condition without neglecting the others altogether. Thus, those who either through age or otherwise
are weak inevitably have a weakness of capacity that prevails when they indulge in sexual intercourse, whereas in
those who are strong and young, just like the young man
we are now assuming in the discussion, the state of the
body is changed to a looseness of texture more than the
capacity is changed to weakness. Accordingly, the restoration is not through those things that loosen the texture,
like for example apotherapeutic exercise, but through
those things that bring together and compress, such as the
preparatory exercise is in such bodies. And certainly, if
some chilling is engendered in bodies after sexual intercourse, what must be used in this is the preparatory exercise, for that stirs up the heat by the speed and vigor of
the movements, and by bringing together and compressing the state (of the body). Also, that this alone of the
exercises cures the looseness of texture is agreed by all   224K
gymnastic trainers who have been taught by experience.
Consequently, there is nothing better for the present purposes than such exercise.

---

35 συνέλθωσι Ko; συναχθῶσι Ku

36 post νῦν: ἡνῖν (Ku) om.

37 ἐν τοῖς τοιούτοις γίνεται σώμασιν add. Ko

τῆς δὲ ὥρας τοῦ ἔτους ἐπιτρεπούσης οὐδὲ τῆς ψυχρολουσίας ἀφεκτέον ἐστίν. ἐδέσματα δὲ τῷ πλήθει
μὲν ἐλάττω, τῇ ποιότητι δὲ ὑγρότερα δοτέον, ἵνα καὶ
πέψῃ καλῶς αὐτὰ καὶ τὴν ἐκ τῶν ἀφροδισίων ἐπανορ-
θώσηται ξηρότητα. χρὴ δ᾽ οὐδὲ ψυχρότερα τὴν κρᾶσιν, ἀλλ᾽ ἤτοι τῆς μέσης ἰδέας ἢ τῶν θερμοτέρων
ὑπάρχειν αὐτά. διότι γὰρ ἐξ ἀφροδισίων ἀραιότερον
καὶ ψυχρότερον ἅμα καὶ ἀσθενέστερον καὶ ξηρότερον
ἀποτελεῖται τὸ σῶμα, χρὴ δήπου τὰ πυκνοῦντα καὶ
θερμαίνοντα καὶ ὑγραίνοντα[38] καὶ τὴν δύναμιν ἀναρ-
ρωννύντα προσφέρεσθαι, καὶ τούτους εἶναι τοὺς σκοποὺς ἐπ᾽ αὐτοῖς.

ὅτι δὲ αὕτη δύναμίς ἐστιν τῶν ἀφροδισίων, οὐ τοῦ
νῦν ἐνεστῶτος λόγου· προὔκειτο γὰρ ἐν αὐτῷ διελθεῖν, ὅπως ἂν κάλλιστα γυμνάζοιτο τὸ προκείμενον
σῶμα μετὰ τὴν τῶν ἀφροδισίων χρῆσιν, ὅπερ οὐκ
ἠδύνατο περανθῆναι καλῶς ἄνευ τοῦ προλαβεῖν ἐξ
225K ὑποθέσεως, ὁποία τίς ἐστιν ἡ ἐν τῷ σώματι γινομένη
διάθεσις ἐπὶ τοῖς ἀφροδισίοις. ἀλλὰ νῦν μὲν ἐξ ὑποθέσεως, αὖθις δὲ μετ᾽ ἀποδείξεως εἰρήσεται, τίς τε ἡ
δύναμις αὐτῶν ἐστι καὶ εἰ χρηστέον ὅλως ἢ μὴ καὶ
τίνες αὐτῶν ὠφέλειαι καὶ βλάβαι κατά τε τὰς τοῦ
σώματος διαθέσεις εἰσὶ καὶ τὰς ὥρας τοῦ ἔτους καὶ
τὰς χώρας ὅσα τ᾽ ἄλλα τοιαῦτα χρὴ προσδιορίζε-
σθαι.

12. Μετὰ μὲν δὴ τὴν τῶν ἀφροδισίων χρῆσιν εἶδος
ἔστω γυμνασίου τὸ παρασκευαστικὸν ὀνομαζόμενον,
ἀγρυπνίας δὲ προσγινομένης ἢ λύπης ἢ ἀμφοτέρων

If the season of the year permits, one must not abstain from cold baths. Foods should be less in quantity but in quality one must give those that are more moist, so you may concoct them well and correct the dryness from sexual intercourse. They must not be colder in respect of *krasis* but either intermediate in kind or be among those that are hotter. Because the body is made looser in texture and colder, and at the same time weaker and drier from sexual intercourse, one must, of course, provide things that are condensing, heating and moistening, and that strengthen the capacity. Such are the objectives after these [activities].

What this potency is in sexual intercourse is not part of the present discussion. What lies before us in it is to go over how best the body presently proposed may exercise after the use of sexual intercourse. This could not be accomplished well without anticipating from the hypothesis what kind of condition exists in the body after sexual intercourse. But now, from hypothesis and again along with demonstration, I shall say what the potency of sexual intercourse is, and if on the whole one should indulge in it or not, and what the benefits and harms are in respect of the conditions of the body, the seasons of the year, the places and other such things that one must determine besides.

225K

12. Certainly, after the use of sexual intercourse, the kind of exercise should be what is termed preparatory. However, when insomnia is added, or grief, or both, it

---

38 καὶ ὑγραίνοντα *add.* Ko

τὸ ἀποθεραπευτικόν, ὅταν γε χωρὶς ἀπεψίας γεννηθῶ-
σιν· ἐπὶ γὰρ ταῖς ἀπεψίαις οὐδ᾽ ὅλως γυμναστέον. ὅτι
δὲ τὸ ἀποθεραπευτικὸν γυμνάσιον ἐπὶ λύπαις τε καὶ
ἀγρυπνίαις ἁρμόττει, δηλοῖ μὲν καὶ ἡ πεῖρα· φαίνονται γὰρ ὑπὸ τῶν ἄλλων γυμνασίων βλαπτόμενοι,
πρὸς τῷ μηδ᾽ ἀνέχεσθαι τῶν ἐπιταττόντων, εἰ τύχοιεν
ἔτι λυπούμενοι. δηλοῖ δ᾽ οὐχ ἧττον τῆς πείρας καὶ ὁ
λόγος· ἐπειδὴ γὰρ ἐπ᾽ ἀγρυπνίαις τε καὶ λύπαις ὁρῶν-
226K ται λεπτότεροί τε καὶ αὐχμηρότεροι καὶ δυσήκοοι γι-
νόμενοι, ξηρότερον ἡγητέον αὐτοῖς εἶναι τὸ σῶμα. τὰς
δὲ τοιαύτας διαθέσεις αἵ τε μαλακώτεραι τρίψεις ἐξ-
ιῶνται σὺν ἐλαίῳ πλείονι γινόμεναι καὶ λουτροῖς
εὐκράτοις αἵ τε κινήσεις αἱ βραδύτεραι καὶ χωρὶς
ἰσχυροτέρας τάσεως ἀναπαύσεσι πλείοσι διειλημμέ-
ναι. τύπος δ᾽ ἦν οὗτος ἀποθεραπευτικοῦ γυμνασίου.

κατὰ δὲ τὸν αὐτὸν τρόπον καὶ τὰς ἐπὶ θυμοῖς ἢ δι᾽
ἔνδειαν ποτοῦ γινομένας ξηρότητας ἐπανορθωτέον
ἐστίν. ἐναντίως δὲ τοῖς εἰρημένοις ἐπανορθοῦσθαι
χρὴ τὰς κατὰ ταύτην τὴν ἕξιν ὑγρότητας, εἴτε διὰ
πόμα πλέον εἴτε δι᾽ ἄλλην τινὰ πρόφασιν ἐγένοντο.
σκοπὸς γὰρ οὖν δὴ καὶ τῶν τοιούτων διαθέσεών ἐστιν
ἡ ξήρανσις. ἀλλὰ τοῦτο μὲν ἁπασῶν κοινόν, ἴδιον δ᾽
ἑκάστης ἐν ταῖς κατὰ μέρος διαφοραῖς. εἰ μὲν οὖν ἐπ᾽
ἀργίᾳ πλείονι καὶ ταῖς τῶν ὑγραινόντων ἐδεσμάτων
ἀμέτροις τε καὶ ἀκαίροις χρήσεσιν ὑγρότης ἐγένετο,
μακροτέρας δεῖται τῆς ἐπανορθώσεως· εἰ δ᾽ ἐπὶ ποτῷ
πλείονι κατὰ τὴν προτεραίαν ἡμέραν γεγονότι χωρὶς
τοῦ πεπονθέναι τι τὴν κεφαλὴν ἢ τὸ στόμα τῆς κοι-

should be apotherapeutic, whenever they are engendered apart from *apepsia*, because, after the *apepsias*, there must be no exercise at all. Experience also makes it clear that apotherapeutic exercise is suitable in the griefs and insomnias, for people are obviously harmed by the other exercises, besides not putting up with those who order them, if they happen to be still grieving. And reason makes this clear no less than experience. Since in the insomnias and griefs, greater thinness and dryness are seen, and people become intractable, one must consider the body to be drier in them. The softer massages dispel such conditions when they are done with plenty of oil and *eukratic* baths, and the movements are slower and without too strong a tension, and they are divided by numerous rests. This is an outline of apotherapeutic exercise.

    In the same way too, one must correct the dryness following anger or lack of drink. In the opposite manner to the things mentioned, it is necessary to correct the moistness in the bodily state produced by either excessive drink or some other cause. Drying, then, is certainly the aim in such conditions. But this is common to all; what is specific to each lies in the individual differences. If, therefore, in excessive idleness and in the immoderate and untimely use of moistening foods, moistness arises, a longer period of correction is needed. If, however, it follows excessive drink on the previous day, without the head or the opening

226K

227K λίας, ἐν μιᾷ δυνατὸν ἡμέρᾳ τελέως ἐξιάσασθαι πλεο-
νάσαντας μὲν ἐν ταῖς ξηραῖς τρίψεσι, γυμνάσαντας
δ᾽ ὀξύτερον, ἐλάττονι δὲ ποτῷ χρησαμένους ἐδέσμασί
τε ξηραντικωτέροις. ὡς, ὅσαι γε μετὰ τοῦ τὴν κεφαλὴν ἢ τὸν στόμαχον ἀπολαῦσαί τι τῆς ἐξ οἴνου βλάβης ὑγρότητές εἰσι κατὰ τὸ σῶμα[39] περιτταί, τοῦ νῦν
ἐνεστῶτος οὐ δέονται λόγου· ῥηθήσεται γὰρ ὑπὲρ
αὐτῶν ἐν τοῖς περὶ τῶν νοσωδῶν συμπτωμάτων.

αἱ δ᾽ ἐπ᾽ ἀργίᾳ πάνυ μακροτέρᾳ τὴν ἀρχὴν μὲν οὐδ᾽
ἂν γένοιντό ποτε κατὰ τὴν προκειμένην διάθεσιν,
ὥσπερ οὐδ᾽ αἱ διὰ πλῆθος ἐδεσμάτων ὑγρῶν τὴν φύσιν, οἷάπερ αἱ πλεῖσται τῶν ὀπωρῶν εἰσι καὶ τῶν
λαχάνων ὅσα μὴ δριμέα· γενομένας δ᾽ αὐτὰς ἀθρόον
μὲν οὐχ οἷόν τε θεραπεύειν· εἰ γὰρ εἰς τοσοῦτον πο-
νήσειεν ὁ ἄνθρωπος, ὡς αὐτάρκως ξηρᾶναι τὴν ἕξιν,
ἁλώσεται κόπῳ καὶ πυρέξει πυρετὸν μὲν ἐφήμερον
πάντως, ἂν δὲ καὶ μοχθηραὶ τύχωσιν αἱ ὑγρότητες
ὑπάρχουσαι, πλειόνων ἡμερῶν· ἐν χρόνῳ δ᾽ ἂν ἐπα-
νορθωθεῖεν, ὡς ὕστερον εἰρήσεται κατ᾽ ἐκεῖνον τὸν
λόγον, ἐν ᾧ τὰς μοχθηρὰς κράσεις ἐπὶ τὸ βέλτιον
ἀλλοιοῦμεν. ὁμοία γὰρ ἡ πρόνοια τῶν ἐπικτήτων
228K διαθέσεών ἐστι καὶ τῶν φυσικῶν δυσκρασιῶν, ὥστ᾽
οὐδὲν χρὴ τό γε νῦν εἶναι περὶ αὐτῶν διεξέρχεσθαι.

13. Λείπεται οὖν ἔτι περὶ τῶν ἑωθινῶν τε καὶ κατὰ
τὴν ἑσπέραν τρίψεων διελθεῖν, οὐ μὰ Δία οὕτως, ὡς

___

[39] κατὰ τὸ σῶμα add. Ko

of the stomach having been affected, it is possible to effect   227K
a complete cure in one day by an increase in the dry massages, exercising more rapidly, and by using less drink and
more drying foods. Those moist excesses involving the
body along with the head and esophagus, "enjoyed" as part
of the harm from wine, are not a required component of
the present discussion. I shall speak about these in the
works on disease symptoms.[29]

The moist superfluities following very prolonged idleness would not occur in the first place in the proposed
condition, just as those due to an abundance of foods that
are moist in nature would not—foods like the majority of
fruits and herbs that are not acrid. When these do occur,
it is not possible to treat them all at once. If the person has
labored to such a degree as to dry the bodily state enough,
he will be seized by fatigue and become febrile with a
fever, which in all cases will be ephemeral. If, however,
abnormal moisture also happens to exist, [the fever] lasts
a number of days. They would be corrected in time, as I
shall recount later in that discussion in which we change
the bad *krasias* to the better. For there is the same care
for the acquired conditions as there is for the natural *dys-*   228K
krasias, so there is no need to go over these now.

13. It still remains, then, to go over morning and evening massages, although not, by the gods, in the way they

[29] *Sympt. Diff.* (VII.42–84K), and *Sympt. Caus.* (VII.85–
272K) (English trans., Johnston, *Galen: On Diseases and Symptoms*).

φασιν ἀποκρίνασθαι Κόιντον ἐρομένῳ τινὶ γυμναστῇ,
τίνα δύναμιν ἔχει τὸ ὑποσυγχρίεσθαι, φάμενον ἀφα-
νίζειν τὰ ἱμάτια. τούτοις γὰρ τοῖς ὀνόμασιν, οἷς ἐγὼ
νῦν ἐχρησάμην, ἐρέσθαι τε λέγουσι τὸν γυμναστὴν
ἀποκρίνασθαί τε τὸν Κόιντον. ὅμοιόν τι τοῦ Κόιντου
περιφέρεται ἀπόφθεγμα τό τε περὶ τῶν οὔρων, ὡς
γναφέως⁴⁰ ἐστὶ καταμανθάνειν αὐτά, καὶ τὸ περὶ θερμοῦ, ψυχροῦ, ξηροῦ καὶ ὑγροῦ, διότι βαλανέων ἐστὶν
ὀνόματα ταῦτα.⁴¹ ἃ ἐγὼ μὲν οὐκ ἂν πεισθείην,⁴² μὴ ὅτι
Κόιντον, ἀλλ᾽ οὐδὲ τῶν ἀπὸ Θεσσαλοῦ τινα φθέγξα-
σθαι· βωμολοχικὰ γὰρ ἅπαντ᾽ ἐστὶ τὰ τοιαῦτα κομ-
ψεύματα καὶ οὐδαμῶς ἀνδρὶ προσήκοντα σεμνῆς
οὕτω τέχνης ἐπιστήμονι.

βέλτιον οὖν ὑπὲρ μὲν τῆς ἑωθινῆς ἀνατρίψεως ὧδέ
πως σκοπεῖσθαι κατά γε τὸ προκείμενον ἐν τῷ νῦν
ἐνεστῶτι λόγῳ σῶμα. τουτὶ γὰρ ἤτοι παντάπασιν
ἄμεμπτόν ἐστι μετὰ τοὺς ὕπνους ἤ τινι τῶν κοπωδῶν
ἐνέχεται διαθέσεων ἢ καὶ τῶν ἄλλων τινί, περὶ ὧν
ἐφεξῆς ταῖς κοπώδεσιν ὀλίγον ἔμπροσθεν ἄχρι τοῦ
δεῦρο διῆλθον. εἰ μὲν οὖν ἄμεμπτον ὑπάρχει, περί-
εργόν ἐστιν ἢ ἀνατρίβειν ἢ ἀλείφειν αὐτό, πλὴν εἴ
ποτ᾽ ἀναγκαῖον εἴη συνενεχθῆναι κρύει κρατερῷ· τηνικαῦτα γὰρ ὡς τοὺς μέλλοντας ψυχρολουτρεῖν οὕτω
καὶ τούτους τῇ τρίψει παρασκευάσομεν. εἰ δέ τις αἴσθησις εἴη κοπώδης, λέλεκται καὶ πρόσθεν, ὡς ἀλεί-

229K

---

⁴⁰ γναφέως Ko; γραφέως Ku
τὰ τοιαῦτα τῶν ὀνομάτων Ku

⁴¹ ὀνόματα ταῦτα Ko;
⁴² πεισθείην Ko; δοίην Ku

say Quintus[30] replied to some gymnastic trainer who asked him what value anointing has, when he said, "it gets rid of the tunic." The words I used just now are those they say the gymnastic trainer asked, and what Quintus replied. This apothegm, put about by Quintus, is rather like that about the urine—it is for the scribe to find out about this—and about hot, cold, dry and moist, that they are the names of baths. These are things I would not be persuaded that Quintus or any of those associated with Thessalus[31] said. All such quibbles smack of ribaldry and are in no way fitting for a man knowledgeable in so serious an art.

Better, then, to consider morning massage in relation to the body proposed in the present discussion as follows. For this is either altogether without fault after sleep, or is   229K subject to one of the fatigue conditions, or one of the others which I went over following the fatigues a little earlier up to this point. If, then, the body is without fault, it is superfluous to massage or anoint it, unless at any time it is necessary to cope with severe cold. Under these circumstances, we shall prepare such people with massage as we would those about to undergo a cold bath. If there is some sensation of fatigue, as has also been said before, it is then

[30] Quintus (fl. AD 115–145) was a student of Marinos and was praised by Galen, who described him as "being like an Empiric but not of that school" (see EANS, 717). The story about the four qualities was that they were just the names of baths and not those of fundamental qualities / components of the body.

[31] Thessalos of Tralleis (ca. AD 20–70) was a leading exponent of Methodism and the subject of very severe criticism by Galen. This is well exemplified in Book 1 of Galen's *MM*, 1–1021K (English trans., Johnston and Horsley, *Galen: Method of Medicine*).

φειν τε χρὴ τηνικαῦτα καὶ ἀνατρίβειν μαλακῶς. οὕτω
δὲ καὶ εἰ ξηρότερον εἴη πλέον τοῦ δέοντος, ἀλειπτέον
μὲν ἐλαίῳ γλυκεῖ τέγγει γὰρ τοῦτο τὸν ξηρὸν χρῶτα,
τριπτέον δ᾽ ἐλάχιστα μέν, ἀλλὰ μήτε σκληρᾷ τρίψει
μήτε μαλακῇ. προτρέψαι γὰρ μόνον δεόμεθα τὴν ἀνάδοσιν, οὔτε δ᾽ ἀλλοιῶσαι τοῦ δέρματος ἢ τῆς σαρκὸς
τὴν ἕξιν οὔτε διαφορῆσαί τι τῶν περιεχομένων ἐν
αὐτοῖς. ἐργάζεται δὲ ἄμφω μὲν ἡ μαλακή, θάτερον δὲ
καὶ ἡ σκληρὰ τρίψις, εἴ γε δὴ πυκνοῖ μὲν αὕτη καὶ
σκληρύνει τὸ δέρμα, διαφορεῖ δ᾽ ἡ μαλακὴ καὶ
230K   ἀραιοῖ[43] καὶ μαλακὸν ἀπεργάζεται τὸ σῶμα.

πύκνωσιν μέντοι τοῦ δέρματος ἐπανορθώσασθαι
βουλόμενοι, τὴν μὲν ἐπὶ ταῖς σκληραῖς ἀνατρίψεσι
καὶ λαβαῖς καὶ σφοδρῷ γυμνασίῳ καὶ κόνει πολλῇ
γεγενημένην ἐλαίῳ δαψιλεῖ καὶ γλυκεῖ χρώμενοι μαλακῶς ἀνατρίψομεν, τὴν δ᾽ ἐπὶ ψύξει πρώτως μὲν ταῖς
ξηραῖς τε ἅμα καὶ ταχείαις ἀνατρίψεσι, δευτέρως δὲ
ταῖς δι᾽ ἐλαίου θερμαίνοντος εἰς τὸ κατὰ φύσιν ἐπανά-
ξομεν. ἀραιότητα δὲ τὴν ἐπί τε λουτροῖς πλείοσι καὶ
τρίψεσι μαλακαῖς ἀφροδισίων τε χρήσεσι γεγενημένην ὀλίγαις μὲν ταῖς ξηραῖς ἀνατρίψεσιν, ὀλίγαις
δ᾽ ἐφεξῆς αὐτῶν ταῖς σὺν ἐλαίῳ τινὶ τῶν στυφόντων
ἰασόμεθα. τὰς δ᾽ ἐπὶ πλέοσι ποτοῖς ὑγρότητας τρίψεις[44] ξηραὶ μόναι θεραπεύουσι διά τε σινδόνων ἢ
χειριδίων ἐπιτελούμεναι καὶ αὐτῶν μόνων ἐνίοτε τῶν
χειρῶν ἢ χωρὶς λίπους παντὸς ἢ σὺν ἐλαίῳ[45] ἐλαχί-
στῳ τινί. ἔστω δὲ τοῦτο τοὔλαιον γλυκύ, ἵν᾽ ᾖ διαφο-

necessary to anoint and massage gently. In this way too, if
the body is drier than it should be, it must be anointed
with sweet oil, for this moistens the dry surface of the
body, and it must be massaged for the shortest time, but
with neither hard nor soft massage. For we only need to
promote the distribution and not to change the state of the
skin or flesh, or disperse any of those things contained in
them. Soft massage effects both, whereas hard massage
effects one of the two, if in fact it condenses and hardens
the skin, while soft massage disperses and makes the body   230K
loose textured and soft.

However, if we wish to correct condensation of the
skin which has arisen following hard massages, wrestling
holds, violent exercise and much powder, we shall massage
gently using sweet oil in abundance. On the other hand,
we shall return to an accord with nature that condensation
which has arisen from chilling primarily with dry and rapid
massages and secondarily to these with oil that is warming.
Loose texture that has arisen after many baths, gentle massages and the use of sexual intercourse, we shall cure with
a few dry massages, and following a few of these, with
those with some oil that has astringent properties. Dry
massages alone treat the moistness after too many drinks.
These massages are done with muslin cloths and the
hands, but sometimes just with the hands alone, either
without oil altogether or with the least amount. This oil

---

43 ἀραιοῖ Ko; ἀραιὸν Ku
44 τρύψεις Ko; αἱ Ku
45 ἐλαίῳ add. Ko

ρητικόν, ἁπάσης ἀπηλλαγμένον στυφούσης ποιότητος. ὧδε μὲν ἔχει περὶ τῆς ἑωθινῆς ἀνατρίψεως.

231K ἡ δ᾽ εἰς ἑσπέραν ἤτοι κοπώδεσιν ἱκανῶς ὑπάρχουσιν ἢ κατεξηρασμένοις ἢ ἀτροφοῦσιν ἐπιτήδειος.
ἀλλὰ τὸ μὲν τῆς ἀτροφίας σύμπτωμα τό γε νῦν ἐξαι-
ρείσθω τοῦ λόγου μετὰ τῶν ἄλλων ἁπάντων νοσωδῶν
συμπτωμάτων ἑξῆς προχειρισθησόμενον. ἐπὶ δὲ τῆς
ὑποκειμένης φύσεως, ὅταν ἤτοι κόπος ἰσχυρὸς ἢ ξηρότης τις ἄμετρος ὑπάρχῃ κατὰ τὸ σῶμα, τὸ μὲν ἄριστον ἔλαττον γινέσθω, πλείων δ᾽ ὁ μεταξὺ χρόνος
ἄχρι τοῦ δείπνου, τὰ πολλὰ δ᾽ ἐφ᾽ ἡσυχίας· ὀλίγον δέ
τι καὶ περιπατείτωσαν, ὡς ὑποκαταβῆναι τὰ σιτία
ταῖς ὀρθίαις κινήσεσι κατασεισθέντα· βέλτιον δὲ καὶ
εἰ ἀποπατῆσαι δυνηθεῖεν. τούτων γὰρ ἁπάντων καλῶς
γενομένων ἀκίνδυνον ἀνατρίβειν ἐλαίῳ γλυκεῖ, μὴ
πάνυ τι τῆς γαστρὸς ἐφαπτόμενον· εἰ δὲ μή, κίνδυνος
αὐτά τε τὰ σιτία πεφθῆναι χεῖρον ἀναδοθῆναί τέ τινα
χυμὸν ἐξ αὐτῶν ἡμίπεπτον ἐπιθολωθῆναί τε τὴν κεφαλὴν ἀνατραπῆναί τε τὸν στόμαχον. ἄριστον μὲν
οὖν ἐστι τὸ μηδ᾽ ὅλως ἅπτεσθαι τῆς γαστρός· εἰ δέ
ποτε τῶν ἀμφ᾽ αὐτῇ μυῶν ἤτοι κοπώδης τις αἴσθησις
ἢ πλείων ἐμφαίνοιτο ξηρότης, ἀλείφειν τὰ μέτρια
232K πράως ἐφαπτόμενον. εἰ δὲ καὶ τὰς αἰτίας τις ἀκοῦσαι
ποθεῖ τῶν εἰρημένων, τὸν ἑξῆς ἀναμεινάτω λόγον, ἐν
ᾧ περὶ τῶν νοσωδῶν συμπτωμάτων διερχόμεθα· νυνὶ
μὲν γὰρ ἤδη μοι δοκῶ μέγεθος αὔταρκες ἔχειν τὸν
ἐνεστῶτα λόγον.

should be sweet so that it is dispersing (diaphoretic) and is free of any astringent quality. Such, then, is the matter of morning massage.

The evening massage is suitable for those who are greatly fatigued, or dried out, or wasting away. But let the symptom of atrophy be set aside from the present discussion along with all the other disease symptoms which will be dealt with in due course. In the case of the assumed nature, whenever there is either strong fatigue or some immoderate dryness involving the body, let the breakfast be less and the time between it and dinner be longer, and for the most part be spent resting, although there should also be a little walking around, so that the foods, shaken by the upright movements, settle down. And it is better if the person is enabled to move the bowels. When all these things occur properly, massage with sweet oil is free of danger, if there is no touching of the abdomen at all. Otherwise, there is a danger that the foods themselves will be concocted badly and that some semiconcocted humor from them will be distributed, making the head turbid and the stomach upset, Therefore, it is best not to touch the abdomen at all. If, at any time, there is either a sensation of fatigue of the muscles on both sides of this, or an excessive dryness appears, anoint moderately with a gentle application. If someone is anxious to learn the causes of the things mentioned, let him await the subsequent discussion in which I go over the disease symptoms. For now, the present discussion seems to me already long enough.

Δ

233K   1. Οὐχ ὡς οἱ πλεῖστοι τῶν νεωτέρων ἰατρῶν ἐν τοῖς σοφιστικοῖς ζητήμασι κατατρίψαντες τὸν χρόνον ἤτοι διὰ βραχέων ἐπιτρέχουσι τὸν περὶ τῶν ἀναγκαιοτάτων λόγον ἢ καὶ παντάπασι παραλείπουσιν, οὕτω καὶ ἡμεῖς ποιήσομεν, ἀλλ᾽, ὅπερ ἐξ ἀρχῆς ἐνεστησάμεθα, τὸ χρήσιμον αὐτὸ διερχόμενοι τὰ λογικωτέραν ἔχοντα τὴν ἐπίσκεψιν εἰς ἕτερον ἀναβαλλόμεθα καιρόν. αὐτίκα γέ τοι περὶ τῶν νοσωδῶν συμπτωμάτων, ὑπὲρ ὧν ἐν τῷδε τῷ λόγῳ πρόκειται διελθεῖν, οὐ μι-
234K   κρὰ ζήτησίς ἐστι, πότερον ἐκ τῆς ὑγιεινῆς ὑπάρχει πραγματείας ἢ ἐκ τῆς θεραπευτικῆς ἢ τούτων μὲν οὐδετέρας, ἄλλης δέ τινος ἀμφοῖν τρίτης, ἣν δὴ καὶ μέσην ὑγείας τε καὶ νόσου τίθενταί τινες οὐδετέραν ὀνομάζοντες.

ἐγὼ δ᾽ ἐπιστάμενος μέν, ὡς, εἴτ᾽ ἐν τοῖς ὑγιεινοῖς τις εἴτ᾽ ἐν τοῖς θεραπευτικοῖς αὐτῶν μνημονεύσειεν, ὁμοίως ὑπὸ τῶν σοφιστῶν ἐπηρεασθήσεται, γινώσκων δ᾽ οὐδὲν ἧττον, ὡς, εἰ καὶ τρίτης τις αὐτοῖς ἀναθείη γενέσθαι πραγματείας ὑπὲρ τῶν οὐδετέρων δια-

---

1 On this issue—what is intermediate between health and disease—see Galen's *Ars M.*, particularly chapters 1–4; Johnston,

BOOK IV

1. I shall not do as the majority of younger doctors do who 233K waste their time in sophistical inquiries, either treating in summary fashion the discussion of the most essential things or leaving them out altogether. Instead, as I set out to do from the beginning, going over what is actually useful, I shall defer to another time those matters that have a more logical investigation. For certainly the immediate inquiry into disease symptoms, which lie before us to consider in this book, is no small matter, whether it belongs 234K to the subject of hygiene, or to therapeutics, or to neither of these, but rather to some third matter different to both which therefore occupies some place between health and disease; some call this "neither" ("neutral").[1]

However, since I feel sure that, whether I make mention of these things under hygiene or therapeutics, they will be similarly spoken of disparagingly by the Sophists,[2] knowing no less that, even if I were to ascribe some third category to them, writing about neither of the other condi-

On the Constitution, 156–77, and von Staden, *The Art of Medicine in Early Alexandria,* 103–7.

[2] It is not clear whether Galen is referring to anyone in particular here or simply using "Sophist" in a general pejorative sense applied to those who indulge in fruitless quibbles about names.

θέσεων ἐπιγράψας, ἔτι καὶ μᾶλλον ἐπιγελάσονταί τε
καὶ τωθάσονται καὶ ἐρήσονται, περὶ τῶν ἀρρενικῶν
καὶ θηλυκῶν ἐν ποίᾳ πραγματείᾳ διδάσκομεν, εἱλόμην
ἐν τῷ νῦν ἐνεστῶτι λόγῳ διελθεῖν ὑπὲρ αὐτῶν. εἰ
γὰρ ἀδύνατον μέν ἐστι διαφυγεῖν[1] τῶν σοφιστῶν τὴν
γλωσσαλγίαν, ἔλαττον δ᾽ ἐπηρεάσουσιν οὕτω δια-
πραξάντων, ἄμεινον ἴσως ἐστὶν τοῦτο ποιεῖν. ἔτι δὲ
μᾶλλον ἄν τις ἐξ αὐτῆς τῆς θεωρίας ἐπιγνοίη τὴν
κοινωνίαν τῆς διδασκαλίας, εἰ προσέχοι τὸν νοῦν
ἀκριβῶς τοῖς λεχθησομένοις, ὧν ἀρκτέον ἐνθένδε.

235K τῆς ὑγιεινῆς ἐπιστήμης οὐ φαυλότατόν ἐστι μόριον
ἡ περὶ τὰ γυμνάσια τέχνη· ταύτης δ᾽ αὐτῆς οὐ μικρὰ
μοῖρα τὸ φυλάξασθαι κόπους. ἐδείχθη δ᾽ ἔμπροσθεν,
ὡς ὁμοία τίς ἐστιν ἥ τε προφυλακὴ τῶν ἐσομένων
κόπων καὶ ἡ ἐπανόρθωσις τῶν ἤδη γεγονότων. οὔκουν
ἑτέρωθι μὲν ἐχρῆν ἐκδιδάσκειν, ὅπως χρὴ φυλάττεσθαι κόπους, ἑτέρωθι δ᾽, ὅπως εἰς τὸ κατὰ φύσιν
ἐπανάγειν προσήκει τοὺς ἤδη γεγονότας. διὰ ταῦτα
μὲν οὖν ἐν τῷ πρὸ τούτου γράμματι, τρίτῳ τῆς ὅλης
πραγματείας ὄντι, περὶ τῶν ἐπὶ γυμνασίοις κόπων ὁ
λόγος ἡμῖν ἐγένετο μετὰ τοῦ τῶν ὁμοίων αὐτοῖς ἐπι-
μνησθῆναι διαθέσεων, ὧν ἔνιαί τινες ἐπὶ γυμνασίοις
ἐγίνοντο μάλιστα. νυνὶ δὲ πρῶτον μὲν ὑπὲρ αὐτῶν
ἐροῦμεν, ἐπειδὰν ἄνευ τῶν γυμνασίων γίνωνται, δεύτερον δὲ καὶ περὶ τῶν ὁμοειδῶν αὐτοῖς.

2. Ὁ μὲν οὖν ἐπὶ γυμνασίοις ἀμέτροις γινόμενος
κόπος ὑγιεινόν τι σύμπτωμά ἐστιν, ὁ δὲ χωρὶς τούτων
νοσώδης. ὥστε καὶ Ἱπποκράτει δοκεῖ κάλλιστα εἰρῆ-

tions, they will laugh and mock even more and will ask under what kind of subject heading I will teach about men and women, which is what I have chosen to consider in the present discussion. If it is impossible to escape the endless talking of the Sophists, but they would be less disparaging if I were to do it in this way, perhaps it would be better if I were to do so. Still more, someone might recognize the generality of the teaching from this same theory, if he directs his attention closely to the things that will be said, of which I must make a start here.

The art concerning exercises is by no means the most trivial part of the knowledge of hygiene. And no small part 235K of the same knowledge is the avoidance of fatigues. It was shown previously that what is prophylactic for fatigues that will occur and what is corrective of those that have already occurred are the same. It is not therefore necessary to teach thoroughly in one place how one must avoid fatigues but in another place how it is a appropriate to return to an accord with nature those fatigues that have already occurred. Because of these things then, in the book prior to this one—the third in the whole work—our discussion was about the fatigues after exercises, together with that of the conditions similar to them that have occurred and were mentioned by me. Some of these occurred particularly after exercises. I shall now speak first about these when they occur without exercise and second, about the conditions of a similar kind to them.

2. The fatigue that occurs after excessive exercises is a healthy symptom whereas that which occurs apart from exercises is a disease symptom. So Hippocrates seems

---

1 διαφυγεῖν Ko; ἐκφυγεῖν Ku

σθαι· "κόποι αὐτόματοι φράζουσι νούσους." ἡ μὲν οὖν
236K ἑλκώδης αἴσθησις σύμπτωμά ἐστι κοπῶδες, ἡ δ' αἰτία, δι' ἣν αὕτη γίνεται, διάθεσις κοπώδης. ἡ δὲ καὶ
ταύτης αὐτῆς αἰτία διττὴν ἔχει τὴν διαφοράν, ἤτοι
κατ' αὐτὸ τοῦ ζῴου τὸ σῶμα περιεχομένη, καὶ ὀνομάζεται τηνικαῦτα προηγούμενον αἴτιον, ἢ μηδ' ὅλως
ἐνυπάρχουσα, καὶ καλεῖται τηνικαῦτα² προκαταρκτι-
κὸν αἴτιον. ὡς εἶναι μὲν τρία τὰ σύμπαντα γένη, περὶ
ὧν ὁ νῦν ἡμῖν ἐνέστηκε λόγος, τό τε σύμπτωμα τὸ
κοπῶδες καὶ τὴν διάθεσιν τὴν κοπώδη καὶ τὴν αἰτίαν
αὐτῆς. ἑκάστου δ' αὐτῶν εἰδικάς τινας ὑπάρχειν διαφοράς, ἐν μὲν ταῖς αἰτίαις, ὡς εἴρηται νῦν δή, τὴν
προηγουμένην καὶ τὴν προκατάρχουσαν, ἐν δὲ ταῖς
διαθέσεσιν, ὡς ἔν γε τῷ πρὸ τούτου λόγῳ δέδεικται,
τὴν ἑλκώδη καὶ τὴν τονώδη καὶ τὴν φλεγμονώδη, καὶ
μέντοι κἂν τοῖς συμπτώμασι τρεῖς τὰς αὐτάς. ὀνομάζειν μὲν οὖν ἔξεστιν, εἰ βούλοιτό τις, ἑτέρως· οὔτε δὲ
πλείω τῶν εἰρημένων λέγειν τε καὶ ποιεῖν ἔξεστιν,
ἀληθεύειν γε βουλομένοις, οὔτε πλείους διαφορὰς τῶν
λελεγμένων.

ὁ μὲν οὖν ἑλκώδης κόπος (ἀρκτέον γὰρ ἀπὸ τοῦδε)
κινουμένοις αἴσθησιν ὀδυνηρὰν³ ὡς ἑλκουμένου φέρει
237K τοῦ σώματος, ἤτοι κατὰ τὸ δέρμα μόνον, ὅταν ᾖ με-
τριώτερος, ἢ κατὰ τὰς ὑποκειμένας αὐτῷ σάρκας,

²τηνικαῦτα add. Ko
³ὀδυνηρὰν Ko; ἀνιαρὰν Ku

342

to have said it best: "spontaneous fatigues announce diseases."[3] Thus, the wound-like sensation is a fatigue symptom; the cause due to which it occurs is a fatigue condition. And the cause of this same condition has a twofold difference. Either it is contained within the actual body of the animal, and under these circumstances is called a *proegoumenic* cause, or it is not at all intrinsic, and under these circumstances is called a *prokatarktic* cause.[4] So there are in all three classes about which our discussion is now established: the symptom of fatigue, the condition of fatigue, and the cause of this condition. And there are certain specific differences of each of these. Among the causes, as I said just now, there are the *proegoumenic* and *prokatarktic;* among the conditions, as has been shown in the discussion prior to this, there are the wound-like, the tensive (tension-like) and the inflammation-like [fatigues], and of course, in the symptoms too, there are the same three. It is possible for someone to name these differently, should he wish to. However, it is not possible for those who wish to speak the truth, to name or make more conditions than those spoken of or more *differentiae* than those that have been stated.

Therefore, the wound-like fatigue (for we must begin from this) produces a sensation that is distressing in those who move, as if the body were wounded, either in the skin alone whenever it is more moderate, in the flesh underly-

236K

237K

---

[3] Hippocrates, *Aphorisms* 2.5, *Hippocrates* IV, LCL 150, 108–9.

[4] For a consideration of these two terms, which might be rendered "internal antecedent" and "external antecedent," see Johnston, *Galen: On Diseases and Symptoms,* 33–35.

ὅταν γένηται σφοδρότερος, ἢ κατὰ τὸ συναμφότερον, ἐπειδὰν ἰσχυρότερος ᾖ. καὶ τοῦτο μέν ἐστι τὸ κοπῶδες σύμπτωμα. διάθεσις δὲ κοπώδης, ἐφ᾽ ᾗ γίνεται τὸ σύμπτωμα, δριμύτης ὑγρῶν ἐστι λεπτῶν καὶ θερμῶν, ὡς διαβιβρώσκειν τε καὶ κεντεῖν καὶ νύττειν τὰ σώματα. γίνεται δὲ αὕτη ποτὲ μὲν ἐπὶ ταῖς ἀμέτροις κινήσεσιν, ὡς ἐν τῷ πρὸ τούτου δέδεικται λόγῳ, ποτὲ δ᾽ ἐπὶ κακοχυμίᾳ τινὶ λεληθότως ὑποτραφείσῃ, καὶ τοὺς τοιούτους κόπους αὐτομάτους ὁ Ἱπποκράτης ὀνομάζει.

τὸ δ᾽ ἕτερον γένος τοῦ κόπου τὸ τονῶδες, ὅταν αὐτόματον συνίστηται, ταῖς καλουμέναις πληθώραις ἕπεται. διατείνεται γὰρ ἐν ταύταις τὰ στερεὰ τοῦ ζῴου μόρια, καὶ μάλιστα ἐν οἷς οἱ χυμοὶ περιέχονται. τὸ δὲ τρίτον τοῦ κόπου γένος τὸ φλεγμονῶδες ἐπὶ πληθώρᾳ τε ἅμα καὶ τῇ προειρημένῃ γίνεται κακοχυμίᾳ. οὐ γὰρ δὴ πᾶν εἶδος κακοχυμίας, ἀλλ᾽ ἐκεῖνο μόνον, ἐν ᾧ δριμύτης ἐστὶ δακνώδης, ἐργάζεται τὸν ἑλκώδη κό-

238K   πον· οὐδ᾽ οὖν οὐδ᾽ αὐτὴ κατὰ τὰς φλέβας ἀναμεμιγμένη τῷ αἵματι λανθάνει γὰρ αὐτῆς ἡ δύναμις τηνικαῦτα διαρρεούσης τε ἅμα καὶ νικωμένης ὑπὸ τῆς τοῦ αἵματος χρηστότητος, ἀλλ᾽ ὅταν εἴς τε τὰς σάρκας καὶ τὸ δέρμα μόνη μεταληφθεῖσα στηριχθῇ, τὴν ἑλκώδη διάθεσίν τε καὶ αἴσθησιν ἐπιφέρει. ὅσοι δὲ πλήθους ἔκγονον ὑπάρχειν οἴονται τὸν κόπον τοῦτον, ἁμαρτάνουσιν·[4] οὔτε γὰρ τοῦ τοιούτου πλήθους, ὃ δὴ καὶ πληθώραν ὀνομάζουσι τὸν τονώδη γὰρ ἐκεῖνο κόπον ἐργάζεται, οὔτε τοῦ τὴν δύναμιν βαρύνοντος· οὐ

ing the skin, when it becomes more severe, or in both together when it is stronger. This is the fatigue symptom. However, the condition of fatigue, in which the symptom occurs, is an acridity of fluids that are thin and hot, such as to corrode, sting and prick the bodies. Sometimes this occurs after excessive movements, as I have shown in the discussion prior to this, and sometimes grows up imperceptibly from some *kakochymia*—Hippocrates termed such fatigues "spontaneous."

The second class of fatigue—the tensive (tension-like)—when it exists spontaneously, follows the so-called *plethoras.* For in these, the solid parts of the animal are stretched to the uttermost, and especially those in which the humors are contained. The third class of fatigue—the inflammation-like—occurs with *plethora* and with the previously mentioned *kakochymia.* For not every kind of *kakochymia,* but only that in which there is a biting acridity, brings about the wound-like fatigue. But even this does not do so, if it is mixed with the blood in the veins, for under these circumstances its potency escapes notice since it flows along and is overcome by the most useful of the blood, but only when it is transferred to the flesh and skin, and is firmly fixed, does it bring the wound-like condition and sensation. However, those who think this fatigue is a product of excess are mistaken, for it is not of the kind of excess which they also term *plethora,* for that brings about the tensive fatigue, nor of that which weighs

238K

---

4 ἁμαρτάνουσιν Ko; σφάλλονται Ku

γὰρ δῆξις οὐδὲ ἑλκώδης αἴσθησις,[5] ἀλλ᾽ ἤτοι βάρος
τε καὶ δυσκινησία τοῦ τοιούτου πλήθους ἐστὶ συμπτώματα τὴν ψυχικὴν βαρύνοντος δύναμιν, ὅταν ὡς
πρὸς ταύτην ὑπάρχῃ πλέον, ἢ κακοσφυξίαι τινές,
ὅταν ὡς πρὸς τὴν ζωτικήν. εἴρηται δ᾽ ὑπὲρ αὐτῶν
αὐτάρκως ἐν τοῖς περὶ σφυγμῶν, ὥσπερ γε κἂν τῷ
Περὶ πλήθους βιβλίῳ τὰ γνωρίσματα τῆς φυσικῆς
δυνάμεως εἴρηται βαρυνομένης ἅμα τοῖς τῶν ἄλλων
ἀμφοτέρων. οὔκουν τὸ πλῆθος αἴτιόν ἐστι τοῦ τὴν
ἑλκώδη φέροντος αἴσθησιν κόπου, ἀλλὰ τῶν ἐν τῷ
δέρματι καὶ τῇ σαρκὶ περιεχομένων χυμῶν ἡ δρι-
239K μύτης. αὕτη γάρ, ἐπειδὰν μὲν ἡσυχάζουσα περικέηται, διαλανθάνει τὴν αἴσθησιν, εἰς κίνησιν δ᾽ ἀφικο-
μένη παραχρῆμα γνωρίζεται. κίνησις δ᾽ αὐτῇ πρώτως
μὲν καὶ μάλιστ᾽ ἐγγίνεται κατά τινας οἰκείους λόγους,
οὓς ἑξῆς ἐροῦμεν, ἑτέρα δὲ κατὰ συμβεβηκός, ἐπειδὰν
ἡμεῖς αὐτοὶ προελόμενοι κινῆσαί τι μέρος ἢ καὶ σύμπαν τὸ σῶμα σὺν ἐκείνῳ καὶ τὰς ἐν αὐτῷ περιεχομέ-
νας κινήσωμεν ὑγρότητας. ἀλλ᾽ ἡ μὲν τοιαύτη κίνησις ἐλαχίστη τέ ἐστι καὶ τὴν κοπώδη μόνην αἴσθησιν
ἐπιφέρει, ἡ δὲ σφοδροτέρα ῥίγους ἐστὶν αἰτία, ἡ μέση
δ᾽ ἀμφοῖν φρίκης. ὅτι δ᾽ οὐδὲν κωλύει, κἂν θερμὸν

---

[5] ἑλκώδης αἴσθησις Ko; ἕλκωσις Ku

[5] The only other use of this term I could find in Galen is in his
Sympt. Diff., VII.63K. In discussing symptoms due to abnormal
nutrition there, among the symptoms he writes, "stoppages and

down the capacity. For neither a biting nor a wound-like
sensation, but either heaviness or difficulty of movement
are the symptoms of such an excess, weighing down the
psychical capacity when it is more in relation to this,
or creating certain *kakosphyxias* (pulse abnormalities),[5]
when it is more in relation to the physical capacity.[6]
Enough has been said about these in the works on the
pulses,[7] just as the signs of the physical capacity being
weighed down have been spoken about in the book, *On
Plethora*,[8] along with those of both the others. The amount
is not therefore a cause which produces the wound-like
sensation of the fatigue; rather, it is the acridity of the
humors contained in the skin and flesh. For this, when it   239K
lies about at rest, escapes perception, but if it comes to
movement it is immediately detected. Movement in it
arises first and foremost in relation to certain specific reasons, which I shall speak about next; others are contingent
when we ourselves choose to move some part or the whole
body, and with that we also move the fluids contained in
it. But such movement, when it is at its least, brings only
the fatigue sensation, whereas a stronger movement is a
cause of rigors, and that in between both, of shivering.
That nothing prevents even heat also being the cause, and

irregularities are symptoms involving the function of the pulses."
(See Johnston, *Galen: On Diseases and Symptoms,* 192 [VII.63K]).

[6] Galen's detailed account of the capacities or faculties (*duna-
meis*) is to be found in his *Nat. Fac.,* II.1–204K (English trans.,
Brock, *On the Natural Faculties*).   [7] There are seven extant
works on the pulses—see Johnston, *On the Constitution,* 312n73.
Relevant passages include VIII.493K ff. and IX.1K ff.

[8] *Plenit.,* VII.513–83K. See particularly VII.513ff.

ὑπάρχῃ τὸ αἴτιον, ῥῖγός τε καὶ φρίκην αὐτὸ ποιεῖν,
ἐν ταῖς τῶν συμπτωμάτων αἰτίαις ἀναδέδεικται. νυνὶ
δ᾿ ἀρκεῖ μόνον αὐτὰ τὰ κεφάλαια τῶν ἐν ἐκείνῳ τῷ
βιβλίῳ[6] δεδειγμένων ὑπόθεσιν ποιήσασθαι τοῖς παροῦσιν.

ὅταν οὖν ἐν τοῖς αἰσθητικοῖς σώμασιν ὑποτραφῇ
περιττώματα δάκνοντα, κατὰ διττὸν τρόπον εἰς κίνησιν ἀφικνεῖται, καθ᾿ ἕνα μὲν ὑπ᾿ αὐτῶν τῶν αἰσθη-
τικῶν σωμάτων ὠθούμενα, δύναμιν ἐχόντων ἀποκρι-
240K τικὴν τῶν ἀλλοτρίων, καθ᾿ ἕτερον δὲ ὑπὸ κινήσεως
σφοδροτέρας, ἣν ἔκ τε γυμνασίων καὶ θυμοῦ ἢ καὶ
τῆς ἐκ τοῦ περιέχοντος θερμασίας ἐπικτᾶται. τὰ μὲν
οὖν ὑπόθερμά τε καὶ σηπεδονώδη περιττώματα, κινη-
θέντα σφοδρότερον, οὐ φρίκην μόνον ἢ ῥῖγος ἐπιφέρει, ἀλλὰ καὶ πυρετὸν ἐξάπτει· τὰ δὲ ψυχρά τε ἅμα
καὶ λεπτομερῆ φρίκην μὲν καὶ ῥῖγος ἐπιφέρει, πυρετὸν δὲ οὐκ ἐξάπτει. προσεῖναι δέ τι καὶ πλῆθος ἀξι-
όλογον ἑκατέροις ἀναγκαῖον, εἰ μέλλοι ταῦτα ποι-
ήσειν. ὅσα δ᾿ ἤτοι παντάπασιν ὀλίγα περιττώματα
δάκνοντα τοῖς αἰσθητικοῖς ἐγγίνεται σώμασιν ἢ
πλείω μέν ἐστιν, οὔπω δ᾿ ἀκριβῶς δακνώδη, τὸν ἑλκώδη κόπον ἐργάζεται. καὶ δὴ καὶ λεκτέον ἡμῖν ὑπὲρ
τούτων[7] ἐν τῷ παρόντι· τὰ γὰρ τοὺς πυρετοὺς ἐπι-
φέροντα τῆς θεραπευτικῆς ἐστι πραγματείας· οὐ μὴν
ἀλλὰ καὶ ὅσα φρίκην μὲν ἐργάζεται, πυρετοὺς δ᾿ οὐκ
ἐξάπτει, καὶ ταῦτα τῆς ἐνεστώσης ἐστὶ πραγματείας.

6 βιβλίῳ Ko; λόγῳ Ku

that this creates rigors and shivering, was shown in the work, *On the Causes of Symptoms*.[9] For the present, this alone is sufficient—to establish the hypothesis for our current purposes, having demonstrated the actual chief points in that book.

Whenever, in perceiving bodies, biting superfluities are produced, they come to movement in a twofold way. In the first case, they are urged on by the perceiving bodies themselves, since these have a capacity which separates alien materials. In the second case, they are urged on by a stronger movement which is gained as an adjunct from exercises and anger, and also from the surrounding heat. Therefore, somewhat hot and putrefying superfluities, when they are moved more violently, bring about not only shivering or rigors, but also kindle a fever. The cold superfluities, that are at the same time fine-particled, bring shivering and rigors but do not kindle a fever. It is, however, necessary for a notable excess to be present in both cases, if they are going to create these effects. Those superfluities that are biting, but only slightly so, or are more so but not yet entirely biting, arising in perceiving bodies, bring about the wound-like fatigue. And actually, we must speak about these in the present discussion, for those things that produce fevers fall within the province of therapeutics, whereas those that produce shivering but do not kindle fevers fall within the scope of the present subject.

240K

---

[9] See *Sympt. Caus.*, 2.5 (VII.179–80K); Johnston, *Galen: On Diseases and Symptoms* 253–54.

---

[7] *post* τούτων: ἐν τῷ παρόντι Ko; ἐστιν ἐν τῷ παρόντι λόγῳ Ku

241K 3. Ἡ ἴασις δὲ ἡ μέν τις εἰς κοινοὺς ἀμφοτέροις
ἀνάγεται σκοπούς, ἡ δέ ἐστιν ἑκατέρων ἴδιος· εἰρήσεται δὲ πρότερον ἡ κοινή. χρὴ τοίνυν, εἴτε θερμὸν εἴτε
ψυχρὸν εἴη τὸ περίττωμα, κενοῦν ἢ ἀλλοιοῦν αὐτό.
δέχεται δ᾽ οὐ πᾶν περίττωμα τὴν ἐκ τῆς φύσεως ἀλλοίωσιν, ὥσπερ οὐδὲ πᾶν ἔδεσμα πᾶσι τοῖς ζῴοις τὴν
ἐν τῇ γαστρὶ πέψιν, ἀλλ᾽ εἶναι χρή τινα συγγένειαν
ἀεὶ τῷ πεττομένῳ πρὸς τὸ πέττον. ὅταν οὖν ἀλλότριον
ᾖ παντάπασιν, οὐδεμία μηχανὴ τοῦτο τὸν ἐκ τῆς φύσεως ἀναδέξασθαι κόσμον, ἀλλὰ χρὴ κενοῦν αὐτὸ
πειρᾶσθαι διὰ ταχέων, ὥσπερ γε καὶ τὰ κατὰ τὴν
γαστέρα διεφθαρμένα τελέως ἢ ἐμέτοις ἢ διαχωρήσε-
σιν ἐκκενοῦσθαι κράτιστον. οὐ μὴν ἐνδέχεται τὴν ἐν
τῇ σαρκὶ καὶ τοῖς ἄλλοις σώμασιν ἀναπεπομένην
κακοχυμίαν ἑτοίμως οὕτως ἐκκενοῦν, ὡς τὴν ἐν ταῖς
αἰσθηταῖς εὐρυχωρίαις περιεχομένην. ἐνίοτε δὲ καὶ ἡ
φύσις αὐτὴ τοῦ κάμνοντος οὐ προσίεται βοήθημα
ταχέως ἐκκενῶσαι δυνάμενον. ἔστιν δ᾽ ὅτε καὶ ἄλλη
242K τις διάθεσις ἀνθίσταταί τε καὶ ἀπαγορεύει τὸν τοιοῦτον τρόπον τῆς κενώσεως, ὑπὲρ ὧν ἐφεξῆς εἰρήσεται, πρότερόν γε τὸ λεῖπον ἐν τῇ πρώτῃ διαιρέσει
προσθέντων ἡμῶν.

οἱ μὲν γὰρ κοινοὶ σκοποὶ τῆς τῶν περιττωμάτων
ἰάσεως εἴρηνται, κένωσίς τε καὶ ἀλλοίωσις. ἰδίους δ᾽
ἑκατέρων προσθετέον ἡμῖν· οὐ γὰρ ἑνὶ τρόπῳ κενω-
τέον οὐδὲ ἀλλοιωτέον ἐστίν,[8] ἀλλὰ τὸν οἰκεῖον ἀεὶ τοῦ
λυποῦντος ζητητέον. οἰκεῖος δὲ συλλήβδην μὲν εἰπεῖν
ὁ διὰ τῶν ἐναντίων ἐστίν, ἐν μέρει δὲ ὁ καθ᾽ ἕκαστον

3. The cure is either what pertains to common objec- 241K
tives in both or is specific for each. I shall speak about the
common first. Thus, it is necessary to evacuate the superfluity, whether it is hot or cold, or to change it. Not every
superfluity is susceptible to change by Nature, just as not
every food is susceptible to concoction in the stomach in
all animals; there must always be a certain relationship
between what is being concocted and what is doing the
concocting. Whenever the superfluity is altogether alien,
there is no mechanism for this to receive the good order
of Nature; rather, it is necessary to attempt to evacuate it
quickly, just as it is best to evacuate completely, by vomiting or defecation, those things corrupted in the stomach.
It is not possible for the reconcocted *kakochymia* in the
flesh and other bodies to be readily evacuated in the same
way as that contained in the perceptible open spaces.
Sometimes too, the actual nature of the one suffering is
not able to contribute assistance for rapid evacuation.
Sometimes too, another condition opposes and prevents 242K
such a kind of evacuation. I shall speak about this next
after first adding what remains in the first division.

The common objectives of the cure of the superfluities
are as stated: evacuation and change. However, we must
add the specific objectives of each, for we must not evacuate or change in one way only, but must always seek what
is proper for the one who is distressed. In short, what is
proper is that which is through opposites, individually in

---

8 ἐστίν *add.* Ko

ἐναντίως. τὰ μὲν κεφάλαια τοῦ λόγου ταῦτα· χρὴ
δ᾽ ἐξηγήσασθαι πλατύτερον αὐτὰ καὶ τὴν οἰκείαν
ἀπόδειξιν ἑκάστῳ προσθεῖναι, τὴν ἀρχὴν ἀπὸ τῆς
ἑλκώδους διαθέσεως ποιησαμένους τῶν αὐτομάτων
κόπων.[9]

4. Ἐπεὶ τοίνυν ἐπὶ κακοχυμίᾳ δριμέων περιττωμάτων ὁ τοιοῦτος ἐγίνετο κόπος, ἐπισκεπτέον πρότερον, εἴτ᾽ ἐν τοῖς στερεοῖς μόνοις σώμασιν εἴτε κἂν
ταῖς κοιλίαις τῶν φλεβῶν ἡ κακοχυμία περιέχεται.
γνώρισμα δ᾽ οὐδὲν ἔχομεν ἐναργὲς οὐδὲ σαφὲς ὑπὲρ
243K τῶν ἐν ταῖς φλεψὶ περιττωμάτων, ὅτι μὴ κατὰ τὰ οὖρα
μόνον, ἀλλὰ στοχάζεσθαι χρὴ διὰ τῶνδε· πρῶτον μὲν
ἐπισκοπουμένους, ᾗτινι κέχρηται διαίτῃ τὸ κάμνον
σῶμα· δεύτερον δ᾽, εἰ καὶ φύσει κακοχυμίαν ἦν ἔθος
ἀθροίζειν αὐτῷ· καὶ πρὸς τούτοις, εἰ αἱ συνήθεις ἐκ-
κρίσεις ἐπέχονται [αἱ φυσικαί]·[10] τέταρτον ἐπὶ τούτοις,
εἰ γυμνασίοις ἢ καθάρσεσιν ἢ ἐμέτοις ἢ αἰωρήσεσιν
ἢ χρήσεσιν αὐτοφυῶν ὑδάτων εἰθισμένος ἐκκενοῦν τὰ
περιττώματα νῦν ὠλιγώρησεν. ἐν μὲν οὖν τῇ διαίτῃ
σκεπτέον, εἰ ἀπεψίαι προήγηνται πολὺ πλείους καὶ
μείζους τῶν συνήθων ἢ κακοχύμων ἐδεσμάτων ἐνεφ-
ορήσατο πλῆθος ἢ οἶνον ἀντὶ μὲν παλαιοῦ γλεύκινον,
ἀντὶ δὲ λεπτοῦ παχὺν ἢ τεθολωμένον ἔπιεν ἢ καὶ παντάπασιν εἰς ὕδατος πόσιν ἐξ οἴνου μετῆλθεν, οὐχ
ἅπαξ ἢ δὶς ἐφ᾽ ἑκάστῳ τῶν εἰρημένων πλημμελήσας,
ἀλλὰ συνεχῶς τε ἅμα καὶ χρόνῳ πολλῷ.

relation to each opposition. These are the chief points of
the discussion. It is necessary to elaborate on these more
broadly and to add the proper demonstration in each case,
making a start from the wound-like condition of the spontaneous fatigues.

4. Accordingly, since such a fatigue arises from a *kakochymia* of the acrid superfluities, we must first consider
whether the *kakochymia* is contained in the solid bodies
alone or is also in the lumina of the veins. As we have no
visible or clear signs about the superfluities in the veins
except from the urine alone, we must make our assess-   243K
ment through these. We must consider first what diet
the suffering body has used. Second, we must consider
whether it was customary for *kakochymia* to naturally
collect in it; and in addition to these, if some of the customary and natural excretions are being held back. Fourth,
in addition to these, we must consider whether the body
is accustomed to evacuate the superfluities with exercises,
purgatives, emetics, or passive exercises, or with the use of
natural waters, and has now paid little attention to [these
measures]. In the regimen, we must consider whether
apepsias precede much more and to a greater degree than
is customary, or whether the person has ingested an abundance of *kakochymous* foods or has drunk wine that is
sweet instead of aged, or is thick or turbid instead of thin,
or if he has changed completely from drinking wine to
drinking water, and has not offended just once or twice in
the case of each of the things mentioned, but continuously
and over a long time.

---

9 τῶν αὐτομάτων κόπων add. Ko
10 [αἱ φυσικαί] Ko; καὶ φυσικαί Ku

δεύτερον δ' ἐπισκεπτέον, ὡς εἴρηται, μή τις τῶν
φύσει κακοχυμίαν ἑτοίμως ἀθροιζόντων ἐστὶν ὁ κάμνων.[11] ἐξευρήσεις δὲ τοῦτο πυθόμενος, εἰ ψωρώδης
ποτὲ διάθεσις ἢ λεπρώδης ἢ ἀλφώδης ἢ κνησμώδης
244K ἐπὶ πλέον αὐτῷ συνέπεσεν ἢ ἐρυσίπελας ἢ ἕρπης ἢ
ἐλέφας ἢ ὀφίασις ἢ ἀλωπεκίασις ἢ φλύκταιναι πλείους ἢ ἑλκώδεις ἐξανθήσεις ἢ ἐπινυκτίδες ἢ ὅλως ὁτιοῦν τῶν ἐπὶ κακοχυμίᾳ γεννωμένων τε καὶ αὐξανο-
μένων συμπτωμάτων. ἐπὶ δὲ τούτοις ἐλέγομεν χρῆναι
σκοπεῖσθαι, μὴ συνήθης τις ἔκκρισις ἐπέσχηται δι'
ἐμέτων ἢ δι' αἱμορροΐδων ἢ σύριγγός τινος ἢ δυσεν-
τερίας ἢ γυναικὶ καταμήνια· εἶθ' ἑξῆς, εἰ αὐτὸς ἐπιτη-
δεύων ἐκκαθαίρειν ἑαυτὸν ἀεὶ πέπαυται νῦν.

ἔνιοι μὲν γὰρ ὑπηλάτοις φαρμάκοις, ἔνιοι δὲ ἐμε-
τηρίοις ἢ οὐρητικοῖς ἢ ἱδρωτικοῖς ἢ χρήσεσιν ὑδάτων
αὐτοφυῶν ἤτοι θειωδῶν ἢ ἀσφαλτωδῶν ἢ νιτρωδῶν
ἐκκενοῦντες ἑαυτῶν τὰ περιττώματα καθ' ἕκαστον ἔαρ
ἢ φθινόπωρον ὠλιγώρησαν νῦν, πολλοὶ δέ, ὡς εἴρηται, καὶ γυμνασίων ἔθους ἀπέστησαν, ἔνιοι δὲ καὶ
τρίψεως ἁπάσης ἢ λουτρῶν ἢ τῶν μετὰ τὸ βαλανεῖον
ἐμέτων ἐπ' οἴνῳ γλυκεῖ. πρόδηλον δέ ἐστιν, ὡς ὁ λόγος οὐ μόνον τῶν τὴν ἀρίστην ἐχόντων κατασκευὴν
245K ἐμνημόνευσεν, ἀλλ' ὑπὲρ τοῦ μηδὲν λείπειν τῷ κατα-
λόγῳ τῶν τῆς κακοχυμίας αἰτίων ἐφήψατο καὶ τῶν
ἀθροιζόντων φύσει κακοχυμίαν σωμάτων, ὑπὲρ ὧν ἐν
ταῖς μοχθηραῖς κατασκευαῖς τοῦ σώματος ἐν τοῖς
ἑξῆς ὑπομνήμασιν ἐπὶ πλέον ἐροῦμεν. ἐκ τούτων μὲν
οὖν στοχάζεσθαι χρὴ τοῦ ποσοῦ τῆς κακοχυμίας, ἵα-

The second thing we must consider, as I said, is whether the affected person is someone who by nature readily collects *kakochymia*. You will discover this by inquiring if sometimes a scabby, leprous, dull-white leprous or pruritic condition befalls him more than usual, or erysipelas, herpes, elephas, ophiasis, alopecia, pustules that are frequent (*phlyktania*), an ulcerous efflorescence, epinoctis or generally any one whatsoever of the symptoms generated and increased by *kakochymia*. In addition to these, we said it was necessary to consider whether or not some customary excretion through vomiting, hemorrhoids, a fistula, dysentery or menstrual flow is held back, and then next, if the person himself always makes a practice of purging himself but has now ceased to do so.

244K

There are some who, each spring or autumn, purge their own superfluities with purging medications, some with emetics, diuretics or diaphoretics, or the use of natural waters (those containing sulfur, asphalt or sodium carbonate), and now pay no attention to this. And many, as was said, give up their customary exercises, while some give up all massage, or baths, or vomiting due to sweet wine after a bath. It is clear that the discussion not only makes mention of those who have the best constitution, but so as nothing is left out in the list of causes of *kakochymia*, it should also include the natural *kakochymia* which collects in bodies; I shall say still more about this in the books to follow on the bad constitutions of the body. From these factors, then, it is necessary to estimate the amount of the *kakochymia* and to discover a cure in pro-

245K

---

11 post κάμνων: ἄνθρωπος *add.* Ku

σιν δὲ τῷ μέτρῳ τῆς ποσότητος ἐξευρίσκειν ἀνὰ λόγον, εἰ μὲν ὀλίγη παντάπασιν ὑπάρχει καὶ κατ᾽ αὐτὸ
μόνον ἠθροισμένη τὸ δέρμα, μετριωτέραν, εἰ δὲ μείζων καὶ διὰ βάθους, ἰσχυροτέραν. εἰρήσεται δὲ πρῶτον μὲν ἡ τῆς ἐπιεικοῦς τε καὶ περὶ τῷ δέρματι μόνον,
ἑξῆς δὲ καὶ ἡ τῶν σαρκῶν ἐμπεπλησμένων, καὶ τρίτη
πρὸς αὐταῖς, ὅταν ὅλον ἀκάθαρτόν τε καὶ περιττωμα-
τικὸν ὑπάρχῃ τὸ αἷμα.

ὑποτίθεμαι[12] πρῶτον μὲν ἐπὶ τὴν ἐξ ἀρχῆς ὑπόθεσιν ἀνελθὼν εὔχυμόν τινα φύσει νεανίσκον, ἔμπροσθεν μὲν ὑγιεινῶς διαιτώμενον κατὰ πάντα, νῦν
δὲ διά τινα χρείαν ἀναγκαίαν ἐν ὁδοιπορίᾳ πλείονι
χρόνῳ διατετριφότα, μήτε γυμνάσασθαι τὰ συνήθη
μήτε λούσασθαι, κεχρῆσθαι δὲ καὶ βρώμασι καὶ πό-
246K μασι μοχθηροῖς καὶ μετὰ τὸ ἄριστον ἢ τὸ δεῖπνον ἢ
καὶ δι᾽ ὅλης τῆς ἡμέρας ἐπ᾽ ὀχήματος ἐνηνέχθαι μηδ᾽
ὕπνου τὰ πολλὰ καλῶς ἀπολαύσαντα· προσυπο-
κείσθω δὲ μηδὲν αὐτῷ πεπλημμελῆσθαι περὶ τὴν
ποσότητα τῶν προσενηνεγμένων καὶ διὰ τοῦτο μηδ᾽
ἀπεψίᾳ τινὶ περιπεπτωκέναι.[13] τὸν γὰρ τοιοῦτον ἄνθρωπον οὐκ ἐνδέχεται κακοχυμίαν ἠθροικέναι πολλήν. οὔκουν οὐδὲ τῆς ἐπανορθώσεως δεῖται μακρᾶς,
ἀλλ᾽ ἀρκεῖ[14] γυμνάσιον ἀποθεραπευτικόν, οἷον ἐν τῷ
πρὸ τούτου γράμματι διήλθομεν. εἴρηται δὲ καὶ περὶ
τῆς ἀκολούθου διαίτης ἐν αὐτῷ. καὶ νῦν οὐδὲν ἔτι χρὴ
μηκύνειν, ἀλλ᾽ ἀναμνῆσαι μόνον, ὡς ὁ σκοπὸς τῶν

portion to the measure of quantity. If the amount is altogether small and collected in the skin alone, the cure is more moderate, whereas if it is greater and in the depths, the cure is stronger. I shall speak first of what is suitable for the skin alone; next the cure when the flesh is also involved; and third, in addition to these, when the blood as a whole is impure and excrementitious.

Let us assume first the hypothetical situation I went over at the start: a young man *euchymous* in nature, who previously followed a healthy regimen in every respect, but now, through some need, has found it necessary to make a journey of long duration and has not carried out his customary exercises, nor bathed, but has used deleterious foods and drinks either with breakfast or dinner, and has been borne in a carriage for the whole day, and has not, for the most part, properly benefitted from sleep. Let us postulate that he has made no error in the amount of things taken and because of this has not fallen into any *apepsia*. It is not possible for such a man to accumulate much *kakochymia*. He does not, therefore, need a long period of correction; it is enough to use apotherapeutic exercise, such as I went over in the book prior to this one. I have also spoken about the regimen he should follow in that book. So now there is nothing I must still delay over, other than to call to mind that the objective for the bodies

246K

---

12 ὑποτίθεμαι Ko; δοίη ἴδη μοι Ku
13 post περιπεπτωκέναι: τονδί Ku
14 ἀρκεῖ Ko; χρὴ Ku

οὕτω διακειμένων σωμάτων κένωσίς ἐστι τῶν κατὰ
τὸ δέρμα περιττωμάτων, ὡς ἂν καὶ τῆς διαθέσεως ἐν
τούτῳ μόνῳ γεγενημένης.

ὑποκείσθω δὲ πάλιν ὁ αὐτὸς ἄνθρωπος ἐπὶ τοῖς
ἄλλοις τοῖς αὐτοῖς ἀπεψίαις πλείοσι περιπεπτωκώς·
ὑποκείσθω δὲ καὶ ἡ ἑλκώδης αἴσθησις αὐτῷ μὴ κατὰ
τὸ δέρμα μόνον, ἀλλὰ καὶ διὰ βάθους, ὡς ὑπονοεῖν
ὅλον ἐμπεπλῆσθαι τὸ σῶμα τῆς κακοχυμίας. οὐκέτι
τὸν τοιοῦτον οὔτ' ἐπὶ γυμνάσιον ἄξομεν οὔτ' ἐπὶ κίνη-
247K σιν ὅλως οὐδεμίαν, ἡσυχάσαι δὲ καὶ ὑπνῶσαι κελεύ-
σαντες ἐν ἀσιτίᾳ διαφυλάξομεν ὅλην τὴν ἡμέραν· εἶτ'
εἰς ἑσπέραν ἀλείψαντές τε λιπαρῶς καὶ λούσαντες
εὐκράτῳ θερμῷ τροφὴν εὔχυμον καὶ ῥοφηματώδη δώσομεν ὀλιγίστην· οὐκ ἀφέξομεν δ' αὐτὸν οὐδ' οἴνου·
συμπέττει γὰρ τοὺς ἡμιπέπτους χυμοὺς ὁ οἶνος, εἴπερ
τι καὶ ἄλλο, καὶ ἱδρῶτας καὶ οὖρα προτρέπει καὶ
ὕπνῳ συντελεῖ. δεόμεθα δ' ἐπὶ τῶν οὕτως ἐχόντων,
ὅσον μὲν ἤδη ἀκριβῶς μοχθηρόν ἐστι, τῆς κακοχυμίας πεφθῆναι μηκέτι δυναμένης, ἱδρῶσί τε καὶ οὔροις ἐκκενῶσαι, τὸ δ' οἷον ἡμίπεπτον ἔτι συμπέψαι τε
καὶ χρηστὸν ἀπεργάσασθαι. τοῦτο δὲ μάλιστα[15] δι'
ἡσυχίας καὶ ὕπνου ἀποτελεῖται.

εἰ μὲν οὖν ἐπὶ τοῖς εἰρημένοις κατασταίη τὸ σύμπτωμα, πρὸς τὰ συνήθη κατ' ὀλίγον ἐπανάγειν χρὴ
τὸν ἄνθρωπον. εἰ δὲ καὶ κατὰ τὴν ἑξῆς ἡμέραν ἔτι
παραμένοι, σκεπτέον ἤδη περὶ βοηθήματος ἰσχυρο-
τέρου, καὶ μάλιστ' εἰ διὰ τῆς νυκτὸς ἤτοι κοπώδης ἐπὶ
πλέον ἢ ἀσώδης ἢ ἄγρυπνος ἢ ἐν ὕπνοις τισὶ φαντα-

in this sort of state is evacuation of the superfluities in the skin, since the condition has occurred in the skin alone.

However, let us assume again this same man who, apart from the other things, has fallen into a number of these same *apepsias*. Let us also assume that the wound-like sensation in him is not in the skin alone, but also in the depths, so we suspect the whole body has been filled with *kakochymia*. We shall no longer lead such a person to exercise or to any movement at all, rather directing him to 247K rest and sleep and to maintain a fast for the whole day. Then, when evening comes, after anointing him with oil and bathing him in *eukratic* warm water, we shall give *euchymous* nutriment and a little porridge. But we shall not keep him away from wine, for the wine helps to concoct the semiconcocted humors, and as well as this, encourages sweating and urination, and contributes to bringing about sleep. However, we need in the case of those so disposed, to eliminate with sweat and urine as much of what is already entirely bad of the *kakochymia* that has not yet been able to be concocted, while in the case of what is still, as it were, semidigested, we need to concoct it and make it useful. This is best brought about by rest and sleep.

If, then, after those things mentioned, the symptom should settle, we ought to lead the man toward customary things a little. If, however, it should still remain on the next day, we must now give consideration to a stronger remedy, and particularly if through the night, either the fatigue is greater, or there is nausea or sleeplessness, or in sleep he

---

15 μάλιστα Ko; κάλλιστα Ku

248K σιώδεσί τε καὶ ταραχώδεσι γένοιτο. τοὺς γὰρ τοιούτους σὺν μὲν ἰσχυρᾷ τῇ δυνάμει δυοῖν θάτερον, ἢ
φλεβοτομεῖν ἢ καθαίρειν προσήκει, διορισάμενον,
ὁποτέρου δεῖ μᾶλλον, ὡς ἐφεξῆς ἐρῶ· σὺν ἀσθενεῖ
γὰρ τῇ δυνάμει[16] φλεβοτομεῖν μὲν οὐδαμῶς, ὑποκαθαίρειν δὲ μετρίως. ὁποῖαι δ᾿ εἰσὶν αἱ μέτριαι καθάρσεις, ἐν τοῖς ἑξῆς εἰρήσεται, πρότερόν γε διορισα-
μένων ἡμῶν τὰ πρότερον.

τῆς γὰρ δυνάμεως ἰσχυρᾶς οὔσης καὶ τοῦ κόπου
παραμένοντος ἐπισκεπτέον, εἴτε μετὰ πλήθους αἵματος ἢ ὠμῶν καὶ ἀπέπτων χυμῶν εἴτε αὐτὴ καθ᾿ ἑαυτὴν
μόνη γέγονεν ἡ τὸν κόπον ἐργαζομένη κακοχυμία. εἰ
μὲν γὰρ μετὰ πλήθους αἵματος, ἤτοι φλεβοτομητέον
ἢ τι τῶν ἀνὰ λόγον[17] πρακτέον. ἀνὰ λόγον δέ ἐστι
τόδε· τοῖς μὲν αἱμορροΐδας ἐπεσχημένοις ἐκείνας ἀνα-
στομῶσαι, ταῖς δὲ γυναιξὶ τὴν τῶν καταμηνίων ἔκ-
κρισιν κινῆσαι, ὥσπερ γε καὶ οἷς τούτων οὐδέν ἐστιν,
ἀποσχάσαι τὰ σφυρά, κἄπειθ᾿ οὕτως ὑποκαθαίρειν
φαρμάκῳ τῷ μάλιστ᾿ οἰκείῳ τῇ κακοχυμίᾳ. μόνης δὲ
249K συστάσης τῆς κακοχυμίας ἄνευ πλήθους αἵματος ἐπὶ
τὴν οἰκείαν τῷ λυποῦντι περιττώματι κάθαρσιν ἔρχεσθαι χρή. λυπεῖ δὲ ποτὲ μὲν ἤτοι πικρόχολον ἢ
μελαγχολικόν, ἔστι δ᾿ ὅτε φλεγματῶδες ἢ ἁλυκὸν ἢ
ὀξύ, καὶ τούτων ἕκαστον ἢ ὀρρωδέστερον ἢ παχύτερον ἢ μέσον πως κατὰ τὴν σύστασιν. ὑπὲρ ὧν τῆς
διαγνώσεως ἤδη λέγωμεν. εἰ μὲν ἅμα τισὶν ἐξαν-

is troubled with dreams and disturbances. In such people   248K
with a strong capacity one of two things is appropriate—
phlebotomy or purging—and determining which of the
two is required more, as I shall speak about in what follows. With a weak capacity, on the other hand, never phle-
botomize but purge downward moderately. The kinds of
things that are moderate cathartics will be spoken of subsequently, after we define the first things first.

When the capacity is strong and the fatigue persists, we
must consider whether the *kakochymia* bringing about the
fatigue has occurred in association with an abundance of
blood or of raw and unconcocted humors, or exists by itself
alone. If it is with an abundance of blood, we must either
carry out phlebotomy or do one of the things analogous to
this. The following things are analogous: to open hemorrhoids in those who have them; to set in motion the excretion through the menstrual flow in woman; in like manner,
in those in whom there are none of these things, to scarify
the ankles and then in this way purge downward with a
medication which is particularly for the specific *kakochymia.* When *kakochymia* exists alone without an abundance of blood, it is necessary to proceed to the evacuation   249K
specific for the distressing superfluity. Sometimes what
brings about distress is either picrocholic or melancholic,
and sometimes phlegmatic, either salty or acidic, and each
of these may be more serous or thicker, or to some degree
intermediate in consistency. Let me now speak about the
diagnosis of these. If the fatigue condition has occurred

---

16 *post* ἀσθενεῖ: γὰρ τῇ δυνάμει Ko; δὲ Ku
17 *post* τῶν: ἀνὰ λόγον Ko; ἀναλόγων τῷδε Ku

θήμασιν ἡ κοπώδης γεγένηται διάθεσις, ἐξ ἐκείνων
ἔτοιμον εὑρίσκειν, ὁποῖόν τι τὸ εἶδός ἐστι τοῦ περιτ-
τώματος· εἰ δὲ τούτων χωρίς, ἐπὶ μὲν εὐχύμου φύσεως
ἔκ τε τῶν προηγησαμένων ἐδεσμάτων καὶ τῶν ἄλλων
ἁπάντων, ὁπόσα συνέπεσεν αὐτῷ, κακοχύμου δ' ὄντος
φύσει κἀντεῦθέν τι ληπτέον. εἰρήσεται δ' ἐπὶ πλέον
αὖθις ὑπὲρ τῶν τοιούτων κράσεων· νυνὶ δὲ περὶ τῶν
ἄλλων λεκτέον, ἐξ ὧν καὶ αὐτῶν ἔνεστι τεκμήρασθαι
τὸ τῆς κακοχυμίας εἶδος.

ἀργότερον μὲν γὰρ διῃτημένου φλεγματωδέστερος
ἀθροίζεται χυμός, ἐν πόνοις δὲ πλείοσιν ἤτοι πικρόχολος ἢ μελαγχολικός, ἐν θέρει μὲν πικρόχολος, ἐν
φθινοπώρῳ δὲ μελαγχολικός. ἀλλὰ καὶ τῶν πόνων τὸ
250K μῆκος ἐπισκεπτέον· ὅσῳ γὰρ ἂν ὦσι πολυχρονιώ-
τεροι, τοσῷδε μᾶλλον ἐπὶ τὸ μελαγχολικὸν ἐκτρέ-
πονται. καὶ τοίνυν, ὅσοι μὲν ἅμα πολλοῖς ἱδρῶσιν
ἐγένοντο, παχύτερον ἐργάζονται τὸ περίττωμα, λεπτότερον δὲ οἱ χωρὶς ἱδρῶτος, ὥσπερ οἱ ἐν χειμῶνι καὶ
ὅλως ταῖς ψυχραῖς καταστάσεσι. συνεπισκεπτέον δ'
ἐν τῷδε καὶ περὶ τῶν οὔρων τι τοῦ πλήθους, ὥσπερ
γε καὶ περὶ τῆς τῶν ἱδρώτων ποιότητος· οἱ μὲν γὰρ
ὀξώδες, οἱ δὲ ἁλμυρόν, οἱ δὲ οἷον βορβόρου τινὸς ἢ
βρώμου σαφῶς ἐξόζουσιν. ἔνεστι δὲ τοῦτο καὶ διὰ
τῆς στλεγγίδος, ὁπότε λούοιντο, σκοπεῖσθαι. πολλάκις γοῦν ἐφάνη πικρόχολος ἀκριβῶς, οἷος ἐπὶ τῶν
ἰκτεριώντων ἀποκρίνεται. διάγνωσις δ' αὐτοῦ ῥαδία
καὶ πρὸ τῆς γεύσεώς ἐστιν ἐκ μόνης τῆς χρόας·
ὠχρὸς γὰρ ὁμοίως τῇ τοιαύτῃ χολῇ φαίνεται. πολ-

along with certain exanthemata, it is easy to discover from those which kind of superfluity it is. If, however, it is without these, in the case of a *euchymous* nature, it is from the previously taken foods, and from all the other things that have befallen him, whereas in one who is *kakochymous* in nature, we must undertake something at the time. More will be said again about such *krasias*. For the present, I must speak about the others, from which it is also possible to gain evidence of the kind of *kakochymia* of these.

When a person's life is quite idle, a more phlegmatic humor collects. In greater exertions, it is either picrocholic or melancholic—in summer, picrocholic and in autumn, melancholic. But one must also consider the length of the labors, for the longer they are in duration, the more the 250K tendency is toward the melancholic. And moreover, those that have occurred along with many sweats create a thicker superfluity, while those without sweat create a thinner superfluity, like those in winter and altogether cold climatic conditions. In this situation, we must also give consideration to the amount of urine, just as also to the quality of the sweats. Some smell acidic, some salty, some like mud, and some clearly smell foul. It is also possible to examine this from the skin scraper (strigil) when they bathe. Anyway, it often appears entirely picrocholic, of the kind that is secreted in those who are jaundiced. The diagnosis of this is easy, and from the color alone rather than the taste, for it appears yellow similar to such bile. In fact, often due

λάκις μὲν ἐπὶ πόνοις ἰσχυροτάτοις καὶ καύμασι σφο-
δροτάτοις ἀκριβῶς ξανθὸς ὤφθη. καὶ μὲν δὴ καὶ μέ-
σος ποτὲ καὶ μικτὸς ἐξ ἀμφοῖν, οἷον ὠχρόξανθός τις,
ὥσπερ γε καὶ ὁ τῆς χολῆς χυμός. ἔστι γὰρ οὖν καὶ
251K τοῦτον ἰδεῖν ἐν ἐμέτοις τε καὶ διαχωρήμασιν ἤτοι γ᾽
ὠχρὸν ἢ ξανθὸν ἢ ἐξ ἀμφοῖν σύνθετον. ὁποῖος δ᾽ ἂν
οὗτος ἐν τῷ σώματι περιέχηται, τοιοῦτον ἀναγκαῖον
αὐτοῦ φαίνεσθαι καὶ τὸν ὀρρόν.

ὁ μὲν οὖν ἱδρὼς τῶν καθ᾽ ὅλον τὸ σῶμα πλεοναζόντων χυμῶν ἐστι γνώρισμα, τὸ δ᾽ οὖρον ἐκείνων
μόνον, ὅσοιπερ ἂν ἐν τοῖς ἀγγείοις περιέχωνται. μηδὲν οὖν παραλιπεῖν ὅλως, ἀλλὰ καὶ τοὺς ἱδρῶτας ἐπισκεπτέον, ὡς εἴρηται νῦν. καὶ ποτ᾽ αὐτὸν τὸν κάμ-
νοντα κελεύειν αὐτῶν ἀπογεύεσθαι πρὸς ἀκριβεστέραν
διάγνωσιν· ὅπερ εἴωθεν ἔξθ᾽ ὅτε καὶ αὐτομάτως γίνεσθαι, παραρρυέντος εἰς τὸ στόμα τοῦ καταφερομένου
πολλάκις ἔκ τε τοῦ μετώπου καὶ τῶν ταύτῃ μορίων.
ἐπισκεπτέον δὲ καὶ τῶν οὔρων τήν τε σύστασιν ἅμα
καὶ τὴν χροιάν. οὐ παραλειπτέον δὲ οὐδέ τι τῶν ἐναι-
ωρουμένων οὐδὲ τῶν παρυφισταμένων ἀνεπίσκεπτον.
δηλοῖ γὰρ ἀκριβῶς τὰ τοιαῦτα πάντα, ὁποῖόν τι τὸ ἐν
τοῖς ἀγγείοις ἐστὶν αἷμα. χολώδους μὲν οὖν ὄντος,
ἀναγκαῖόν ἐστι καὶ τὸν ὀρρὸν αὐτοῦ χολώδη φαίνεσθαι καθ᾽ ἑκατέραν τῆς χολῆς τὴν ἰδέαν, ὁμοίως δ᾽
ἔτι καὶ φλεγματώδους ὑπάρχοντος.

252K   ὅταν μὲν οὖν ἀκριβῶς ἄπεπτον ὑπάρχῃ, λεπτὸν καὶ
ὑδατῶδές ἐστι τὸ οὖρον, οὔθ᾽ ὑπόστασιν ἴσχον οὔτε
ἐναιώρημά τι· πεττομένου δὲ ταῦτα φαίνεται καί τινες

to the strongest exertions and the most severe heatstrokes, it is seen to be completely yellow. Furthermore, it is also sometimes intermediate and a mixture of both—a certain pale yellow as it were—just like the humor of bile. This is seen in both vomitus and feces, either as pale, or yellow, or a combination of both. When this kind of thing is contained in the body, its serum inevitably also has such an appearance. 251K

Thus the sweat is a sign of excessive humors in the whole body, while the urine is only a sign of those which are contained in the blood vessels. Nothing at all must be neglected; the sweats must also be examined, as I stated just now. And on occasion, we direct the patient himself to taste his own sweat toward a more precise diagnosis. This is also wont to occur sometimes spontaneously, if it flows into the mouth, often being carried down from the face and the parts thereof. We must also examine the urine, both its consistency and its color. And we must not overlook the suspended matter or be inattentive to the sediment—all such things show precisely what kind of blood is in the vessels. Thus, when it is bile-containing, its serum inevitably appears bilious in relation to each kind of bile, and similarly when it is phlegmatic.

Therefore, whenever it is completely unconcocted, the 252K urine is thin and watery and has neither sediment nor any suspended matter. However, when it is concocted, these

ἄνωθεν ἐφίστανται νεφέλαι λεπταί, καθάπερ καὶ ἡ
καλουμένη γραῦς ἢ καὶ ἐπίπαγος ἐπὶ τῶν ἀποψυχο-
μένων ζωμῶν. εἰ δὲ θολερόν, οἷον καὶ τὸ τῶν ὑπο-
ζυγίων, φαίνοιτο, δηλώσει μὲν ἐμπεπλῆσθαι τῶν κα-
λουμένων ὠμῶν χυμῶν τὰς φλέβας, οὐ μὴν ἡσυχάζειν
γε περὶ αὐτοὺς τὴν φύσιν, ἀλλὰ πέττειν ἐρρωμένως.
εἰ δὲ διακρίνοιτο ταχέως καὶ τὸ ὑφιστάμενον εἴη λευκόν τε καὶ λεῖον καὶ ὁμαλόν, ὅσον οὔπω δηλοῖ κρα-
τήσειν ἁπάντων αὐτῶν τὴν φύσιν. εἰ δ' οὐρησάντων
μὲν εἴη καθαρόν, ἀναθολωθείη δ' εὐθέως, ἐπιχειρεῖν
τῇ πέψει τῶν ὠμῶν χυμῶν ἐνδείκνυται τὴν φύσιν, εἰ
δὲ μετὰ πλείονα χρόνον, οὐκ εὐθύς, ἀλλ' ὕστερον ἐπι-
χειρήσειν. κοινὸν δ' ἐπὶ πάντων οὔρων θολερῶν ἔστω
σοι γνώρισμα ἡ διάκρισις, ἤτοι ταχέως ἢ βραδέως ἢ
μηδ' ὅλως γινομένη. εἰ μὲν οὖν ταχέως τε γίνοιτο καὶ
τὸ ὑφιστάμενον εἴη λευκὸν καὶ λεῖον καὶ ὁμαλόν,
253K ἰσχυροτέραν ἐνδείκνυται μακρῷ τὴν φύσιν ὧν πέττει
χυμῶν. εἰ δ' ἀγαθὴ μὲν ἡ ὑπόστασις, ἐν χρόνῳ δὲ
γίνοιτο πλείονι, καὶ τὴν φύσιν ἐν χρόνῳ πλείονι κρα-
τήσειν τῶν χυμῶν ἐπαγγέλλεται. εἰ δ' ἤτοι μὴ διακρίνοιτο παντάπασιν ἢ σὺν μοχθηραῖς ὑποστάσεσιν,
ἀσθενὴς ἡ φύσις ἐστὶ καὶ δεῖται βοηθείας τινὸς εἰς
τὸ πέψαι τοὺς χυμούς.

ὥσπερ δὲ τὰ οὖρα τῶν ἐν τοῖς ἀγγείοις χυμῶν
ἐνδείκνυται τὴν διάθεσιν, οὕτως οἱ ἱδρῶτες καὶ τἆλλα
τὰ περὶ τὴν σύμπασαν ἕξιν τοῦ ζῴου φαινόμενα τῶν
κατ' ἐκείνην ἐστὶ δηλωτικά. θερμότητος μὲν γὰρ αἴσθησις ἀήθης ἐν αὐτῇ γίνεται, τῶν θερμῶν ἐπικρα-

things appear and certain thin clouds separate above, just like the so-called scum and congealed surface of cooled soups. If, however, it should appear turbid, like that of asses, it will show the veins are filled with the so-called raw humors, and that the nature[10] is not at ease with them, but is striving to concoct them. Also, if the sediment separates quickly and is white, thin and uniform, it shows that the nature hasn't yet overcome all these. If, however, the urine passed is clear but is immediately made turbid, it shows that the nature is attempting to concoct the unconcocted humors, and if after a longer time—not immediately but later—that it will succeed. In all the turbid urine, the separation should be a general sign for you, whether it occurs quickly, slowly or not at all. Thus, if it occurs quickly, and the sediment is white, thin and uniform, this 253K indicates the nature is far stronger than the humors it is concocting. If, on the other hand, the sediment is good but arises over a longer time, it proclaims that the nature will prevail over the humors in a longer time. If, however, there is no separation at all, or with bad sediments, it shows the nature is weak and needs some help to concoct the humors.

Just as the urine indicates the condition of the humors in the vessels, so the sweats and the other phenomena concerning the whole state of the organism are indicative of things in relation to that. Thus, an unusual sensation of heat occurs in it when hot humors prevail, and of cold

[10] In this passage, φύσις is taken to be the nature of individual, rather than Nature herself, although the latter is a possible rendering.

τούντων χυμῶν, ψυχρότητος δέ, τῶν ψυχρῶν. καὶ λευ-
κότεροι μὲν ἐπὶ ταῖς τοῦ φλέγματος, ὠχρότεροι δὲ ἐπὶ
ταῖς τῆς χολῆς φαίνονται πλεονεξίαις, εἰ δὲ καὶ ἀκρα-
τεστέρα ποτ' εἴη, ξανθότεροι. τὸ γὰρ χρῶμα τῶν χυμῶν ἐστιν, οὐ τῶν στερεῶν τοῦ ζῴου μορίων, ὅταν γε
μὴ ὑποχωρήσωσιν εἰς τὸ βάθος οἱ χυμοί. συμβαίνει
δ' αὐτοῖς τοῦτο διὰ κρύος ἢ ῥῖγος ἢ πάθος ψυχικόν,
254K οἷον φόβον ἢ λύπην ἰσχυρὰν ἢ ἀρχομένην αἰδώ· μηδενὸς δὲ τούτων παρόντος, οὐκ ἄν ποθ' ὑπονοστή-
σειαν εἰς τὸ βάθος οἱ χυμοί, ὥσπερ οὐδ' ἐπικαύσαιεν
ἄν ποτε τὸ δέρμα βιαιότερον ὁρμήσαντες ἐπ' αὐτὸ
χωρὶς τοῦ παθεῖν τι τὴν ψυχὴν ἢ θάλπος ἄμετρον
ἔξωθεν περιστῆναι τῷ ζῴῳ. ὀργισθέντων οὖν ποτ'
ἰσχυρῶς ἢ θυμωθέντων ἢ τὴν ἐκ τῆς αἰδοῦς οἷον ἄμ-
πωτιν τῶν χυμῶν ἀναφερόντων, μὴ προσέχειν τὸν
νοῦν τῇ χροιᾷ· χωρὶς δὲ τοῦ βιάζεσθαι τὸ περιϊστά-
μενον ἔξωθεν ἤτοι θερμὸν ἢ ψυχρὸν ἤ τι πάθος ὧν
ἀρτίως εἴρηται γεγενημένον, ἀψευδής ἐστιν ἡ ἐκ τῆς
χροιᾶς τοῦ ζῴου διάγνωσις τῶν χυμῶν.

ὡς οὖν τὸ μὲν λευκότερον ἑαυτοῦ γεγονὸς σῶμα
τὸν φλεγματικὸν ἐπικρατεῖν ἐνδείκνυται χυμόν, τὸ
δ' ὠχρότερον ἢ ξανθότερον τὸν χολώδη, κατὰ τὸν
αὐτὸν τρόπον καὶ ἡ ἐπὶ τὸ ἐρυθρότερον ἐκτροπὴ τοῦ
κατὰ φύσιν αἷμα πλεονάζειν, ἡ δὲ ἐπὶ τὸ μελάντερον
τὴν μέλαιναν χολὴν δηλοῖ. δόξειε δ' ἄν σοί ποτε καὶ

when cold humors prevail. And people appear paler with excesses of phlegm, pale yellow with excesses of bile, and sometimes, if it is more pure, more yellow. For the color is from the humors and not from the solid parts of the organism, whenever the humors do not retreat to the depths. This happens to them due to severe cold, rigors, or a psychical affection, such as fear, strong grief or incipient shame. But if none of these were to be present, the humors would never go down into the depths, just as they would never at any time make the skin burn by rushing into it too strongly, apart from some psychical affection or excessive heat surrounding the organism externally. If then at some time, people are strongly provoked or angered, or there is a kind of inward return of the humors[11] brought forth by shame, pay no attention to the color. However, apart from what is surrounding externally being overpowering—either heat or cold—or some affection among those spoken of just now having occurred, the diagnosis from the color of the humors of the organism is thoroughly reliable.

254K

Therefore, as a body that has become whiter than usual indicates the predominance of the phlegmatic humor, and one that is more pale yellow or yellow, the predominance of the bilious humor, in the same way too, the change to a greater redness than accords with nature indicates blood in excess, while a change to a greater darkness shows black

---

[11] The term ἄμπωτις can mean the ebb and flow of the tides, but a specific medical / physiological meaning is also listed in LSJ as, "return of humors inward from the surface of the body," citing Hippocrates, *Humors* 1. Jones' translation of the opening sentence is: "The colour of the humors, when there is no ebb of them, is like that of flowers" (*Hippocrates* IV, LCL 150, 63).

οἷον[18] μολίβδῳ τινὶ τὴν χροιὰν ἐοικέναι καὶ αὖθις
οἷόν τις μίξις εἶναι λευκοῦ τε ἅμα καὶ πελιδνοῦ καί
255K ποτε τὸ πελιδνὸν αὐτὸ μόνον ἐπικρατεῖν ἄνευ τοῦ λευκοῦ. τὰ τοιαῦτα οὖν χρώματα τὸν ὠμὸν ἐπικρατεῖν
ἐνδείκνυται χυμόν, ἐν εἴδει μὲν ὑπάρχοντα φλέγματος, ἧττον δ' ὑγρὸν ὄντα τοῦ συνήθους ὀνομαζομένου
φλέγματος. ὡς τὰ πολλὰ δὲ οὐδὲ γλισχρότης αὐτῷ
πρόσεστιν· ὡς, εἴ γε προσείη,[19] τὸν τοιοῦτον χυμὸν ὁ
Πραξαγόρας ὑαλώδη καλεῖ, ψυχρὸν μὲν ἱκανῶς ὑπάρχοντα, παχύτερον δ' ἧττον ὄντα τοῦ κατ' ἐξοχὴν
ὠμοῦ προσαγορευομένου. κοινῇ μὲν γὰρ οἱ τοιοῦτοι
χυμοὶ λευκοί τε καὶ ὠμοὶ πάντες εἰσί, προσαγορεύεται δ' αὐτῶν ἄλλος ἄλλῃ προσηγορίᾳ.

καὶ οὐ τοῦ νῦν ἐνεστῶτος καιροῦ διορίσασθαι πάντας αὐτούς· μόνου γὰρ εἰς τὰ παρόντα τοῦ κοινῇ πᾶσι
συμβεβηκότος δεόμεθα, τοῦ μηδέπω κατειργάσθαι
τελέως αὐτοὺς ὑπὸ τῆς φύσεως, ἀλλ' ἔθ' ὑπάρχειν
ὠμούς. ἐν μεθορίῳ γάρ ἐστι τὸ αἷμα τῶν τε χολωδῶν
χυμῶν καὶ τούτων, ὧν τὸ γένος ἑνὶ προσρήματι
καλεῖν ἔξεστιν ἢ ὠμὸν χυμὸν ἢ φλέγμα. ἐκεῖνοι μὲν
γὰρ ὑπερκατεργασθέντος ἀποτελοῦνται τοῦ αἵματος,
οὗτοι δ' οὐδέπω γεγονότος. ἔστι δ' ἑκατέρων ἄπειρος
256K μὲν ἡ κατὰ μέρος διαφορά, διώρισται δέ πως ἤδη
πρὸς τῶν περὶ τὰ τοιαῦτα δεινῶν εἴδεσιν εὐαριθμή-
τοις· ὧν οὐδ' αὐτῶν ἀναγκαῖόν ἐστιν μεμνῆσθαι νῦν

[18] post οἷον: μολίβδῳ τινὶ τὴν χροιὰν ἐοικέναι Ko; μολί-
βδου τὴν χρόαν ἔχειν Ku
[19] προσείη Ko; προσήκοι Ku

370

bile to be in excess. Sometimes it might seem to you as if the skin has a kind of livid color, or again there is some mixture of white and livid, and sometimes livid itself alone 255K prevails without the white. Such colors show the unconcocted humor prevailing is phlegm-like in kind, but less moist than what is customarily called phlegm, as for the most part there is no viscidity present in it, while if the kind of humor that Praxagoras calls hyaloid (green)[12] is present, it is strongly cold, being less thick than the named humor to a great degree. For in general, such humors are white and all are unconcocted, some being called by one name and others by another.

But now is not the appropriate time to distinguish all these; for the present we need only what is contingent to all in common, which is that they have not yet been prevailed upon completely by the nature but are still crude (unconcocted). For blood lies in the boundary zone between the biliary humors and these, the class of which can be called in one word either unconcocted humor or phlegm. Those are produced when the blood is overconcocted while these are produced when concoction has not yet occurred. The individual differences of each are very numerous, but somehow they have already been differentiated into easily enumerated kinds by those skillful in such matters. It is not necessary to mention all these now.

[12] See Fritz Steckerl, *The Fragments of Praxagoras of Cos and His School* (1958), and particularly 22 and 53–55 (which include the present statement). See also Manetti's entry in EANS, 694–95.

ἁπασῶν, ἀλλ᾽ ἀρκεῖ μόνον εἰς ἓν ἀγαγεῖν ἅπαντα τὰ κεφάλαια, ὅπερ οἷον σκοπόν τινα ποιήσασθαι προσήκει τῶν πρακτέων.

ἐπειδὴ γὰρ οἱ μέν τινές εἰσι, πρὶν ἀκριβῶς αἱματωθῆναι τὴν τροφήν, οἷον ἡμίπεπτοι τινες, οἱ δ᾽ ἄπεπτοι παντάπασιν, οἱ δ᾽ ὀλίγον ἀποδέοντες αἵματος ἰδέας, ἕτεροι δ᾽ ἔσχατοι τῆς αἱματώσεως, ἀμετρίᾳ θερμότητος ἑπόμενοι, καὶ τούτων αὐτῶν οἱ μὲν ὀλίγον ἀποκεχωρηκότες τοῦ αἵματος, οἱ δὲ πλέον, οἱ δὲ πλεῖστον, ἐπὶ μὲν τῶν ὀλίγον ἀπεχόντων ἐφ᾽ ἑκάτερα θαρρούντως χρῆσθαι φλεβοτομίᾳ, ἐπὶ δὲ τῶν πλέον εὐλαβέστερον, ἐπὶ δὲ τῶν πλεῖστον οὐδ᾽ ὅλως. συνεπισκοπεῖσθαι δὲ πειρᾶσθαι καὶ τὸ ποσὸν ἐν αὐτοῖς, οἷον, εἰ οὕτως ἔτυχεν, ἂν μὲν ὀλίγον ὑπάρχῃ τὸ χρηστὸν αἷμα, πλεῖστος δ᾽ ἄλλος τις χυμός, ἀφίστασθαι τῆς φλεβοτομίας· ἂν δ᾽ οὗτος μὲν ὀλίγος ᾖ, δαψιλὲς δ᾽ ὑπάρχῃ τὸ αἷμα, θαρρούντως χρῆσθαι φλεβοτομίᾳ.

257K   εἶθ᾽ οὕτως, ὡς εἴρηται πρόσθεν, ὑπάγειν γαστέρα πρός τε τὸ πλῆθος ἀφορῶντα καὶ τὴν ἰδέαν τοῦ πλεονάζοντος χυμοῦ. εἰ δέ τις ἤτοι δι᾽ ἡλικίαν ἢ διὰ δειλίαν οὐκ ἐθέλοι παρέχειν αὐτὸν τῷ ἰατρῷ πρὸς οὐδένα τρόπον αἵματος ἀφαιρέσεως, ἐπὶ κάθαρσιν ἰσχυροτέραν ἄγειν αὐτόν. εἰ δὲ καὶ ταύτην ὑποπτεύοι, δι᾽ ἑτέρων ἐκκενοῦν τὸ περιττόν. ἐπὶ μὲν οὖν τῆς ὑποκειμένης ἐν τῷ λόγῳ φύσεως οὐδὲν χαλεπὸν ἐξευρεῖν ἑτέρας κενώσεις· ἐπ᾽ ἄλλων δὲ μετὰ διορισμῶν ἀκριβεστέρων ἐξευρίσκειν αὐτὰς προσῆκεν, οὓς αὖθις

It will suffice simply to reduce all the chief points to one, which it is appropriate to make a kind of objective of the things to be done.

Since there are some [humors] before the nutriment has been made entirely into blood, some that are, as it were, partially concocted, some that are completely unconcocted and some that lack to a slight extent the form of blood, and others in the last stage of blood formation, following an excess of heat, and some of these only a little distance away from blood, and some more distant and some greatly so, in the case of those differing a little in either direction, use phlebotomy with confidence; in the case of those differing more, use it more prudently; and in the case of those that differ the most, do not use it at all. Also attempt to estimate the amount in these. For example, should it so happen that the useful blood is small in amount, while some other humor is most abundant, keep away from phlebotomy. But if the other humor is small in amount, while the blood is abundant, use phlebotomy with confidence.

If it is like this, as I said earlier, empty the stomach 257K downward, keeping the primary focus on the amount and kind of the excess humor. If, however, someone, due either to age or timidity, doesn't wish to present himself to the doctor for any kind of blood removal, bring him to a stronger purging. If he is also suspicious of this, evacuate the superfluity by other means. In the case of the nature assumed in the discussion, it is not difficult to discover other evacuations. In other cases it is appropriate to discover them with more precise distinctions, which I shall

ἐροῦμεν, ἐπειδὰν πρότερον ὑπὲρ τῆς εὐχύμου φύσεως
εἴπωμεν. ὑποκείσθω γὰρ ὁ τοιοῦτος ἄνθρωπος ἐπὶ
μοχθηρᾷ διαίτῃ κοπώδης γεγενημένος, εἶτ' ἐξ ὧν εἰ-
ρήκαμεν σημείων ἐν μὲν τῷ φλεβώδει γένει τῶν ἀγγείων ἐμφαινέσθω τι πλῆθος αὐτῷ χυμῶν ἡμιπέπτων,
ἐν δὲ τῷ παντὶ σώματι τούτων δὴ τῶν δακνωδῶν, οἷς
ὁ κόπος εἵπετο, συνηυξῆσθω δέ πως αὐτῷ καὶ τὸ
αἷμα. μάλιστα μὲν οὖν, ὡς εἴρηται, ἐχρῆν τοῦ αἵμα-
258K τος ἀφελόντα καθῆραι τοὐντεῦθεν ἐκεῖνον τὸν χυμόν,
ὃς ἂν ἐπικρατεῖν φαίνηται.

μὴ προσιεμένου δὲ τὴν τοῦ αἵματος ἀφαίρεσιν,
αὐξῆσαι τὴν κάθαρσιν. εἰ δὲ μηδέτερον ὑπομένοι,
σκοπεῖσθαι τὴν ἑτέραν ὁδόν, ᾗ μάλιστα ἄν τις, εἰ καὶ
μὴ διὰ ταχέων, ἀλλ' ἐν χρόνῳ τε πλείονι πρὸς τὴν
ἀρχαίαν τοῦ σώματος ἕξιν ἐπανάγοιτο τὸν ἄνθρωπον.
ἐπεὶ οὖν οἱ πρῶτοι δύο σκοποὶ τῆς ἐπανορθώσεως ἐν
ἁπάσαις ταῖς τοιαύταις διαθέσεσίν εἰσι, πέψις μὲν
τῶν ἀπέπτων ἢ καὶ ἡμιπέπτων χυμῶν, ὁπόσοι πρότε-
ροι τοῦ αἵματος γεννῶνται, κένωσις δὲ τῶν δριμέων
τε καὶ δακνωδῶν, ὁπόσοι δεύτεροί τ' εἰσὶ καὶ ὕστεροι
τοῦ αἵματος, ἀπέχειν μὲν αὐτοὺς χρὴ κινήσεως ἁπάσης ἰσχυρᾶς, ἀτρέμα δ' ἀλείφοντας καὶ ἀνατρίβοντας
ἐλαίῳ λούειν ὅτι μάλιστα προσηνεστάτοις λουτροῖς,
εἶτ' ἐφ' ἡσυχίας τε καὶ ἀσιτίας διάγειν, εἰ δ' οἷόν τ'
εἴη, καὶ ὕπνου, εὖ εἰδότας, ὡς οὐδὲν οὕτω πέττει μὲν
τὰ πεφθῆναι δυνάμενα, διαφορεῖ δὲ τοὺς μοχθηροὺς
χυμοὺς ὡς ὁ μετὰ τὸ βαλανεῖον ὕπνος. ὅταν οὖν, ὡς
ὀλίγον ἔμπροσθεν εἴρηται, τῇ πρώτῃ τῶν ἡμερῶν

speak about again when I have first spoken about the *eu-chymous* nature. Let us suppose such a person has become fatigued due to a bad regimen; then, from the signs I have spoken about, there should be displayed in the venous class of vessels some excess of semiconcocted humors in him, and in the whole body an excess of those that are biting; the fatigue follows the latter. And let the blood somehow be increased in him. Especially then, as I said, it is necessary, by removing blood, to purge from the source that humor which appears to prevail.   258K

If the person doesn't allow the removal of blood, increase the purging. However, if he submits to neither, consider the other path by which particularly someone might restore the person to the original state of the body, even if not quickly, at least over a longer time. Since there are two primary objectives of correction in all such conditions—concoction of the unconcocted and semiconcocted humors that are generated prior to the blood and evacuation of the acrid and biting humors that are secondary and subsequent to the blood—you must keep them away from all strong movement, anointing them gently, massaging with oil, and bathing them especially with the mildest baths, then getting them to spend time resting and fasting, and if possible also sleeping, knowing full well that nothing concocts those things that can be concocted and disperses the bad humors like sleep after bathing. Therefore whenever, as I said a little earlier, having made an attempt

ἀποπειραθέντες, εἰ καθίσταται ῥᾳδίως ὁ κόπος,
259K ὁμοίως ἐνοχλούμενον ὁρῶμεν τὸν ἄνθρωπον, ἐπί τε
λουτρὸν ἄγειν αὐτὸν καὶ τῇ δευτέρᾳ τῶν ἡμερῶν ἡσυχάζειν τε καὶ ἀσιτεῖν ἀναγκάζειν· ὡς[20] καὶ τὸ δεύτερον
ἔτι τε καὶ τρίτον <ἐπιτήδειον>[21] γνόντας λοῦσαι νή-
στιν, ἡσυχίᾳ τε καὶ ὕπνῳ τὰ μεταξὺ τῶν βαλανείων
διαλαμβάνοντα. συντελεῖ δὲ καὶ εἰς τὸν ὕπνον οὐχ
ἥκιστα καὶ αὐτὸ τὸ βαλανεῖον αὐτοῖς. ὑπνωδέστεροι
γὰρ οἱ λουσάμενοι γίνονται πάντες, εἰ μηδὲν ἄλλο
κωλύσει μεῖζον. ὥστε σοι τὸν ὕπνον αἴτιόν τε καὶ
σημεῖον ἀγαθὸν γίνεσθαι τῆς ἐλπιζομένης ὠφελείας,
ὥσπερ γε καὶ τὸ μὴ δυνηθῆναι καθεύδειν ἐπὶ τοῖς
βαλανείοις οὐκ ἀγαθὸν αἴτιον ἅμα καὶ σημεῖον.

ὡς τὰ πολλὰ μέντοι καὶ τῶν πλεοναζόντων χυμῶν
ἐξ ὕπνου τε καὶ ἀγρυπνίας ἔνεστί σοι λαβεῖν διάγνωσιν. ἐπὶ μὲν γὰρ τοῖς ψυχροῖς τά τε κώματα καὶ οἱ
μακρότεροι τῶν ὕπνων, ἐπὶ δὲ τοῖς θερμοῖς καὶ
δακνώδεσιν ἀγρυπνία, καὶ εἰ καθυπνώσειέ γέ ποτε,
φαντασιώδεις τε καὶ θορυβώδεις ὕπνοι ὑποπίπτουσιν,
ὡς ἐξανίστασθαι ταχέως αὐτούς. ὥσπερ δὲ κατὰ τὴν
πρώτην ἡμέραν, οὕτω καὶ κατὰ τὴν δευτέραν ἐλά-
260K χιστά τε καὶ ἁπλᾶ ῥοφήματα προσοίσομεν αὐτοῖς· τὸ
μὲν γὰρ πλείω διδόναι τοῖς κενώσεως δεομένοις ἄντικρυς ἐναντίον, τὸ δ' αὖ μηδ' ὅλως τρέφειν ἀσῶδές
ἐστι καὶ κακωτικὸν τοῦ στομάχου καὶ τῆς δυνάμεως
καταβλητικὸν καὶ τῆς κακοχυμίας αὐξητικόν. ἐλάχιστον οὖν αὐτοῖς διδόναι, μάλιστα μέν, εἰ οἷόν τε, χυλοῦ πτισάνης ἁπλῶς ἠρτυμένης,[22] εἰ δὲ μή, ἀλλὰ τοῦ

on the first day [to determine] if the fatigue is easily set-
tled, we see the person to be similarly troubled, bring him 259K
to the bath and on the second day compel him to rest and
fast. If you discern that on the third day, he is still as he
was on the second day, it is useful to bathe him while fasting, keeping the periods between the baths for rest and
sleep. The bath itself contributes not least to sleep for
them. For all those who are bathed become more drowsy,
if nothing else greater prevents this. As a result, sleep becomes for you the best cause and sign of the expected
benefit, just as, in fact, not being able to lie down to sleep
after the baths is not a good cause and sign.

However, for the most part it is possible for you to take
the diagnosis of the excessive humors from sleep and
wakefulness. For sleeps that are deep and unduly prolonged are due to cold humors, while sleeplessness is due
to hot and biting humors; in the latter case, if people do
in fact fall fast asleep at some time, dream-filled and disturbed sleep is their lot, such that they are quickly awak-
ened. Just as on the first day, so too on the second, we shall 260K
offer them very small amounts of simple gruel; to give
more to those requiring evacuations is utterly contraindi-
cated, whereas not to nourish them at all is accompanied
by nausea, is harmful to the esophagus, overthrows the
capacity and increases the *kakochymia*. Therefore, we
give them the least amount, and particularly, if possible,
barley water simply prepared. If not, give another kind of

---

20 *post* ὡς: εἰ Ku

21 ἐπιτήδειον *om.* Ku

22 ἠρτυμένης . . . ἠρτυμένου (ἀρτύω) Ko: ἠρτημένον . . .
ἠρτημένου (ἀρτάω) Ku

GALEN

χόνδρου τὸν αὐτὸν τρόπον ἠρτυμένου τῇ πτισάνῃ,[23]
καὶ μάλισθ᾽ ὅταν ὠμῶν χυμῶν πλῆθος ὑποπτεύωμεν
ἢ ἐν ταῖς φλεψὶν ἢ καθ᾽ ὅλον ὑπάρχειν τὸν ὄγκον.
εἴπερ γὰρ μηδὲν ὅλως ὄξους ὁ χόνδρος προσλάβοι,
γλισχότερός ἐστιν ἢ πρέπει τοῖς παροῦσιν, ὥστ᾽ ἐμφράξει μᾶλλον, οὐ διαρρύψει τοὺς πόρους, οὗ μάλιστα χρῄζουσιν ἐπὶ τοῖς παχέσι καὶ γλίσχροις χυμοῖς, οἷοίπερ εἰσὶ τοὐπίπαν οἱ φλεγματώδεις ἅπαντες.
εὔχυμος μὲν οὖν ἐστι καὶ διὰ τοῦτο κακοχυμίας ἐπι-
κεραστικός. ἀλλ᾽ εἰ μὴ κολασθείη τὸ γλίσχρον ἐν
αὐτῷ, προσλαβὼν ὄξους τε καὶ πράσου τὸ μέτριον
ἐμφράξει τε καὶ θρέψει μειζόνως ἢ συμφέρει τοῖς ἐν-
εστῶσι. διὰ ταῦτ᾽ ἄρα καὶ ὁ τῆς πτισάνης χυλὸς
261K   ἀμείνων ἐστὶν εἰς τὰ τοιαῦτα καὶ τρέφων συμμέτρως
καὶ μηδαμόσε κατὰ τὰς στενοτέρας ὁδοὺς ἰσχόμενος,
ὥσπερ ὁ χόνδρος, ἀλλ᾽ αὐτός τε διεξερχόμενος προσ-
απορρύπτων τε[24] τοὺς πόρους ἅμα τῷ τέμνειν τε καὶ
διαλύειν, ὁπόσον ἂν ἐν τοῖς ἡμιπέπτοις τε καὶ ἀπ-
έπτοις χυμοῖς ὑπάρχῃ παχύ.

διὰ ταῦτά γε καὶ τὸ μελίκρατον ἐπιτήδειόν ἐστιν
αὐτοῖς, ὀξύμελί τε καὶ ἀπόμελι καὶ πέπερι καὶ ζιγγίβερι καὶ πάνθ᾽ ὅσα τέμνει τε καὶ διαλύει τὰ παχέα
χωρὶς τοῦ κακοχυμίαν ἐργάζεσθαι. λεχθήσεται δὲ
ὑπὲρ τῆς ὕλης αὐτῶν ἐπὶ πλέον ἐν τοῖς ἑξῆς· εἰς δὲ
τὸν ἐνεστῶτα λόγον ὥσπερ τινὰ παραδείγματα τά τε
προειρημένα λελέχθω μοι καὶ τὰ μέλλοντα λεχθήσε-

gruel moistened in the same way with ptisane, having a
little vinegar in it,[13] and especially whenever we suspect
an excess of unconcocted humors exists in the veins or in
the whole mass of the body. For if the gruel were to receive in addition no vinegar at all, it is more viscid than
befits the cure, so that it obstructs more and doesn't
cleanse the pores thoroughly, which is what they particularly need for the thick and viscous humors, as in general
do all those who are phlegmatic. Thus, it is *euchymous* and
because of this tempers the *kakochymias*. But if the viscidity in it is not corrected by it receiving vinegar and leek in
moderation, it will obstruct and nourish more than is beneficial in the existing circumstances. Because of this, then,
the juice of ptisane is better for such things, is moderately   261K
nourishing, and in no way stops up the narrower passages,
like gruel does, but going through itself, washes out and
thoroughly cleanses the channels at the same time as cutting and dissolving whatever thickness exists in the semiconcocted and unconcocted humors.

Because of these things, melikratos is also in fact useful
for them, as are oxymel, apomel, pepper, ginger and all
such things that cut and dissolve the thick substances without creating *kakochymia*. More will be said on the material of these things in what follows. For the present discussion, let the things I have previously said and those that
are going to be said serve as examples. Of the pulses,

---

[13] The Kühn text is followed here in the light of the subsequent sentence.

---

23 *post* πτισάνῃ,: ὀλίγον ὄξους ἔχοντος Ku
24 *post* τε: *add.* καὶ διαρρύπτων Ku

σθαι. ὀσπρίων μὲν γὰρ ἐπιτηδειοτάτη ἡ πτισάνη,
λαχάνων δ᾽ ἡ θριδακίνη, τῶν δ᾽ ἰχθύων οἱ πετραῖοι,
καὶ τῶν ἄρτων οἱ κριβανῖται καὶ ζυμῖται καὶ καθαροὶ
συμμέτρως, ὀρνίθων δὲ οἱ ὄρειοι, τῶν δὲ ποτῶν ὀξύμελι, μελίκρατον, οἶνος λεπτὸς καὶ λευκός, ἁπλῶς δ᾽
εἰπεῖν, ὅσαπερ εὔχυμά τ᾽ ἐστὶ καὶ ῥυπτικὰ καὶ μὴ
γλίσχρα μηδὲ παχύχυμα μηδ᾽ ἱκανῶς πολύτροφα.

262K   τὰ δ᾽ οὐρητικὰ προσαγορευόμενα κατὰ τὸν ἐνεστῶτα καιρὸν οὐκ ἐπαινῶ, καὶ μάλισθ᾽ ὅσα σφοδρότερον θερμαίνει τε καὶ κατατήκει τὸ αἷμα· τῷ γὰρ
μέλλοντι καλῶς πεφθήσεσθαι τοσαύτης ταραχῆς οὐδέπω χρεία. ταῦτά τε οὖν ἅπαντα πρακτέον οὕτως
ἐστὶ κατὰ τὴν δευτέραν ἡμέραν, οὐχ ἥκιστα δὲ κατὰ
τὴν τρίτην τε καὶ τὴν τετάρτην. ἔτι τε πρὸς τούτοις,
εἰ πραΰνοιτο μὲν ἡ κοπώδης διάθεσις, εὔχρουν δὲ γίνοιτο τὸ σῶμα καὶ οὖρα πέπονα καὶ ὕπνοι χρηστοί,
τρῖψαι μὲν ἐπὶ πλέον αὐτὸν ἀποτολμήσαντα, γυμνά-
σαι δ᾽ ὀλίγον. πράξαντος γὰρ οὕτως, εἰ μὲν μηδεμία
κοπώδης αἴσθησις ἐπιγίνοιτο, πρὸς τὰ συνήθη γυμνάσια διὰ ταχέων ἐπανάγειν· εἰ δ᾽ ἐπιφανείη τι τῶν
ἔμπροσθεν ἤτοι συμπτωμάτων ἢ σημείων, αὖθις οὖν
καὶ σὺ πρὸς ἐκεῖνο βλέπων ἐξαλλάττειν πειρῶ τὰ
κατὰ μέρος. εἰ μὲν οὖν τῆς κοπώδους αἰσθήσεως ἀνά-
μνησις γένοιτο μόνης ἐπὶ τοῖς ἄλλοις ἅπασι σημείοις
ἀγαθοῖς διαμένουσιν, ἀποθεραπευτικῶς ἐπανορθοῦμεν
263K τὸν κόπον· εἰ δὲ τὰ μὲν σημεῖα ταραχθείη τε καὶ οἷον
χυθείη, μὴ παρείη δ᾽ ὁ κόπος, ἐν ἡσυχίᾳ πλείονι διαφυλάττειν τὸν ἄνθρωπον· εἰ δ᾽ ἄμφω συνέλθοι, διὰ

ptisane is the most suitable; of vegetables, lettuce; of fish, those living among rocks; of breads, those that are oven baked, leavened and moderately pure; of birds, those that are mountain dwellers; and of drinks, oxymel, melikraton, wine that is thin and white, and in short, those things that are *euchymous* and cleansing but neither viscous nor with thick juices, nor excessively nutritious.

During this time, I do not recommend the so-called 262K diuretics, and particularly not those that heat and dissolve the blood too strongly; something that is going to be well-concocted has not as yet need of such disturbance. These, then, are all the things that must be done in this way on the second day, and no less on the third and fourth days. In addition to these, if the fatigue condition were to become milder, the body would become a good color, the urine concocted, and sleep beneficial, confidently go ahead to massage the person more and exercise him a little. Having done this, if no fatigue sensation supervenes, quickly bring him back to his customary exercises. If, however, any of the previous symptoms or signs should come to light, reconsider the situation and endeavor to change them individually. But if only the recollection of the fatigue sensation exists, while all the other good signs remain, we correct the fatigue with apotherapeutic measures. On the other hand, if the signs are disordered and, 263K as it were, mixed, although the fatigue is not present, keep the person at rest for a longer time. But if both should

τῆς αὐτῆς ἐπιμελείας ἄγειν, ᾗ χρώμενος ἔμπροσθεν
εἰς τοσοῦτον προσήγαγες ὠφελείας αὐτόν, ὡς τολμῆ-
σαί τι καὶ περὶ γυμνασίων. οὕτω μὲν οὖν ἐπανορθοῦσθαι προσήκει τὴν εἰρημένην διάθεσιν.

5. Εἰ δὲ τἄλλα μὲν εἴη ταὐτὰ κατὰ τὸν εἰρημένον
ἄνθρωπον, ὑπάρχοι δ' ἐν τῷ κοπώδει σώματι τὸ μὲν
χρηστὸν αἷμα ὀλίγον, οἱ δ' ὠμοὶ χυμοὶ πάμπολλοι,
μήτε φλεβοτομεῖν μήτε καθαίρειν μήτε γυμνάζειν,
ἀλλὰ μηδὲ κινεῖν ὅλως μηδὲ λούειν. αἱ μὲν γὰρ φλε-
βοτομίαι τὸ μὲν χρηστὸν αἷμα κενοῦσι, τὸ δὲ μοχθηρόν, ὅπερ ἐν ταῖς πρώταις μάλιστα φλεψὶ ταῖς καθ'
ἧπάρ τε καὶ μεσάραιον ἀθροίζεται, πρὸς ὅλον ἐπι-
σπῶνται τὸ σῶμα. κάθαρσις δὲ ἐπὶ τῶν τοιούτων
στρόφους τε καὶ δήξεις ἐργάζεται καὶ λειποψυχίας
σὺν τῷ μηδὲ κενοῦν ἀξιολόγως· οἱ γὰρ ὠμοὶ χυμοὶ
πάντες ἀργοὶ καὶ δυσκίνητοι διὰ τὸ πάχος εἰσὶ καὶ
τὴν ψυχρότητα. προσεμφράττουσι γοῦν ἁπάσας τὰς
264K στενὰς ὁδούς, δι' ὧν χρὴ τὸ κενούμενον ἐν ταῖς καθάρσεσιν ἐπὶ τὴν γαστέρα παραγίνεσθαι, καὶ διὰ
ταύτην τὴν αἰτίαν οὔτ' αὐτοὶ κενοῦνται καὶ τοὺς ἄλλους ἐμποδίζουσι. τοῦτο μὲν οὖν καὶ ὑφ' Ἱπποκράτους
διὰ βραχυτάτων παρήνηται ῥημάτων εἰπόντος· "πέπονα φαρμακεύειν καὶ κινέειν,[25] μὴ ὠμά."

25 καὶ κινέειν add. Ko

14 LSJ lists, as the second of two groups of meanings of
στρόφος, "twisting of the bowels, colic" referring to Hippocrates,

come together, act with the same care and using what you previously used to bring him to such a degree of benefit, you may be bold as regards exercise. This, then, is how it is appropriate to correct the condition spoken of.

5. If all the other things are the same in the aforementioned person, but also in the fatigued body, the useful blood is small in amount, while the unconcocted humors are large in amount, do not carry out phlebotomy, purge or exercise; allow no movement at all and do not bathe. For phlebotomies evacuate the useful blood, while what is bad is collected particularly in the primary veins—that is, those in relation to the liver and mesentery—and is drawn toward the whole body. Evacuation in such cases produces twisting of the bowels (colic),[14] gnawings and fainting (*lypopsychia*) without any worthwhile evacuation. For all the unconcocted humors are sluggish and difficult to move due to their thickness and coldness. Therefore, they contribute to the blockage of all the narrow channels through which what is being evacuated in the purgings 264K must reach the stomach. And for that very reason, they themselves are not evacuated and they block the others. This, then, was recommended by Hippocrates in his very terse statement when he said: "medicate and move what has been concocted—not what is unconcocted."[15]

Aphorisms 4 11, and *Ancient Medicine* X, where Jones in both instances translates it as "colic" (LCL 150 and 147, respectively), and to the present passage. Presumably, Galen is attributing colicky abdominal pain to twisting of the bowel—it is highly improbable that the actual condition of volvulus was recognized!

[15] Hippocrates, *Aphorisms* 1.22, *Hippocrates* IV, LCL 150, 72–75.

διὰ δὲ τὴν αὐτὴν αἰτίαν οὐδὲ γυμνάζειν οὐδὲ κινεῖν
ὅλως ἀλλ' οὐδὲ λούειν προσήκει τοὺς ἐν ταῖς πρώταις
φλεψὶ τὸ πλῆθος τῶν ὠμῶν χυμῶν ἔχοντας. ἅπασαι
γὰρ αἱ τοιαῦται κινήσεις εἰς ὅλον τὸ σῶμα ποδη-
γοῦσι τοὺς χυμούς. φυλακτέον οὖν αὐτοὺς ἐν ἡσυχίᾳ
πάσῃ καὶ δοτέον ἐδέσματά τε καὶ ποτὰ καὶ φάρμακα
λεπτύνοντά τε καὶ τέμνοντα καὶ κατεργαζόμενα τὸ
πάχος τῶν χυμῶν, ἄνευ τοῦ θερμαίνειν ἐπιφανῶς· οἱ
γὰρ θερμανθέντες ἰσχυρότερον χυμοὶ πανταχόσε τοῦ
σώματος ἴασιν. διαιτᾶν οὖν αὐτοὺς ἐπ' ὀξυμέλιτι
μάλιστα βραχύ τι καὶ πτισάνης ἐνίοτε καὶ μελι-
κράτου διδόντα. καὶ γὰρ φέρουσι τὴν λεπτὴν δίαιταν,
εἴπερ τινὲς ἄλλοι, καταχρώμενοι τῷ πλήθει τῶν ὠμῶν
χυμῶν εἰς τροφὴν τοῦ σώματος ἐν τῷ κατὰ βραχὺ
265K   πέττειν αὐτούς. ἐπεὶ δὲ καὶ τὸ ὑποχόνδριον ἅπασι τοῖς
τοιούτοις ἐπῆρταί τε καὶ διαπεφύσηται καὶ ῥᾳδίως, ὅ
τι περ ἂν προσάρωνται, πνευματοῦται, βέλτιον ἂν εἴη
διδόναι σὺν τῇ τροφῇ πεπέρεως μακροῦ. καὶ γὰρ διαλύει τοῦτο παχύτητα φυσώδους πνεύματος, ἀπωθεῖται
δὲ καὶ πρὸς τὴν κάτω γαστέρα τὰ καθ' ὑποχόνδριον
ἀργῶς συνεστῶτα καὶ τῇ πέψει τῶν ληφθέντων συν-
αίρεται κατὰ τὸν αὐτὸν[26] λόγον ἁπάντων πεπέρεων. εἰ
δὲ μὴ παρείη τοῦτο, τῷ λευκῷ χρηστέον· ἔστι γὰρ
στομάχου τονικώτερον ἀμφοῖν τοῖν ἄλλοιν πεπερέοιν.
εἰ δὲ μηδὲ τοῦτο παρείη, χρῆσθαι τῷ καλλίστῳ μέ-
λανι· τοῦτο δὲ τὸ βαρύσταθμον.

[26] post αὐτὸν: κοινὸν (Ku) om.

For this same reason, it is not appropriate for those who have an excess of unconcocted humors in the primary veins to exercise or move at all, or even bathe. All such movements lead the humors to the whole body. Therefore, one must keep those affected at complete rest, give them foods, drinks and medications that are thinning, and cut and prevail over the thickness of the humors without manifestly heating. For humors that are strongly heated go everywhere in the body. Therefore, feed these patients chiefly with oxymel, also giving a little ptisane and sometimes melikraton. Also, they tolerate the thin diet, even though there are certain others who use up the excess of unconcocted humors for the nourishment of the body in which they gradually concoct them. And since the hypo-   265K
chondrium in all such patients is raised and distended easily, and what is administered turns into gas, it would be better to give the long pepper[16] with the food, for this disperses the thickness of the flatulent *pneuma* and urges on toward the lower abdomen what is sluggishly arrested in the hypochondrium, and it combines in the concoction of what has been taken in the manner common to all the peppers. If this is not available, you must use white pepper, as it is more contracting for the stomach than both the other peppers. If this not available, use the best black pepper; this weighs heavily.

16 LSJ lists two kinds of pepper: *Piper nigrum* and *Piper officinarum.* See Theophrastus, *History of Plants* 9.20, and Dioscorides 2.189.

ἄμεινον δὲ καὶ τῷ Διοσπολιτικῷ προσαγορευομένῳ
χρῆσθαι φαρμάκῳ. συντίθεται δὲ διττῶς, ἐνίοτε μὲν
ἐξ ἁπάντων ἴσων, κυμίνου τε καὶ πεπέρεως καὶ πηγάνου καὶ νίτρου, καὶ ἔστιν οὕτω μᾶλλον ὑπακτικώτερον
γαστρός· ἐνίοτε δὲ τῶν μὲν ἄλλων ἴσον ἑκάστου
μίγνυται, τοῦ νίτρου δ᾽ ἥμισυ. κάλλιον δὲ κύμινον μὲν
ἐμβάλλεσθαι τὸ καλούμενον Αἰθιοπικόν, πεπέρεως δὲ
ἤτοι τὸ μακρὸν ἢ τὸ λευκόν. ἐμβρέχεσθαι δὲ τὸ κύμι-
266K νον ὄξει δριμυτάτῳ· κἄπειτ᾽ εὐθέως τριβέσθω ἢ πρότερον φρυγέσθω μετρίως ἐν ἀγγείῳ κεραμέῳ, τελέως
ὠπτημένῳ κατὰ τὴν κάμινον. ὅσα γὰρ ἐνδεῶς ἐξη-
ράνθη, πηλώδη μᾶλλόν ἐστιν ἢ κεράμεα καὶ τοῖς
φαρμάκοις τι προστρίβεται τῆς ἑαυτῶν ποιότητος.
ἔστω δὲ καὶ τὰ τοῦ πηγάνου φύλλα προανεξηραμμένα
συμμέτρως. εἰ μὲν γὰρ ἐπὶ πλέον ξηρανθείη, δριμέα
τε γίνεται καὶ πικρὰ καὶ περαίτερον τοῦ προσήκοντος
θερμά, μηδ᾽ ὅλως δὲ προξηρανθέντα περιέχει τινὰ
περιττωματικὴν ὑγρότητα μηδέπω κατειργασμένην
ἀκριβῶς, δι᾽ ἣν οὐ γίνεται παντάπασιν ἄφυσα.

τούτοις τοῖς τέσσαρσιν ἐνίοτε μὲν ἀναμίγνυται
μέλι προαπηφρισμένον, ἐνίοτε δ᾽ οὐδέν, ἀλλὰ μόνα
χωρὶς τοῦ μέλιτος ἀποτεθέντα πτισάνης ἐμβάλλεται
χυλῷ καὶ ὅτῳπερ ἂν ἄλλῳ τῶν ἐδεσμάτων μάλιστα
πρέπειν δοκῇ. λαμβάνεται δὲ καὶ καθ᾽ ἑαυτὸ τὸ φάρμακον τοῦτο πρὸ τροφῆς τε καὶ μετὰ τροφήν. καὶ ἔστι
κάλλιστον ἐπὶ τῇ τοιαύτῃ χρήσει τὸ τῷ μέλιτι μιγνύ-

It is better to use the so-called Diospoliticum[17] medication which is compounded in two ways. Sometimes, it is mixed from equal parts of cumin, pepper, rue and niter, and in this way is more aperient for the stomach. Sometimes it is mixed from equal parts of the first three but only half the amount of niter. It is better, however, to put in the so-called Ethiopian cumin while the pepper should be either long or white. Soak the cumin in very sharp vinegar and then immediately rub it or first roast it moderately in   266K
a ceramic vessel that has been completely baked in the oven. Vessels that are insufficiently dried out are more clay-like than the ceramic and something of their own quality is imparted to the medications. The leaves of rue should have been dried moderately beforehand. If they are dried up still more, they become acrid and bitter, and hot beyond what is appropriate, whereas, if they have not been previously dried at all, they contain a certain excrementitious moisture which is not yet completely overcome; due to this, they are not altogether free of gas.

Sometimes despumated honey is mixed with these four [ingredients] and sometimes nothing, but they are kept separate without the honey and are thrown into the juice of ptisane and into whatever other of the foods seems to be particularly suitable. This medication is taken by itself before food and also after food. And it is best in such a use

17 Diospoliticum presumably refers to a city of that name, probably in Egypt (there were several cities so named). According to Ackermann's index to Galen's *Opera Omnia*, there are five references to this medication, all in the *Hygiene* (265, 283, 413ff, 430 and 431K). The compound medication is clearly described here.

μένον, ἀκριβῶς προαπηφρισμένῳ· τοῦτο γὰρ ἀφυσώτατον.[27] ἔστω δὲ δηλονότι καὶ αὐτὸ τὸ μέλι κάλλιστον, εἴπερ ἀφυσώτατόν τε καὶ τμητικώτατον ἔσεσθαι
μέλλει τὸ φάρμακον.

267K

ἐπιτήδειον δὲ τοῖς οὕτω διακειμένοις ἐστὶ καὶ τὸ
διὰ τριῶν πεπέρεων, ὅταν μὴ πάνυ φαρμακῶδες καὶ
ποικίλον κατασκευασθῇ καθάπερ οἱ πολλοὶ τῶν ἰατρῶν συντιθέασιν αὐτό, δίκην καρυκείας τινὸς ἢ συὸς
ἀγρίου πρὸς δὴ τούτων τῶν θαυμασίων ὀψοποιῶν
ἐσκευασμένον. ὥστ' ἔργον οὐ σμικρὸν τοῖς ἰατροῖς
ἐστιν ἢ τοῖς προσαραμένοις αὐτὸ τὰ ἐπεμβαλλόμενα
φάρμακα ἀποπέψαι, τὸ ἄμμι καὶ τὸ σέσελι καὶ τὸ
λιβυστικὸν ὅσα τ' ἄλλα τοιαῦτα· διαγινώσκεται γοῦν
λαμβανόντων ὠμὰ καὶ ἀμετάβλητα μέχρι πλείστου
κατὰ τὴν γαστέρα μένοντα. ταῦτά τε οὖν ἀφελεῖν χρὴ
τοῦ φαρμάκου καὶ προσέτι τὸ ἐλένιόν τε καὶ τὸν τῆς
νάρδου στάχυν, ὡς δὲ ἔνιοι συντιθέασι, καὶ τὴν κα-
σίαν. ἔχειν δ' ἐσκευασμένον ἕτοιμον διττόν, ὡς ἡμεῖς
εἰώθαμεν· ἁπλοῦν μὲν τὸ ἕτερον, ὅπερ ἐπί τε τῶν
ἠπεπτηκότων χρὴ διδόναι καὶ τῶν πρόσφατον ψύξιν
ἐχόντων κατὰ τὴν γαστέρα καὶ φλεγματώδη χυμόν·
ἕτερον δὲ φαρμακῶδες, ᾧ μάλιστα ἐπὶ τῶν ἀπὸ τῆς

268K   κεφαλῆς εἰς τὸν θώρακα ῥευμάτων χρώμεθα. ἀλλὰ
τοῦτο μὲν ὅπως χρὴ σκευάζειν, αὖθις εἰρήσεται.

τὸ δ' ἁπλοῦν, ᾧ καὶ πρὸς τὸ πλῆθος τῶν κατὰ τὰς
πρώτας φλέβας ὠμῶν χρώμεθα, τοιόνδ' ἐστίν· εἰς
πεντήκοντα δραχμὰς ἑκάστου τῶν τριῶν πεπέρεων
ἀρκεῖ μιγνύειν ἀνίσου τε καὶ θύμου καὶ ζιγγιβέρεως

that it be mixed with honey that is entirely despumated, as this is most free of gas. Obviously the honey itself must also be the best, if the medication is going to be most free of gas and most cutting. 267K

Suitable also for those in this state is the mixture of three peppers, whenever it is not unduly medicinal and is variably prepared, as many doctors compound it, after the manner of rich cooking or of wild pig prepared by those wondrous cooks. As a result, it is no small task for doctors or their assistants to concoct these same compound medications with anise, hartwort, libustikon and other such things added in. Anyway, it is recognized that, when taken, these things remain unconcocted and unchanged by concoction in the stomach for a long time. Therefore, it is necessary to keep these away from the medication, and besides these, the catmint and spikenard, and as some compound it, also the cassia. Have a double preparation ready, as I am accustomed to do—one simple one which should be given to those who are partially digesting and have a recent chill involving the stomach and phlegmatous humor, and another which is medicinal, and which we use 268K particularly in the case of the fluxes from the head to the chest. I shall speak again about how this ought to be prepared.

The simple preparation which we also use for the excess of unconcocted humors in the primary veins, is as follows: it is enough to mix 8 drachms each of anise, thyme and ginger into 50 drachms of each of the three peppers.

---

[27] post ἀφυσώτατον: add. ἐστιν (Ku)

ἑκάστου δραχμὰς ὀκτώ. τὸ μὲν ἁπλούσταστον τοῦτο,
καὶ τούτου μᾶλλον ἔτι τὸ χωρὶς ζιγγιβέρεως. ἕτερον
δὲ τὸ διὰ τῶν αὐτῶν συγκείμενον, ἀλλ' εἰς τὰς πεντήκοντα δραχμὰς ἑκάστου τῶν τριῶν πεπέρεων ἑκκαίδεκα δραχμῶν ἐμβαλλομένων ἑκάστου τῶν τριῶν,
ἀνίσου καὶ θύμου καὶ ζιγγιβέρεως, ᾧ καὶ μάλιστα
χρώμεθα πρὸς τὰ παρόντα. χρὴ δ', εἴπερ οἷόν τ' εἴη,
τὸ μὲν ἄνισον εἶναι Κρητικόν, Ἀττικὰ δὲ τὰ θύμα ἢ
πάντως γε ἐκ χωρίων ὑψηλῶν τε καὶ ξηρῶν. ἐμβάλλειν δ' αὐτῶν τὴν κόμην ἅμα τοῖς ἄνθεσιν, ἀποκρί-
νοντα τὸ ξυλῶδες. ἔστω δὲ καὶ τὸ πέπερι τὸ μὲν μακρὸν ἄτρητόν τε καὶ ὑγιές, ὥσπερ οὖν καὶ τὸ ζιγγίβερι·
τάχιστα γὰρ ἀμφότερα τιτρᾶται. καὶ πρὸς τῷ μὴ τε-
269K τρῆσθαι τὸ ἀληθινὸν δηλονότι αὐτὸ τὸ πέπερι τὸ ἀπὸ
τῆς βαρβάρου κομιζόμενον ἔστω. διττῶς γὰρ ἐνταῦθα
πανουργεῖται, σκευαζόμενον μὲν τὸ ἕτερον ἐπὶ τῆς
Ἀλεξανδρείας μάλιστα, βοτάνης δέ τινος ἐκβλά-
στημα θάτερον ὑπάρχει.

ὡς δ' ἄν τις μάλιστα γνωρίζοι τὸ πεπανουργημέ-
νον, ἐγὼ διηγήσομαι, τοσοῦτον πρότερον εἰπὼν ὑπὲρ
τοῦ μὴ θαυμάζειν τινὰ μηδὲ ζητεῖν τὴν αἰτίαν, δι' ἣν
εἴτε συνθέσεις φαρμάκων ἢ δοκιμασίας ἐπῆλθέ μοι
γράφειν ἐνταῦθα, μὴ πάνυ τι πράττειν τοῦτο εἰθι-
σμένῳ κατὰ τὴν θεραπευτικὴν πραγματείαν. ἐν ἐκείνῃ
μὲν γὰρ αὐτοῖς μόνοις διαλέγομαι τοῖς ἰατροῖς, ἐνταυ-
θοῖ δὲ καὶ τοῖς ἄλλοις ἅπασιν, οὓς ὀνόματι κοινῷ
προσαγορεύουσιν ἔνιοι φιλιάτρους, ἐν τοῖς πρώτοις
δηλονότι μαθήμασι γεγονότας, ὡς γεγυμνάσθαι τὴν

This is the simplest and is more simple still without the ginger. The other preparation is compounded of the same things, but to the 50 drachms of each of the three peppers put in 16 drachms of each of the three ingredients—anise, thyme and ginger. This is what we particularly use in present circumstances. And if possible, the anise should be Cretan and the thyme Attic, or at all events, from high and dry regions. Put in the foliage of these along with the flowers, separating that which is woody. Also let the pepper be long, intact and without holes, just as the ginger should be, as both are perforated very quickly. And in addition to being intact, let it clearly be genuine pepper brought from abroad, for here it is adulterated in two ways: the one is prepared particularly at Alexandria and the other is the new shoots of some herb. 269K

How you might especially recognize what is adulterated, I shall explain after first saying this: no one should wonder about or seek the reason why it came to me to write here of the compounding of medications or their assay—I who am not particularly accustomed to do this in relation to a therapeutic matter. For in that case I am conversing only with doctors themselves, whereas here the discussion is also with all those others whom some call by the general name "friends of medicine"—people who have obviously been involved in the primary teachings, so

διάνοιαν. οὔκουν ἀναγκαῖόν ἐστι τοῖς τοιούτοις οὔτ᾽
ἐν τῇ περὶ τῶν ἁπλῶν φαρμάκων οὔτ᾽ ἐν τῇ περὶ συνθέσεως αὐτῶν γεγυμνάσθαι πραγματείᾳ, πολὺ δὲ δὴ
μᾶλλον²⁸ οὔθ᾽ ὡς σκευάζειν ἐπίστασθαι οὔθ᾽ ὡς χρὴ
δοκιμάζειν ἕκαστον. τούτοις οὖν ἄμεινόν ἐστι γράφε-
σθαι τὰ τοιαῦτα πάντ᾽ ἀκριβῶς, ὥσπερ ἀρτίως τὰ
270K περὶ τοῦ μακροῦ πεπέρεως ἐπεχείρησα διηγεῖσθαι.
χρὴ γὰρ ἀπογεύεσθαι μὲν αὐτοῦ πρῶτον, ἀκριβῶς
ἐπισκοπούμενον, εἰ πεπέρεως ἀποσῴζει ποιότητα,
μετὰ δὲ τοῦτ᾽ ἐμβαλεῖν ὕδατι. τὸ γὰρ ἐσκευασμένον,
εἰ βραχείη δι᾽ ὅλης τῆς ἡμέρας, αὐτίκα διαλύεται τη-
κόμενον. εἴπερ οὖν ἥ τε ποιότης ἀκριβῶς αὐτῷ πεπέρεως ὑπάρχει καὶ μὴ διαλύεται βρεχόμενον, εἶθ᾽,
ὡς εἴρηται, καὶ ἄτρητον εἴη, ἐπιτήδειον νόμιζε τὸ τοιοῦτον μακρὸν πέπερι. τὸ δέ γε μέλαν μήτε μικρὸν
ἔστω μήτε ῥυσσὸν μήτε παχύφλοιον, ἀλλ᾽ ἐκ τοῦ βα-
ρυστάθμου καλουμένου τὸ μέγιστόν τε ἅμα καὶ εὐτραφέστατον ἐκλεγέσθω. καὶ τοῦ λευκοῦ δ᾽ ὁμοίως
ἐκλεγέσθω τὸ μέγιστόν τε καὶ εὐτραφέστατον. εἶτα
πάντων ἅμα κοπέντων καὶ λεπτῷ κοσκίνῳ διαττηθέν-
των ἀπηφρισμένον ἐπιμελῶς μιγνύσθω μέλι τὸ κάλλιστον, εὐῶδες δήπου τοῦτο καὶ ξανθὸν ὑπάρχον γλυ-
κύτατόν τε ἅμα καὶ δριμύτατον καὶ τῇ συστάσει μήτε
παχὺ μήθ᾽ ὑγρόν, ὡς ἀποσπᾶσθαι τῆς συνεχείας,²⁹
ἀλλ᾽ ὥστε καθέντα τὸν δάκτυλον εἰς αὐτό, κἄπειτα
καταστήσαντα μετέωρον, ἀπορρέον ὁρᾶν αὐτοῦ τὸ
271K μέλι μέχρι πλείστου συνεχὲς ἑαυτοῦ.

ἄμεινον δ᾽ ἐπ᾽ ἀνθράκων ἕψειν ἢ ξύλων ἀκριβῶς

as to have become practiced in the concept. But it is
not necessary for such people to be practiced either in
the matter of simple medications or in the compounding
of these; much more so, it is not necessary to know how
each is prepared or how each must be assayed. For them,
then, it is better to write all such things precisely, as I did
just now, in attempting to explain about the long pepper.   270K
It is necessary to taste this first, considering accurately
whether it preserves the quality of pepper, and after this,
to throw it into water. For what is prepared, if it soaks for
a whole day, immediately dissolves and is liquefied. If the
quality of pepper is entirely in this and it doesn't dissolve
when soaked, or, as was said, it is also unperforated, regard
such long pepper as suitable. However, the so-called black
pepper should not be small or wrinkled, or have thick
bark; choose from the so-called heavy pepper the biggest
and most well-grown. And of the white pepper, similarly
choose the biggest and most well-grown. Then, when all
have been pounded together at the same time and sifted
through a fine sieve, carefully mix the best despumated
honey. This, of course, should be fragrant and yellow and
at the same time very sweet and very acrid, but neither
thick in consistency nor liquid so as to take away its continuity. Rather, it should be such that, if you put a finger into
it and then raise the finger to a height, you see the honey
flow for the most part in continuity with itself.   271K

It is better to cook [the mixture] over charcoal or wood

---

[28] *post μᾶλλον:* οὔθ ὡς σκευάζειν ἐπίστασθαι οὔθ᾽ ὡς χρὴ
δοκιμάζειν ἕκαστον. Κο; οὐδ᾽ ὡς χρὴ δοκιμάζειν ἕκαστον ἐπίστασθαι. Ku

[29] τῆς συνεχείας Κο; αὐτοῦ τὸ συνεχὲς Ku

ξηρῶν, ἃ δὴ καὶ καλοῦσιν ἄκαπνα. τοῦτό τε οὖν διδόναι χρὴ τὸ φάρμακον οὐχ ἅπαξ μόνον ἢ δίς, ἀλλὰ
καὶ πλεονάκις ἑκάστης ἡμέρας· καὶ γὰρ ἔωθεν καὶ
πρὸ τροφῆς καὶ μετὰ τροφὴν καὶ καθυπνοῦν μελλόντων ἐπιτήδειον ὑπάρχει· τὸ δὲ πλῆθος ἑκάστης δόσεως ἔστω κοχλιάριον μεστόν, μικρὸν μὲν ἐπὶ τῶν
μικρῶν σωμάτων, μέγιστον δὲ ἐπὶ τῶν μεγίστων, ἀνὰ
λόγον δ᾽ ἐπὶ τῶν μεταξύ. καὶ μέντοι καὶ τὸ ζιγγίβερι,
τὸ κομιζόμενον ἐκ τῆς βαρβάρου, διάβροχον ὄξει,
συμφέρει λαμβάνειν. ἔστι δὲ τοῦτο ῥίζα χλωρᾶς τῆς
πόας ἐμβαλλομένη τῷ ὄξει μετὰ τὴν ἀναίρεσιν εὐθέως. κατασβέννυται γὰρ ἡ τῶν τοιούτων φαρμάκων
θερμότης αὐτόθι που περὶ τὰς πρώτας φλέβας, οὐκέτ᾽
ἀναφερομένη πρὸς ὅλον τὸ σῶμα, καθάπερ ἑτέρων
τινῶν, οἷον καὶ τὸ διὰ τῆς καλαμίνθης, ὑπὲρ οὗ μετ᾽
ὀλίγον ἐπὶ πλέον εἰρήσεται.

6. Ταῦτ᾽ ἄρα καὶ ὀξύμελι χρησιμώτατον αὐτοῖς
ἐστιν, ὡς καὶ πρότερον εἴρηται. σκευάζειν δὲ καὶ
272K τοῦτο προσήκει κατὰ τάδε. μέλι τὸ κάλλιστον ἐπ᾽ ἀνθράκων ἀπαφρίσαντας ἐπεμβάλλειν αὐτῷ τοσοῦτον
ὄξους, ὡς γευομένῳ μήτ᾽ ἄγαν ὀξὺ φαίνεσθαι μήτε
γλυκύ· καὶ τότ᾽ αὖθις ἑψεῖν ἐπ᾽ ἀνθράκων, ὡς ἑνωθῆ-
ναί τε τὰς ποιότητας αὐτῶν ἀκριβῶς καὶ μὴ φαίνεσθαι γευομένοις ὠμὸν τὸ ὄξος· εἶτ᾽ ἀποθεμένους
τούτῳ μιγνύειν ὕδωρ ἐπὶ τῆς χρήσεως, οὕτω κεραννύν-
τας ὡς οἶνον. εἰ μὲν οὖν ὁ πίνων αὐτὸ μήθ᾽ ὡς ὀξὺ
μήθ᾽ ὡς γλυκὺ μέμφοιτο, χρηστέον ἄχρι παντός· εἰ
δὲ μή, τὸ λεῖπον[30] ἐπεμβάλλοντας ἀφεψεῖν αὖθις. οὐ

that is completely dry, which they also call smokeless. It is then necessary to give this medication not once only, or twice, but several times each day. At dawn, before food, after food, and when about to go to sleep are suitable times. Let the amount of each dose be a full spoonful—a small spoon for small bodies, a large spoon for large bodies, and in proportion for those bodies in between. Of course, the ginger brought from abroad and soaked in vinegar is beneficial to take. This root of the fresh herb is thrown into the vinegar immediately after picking it, for the heat of such medications is quenched on the spot in the primary veins, and not carried to the whole body, like that of some others—for example, that made with catmint, which I shall say more about shortly.

6. In this way, then, oxymel also is most useful for such patients, as I said previously. It is appropriate to prepare 272K this as follows: when you have skimmed the best honey over charcoal, put into it as much vinegar as to make the taste neither obviously very acidic nor sweet, and at that time, again boil it over charcoal so as its qualities are entirely mixed into one and the vinegar does not obviously seem raw to those tasting it. Then, laying it aside, mix water with it according to the use, mixing it in this way like wine. If, then, the one drinking it finds no fault with it, either as acidic or as sweet, it must be used assiduously. If not, put in what is lacking and boil it again. I do not ap-

---

30 *ante* τὸ λεῖπον: τότε (Ku) *om.*

γὰρ ἐπαινῶ τοὺς κατὰ μίαν συμμετρίαν σκευάζοντας αὐτό· παραπλήσιον γάρ τί μοι δοκοῦσιν οἱ τοιοῦτοι ποιεῖν τοῖς ἀξιοῦσιν ἅπαντας τοὺς πίνοντας ὡσαύτως κεραννύναι τὸν οἶνον τῷ ὕδατι, μὴ γινώσκοντες, ὡς ἔνιοι μὲν ὑδαρέστερον εἰθισμένοι πίνειν εὐθέως πλήττονται τὴν κεφαλήν, εἰ καὶ βραχύ τις αὐτοῖς ἀκρατέστερον κεράσειεν, ἔνιοι δ᾽ ἀκρατεστέρῳ χαίροντες ἀνατρέπονται τὸν στόμαχον ὑδαρέστερον πιόντες.

ὁπότ᾽ οὖν ἐπὶ οἴνου ταῦτα συμπίπτει, συνήθους οὕτω ποτοῦ, πολὺ δὴ μᾶλλον ἐπ᾽ ὀξυμέλιτος, ὅσῳ καὶ
273K ἀηθέστερον οἴνου καὶ ἰσχυρότερόν ἐστιν, εἰκὸς ἀκολουθήσειν αὐτά. βέλτιον οὖν ἐστιν ταῖς τῶν λαμβανόντων αἰσθήσεσιν κρίνειν τὸ σύμμετρον, οὐ ταῖς ἡμετέραις,[31] οἰκειότατον μὲν εἶναι νομίζοντας τῇ φύσει τοῦ λαμβάνοντος τὸ ἥδιστον ὀξύμελι καὶ διὰ τοῦτο καὶ ὠφέλιμον, ἐναντιώτατον δὲ τὸ ἀηδέστατον. αὐτὴν δὲ τὴν πρώτην κρᾶσιν αὐτοῦ, ὡς ἂν μάλιστα τοῖς πλείστοις ἁρμόσειε, κατὰ τάδε χρὴ ποιεῖσθαι· ὄξους ἑνὶ μέρει διπλάσιον μιγνύσθω τοῦ τὸν ἀφρὸν ἀφῃρημένου μέλιτος, εἶθ᾽ οὕτως ἐπὶ μαλακοῦ πυρὸς ἑψείσθω, μέχρις ἂν ἑνωθῶσιν αὐτῶν αἱ ποιότητες· οὕτω γὰρ ἂν οὐδὲ τὸ ὄξος ὠμὸν ἔτι φαίνοιτο. δι᾽ ὕδατος δ᾽ εὐθέως ἐξ ἀρχῆς ὧδε σκευάζειν ὀξύμελι. τῷ μέλιτι μιγνύσθω τετραπλάσιον ὕδατος καλλίστου κἄπειθ᾽ ἑψείσθω μετρίως, μέχρις ἂν ὁ ἀφρὸς ἐφιστῆται.

[31] post σύμμετρον: οὐ ταῖς ἡμετέραις Ko; αὐταῖς, ἢ ἡμετέραις, Ku

prove of those who prepare it according to one single set of proportions. Such people seem to me to be acting like those who think it right for all who drink wine to mix it with water in the same way, not realizing that some who are accustomed to drinking a more watery wine are immediately overpowered,[18] if someone has mixed it a little too neat for them. However, some who enjoy more neat wine are upset in the stomach, if they drink wine that is too watery.

Therefore, when these things happen in the case of wine, so customary a drink, how much more are they likely to follow in the case of oxymel, which is less customary than wine and stronger. It is better then to judge the proportion by the perceptions of those taking [the oxymel] rather than by our own, knowing that the most pleasant oxymel is most suited to the nature of the person taking it, and because of this is beneficial, while the very opposite applies to that which is most unpleasant. However, the actual primary mixture of this, so it would be most suitable for most people, must be made as follows: Mix a double quantity of despumated honey with one part of vinegar, then boil over a gentle flame until their qualities are combined into one, for in this way the vinegar does not still seem to be raw. Oxymel may be prepared with water in this way right from the start. Mix with the honey four times as much of the best water and then boil moderately until

273K

---

[18] Literally, "struck in the head"—see also Galen, *HVA*, XV.672K, and *Hipp. Fract.*, XVIII(2).568K.

τὸ μὲν οὖν φαῦλον μέλι πάμπολυν ἐξερεύγεται[32] τὸν
ἀφρόν, ὥστε καὶ ἡ ἕψησις αὐτοῦ πολυχρονιωτέρα γί-
νεται· τὸ δὲ ἄριστον ἐλαχιστοτέρῳ χρόνῳ καὶ βρα-
χύτατον ἀφίησιν, ὥστ᾽ οὐδὲ ἴσης αὐτῷ δεῖ τῆς ἑψήσεως. ἡ δ᾽ οὖν πλείστη τὸ τέταρτον ἀπολείπει μέρος
274K τοῦ κραθέντος ἐξ ἀρχῆς. μίξαντας δ᾽ ὄξους ἥμισυ τοῦ
μέλιτος[33] πάλιν ἑψητέον ἄχρι τοῦ τὰς ποιότητας αὐτῶν ἀκριβῶς ἑνωθῆναι καὶ μηκέτ᾽ ὠμὸν φαίνεσθαι τὸ
ὄξος. σκευάζεται δὲ καὶ κατ᾽ ἀρχὰς εὐθέως τῶν τριῶν
μιχθέντων. ἔστω δὲ ἐν μὲν ὄξους μέρος, δύο δὲ μέλιτος, ὕδατος δὲ τέτταρα· καὶ ταῦθ᾽ ἑψείσθω μέχρι τοῦ
τρίτου μέρους ἢ τετάρτου, τὸν ἀφρὸν ἀφαιρούντων
ἡμῶν. εἰ δὲ ἰσχυρότερον αὐτὸ ποιῆσαι βούλοιο, τοσοῦτον ἐμβαλεῖς ὄξους, ὅσον καὶ μέλιτος.

ἀπόμελι δὲ κάλλιστον ἐν ὕδατι σκευάζεται, καὶ
πίνουσιν αὐτὸ δι᾽ ὅλου τοῦ θέρους, ὡς ἐμψῦχον ποτόν.[34] ἔνεστι δὲ τῷ βουλομένῳ καὶ πρὸς τὴν ὑποκει-
μένην ἐν τῷ παρόντι λόγῳ διάθεσιν ὠφελίμως χρῆσθαι, καὶ μάλισθ᾽ ὅταν ὀξυνθῇ· πάσχει δὲ πλειστάκις
τοῦτο, τὸ μὲν μᾶλλον, τὸ δὲ ἧττον, ὡς ἂν δι᾽ ὕδατος
σκευαζόμενον, οὐ τοῦ ὀμβρίου, καθάπερ τὸ ὑδρόμελι,
ἀλλὰ τοῦ ἐπιτυχόντος. ἔνεστι δ᾽, εἰ βούλοιτό τις, καὶ
δι᾽ ὀμβρίου συντιθέναι. καὶ ἔγωγ᾽ ἂν οὕτω ποιεῖν συνεβούλευον, εἰ ἐπῄνουν τὸ ὄμβριον· ἀλλὰ γὰρ οὔτε
τοῦτο ἐπαινῶ καὶ οὐδὲν χεῖρον ὀξυνόμενον ἀποτε-
275K λεῖται καὶ μάλιστα εἰς τὰ παρόντα. καὶ γὰρ οὖν καὶ
μετρίως ὀξύνεται τό γε μὴ παντάπασιν ἀμελῶς
ἐσκευασμένον. ἡ δ᾽ ἐπιμέλεια τῆς σκευασίας ἐστὶν ἐν

the foam is removed. Poor quality honey spews forth a
large amount of foam, so the boiling of this also takes a
much longer time, whereas the best honey gets rid of the
foam, which is very little, in less time. Consequently, it
doesn't need to be boiled for an equal time. At the most,
a quarter of the initial mixture leaves. When you mix in   274K
half as much vinegar as honey, you must boil it again until
their qualities are perfectly combined into one and the
vinegar no longer seems raw. Oxymel is also prepared by
mixing the three ingredients right from the start. Let there
be one part of vinegar, two of honey, and four of water.
Boil these down to a third or a quarter and remove the
foam. If you wish to make this stronger, put in as much
vinegar as there is honey.

Apomel is best prepared in water; people drink this
throughout the summer as a cooling drink. For someone
who wishes to do so, it is also possible to use it beneficially
for the condition assumed in the present discussion, in
particular when it is made acidic. And it is affected in this
way frequently to a greater or lesser degree, being prepared with water, not rain water like hydromel, but any
water. On the other hand, it is possible, if someone so
wishes, to compound it with rain water. And I myself
would do this, if I approved of rain water, but I do not
recommend this; there is nothing worse than making it
acidic, particularly for present purposes. And in conse-   275K
quence, it is also moderately acidic, if it is not altogether
carelessly prepared. The care of the preparation is in the

---

32 ἐξερεύγεται Ko; ἐξεργάζεται Ku
33 τοῦ μέλιτος add. Ko
34 ποτόν add. Ko

τῷ τὸ κηρίον εἶναι μὴ πάνυ φαῦλον ἑψεῖσθαί τ' ἐπὶ
πλέον ἐν ὕδατι πηγαίῳ, καθαρῷ τε καὶ ἡδεῖ. χρὴ γὰρ
ἐκπιέσαντα τῶν κηρίων τὸ μέλι μέχρι τοσούτου καθ-
εψεῖν ἐν ὕδατι, ἄχρις ἂν μηκέτι μηδεὶς ἀφρὸς ἐπαν-
ιστῆται. τούτῳ τε οὖν χρῆσθαι ποτῷ καὶ τῶν οἴνων
τοῖς ὠξυσμένοις ἀτρέμα καὶ τῶν ἐδεσμάτων τοῖς λεπτύνουσιν ἄνευ τοῦ θερμαίνειν, οἷάπερ ἐστὶ καὶ ἡ
κάππαρις, εἰ δι' ὀξυμέλιτος ἢ δι' ὀξελαίου λαμβά-
νοιτο.

μέχρι μὲν δὴ δυοῖν ἢ τριῶν ἡμερῶν, ὡς εἴρηται,
διαιτᾶν· εἰ δ' ἐλπίζοις ἱκανῶς ἤδη λελεπτύσθαι τοὺς
ὠμοὺς χυμούς, οἶνον προσφέρειν, λεπτὸν μὲν τῇ συστάσει, κιρρὸν δὲ ἢ λευκὸν τῇ χροιᾷ· ὁ μὲν γὰρ εἰς
εὐχυμίαν τε καὶ πέψιν, ὁ δ' εἰς οὔρησιν ἀγαθός. εἰσὶ
δ' ἐπὶ μὲν τῆς Ἰταλίας ὅ τε Φαλερῖνος καὶ ὁ Σουρεν-
τῖνος ἐκ τοῦ προτέρου γένους, ὥσπερ οὖν ἐκ τοῦ δευτέρου ὁ Σαβῖνός τε καὶ ὁ Ἀλβανὸς καὶ Ἀδριανός, ἐπὶ
δὲ τῆς Ἀσίας ἐκ μὲν τοῦ προτέρου Λέσβιός τε καὶ
276K Ἀριούσιος, ἐκ δὲ τοῦ δευτέρου Τιτακαζηνός τε καὶ
Ἀρσυηνός. οὗτοι μὲν οὖν ὡς παραδείγματα εἴρηνται
τοῦ λόγου. πολλοὶ δὲ καὶ ἄλλοι κατὰ τὴν Ἰταλίαν
εἰσὶ καὶ τὴν Ἀσίαν, οὐχ ἥκιστα δὲ κἀν τοῖς ἄλλοις
ἔθνεσιν, ὅμοιοι τοῖς εἰρημένοις, οὓς αὐτὸν ἕκαστον
ἐκλέγεσθαι χρὴ πρὸς τοὺς εἰρημένους ἀποβλέποντα
σκοπούς, τήν τε χρόαν καὶ τὴν σύστασιν, ἀποδοκι-
μάζειν τε τούς τε παχεῖς καὶ τοὺς μέλανας, ὡς κακο-
χύμους τε καὶ βραδυπόρους.

εἰ δ' ἐπὶ τοῖσδε βελτίων ὁ ἄνθρωπος γίνοιτο, καὶ

honeycomb not being of very poor quality, and boiling it more in spring water, which is pure and sweet. It is necessary to squeeze the honey of the honeycombs and boil this in water until no foam rises up any more. Use this drink, then, and the mildly acidic of the wines, and foods that are thinning without being heating, like capers are, if someone takes them with oxymel or a sauce of vinegar and oil.

Follow this diet for two or three days, as I said. If, however, you have reason to believe the unconcocted humors are already sufficiently thinned, give wine which is thin in consistency and tawny or white in color; the one is good for *euchymia* and concoction and the other as a diuretic. Of the Italian, the Falernian and Sorrentian are of the first class, and the Sabine, Albanian and Adrianian are of the second class. Of the Asian wines, the Lesbian and Ariusian are of the first class and the Titacazine and Arsynian of the second class.[19] I mention these as examples in the discussion. And there are many others, both in Italy and in Asia, and no less even in other nations, like those mentioned which you must choose yourself in each case, looking toward the stated objectives in respect of color and consistency, and rejecting those that are thick and black as being *kakochymous* and slow in passing. 276K

If, after these [measures], the person should become

[10] On the various wines, see Galen's *MM,* X.829–37K and elsewhere. For further information in modern works, see C. Sellman, *Wines in the Ancient World* (1957), and McGovern, *Ancient Wines.*

λούειν ἤδη προσήκει καὶ ἀλείφειν καὶ ἀνατρίβειν μαλακῶς, κἀπειδὰν πρῶτον ὑπόστασιν ἴσχῃ τὰ οὖρα,
τάς τε τρίψεις αὐξῆσαι καὶ πρὸς τὰ συνήθη γυμνάσια
κατ᾽ ὀλίγον ἐπανάγειν. ἐν τούτῳ δὲ τῷ καιρῷ καὶ τοῖς
διαφορητικοῖς ἀλείμμασι χρηστέον, ὧν καὶ πρόσθεν
μὲν ἐμνημόνευσα, καὶ αὖθις δ᾽ ὑπὲρ αὐτῶν ἐρῶ. οὐ
μὴν ἐμέτοις χρῆσθαι συμβουλεύω κατὰ τὰς τοιαύτας
διαθέσεις, ὥσπερ ἐνίοις ἔδοξεν ἰατροῖς τε καὶ γυμνασταῖς, ἐξαπατηθεῖσιν, οἶμαι, πρὸς τῶν ὑπὸ Φιλοτίμου
τε καὶ Πραξαγόρου γεγραμμένων ἐπὶ ταῖς τῶν τοι-
277K ούτων χυμῶν θεραπείαις. οὐ γάρ, ὅταν ἅμα κοπώδεις
διαθέσεις πλεονάζωσιν, ἀλλ᾽ ἐπειδὰν μόνοι λυπῶσιν,
ἐμέτοις αὐτοὺς ἐκκενοῦσιν· οὐδεὶς γὰρ κίνδυνος ἀντι-
σπασθῆναι τηνικαῦτα βιαιότερον ἔσω τι τῶν κατὰ
τὰς σάρκας περιττωμάτων, ὥσπερ ὅταν ἄμφω πλεο-
νάζῃ, τὰ μὲν ὠμὰ κατὰ τὰς πρώτας φλέβας, ἐν δὲ τοῖς
στερεοῖς τὰ δακνώδη. φυλάττεσθαι γὰρ τῶν τοιούτων
ἄμεινόν ἐστιν ἑκατέρας τὰς ἀντισπάσεις, ἔξω μὲν τὴν
τῶν ὠμῶν, ἔσωθεν δὲ τὴν τῶν δακνωδῶν.

ὥσπερ οὖν ἐφυλαξάμεθα κατὰ τὸν ἔμπροσθεν λόγον ἔξω τοὺς ὠμοὺς ἐπισπᾶσθαι χυμούς, οὕτω χρὴ
φυλάττεσθαι τοὺς δακνώδεις ἀντισπᾶν εἴσω. κεφάλαια δὲ τῆς μὲν ἔξω φορᾶς αὐτῶν εἰσι γυμνάσια καὶ
τρίψεις καὶ λουτρὰ καὶ θάλπος ἀλείμματά τε θερμαί-

---

20 Philotimus (330–270 BC) was a Greek doctor and pupil of
Praxagoras, with whom he is usually linked in terms of his views.

better, it is now appropriate to bathe, anoint and massage gently, and when the urine first has a sediment, increase the massages and return him gradually to his customary exercises. At this time, you must also use the diaphoretic ointments which I mentioned previously: I shall speak about these again. I do not advise the use of emetics in such conditions, as seemed good to some doctors and gymnastic trainers, deceived, in my opinion, by the writings of Philotimus and Paraxagoras,[20] for the treatments of such humors. For it is not, when fatigue conditions become more severe, but when they alone cause distress, that [doctors] evacuate them with emetics. For there is no danger under these circumstances of an overly forceful revulsion inward of the superfluities in the flesh, as there is when both increase—that is, both the unconcocted humors in the primary veins and those that are biting in the solid parts. It is better to guard against each of the revulsions of such humors—outward in the case of those that are unconcocted and inward in the case of those that are biting.

277K

Therefore, just as in the previous discussion, we were on guard against drawing the unconcocted humors outwardly, so too is it necessary to guard against revulsing the biting humors inwardly. The chief agents of the outward passage of these are exercises, massages, baths, warm un-

He is regarded as a Dogmatic. Praxagoras of Cos (325–275 BC) is one of the most renowned of ancient doctors and was the teacher of Herophilus as well as Philotimus and others. He is credited with many writings, all lost. Fragments of the writings of both men (and others of the school) are collected in the work by Steckerl, *The Fragments of Praxagoras*.

νοντα, καὶ τῶν ψυχικῶν παθῶν ἡ ὀξυθυμία καὶ ἁπλῶς
εἰπεῖν ἅπανθ᾽ ὅσα τοὺς ἐν τῷ βάθει τοῦ ζῴου χυμοὺς
εἰς τὴν πανταχόθεν κίνησιν ἐξορμᾷ, τῆς δ᾽ εἴσω τὰ
ἀποτρέποντα τῆς ἔξω φορᾶς. ἔστι δὲ δήπου ταῦτα
τῶν ὁμιλούντων τῷ δέρματι τά τε ψυχρὰ καὶ τὰ στύφοντα καὶ ὅσα πρὸς τούτοις ἐπισπᾶται τοὺς χυμοὺς
278K ἢ ἄλλως ὁπωσοῦν ἐπεγείρει τὴν ἔσω κίνησιν αὐτῶν,
ἐξ ὧν ἐστι λύπη καὶ φρίκη καθ᾽ ἡντινοῦν αἰτίαν γι-
νομένη· καὶ γὰρ καὶ διὰ ψυχρὸν αἴτιον καὶ διὰ[35] θερμὸν ἐδείχθη φρίκη τε καὶ ῥῖγος γινόμενον καὶ διὰ
τῶν ἐκπληττόντων τε καὶ φοβούντων τὴν ψυχὴν ἤτοι
ἀκουσμάτων ἢ θεαμάτων. ἅπαντ᾽ οὖν τὰ τοιαῦτα φυλακτέον ἐστίν, ἐπειδὰν ἅμα τε κοπώδης ὁ αὐτὸς ἄνθρωπος ᾖ καὶ τὰς φλέβας ἔχῃ μεστὰς ἀπέπτων χυμῶν. οὔτε γὰρ τοὺς ἔξω χυμοὺς ἀντισπᾶν ἔσω καλὸν
οὔτε τοὺς ἔνδον ἔξω, ἀλλὰ τοὺς μὲν ἔξω διαφορεῖν
ἀτρέμα (τὰ γὰρ ἰσχυρότερον τοῦτο δρῶντα καὶ τῶν
ἔνδον ἐπισπᾶταί τι), τοὺς δ᾽ ἔνδον λεπτύνειν τε καὶ
συμπέττειν. εἰ δ᾽ ἤτοι δι᾽ ἐμέτων ἢ διὰ γαστρὸς ὑπ-
αγωγῆς ἐπιχειρήσειας αὐτοὺς ἐκκενῶσαι σφοδρότερον, ἐπισπάσῃ τινὰς ἐκ τῶν ἔξωθεν εἴσω.

τοὺς δὲ περὶ τὸν Φιλότιμον οὐ χρὴ μέμφεσθαι τῇ
τοιαύτῃ κενώσει χρωμένους, ὅταν ἐν ταῖς πρώταις
φλεψὶ πλῆθος ὠμῶν χυμῶν[36] περικέηται χωρὶς ἑτέρας
διαθέσεως, ἀλλὰ μᾶλλον ἡμᾶς αὐτοὺς ἐθιστέον ἀκρι-
279K βέστερον ἕπεσθαι παλαιοῖς γράμμασιν. εἰς αὐτὸ μὲν
οὖν τὸ μελίκρατον ὕσσωπον ἐναφεψοῦντες, οὐκ εὐθέως
μὲν οὐδ᾽ ἐν τῇ πρώτῃ τῶν ἡμερῶν, ἐν δὲ ταῖς ἐχομέ-

guents and heating agents. And of the psychical affections,
an unstable temper, and in short, all those things that stir
up the movement in the humors in the depths of the organism in all directions, turn away the things within from
their outward passage. These are, of course, the things
which, when in contact with the skin, are cold and astringent, and those things that draw the humors toward them,
or otherwise, in any way whatsoever, stir up their move-   278K
ment inward; among these are grief and shivering arising
from any cause at all. For truly, it was shown that, through
both cold and heat as a cause, shivering and rigors occur,
and through the things that shock and terrify the soul,
whether heard or seen. Therefore, one must guard against
all such things when, at the same time, the patient himself
is fatigued and has unconcocted humors filling the veins.
For it is not good to revulse the outer humors inward or
the inner humors outward, but to gently disperse those
that are outer (for the things which do this more strongly
are also those which draw something inward), and to thin
and help concoct those within. If, however, you attempt to
evacuate these more strongly, either by vomiting or by a
downward purging through the abdomen, you will draw
some of those that are external inward.

You must not blame the followers of Philotimus for
using such an evacuation, whenever an excess of unconcocted humors is enclosed within the primary veins apart
from another condition. Rather, we must particularly accustom them to follow the ancient writings more accurately. To this end then, when we boil up melikraton and   279K
hyssop, we should not give this immediately, nor on the

---

35 διὰ *add.* Ko   36 χυμῶν *add.* Ko

405

ναις δώσομεν, ἐφ᾿ ὧν ἅμα τε τὸ πλῆθός ἐστι τῶν
ὠμῶν χυμῶν ἔνδον ἥ τε κοπώδης αἴσθησις ἔξω. τοὺς
δ᾿ ἐμέτους παραιτησόμεθα δεδιότες, ὡς εἴρηται, τῶν
ἔξωθέν τι περιττωμάτων εἴσω παλινδρομῆσαι. Πραξαγόρας δὲ καὶ Φιλότιμος εὐλόγως ἔμετον ἐπὶ τοιούτῳ
μελικράτῳ παραλαμβάνουσιν, ὠμοὺς χυμοὺς θερα-
πεύοντες ἄνευ κοπώδους διαθέσεως.

7. Ἀλλ᾿ ἐπειδὴ καὶ περὶ τούτων αὐτάρκως εἴρηται,
καιρὸς ἂν εἴη τῆς ὑπολοίπου διαθέσεως ὑπάρξασθαι,
καθ᾿ ἥν[37] αἱ μὲν φλέβες αἷμα χρηστὸν περιέχουσι
σύμμετρον τῇ ποσότητι, τὸ δὲ τῶν ὠμῶν πλῆθος εἰς
τὴν ἕξιν ἀνελήφθη. γίνεται δὲ ταῦτα κατ᾿ ἐκείνας
μάλιστα τὰς περιστάσεις τῶν πραγμάτων, ἐν αἷς ἤτοι
θάλπος ἢ γυμνάσιον ἄμετρον ἐκ τῶν φλεβῶν ἀναρπάζει τοὺς ὠμοὺς χυμοὺς εἰς τὰς σάρκας οὐδεμίαν ἐν
τῷ παρόντι πρόσφατον ἀπεψίαν ἡπεπτηκότων, ὡς, εἴ
γε καὶ τοῦτο συνέλθοι, σύμπαν οὕτως ἐμπλησθήσεται
τῶν ὠμῶν χυμῶν τὸ σῶμα. καὶ λεχθήσεται μὲν ὀλίγον ὕστερον, ὡς χρὴ τὴν τοιαύτην ἐπανορθοῦσθαι
280K διάθεσιν. ἀλλ᾿ ἐπεὶ τὸ σύνθετον ὕστερόν τέ ἐστι καὶ
δεύτερον τῶν ἁπλῶν, ἄμεινον ἂν εἴη περὶ τῆς ὑπολοίπου διαθέσεως ἁπλῆς διελθόντας ἐπὶ τὰς συνθέτους
αὖθις ἰέναι. καίτοι γε οὐδὲ ταύτην ἀκριβῶς ἁπλῆν
χρὴ νομίζειν, ἀλλ᾿ ὡς ἐν αὐτομάτοις κόποις ἁπλῆν.
ὑποκειμένης γὰρ τῆς τὸν κόπον ἐργαζομένης διαθέσεως, ἐπιμίγνυμεν αὐτῇ τὰς ἄλλας. ἀλλ᾿ ὅτι καὶ κατὰ

first day, but on the days following, in the case of those in whom there is an excess of unconcocted humors within or a fatigue sensation without. However, we shall reject the use of vomiting, fearing, as I said, lest some of the super-fluities that are external run back inward. Paraxagoras and Philotimus reasonably employed vomiting caused by such a melikraton preparation, when treating unconcocted humors apart from a fatigue condition.

7. But since enough has been said about these matters, it would be an appropriate time to make a start on the remaining condition, in which the veins contain useful blood in a moderate amount, while the excess of uncon-cocted humors has been taken into the system. This arises particularly in those states of affairs in which either heat or excessive exercise carries off the unconcocted humors from the veins to the flesh and there is, at the time, no new failure of concoction (*apepsia*) of those humors that are semiconcocted, as, if this should also occur together, the whole body would be filled with unconcocted humors in this way. I shall speak a little later about how such a condi-tion must be corrected. But since the combination is later 280K and secondary to the simple components, it would be bet-ter to go over the remaining condition that is simple and come in turn to the compound ones. And yet this condi-tion must not be thought of as simple in an absolute sense, but as simple among the spontaneous fatigues. For the underlying condition creating the fatigue has other [con-ditions] mixed with itself. But because in the mixing also,

---

37 *post* καθ᾽ ἥν: αἱ μὲν φλέβες αἷμα χρηστὸν περιέχουσι σύμμετρον τῇ ποσότητι, τὸ δὲ Κο; ἅμα τῇ ἑλκώδει διαθέσει καὶ τὸ Κu

τὴν μίξιν ἐνίοτε μὲν ἀπλῆ καὶ μία μίγνυται διάθεσις, ἐνίοτε δὲ σύνθετος, οὕτως ὠνομάσαμεν ἀπλῆν, ὑπὲρ ἧς ὁ λόγος ἐνέστηκεν.

ἔστω δὲ τὸ πλῆθος τῶν ὠμῶν χυμῶν ἐν τοῖς στερεοῖς τοῦ ζῴου μέρεσιν, οὐκ ἐν ταῖς φλεψίν, ἅμα τῷ καὶ τὴν ἑλκώδη τοῦ κόπου διάθεσιν ἐν τοῖς αὐτοῖς ὑπάρχειν· ἡ γὰρ ἐξ ἀρχῆς ὑπόθεσις τοῦ λόγου τοιαύτη τις ἦν. ὡς οὖν ἐφ' οἷς κατὰ τὰς φλέβας ἦν, καὶ μάλιστα τὰς πρώτας, τὸ πλῆθος τῶν ὠμῶν ἐφυλαττόμεθα τὰ θερμαίνοντα, δεδιότες εἰς τὴν ἕξιν ἀναληφθῆναι τοὺς τοιούτους χυμούς, οὕτω νῦν οὐδὲν χρὴ δεδιέναι. δοτέον οὖν αὐτοῖς, ὅσα μέχρι τοῦ δέρματος
281K ἐκτείνει τὴν θερμότητα, καὶ ἀνατριπτέον ἐπὶ πλέον ἐλαίῳ χαλαστικῷ, καὶ μάλιστα μετὰ τὸν ὕπνον ἕωθεν ἐξαναστάντας· ἡ γὰρ τοιαύτη τρίψις ἅμα τε πέττει τοὺς ὠμοὺς χυμοὺς καὶ τρέφει τὸν ὄγκον τοῦ σώματος.[38] χρὴ δ' ἡσυχάσαι μετὰ ταῦτ' ἄχρι πλείονος, εἰ μέλλοι καλῶς γενήσεσθαι ταῦτα.

πολὺ δ' ἂν ἐνεργέστερον ἀνύσειεν, οὗ χάριν γίνεται, βραχέα τε σιτία προσενηνεγμένου κατὰ τὴν προτεραίαν τοῦ ἀνθρώπου καὶ μηδὲν ἐπὶ τῷ δείπνῳ πίνειν πλὴν οἴνου κιρροῦ καὶ λεπτοῦ. καὶ γυμνάζεσθαι δ' οὐ συμφέρει[39] συνεχῶς καὶ σφοδρῶς αὐτίκα, μή πῃ λάθωμεν ἀπέπτων ἔτι χυμῶν ἀναγκάζοντες τρέφεσθαι τὸ σῶμα. βέλτιον οὖν ἕωθεν ἐπὶ πλέον ἀνατρῖψαί τε καὶ μετὰ ταῦτα ἡσυχάσαι περιπατῆσαί τε τὰ μέτρια καὶ αὖθις χρήσασθαι τρίψει πολλῇ καὶ λουτρῷ συμ-

sometimes a simple and single condition is mixed and sometimes a compound one, the discussion is set up in this way about what we called simple.

Let there be an excess of unconcocted humors in the solid parts of the organism, and not in the veins; together with this the wound-like condition of fatigue also exists in these parts. For the initial hypothesis of the discussion was just this. Therefore, as the excess of unconcocted humors was in addition to those in the veins, and particularly the primary veins, we avoided heating agents, fearing lest such humors would be taken up again into the system, now there is no necessity to be fearful. Therefore, we must give them those things that extend the heat even as far as the skin, and we must massage more with oil that is relaxing, 281K particularly in the early morning when they rise after sleep. Such massage simultaneously concocts the unconcocted humors and nourishes the mass of the body. After this, they ought to rest for a long time, if these things are going to turn out well.

However, it would accomplish this much more effectively and would be for the patient's sake, if he were to be offered little in the way of food on the previous day and to drink nothing after dinner apart from tawny, thin wine. It would be of benefit not to exercise him continuously and violently straightway, lest we unwittingly force the body to be nourished by still unconcocted humors. Therefore, it would be better to massage him more early in the morning, and after this, for him to rest and walk around moderately, and again to use a lot of massage and a moderately

---

³⁸ τοῦ σώματος Ko; τοῦ ζῴου Ku   ³⁹ ante συνεχῶς: καὶ γυμνάζεσθαι δ᾽ οὐ συμφέρει Ko; μὴ γυμνάζεσθαι Ku

μέτρως θερμῷ καὶ τροφαῖς εὐχύμοις τε ἅμα καὶ μὴ
γλίσχροις. εἴρηται δ' ἔμπροσθεν αὐτῶν ἡ ὕλη, πτισάνην ἐπαινούντων ἡμῶν εἰς τὰ τοιαῦτα καὶ τοὺς πε-
τραίους ἰχθύας καὶ τῶν ὀρνίθων τοὺς ὀρείους. ἐπιτή-
δεια δὲ καὶ τὰ λεπτύνοντα τῶν ἐδεσμάτων, ὑπὲρ ὧν
ἐν ἰδίᾳ γέγραπται βιβλίον.

282K   εὐλαβεῖσθαι δὲ χρὴ μηδέν, εἰ καὶ θερμαίνοι σφοδρῶς, ἀλλὰ καὶ τὸ διὰ τῆς καλαμίνθης φάρμακον
ἀδεῶς λαμβάνειν. ἔστι δὲ ἡ σύνθεσις αὐτοῦ τοιαύτη.
καλαμίνθης καὶ γλήχωνος καὶ πετροσελίνου καὶ σε-
σέλεως ἑκάστου δραχμὰς δώδεκα, σελίνου σπέρματος, κορύμβων θύμου ἀνὰ δραχμὰς τέτταρας ἑκατέρου, καὶ πρὸς τούτοις ἔτι λιβυστικοῦ μὲν δραχμὰς
ἑκκαίδεκα, πεπέρεως δὲ ὀκτὼ καὶ τετταράκοντα. πεπέρεως μὲν οὖν ἔστω τὸ βαρύσταθμον ὀνομαζόμενον,
σέσελι δὲ τὸ Μασσαλεωτικόν, πετροσέλινον δὲ τὸ
Μακεδονικόν, καὶ τούτου μάλιστα τὸ Ἀστρεωτικόν,
καλαμίνθη δὲ καὶ γλήχων μάλιστα μὲν ἐκ Κρήτης, εἰ
δὲ μή, ἐκ χωρίων ὑψηλῶν τε καὶ ξηρῶν, ὡσαύτως δὲ
καὶ τὰ θύμα. τὰ μὲν οὖν σκληρὰ καὶ ξυλώδη τῶν
βοτανῶν ἀπορρίπτειν, λαμβάνειν δ' εἰς τὸ φάρμακον
τὰ φύλλα, καὶ τούτων μάλιστα τὰ λεπτότατά τε καὶ
εὐθαλέστατα καὶ ἐπ' ἄκραις ταῖς βοτάναις, καὶ πρὸς
αὐτοῖς ἄνθη τε καὶ τὰ σὺν αὐτοῖς λεπτότατα κάρφη.
κόπτειν δ' ἅμα σύμπαντα καὶ διαττᾶν χρὴ διὰ λεπτο-
τάτου κοσκίνου· μάλιστα γὰρ εἰς ὅλην τοῦ ζῴου τὴν
ἕξιν τὰ τοιαῦτα τῶν φαρμάκων ἀναδίδοται, καθάπερ
γε τὰ παχύτατα κατὰ τὴν γαστέρα μένει, τοῖς στενοῖς

hot bath along with *euchymous* nutriments that are at the same time not viscous. We spoke of the material of these before, when we recommended for such purposes ptisane, fish that live among the rocks, and birds from the mountains. Thinning foods are also suitable—I have written one book about these specifically.[21]

There is no need to be cautious, even if something heats strongly; the medication made from catmint can also be taken without fear. The composition of this is as follows: Twelve drachms each of catmint, pennyroyal, parsley and hartwort; four drachms each of the seed of parsley and the fruit of thyme; and in addition to these, sixteen drachms of libustikon and forty-eight drachms of pepper. Let the pepper be what is called heavy pepper, the hartwort, Massaleotic, the parsley, Macedonian, and of this especially the Astreotic, and the catmint and pennyroyal particularly from Crete, but if not, from high and dry places, and likewise with the thyme. Throw away the hard and woody parts of the herbs, taking for the medication the leaves, and of these, particularly the thinnest and most thriving, and from the tips of the plants; in addition to those, take the flowers, and with these the thinnest twigs. It is necessary to pound and sieve all these at the same time through a very fine sieve, for such medications particularly are distributed into the whole system of the organism, just as the thickest remain in fact in the stomach,

282K

[21] *Vict. Att.*, CMG V.4.2 (English trans., Singer, *Galen: Selected Works*).

33K στόμασι τῶν φλεβῶν ἐναρμοσθῆναι μὴ δυνάμενα.
διὰ τοῦτο καὶ τὸ Διοσπολιτικὸν ὀνομαζόμενον φάρμακον, οὗ κατὰ τὸν ἐνεστῶτα λόγον ἔμπροσθεν ἐμνημο-
νεύσαμεν, παχυμερέστερον εἰώθαμεν σκευάζειν, ἐπειδὰν ὑπαχθῆναι τὴν γαστέρα διὰ τούτου δέῃσῃ. καί
τις οὐκ εἰδὼς τοῦτο, λεπτότατόν τε καὶ χνοωδέστατον
ἐργασάμενος αὐτό, τὴν μὲν ὑποχώρησιν οὐδέν τι
προὔτρεψεν, οὖρα δ' ἐκίνησεν οὐκ ὀλίγα· καὶ ἡμῖν
ἐκοινοῦτο θαυμάζων τε ἅμα καὶ ζητῶν τοῦ γεγονότος
τὴν ἀληθινὴν αἰτίαν. αὐτὸς μὲν γὰρ ἔφη νομίζειν
ἰδιοσυγκρισίαν τινὰ τοῦ ἀνθρώπου καὶ γὰρ ἐκάλεσεν
οὕτως αἰτίαν εἶναι τοῦ συμβεβηκότος. ὡς δ' ἔμαθεν,
ὅτι τὸ τῆς συνθέσεως εἶδος αἴτιον ἴδιον ὑπῆρχεν,
αὖθις ἑτέρως σκευάσας ἔτυχε τοῦ σκοποῦ. τούτου μὲν
οὖν τοῦ παραγγέλματος ἐπὶ πάσης συνθέσεως φαρμάκων ἄμεινον μεμνῆσθαι. τὸ δὲ προκείμενον ἐν τῷ
παρόντι λόγῳ φάρμακον ἀκριβῶς ἅπαντα λεπτὰ λαμβανέτω χάριν τοῦ ῥᾳδίως ἀναδίδοσθαί τε καὶ φέρεσθαι πάντη. μιγνύσθω δὲ τοῖς οὕτω παρεσκευασμέ-
284K νοις μέλι τὸ κάλλιστον, ἀκριβῶς ἀπηφρισμένον. ἡ δὲ
χρῆσις αὐτοῦ γινέσθω μετὰ τὴν ἑωθινὴν ἀνάτριψιν
πρὸ τῶν γυμνασίων τε καὶ λουτρῶν.

ἔξεστι δὲ καὶ χωρὶς τοῦ μῖξαι τὸ μέλι ξηρὸν τὸ
φάρμακον φυλάξαντα χρῆσθαι παραπλησίως ἁλσὶ
τοῖς εἰς τὰ ὄψα παρεσκευασμένοις. ἔξεστι δὲ καὶ εἰς
πτισάνην ἐμβάλλειν ἢ ὄξος ἤ τι τοιοῦτον ἀντὶ πεπέρεως. οὐ μόνον δὲ τὸ ξηρὸν ἁλῶν δίκην ἐστὶ πολύ-
χρηστον, ἀλλὰ καὶ τὸ σὺν τῷ μέλιτι. καὶ γὰρ καὶ

since they cannot be adapted to the narrow openings of   283K
the veins. And because of this too, I am accustomed to
prepare the thicker-particled form of the medication
called Diospoliticum, which I mentioned earlier in the
present discussion, whenever there is need for the bowels
to be opened by it. Someone who didn't know this made
it very thin and like very fine powder, so it didn't stimulate
defecation, although it did promote urination to no small
extent. He imparted this to me, since he was amazed and
was seeking the real cause of this occurring. He himself
said he thought it was some idiosyncrasy of the patient,
and this is what he called it, taking this to be cause of what
had happened. However, when he learned that the specific cause was the manner of compounding, he prepared
it again in a different way and gained his objective. It is
better to keep this particular example in mind in the case
of every compounding of medications. Let the medication
proposed in the present discussion consist of all the things
in a very fine state for the sake of easy distribution and
being carried everywhere. Mix with the things prepared
in this way, the best honey entirely despumated. The use   284K
of this medication should be after the early morning massage and before exercises and bathing.

It is also possible, without mixing the honey to keep the
medication dry and use it like salts prepared for cooked
meats. It is also possible to put it into ptisane or vinegar,
or some such thing instead of pepper. It is much used—
and not only that which is dry after the manner of salts,
but also that with honey. For truly, this can be mixed with

τοῦτο δυνατὸν ἀναμιγνύντα τοῖς ἐδέσμασιν ἀποπειρᾶσθαι[40] μετά τινος ἐξ αὐτῶν, ὡς ἡδὺ τῷ χρωμένῳ
φαίνεσθαι. μετὰ μέντοι τὴν τροφὴν μηδέποτε λαμβάνειν μήτε τοῦτο μήτ᾽ ἄλλο τι φάρμακον ἀνάδοσιν
ἰσχυροτέραν ἐργαζόμενον. οὐ γὰρ ἀναδίδοσθαι τηνικαῦτα βέλτιον, ἀλλὰ πέττεσθαι τοῖς ληφθεῖσιν. ἄμεινον οὖν ἐστι τοῖς ἐπικουρίας τινὸς ἔξωθεν εἰς πέψιν
δεομένοις ἐν τούτῳ τῷ καιρῷ δίδοσθαι φάρμακον, οὗ
τὴν σύνθεσιν ἔμπροσθεν ἐδήλωσα. λέγω δὲ τὸ διὰ
τριῶν πεπέρεων ἁπλοῦν. καὶ αὐτὸ δὲ τὸ πέπερι μόνον
ἐπιπαττόμενον τῷ ποτῷ χρηστὸν εἰς τὰ παρόντα, καὶ
285K εἴπερ ἄρα μείζων τις εἴη χρεία, καὶ τὸ διὰ χυλοῦ τῶν
κυδωνίων μήλων, οὗ τὴν σύνθεσιν ἅπασαν ἐρῶ καὶ
τὴν δύναμιν ἀκριβῶς ἐξηγήσομαι κατὰ τὸν ἑξῆς λόγον. εἰς δὲ τὸ παρακείμενον νῦν ἁπάντων ἄριστον
φάρμακον ὧν ἐγὼ γινώσκω τὸ διὰ τῆς καλαμίνθης
ἐστί. καὶ γὰρ λεπτύνει τὰ παχέα καὶ γλίσχρα καὶ
διαφορεῖ καὶ οὔρησιν κινεῖ καὶ καταμήνια γυναιξίν.
ἔστι δὲ καὶ ἥδιστον ἐν τῇ χρήσει, καὶ μάλισθ᾽ ὅταν
λάβῃ πλείονος τοῦ μέλιτος. ἑψεῖσθαι δ᾽ ἐπὶ πλέον
αὐτὸ χρὴ τηνικαῦτα.

τοῖς δ᾽ ἀπεστραμμένοις τὰ γλυκέα καὶ φεύγουσι τὸ
μέλι καὶ γὰρ καὶ τοιαῦταί τινες εὑρίσκονται φύσεις
ὀλίγον ἐν τῇ συνθέσει τὸ μέλι μιγνύσθω. βέλτιον δὲ
καὶ μᾶλλον ἑψεῖσθαι τούτοις· καὶ γὰρ ἧττον οὕτως
ἐστὶ γλυκὺ καὶ ἧττον ἀνατρέπει τὸν στόμαχον ἐπὶ
τῶν ἀπεστραμμένων φύσει τὸ μέλι. τὰ μὲν δὴ τοιαῦτα
πάντα καὶ αὐτός τις ἐπινοείτω πρὸ τῆς ἡμετέρας

414

foods, trying it out with one of these that seems sweet to
the person using it. But never take it after food—neither
this nor any other medication which brings about a stronger distribution. Under these circumstances, it is better
for the things taken not to be distributed but to be concocted. Therefore, it is better for those needing some external assistance toward concoction to give at this time a
medication, the compounding of which I showed previously. I speak of the simple medication made with three
peppers. Also, pepper itself alone dispersed in a drink is
useful in the present circumstances, and if there were to
be greater use of it, that made with the juice of quinces. I   285K
shall speak about the whole composition of this and explain the potency precisely in the discussion that follows.[22]
However, for those matters now under consideration, the
best medication of all those I know is that made with
catmint, for it thins those things that are thick and viscous,
disperses, and sets in motion the urine and the menstrual
flow in women. It is also very pleasant to use, and especially when it contains more honey. Under these circumstances it is necessary to boil it more.

However, for those who are averse to sweet things and
avoid honey—for such natures are found—mix only a little
honey in the compounding. For them, it is also better to
boil it more. In this way it is less sweet and upsets the
esophagus less in those who are by nature averse to honey.
Someone might think of all such things for himself, prior

[22] This is in fact deferred to the final chapter of the treatise—
Book 6, chap. 13.

___

[40] ἀποπειρᾶσθαι Κο; ἀποχεῖσθαι Κυ

συμβουλῆς, ἕνα κοινὸν ἐν ἅπασι διαφυλάττων τὸν
σκοπόν, ἥδιστον γενέσθαι τὸ φάρμακον, εἰς ὅσον ἐγχωρεῖ, φυλαττομένης αὐτοῦ τῆς εἰς τὴν ὠφέλειαν δυ-
νάμεως· οὐ γὰρ δὴ ταύτην γ' ἐκλῦσαι κελεύω τῆς

286K ἡδονῆς στοχασάμενον. ὅταν οὖν ἐπὶ δύο που τὰς
πρώτας ἡμέρας ἢ τρεῖς οὕτως ᾖ παρεσκευασμένος ὁ
ἄνθρωπος, οὐδὲν ἂν εἴη χεῖρον αὐτὸν καὶ γυμνάσασθαι διὰ τῶν συνήθων αὐτῷ, ἀποπειρώμενον εὐχροίας
τε καὶ τῶν ἄλλων σημείων, ἃ κατὰ τὸν ἔμπροσθεν
εἴρηται λόγον. εἰ μὲν γὰρ ἅπαντά σοι χρηστὰ φαίνοιτο, τελέως γυμνάζειν αὐτόν· εἰ δὲ μή, προκατα-
παύειν τε τοῦ συμμέτρου κἂν τοῖς προειρημένοις δι-
αιτήμασί τε καὶ φαρμάκοις φυλάξαντα κατ' ἐκείνην
τὴν ἡμέραν αὖθις ἐπιχειρεῖν τῷ γυμνασίῳ κατὰ τὴν
ὑστεραίαν ἐπὶ τοῖς αὐτοῖς σημείοις τε καὶ σκοποῖς,
ἵν', ὅταν ἤδη πάντα ἄμεμπτα φαίνηται, πρὸς τὴν αὐτὴν δίαιταν ἐπανάγῃς αὐτόν, ἢ καὶ πρὶν ἁλῶναι τῷ
κοπώδει συμπτώματι συνήθης ἦν.

8. Ἐπεὶ δὲ καὶ τῷ διὰ τῆς ἐλάτης ἀκόπῳ φαρμάκῳ
χρῆσθαι συμφέρει τοῖς κατὰ τὴν σάρκα τε καὶ τὸν
ὄγκον ἅπαντα τοῦ ζῴου χυμοὺς ἠθροικόσιν, ἤτοι
πεφθῆναι δεομένους ἢ διαφορηθῆναι, οὐδὲν ἂν εἴη
χεῖρον εἰπεῖν τι καὶ περὶ τῆς ἐκείνου συνθέσεως. ἔστι

287K μὲν οὖν ὡραιότατον αὐτῆς τὸ σπέρμα περὶ τὴν ἐπιτο-
λὴν τοῦ ἀρκτούρου, ὅστις καιρὸς ἐν Ῥώμῃ μὲν ὁ
καλούμενος μήν ἐστι Σεπτέμβριος, ἐν Περγάμῳ δὲ
παρ' ἡμῖν Ὑπερβερεταῖος, Ἀθήνησι δὲ μυστήρια.
ἐλαίῳ δ' ἐμβάλλειν αὐτὸ χρή, καθ' ἣν ἂν ἐθέλῃς ὥραν

416

season of the year you wish, for it makes no difference in regard to this. It is better if you put this in after crushing it, for in this way you will fill the oil quicker with the specific quality and potency. This time [in the oil] should be at least forty days; often I shall let it soak for three, or four, or many more months, and then, having treated it in this way, after squeezing the moisture from the seed, I next discard the seed itself and filter the fluid through fine linen. The oil should be one of those that are relaxing, like the Sabine in Italy. Into twenty-five kotyles of this, put in an Italian *modius* of the seed of the silver fir, these being the Italian kotyles which they also call liters.[24] When the seed is soaked in this, it obviously becomes much less. It is then commensurate to put into what is left remaining four liters of wax and a third part of a liter of the pine resin. These amount to thirty-two drachms. Put in also an equal amount of the pine resin. If these are not available, use terebinth instead of them. It is better to boil in a double vessel, or at all events in a weak fire, like one consisting of charcoals, but these should not be many. This medication is very suitable for all fatigues, whether spontaneous or nonspontaneous.

288K

Good also is the medication compounded from the blossoms of the black poplar. This is prepared as follows: Add an Italian *modius* of the still-closed blossoms of the

[24] As a liquid measure a *kotyle* was 0.475 pint or 270 milliliters. A *modius* was equivalent to a peck or approximately 9 liters.

[41] *post* κἄπειθ᾽ οὕτως: *add.* λαβόντες, εἶτα τὴν ὑγρότητα τοῦ σπέρματος ἐκθλίψαντες αὐτὸ μὲν Κο

λίτραις ἐλαίου Σαβίνου πεντεκαίδεκα ἢ εἴκοσι. μὴ
παρόντος δὲ τοῦ Σαβίνου, τῶν ὁμοίων τι παρασκευα-
στέον ἐλαίων. εἴρηται δ᾽ ἐν τοῖς ἔμπροσθεν, ὡς ὅμοια
πάντ᾽ ἐστὶ τὰ λεπτομερῆ θ᾽ ἅμα καὶ μὴ στύφοντα.
κάλλιον δ᾽, εἰ καὶ μετρίως θλασθὲν ἐμβληθείη τῷ
ἐλαίῳ τὸ ἄνθος. εἰ δὲ καθ᾽ ἑκάστην ἡμέραν[42] διακι-
νοῖτο, καὶ μάλισθ᾽ ὅταν ἥλιος ᾖ θερμός, ἐν οἰκήματί
τε θερμῷ τὴν ἀπόθεσιν ἴσχοι, θᾶττόν τε ἂν οὕτω καὶ
289K   μᾶλλον ἡ τῆς αἰγείρου ποιότης τε καὶ δύναμις εἰς
τοὔλαιον μετέλθοι, ὥστε σε μετὰ δύο που καὶ τρεῖς
μῆνας δύνασθαι ἐκθλίψαντα τὸ ἄνθος αὐτὸ μὲν ἀπορ-
ρῦψαι, τὸ δ᾽ ἔλαιον ἔχειν ἀδήκτως διαφορητικόν, οὐδέν
τι μεῖον τοῦ ἐλατίνου. καὶ μέντοι καὶ κηρὸν καὶ ῥητίνην ἔξεστιν ἐμβάλλειν αὐτῷ τοσοῦτον, ὅσον ἀρτίως
ἐν τῇ συνθέσει τοῦ διὰ τῆς ἐλάτης εἴρηται φαρμάκου.
παχύτερον δ᾽ εἴ ποτε[43] γένοιτο τὸ ἐκθλιβὲν ὑγρόν,
ἱκανὸν οὐ μόνον τὸ τέταρτον μέρος, ἀλλὰ καὶ τὸ πέμ-
πτον μίγνυσθαι κηροῦ. ἐγὼ δ᾽ οἶδά ποτε καὶ τὸ ἕκτον
ἐμβαλών, παχυτέρου τε τοῦ ἐλαίου γινομένου καὶ τοῦ
μέλλοντος αὐτῷ χρῆσθαι χαίροντος ὑγροτέρῳ φαρμάκῳ.

τινὲς δ᾽, οὐκ ἀναμένοντες ἐν χρόνῳ πλείονι διαβρέ-
χεσθαι τά τε ἄνθη καὶ τὰ σπέρματα, τοῦτο μὲν ἑψεῖν
αὐτὰ δέονται, τοῦτο δ᾽ ὕδωρ ἐμβάλλειν, ὅσοι γε προ-
μηθέστεροι, χάριν τοῦ μήτε φρυγῆναι τὰ ἐμβληθέντα
μήτε κνισῶδες γενέσθαι τοὔλαιον. ὅσοι δ᾽ ἔτι τούτων
ἐπιμελέστεροι, τὴν ἕψησιν ἐν ἀγγείοις διπλοῖς ποιοῦν-
290K   ται· καλεῖται δ᾽ οὕτως, ἐπειδὰν προϋποκειμένου λέβη-

black poplar into the fifteen or twenty liters of Sabine oil.
If the Sabine is not available, you must prepare one of the
similar oils. I said in what has gone before that similar oils
are all those that are fine-particled and at the same time
not astringent. It is better if the blossom is moderately
crushed and soaked in the oil. If you also stir it thoroughly
every day, and particularly whenever the sun is hot, and
keep it stored in a house that is hot, the quality and po- 289K
tency of the black poplar transfers into the oil quicker and
to a greater extent, so that after perhaps two or three
months, having squeezed out the blossom itself, discard it,
having the oil as a nonbiting diaphoretic no less than the
pine. And indeed, it is possible to throw wax and resin into
it, to an amount I stated just now in the composition of the
medication made from the silver fir. If ever the squeezed-
out fluid should become thicker, it is sufficient to mix in
not only a fourth part but also a fifth part of wax. And I
know also on one occasion, I put in a sixth part, when the
oil became too thick and the person going to use it was
pleased with a more liquid medication.

Some people, who do not remember to soak the blossoms and seeds for a long time, need on the one hand to
boil them and on the other hand to put in water, while
those who take more care in regard to this, neither roast
the things put in nor let the oil become greasy. Some who
are even more careful than they are, do the boiling in
double vessels—it is described in this way when a pan 290K

---

[42] *add.* ἡμέραν Ko
[43] *ante* γένοιτο· παχύτερον δ᾽ εἴ ποτε Ko; παχύ τι δ᾽ εἴποτε
Ku

τος ὕδωρ ἔχοντος ζέον ἐνιστῆταί τι τούτῳ μικρότε-
ρον⁴⁴ ἀγγεῖον, εἰς ὃ μέλλει τὸ ἔλαιον ἐγχεῖσθαι. τοῦτο
μέν γε καὶ ἡμεῖς ποιοῦμεν, ἀλλ' οὐκ εἰς τὴν τῶν σπερ-
μάτων ἢ ἀνθῶν ἕψησιν οὐδὲν γὰρ ταύτης δεόμεθα
πολυχρονίως αὐτὰ ἀποβρέχοντες, ἀλλ' ὁπόταν δια-
τήκειν ἐν τῷ λίπει⁴⁵ τάς τε ῥητίνας καὶ τὸν κηρὸν
ἐπιχειρῶμεν ἐν τῇ τοῦ φαρμάκου σκευασίᾳ. καταναγ-
καζόμενος δέ τις ἐν τάχει σκευάζειν αὐτὰ δεήσεταί τε
προαφεψεῖν, ὡς εἴρηται, καὶ ὕδατος ἐγχεῖν ἢ οἴνου
χάριν τοῦ μὴ φρύγεσθαι. τὸ μὲν οὖν ὕδωρ εἰς τὰ
παρόντα χρησιμώτερον, ὁ δ' οἶνος εἴς τε τὰς ποδα-
γρικὰς διαθέσεις καὶ ὅλως ἀρθρίτιδας. ἀρκεῖ δὲ μιγνύναι τοσοῦτον ὕδατος, ὡς ἑψώντων ἐκδαπανηθῆναι
τὸ πᾶν. εἴη δ' ἂν οὕτως ὀλίγον, ὡς τετραπλάσιον ἢ
πενταπλάσιον αὐτοῦ τὸ ἔλαιον ὑπάρχειν.

ταῦτά τε οὖν τὰ φάρμακα διαφορεῖ τοὺς κατὰ τὰς
σάρκας τε καὶ τὸ δέρμα μὴ πάνυ παχεῖς μηδὲ γλί-
σχρους χυμούς, καὶ πρὸς τούτοις ἔτι τὸ ἐκ τοῦ χαμαι-
μήλου ἔλαιον, αὐτό τε καθ' ἑαυτὸ μόνον, ἐμβληθέντων
291K τε κηροῦ καὶ ῥητίνης αὐτῷ. ἱκανὸν δὲ κἀνταῦθα τοῦ
μὲν κηροῦ τὸ τέταρτον μέρος, τῆς δὲ ῥητίνης τὸ δωδέκατον. εἰ δὲ μηδενὸς τῶν εἰρημένων ἐλαίων εὐπο-
ροίης, ἀνήθινον ἔλαιον ποιήσασθαί σοι ῥᾷστον. ἔστι
δὲ καὶ τοῦτο διαφορητικὸν φάρμακον, εἴτε καταμόνας
τις εἴτε σὺν κηρῷ τε καὶ ῥητίνῃ χρῷτο. πειρᾶσθαι δ'
ἐπ' ἀγγείου διπλοῦ τὸ ἀνήθινον ἑψεῖν. ἄμεινον δὲ καὶ
χλωρὸν εἶναι τὸ ἄνηθον. ὅταν δὲ ᾖ τοῦτο, καθ' ὃν
χρῄζεις καιρόν, ὥραιον ὑπάρχει.⁴⁶ τὸ σαμψύχινον δ'

containing boiling water is placed beforehand under a
smaller vessel, into which the oil is going to be poured. In
fact, I also do this, but not for the boiling of the seeds or
blossoms, for we do not need this when they are soaked
for a long time, but when we attempt to soften in the pan
the resin and the wax in the preparation of the medication.
If someone is forced to prepare these quickly, he will need
to boil them beforehand, as I said, and pour in either
water or wine, so as not to roast them. Water is more useful for the present purposes, whereas wine is more useful
for the gouty conditions and for the arthritidies generally.
It is sufficient to mix in as much water as will be altogether
consumed in the boiling. And the water should be so small
in amount that the oil is four or five times more than it.

These medications, then, disperse the humors that are
not very thick or viscous in the flesh and skin, as does the
oil made from chamomile in addition to these—either the
oil itself alone or when wax and resin are put into it. And
even here a fourth part of wax and a twelfth part of resin
are sufficient. If you do not have a plentiful supply of any
of the oils mentioned, dill oil is very easy for you to make.
This is also a dispersing (diaphoretic) medication, either
by itself alone or used with wax and resin. Attempt to boil
the dill in a double vessel. It is also better for the dill to
be green. Whenever it is, it should be seasonable at the

291K

---

44 *ante* ἀγγεῖον: ἐνιστῆταί τι τούτῳ μικρότερον Ko; ἔνεστι
τέ τι τούτῳ μικρὸν Ku

45 λίπει Ko; λέβητι Ku

46 ὑπάρχει Ko; ὑπαρχέτω Ku

ἔλαιον ἐπιτήδειον ἐν ὥρᾳ χειμερινῇ καὶ χωρίῳ ψυχρῷ
καὶ καταστάσει παραπλησίᾳ. λαμβανέτω δὲ καὶ
τοῦτο κηροῦ τε καὶ ῥητίνης, εἰ παραμένειν αὐτό γε
βούλοιο τοῖς ἀλειφομένοις ὑπ' αὐτοῦ σώμασιν.

ὁμοίως δὲ καὶ τὴν λιβανωτίδα πόαν ἐναφεψεῖν
ἐλαίῳ. καὶ εἰ μηδὲ ταύτην ἔχεις, ῥίζα τεύτλου λευκοῦ
καὶ ἡ τοῦ σικύου δὲ τοῦ ἀγρίου ῥίζα καὶ ἡ τῆς ἀλ-
θαίας καὶ ἡ τῆς βρυωνίας ἱκανῶς διαφοροῦσιν ἐναπο-
τιθέμεναι καὶ αὗται τὴν ἑαυτῶν ποιότητά τε καὶ δύναμιν ἐλαίῳ τινὶ τῶν διαφορητικῶν. ἐς ὅ τι δ' ἂν ἐθέλῃς
ἔλαιον οὕτω παρασκευασθὲν ἐμβάλλειν ἤτοι κηρὸν
292K μόνον ἢ καὶ ῥητίνην τινὰ σὺν αὐτῷ, καλλίστην τε καὶ
παραμόνιμον ἀλοιφὴν κατασκευάσεις. ἀρκεῖ δὲ τοῦ-
πίπαν ἐμβάλλειν τοῦ μὲν κηροῦ τὸ τέταρτον μέρος,
τῆς δὲ ῥητίνης τὸ δωδέκατον. εἰ δὲ καὶ δύο ῥητίνας
ἐμβάλλοις ἐλατίνην τε ἅμα καὶ στροβιλίνην ἢ καὶ
τρίτην ἐπ' αὐταῖς τὴν τερμινθίνην ἢ καὶ μὴ παρου-
σῶν τούτων τὴν ὑγρὰν πιτυΐνην, ἔσται σοι καὶ οὕτω
διαφορητικὸν τὸ φάρμακον. ἀπορῶν δὲ καὶ ταύτης
ἤτοι τὴν ἐκ τῶν κεραμείων μιγνύναι πιτυΐνην ἢ τὴν
φρυκτὴν ὀνομαζομένην, εἰδὼς[47] μέν, ὡς ἀποδέουσιν
αὗται τῶν προειρημένων, οὐ μὴν παντάπασίν εἰσιν
ἀπόβλητοι.

9. Λοιπῆς δ' οὔσης κοπώδους διαθέσεως, ἐν ᾗ πρὸς
τῷ τῆς αἰσθήσεως ἑλκώδει πλῆθος ὠμῶν χυμῶν ἐν
ὅλῳ τῷ σώματι περιέχεται, λεκτέον ἂν εἴη καὶ περὶ
τῆσδε. χαλεπὸν δ' οὐκέτ' οὐδὲν ἐξευρεῖν αὐτῆς τὴν
ἐπανόρθωσιν, ἀπὸ τῶν εἰρημένων ὁρμώμενον. εἰ γάρ,

time of use. And marjoram oil is suitable in the winter, in a cold place and in similar climatic conditions. Let this also contain wax and resin, if you wish it to remain a longer time on the bodies being anointed with it.

Similarly also, boil the frankincense herb in oil. And if you do not have this, the root of white beet, or wild cucumber, or marshmallow, or bryony disperse adequately when they produce their own quality and potency in one of the diaphoretic oils. And whatever oil, prepared in this way, you wish to put them into—whether wax alone or with some resin in it—you will prepare a very good and very reliable unguent. In general, it is enough to put in a fourth part of wax and a twelfth part of resin. If also you put in two resins (fir and pine) or a third in addition to these (terebinth) and also the liquid pine resin, you will also have a medication capable of dispersing. If these are not available, mix in either the pine resin from potters' workshops or even the so called *phrykte*.[25] I know these are inferior to those previously mentioned, but they are not altogether worthless.

292K

9. There is a remaining fatigue condition, in which, in addition to the wound-like sensation, an excess of raw (unconcocted) humors is contained in the whole body. I must also speak about this. It is not difficult to discover its correction, starting from the things spoken of. When the

---

[25] Φρυκτή is described in LSJ as "a kind of resin"—see also Galen, *Comp. Med. Gen.*, Book 3, chap. 3, XIII.589.

---

[47] εἰδὼς Ko; οἶδα Ku

ὁπότε μὲν ἐν ταῖς πρώταις φλεψὶ τὸ πλῆθος τῶν ὠμῶν
ἐστι, τέμνειν αὐτὰ καὶ πέττειν ἔφαμεν χρῆναι, φυλατ-
293K τομένους τὴν εἰς τὸν ὄγκον ἀνάδοσιν, ὁπότε δ' ἐν ταῖς
ἐσχάταις φλεψὶ καὶ κατὰ τὴν ἕξιν τοῦ ζῴου, πέττειν
τε ἅμα καὶ διαφορεῖν, ἄμφω μικτέον ἐστίν, ἐφ' ὧν καὶ
ἄμφω συμβέβηκεν· εἰ μὲν ἰσοσθενῶς σοι δόξειεν ἐν-
οχλεῖν, ὁμοτίμως ἀμφοτέρων στοχαζόμενον, εἰ δὲ εἴη
θάτερον ἐπικρατέστερον, εἰς ἐκεῖνο μὲν ἀναφέρειν τῆς
ὅλης θεραπείας τὸ κῦρος, ἀμελεῖν δὲ μηδὲ θατέρου
τοῦ μικροτέρου. μιχθήσεται μὲν οὖν ἡ πρὸς ἀμφότερα
θεραπεία κατὰ τόνδε τὸν τρόπον· οὐδὲν γὰρ χεῖρον
ἐπὶ παραδειγμάτων ὀλίγον εἰπεῖν τι καὶ περὶ τοῦδε.

τῷ διὰ τῶν τριῶν πεπέρεων ἁπλῷ φαρμάκῳ χρῆσθαι συνεβούλευον, ἐφ' ὧν ὠμῶν χυμῶν πλῆθος ἐν
ταῖς φλεψίν ἐστι, καὶ μάλιστα ταῖς πρώταις. εἰ τοίνυν
μὴ μόνον ἐν αὐταῖς, ἀλλ' ἐν πάσαις ταῖς φλεψὶν εἴη,
ἤδη δὲ καὶ κατὰ τὰς σάρκας, ἐν ἀρχῇ μὲν τῆς ἐπιμελείας τῷ διὰ τριῶν πεπέρεων φαρμάκῳ χρηστέον,
ἔχοντι καὶ πετροσελίνου τοσοῦτον, ὁπόσον ἂν εἶχεν
ἀνίσου καὶ θύμου καὶ ζιγγιβέρεως· μετὰ δὲ τὴν πρώτην ἡμέραν καὶ μᾶλλον ἔτι τὴν δευτέραν ἐπιμιγνύναι
αὐτῷ τοῦ διὰ τῆς καλαμίνθης· εἶθ' ἑξῆς ἴσα μικτέον·
294K εἶτ' ἐπὶ προήκοντι τῷ χρόνῳ πλέον τοῦ διὰ τῆς κα-
λαμίνθης· εἶτ' ἐπὶ τελευτῇ καὶ μόνον. κατὰ δὲ τὸν
αὐτὸν τρόπον ἐπὶ τῆς ἄλλης ἀπάσης διαίτης, ὅταν
ἰσοκρατεῖς αἱ διαθέσεις ὑπάρχωσι, μιγνύναι μὲν αὐτῶν τῆς ἐπανορθώσεως τοὺς σκοπούς. ἀλλ' ἐν ἀρχῇ
μὲν ἐπικρατείτω τὰ τῶν ἐν ταῖς πρώταις φλεψὶν

excess of unconcocted humors is in the primary veins, I
said it is necessary to cut these and concoct them, guarding against the distribution to the mass of the body. When 293K
they are in the terminal veins and the system of the organism, it is necessary to concoct and at the same time disperse them. Both must be mixed in those in whom both
have occurred. If the disturbance seems to you equal in
force, endeavor to determine if both are of equal rank. If
either is more dominating, the main focus of the whole
treatment is directed at that, while not neglecting the
other lesser condition. Therefore, the treatment will be
combined in regard to both in the following manner—for
it is not a bad idea to say a little about this by way of an
example.

I am in the habit of advising the use of a simple medication made from the three peppers in the case of those
in whom there is an excess of unconcocted humors in the
veins, and particularly in the primary veins. However, if it
is not only in these, but in all the veins, and already in the
flesh, in the beginning of the care, you must use the medication made from the three peppers, having as much
parsley as it has anise, thyme and ginger. After the first day,
and even more after the second day, mix with it the medication made from catmint. Then, next in order, you must
mix equal amounts, and then, as time goes on, more of the 294K
medication made with catmint, and then finally this alone.
And the same method applies in the case of every other
regimen. Whenever the conditions are of equal strength,
combine the objectives of their correction. But in the beginning, let the remedies of humors in the primary veins

427

ἰάματα, κατὰ δὲ τὴν τελευτὴν τὰ τῶν ἐν σαρκί, με-
σοῦντος δὲ τοῦ χρόνου, μιγνύσθω κατ' ἴσον ἀμφότερα. ταῦτά τε οὖν εἴρηταί μοι καὶ ἤδη δῆλον, ὅπως
ἐπανορθοῦσθαι χρὴ τὰ κατὰ τοὺς χυμοὺς ἁμαρτή-
ματα, πρὶν νοσῆσαι τὸν ἄνθρωπον. ἐξ ὧν γὰρ ἐπὶ τῆς
ἑλκώδους διαθέσεως εἴπομεν, ὅταν ἐπιμίγνυταί τινι
κακοχυμία, πάρεστι συλλογίσασθαι καὶ περὶ τῶν ἄλλων ἑκάστης, ἐπειδὰν μόνη ποτὲ συνιστῆται.

10. Περὶ μὲν οὖν τοῦ τε πρώτου γένους τῶν κόπων,
ἐφ' ὧν νυγματώδης ἐστὶ κατὰ πάντα τοῦ ζῴου τὸν
ὄγκον[48] αἴσθησις, ὅσαι τ' ἄλλαι μοχθηρῶν χυμῶν ἐν
τῷ σώματι γίνονται πλεονεξίαι, καθ' ἑαυτάς τε καὶ
σὺν κόποις, σχεδὸν ἤδη λέλεκται πάντα. περὶ δὲ τοῦ
τονώδους ὑφ' ἡμῶν κληθέντος κόπου λέγω ἐφεξῆς.
ὅτι μὲν οὖν ὁ τοιοῦτος κόπος, ὅταν ἄνευ γυμνασίων
συνιστῆται, πλῆθος ἐνδείκνυται διατεῖνον τὰ στερεὰ
μόρια τοῦ ζῴου, καὶ ἄλλοις μέν τισι τῶν εὐδοκίμων
ἰατρῶν ἔδοξεν, οὐχ ἥκιστα δὲ καὶ τοῖς περὶ τὸν Ἐρασίστρατον. ὅτι δ', ὅταν αἵματος ᾖ πλῆθος, ἄριστον
ἤτοι φλέβα τέμνειν ἢ ἀποσχάζειν τὰ σφυρά, λέλεκται
μέν που καὶ πρόσθεν, ἀναληπτέον δ' ἔτι καὶ νῦν τὸν
λόγον Ἐρασιστράτου χάριν, ὃς οὔτ' ἐπ' ἄλλης ὅλως
οὐδεμιᾶς οὐδὲ ἐπὶ τῆσδε τῆς διαθέσεως ἐχρήσατο
φλεβοτομίᾳ. ὅτι μὲν οὖν ἐπὶ τῆς ὑγιεινοτάτης φύσεως, ὑπὲρ ἧς ὁ λόγος ἐνέστηκεν, ἐγχωρεῖ καὶ
κατ' ἄλλον τρόπον ἐκκενοῦν τὸ τοιοῦτον πλῆθος, ἔμπροσθεν εἴρηται. κατὰ μὲν γὰρ τὰς μοχθηρὰς φύ-

prevail, and at the end, the remedies of those in the flesh. In the intervening time mix both equally. These things being said, it is, to me, already clear how one must correct the faults in the humors before the person becomes diseased. For from the things I said in the case of the wound-like condition, when it is mixed with some *kakochymia*, it is possible to also draw inferences about each of the others, when at anytime it exists alone.

10. I have already said almost everything about the first class of the fatigues in which there is a pricking sensation in the whole mass of the organism, and the many other excesses of bad humors that arise in the body, by themselves and with fatigues. I speak next about the tensive fatigue, as I call it. That there is such a fatigue, when it exists apart from exercise, indicates an excess stretching of the solid parts of the organism, and this seemed so to certain other, famous doctors, not least the followers of Erasistratus. On the other hand, that whenever there is an excess of blood, it is best to open veins or scarify the ankles, has also been said before somewhere. What must still be taken up now is the argument with regard to Erasistratus,[26] who did not use phlebotomy in any other condition at all or in this condition. Therefore, in the case of the healthiest nature, which is what the discussion is about, it is also possible to evacuate such an excess in another way, as I said before. For in relation to the bad na-

295K

[26] Only fragments of Erasistratus' writings remain—see Garofalo, *Erasistrati Fragmenta*. The fragments on hygiene are 115–67. For this particular issue, see *Venae Sect.*, XI.281K ff.

48 τὸν ὄγκον add. Ko

429

σεις,⁴⁹ ἐφ᾽ ὧν τὸ περιττὸν κατὰ τὸν ἐγκέφαλον ἢ τὰ
τῆς ἀναπνοῆς ὄργανα φέρεται, πηλίκον ἐστὶ κακὸν
ἑτέρῳ τρόπῳ κενώσεως χρῆσθαι, παραλιπόντα φλε-
βοτομίαν, εἰρήσεται μέν που καὶ διὰ τῶν ἑξῆς ὑπομνημάτων, ὅταν ὑπὲρ τῶν μοχθηρῶν κατασκευῶν ὁ
296K λόγος περαίνηται, λέλεκται δ᾽ ἤδη κἂν τῷ Περὶ φλεβοτομίας πρὸς Ἐρασίστρατον.

ὅθεν οὐδὲν ἔτι⁵⁰ μηκύνειν δεῖ περὶ αὐτῶν, ἀλλ᾽ ἐπὶ
τὸν λοιπὸν καὶ τρίτον κόπον, ὅταν αὐτομάτως συνιστῆται, τὸν λόγον ἄγειν, ὃν ἐν τοῖς ἔμπροσθεν ἐκα-
λέσαμεν φλεγμονώδη διά τε τὸ τῆς ὀδύνης μέγεθος
καὶ ὅτι μετὰ θερμότητος ἐπιφανοῦς συνίσταται συν-
εξαίρων εἰς ὄγκον τοὺς μῦς. οὗτος ὁ κόπος οὐδ᾽ ὡρῶν
ὀλίγων, μή τί γε δυοῖν ἢ τριῶν ἀνέχεται ἡμερῶν⁵¹ τῆς
Ἐρασιστράτου βραδυτῆτος, ἀλλ᾽ αὐτίκα πυρετὸν ἐπιφέρει σφοδρότατον, ἢν μή τις φθάσας ἀποχέῃ τοῦ
αἵματος. καὶ γὰρ οὖν θερμότατόν ἐστι τὸ τῶν τοιούτων
κόπων αἷμα, καὶ πλείστης αὐτοῦ δέονται τῆς κενώσεως ἅπαντες σχεδὸν οἱ καταληφθέντες τῷ κόπῳ. καὶ
οἱ πολλοί γ᾽ αὐτῶν πυρέττουσι, κἂν ἀποχέῃς τοῦ
αἵματος. ὅθεν οὔτε βλακεύειν οὔτε ὀλίγον ἀφαιρεῖν
προσῆκεν, ἀλλὰ καὶ διὰ ταχέων ἐκκενοῦν καὶ μέχρι
λειποθυμίας ἄγειν, εἰ μηδὲν ἕτερον κωλύοι. κάλλιον
δέ, εἰ ἐγχωρεῖ, δὶς ἀφελεῖν ἐν ἡμέρᾳ μιᾷ, τὸ μὲν
πρότερον οὕτω κενοῦντας, ὡς μὴ λειποθυμῆσαι τὸν

⁴⁹ φύσεις add. Ko   ⁵⁰ ἔτι Ko; ἔστι Ku
⁵¹ post ἡμερῶν: ἴσον ὄγκον (Ku) om.

tures, in which the superfluity is carried to the brain or the organs of respiration, how great an evil it is to use another form of evacuation, leaving aside venesection, will be spoken of in the books to follow, when the discussion about bad constitutions is completed. However, I have already spoken of this in the work *On Phlebotomy, against Erasistratus.*[27]

296K

For which reason, nothing need still delay us about these things but to take the discussion to the remaining and third fatigue, whenever it exists spontaneously—the one which, in what has gone before, we called inflammation-like due to the magnitude of the pain and because it exists with evident heat while swelling the muscles into a mass. This fatigue is not of a few hours' duration, nor let me tell you, does it maintain the slow course over two or three day, as *per* Erasistratus, but immediately brings a very severe fever, unless someone anticipates this with a withdrawal of blood. For truly, the blood of such fatigues is very hot, and almost all of those seized by this fatigue need the evacuation of the greatest part of this. And many of them are febrile, even if you drain off the blood. For this reason, it is not appropriate to be tardy about this or to take a small amount, but to evacuate quickly and to bring them to the point of swooning (*leipothumia*), if there is nothing to contraindicate this. However, it is better if possible to remove blood twice in one day, on the first occasion evacuating in such a way as not to cause the patient

[27] *Venae Sect.,* XI.147–86K.

431

297K ἄνθρωπον, τὸ δὲ δεύτερον οὐδὲ τὴν λειποθυμίαν φο-
βεῖσθαι προσήκει. τῇ μὲν γὰρ προτέρᾳ κενώσει κατα-
λυθεὶς οὐκ ἂν ὑπομείναι τὴν δευτέραν· ἐν ταύτῃ δ᾽ εἴ
τι πάθοι τοιοῦτον, εὐανακόμιστος γίνεται. μὴ φλεβο-
τομηθέντες δὲ οἱ οὕτω διακείμενοι τύχης ἀγαθῆς εἰς
τὸ σωθῆναι δέονται, καὶ οὐδὲ σῴζονται, εἰ μὴ καθ᾽
ἕτερόν τινα τρόπον, ἢ αἱμορραγήσαντες ἐκ ῥινῶν ἢ
λάβρων ἱδρώτων αὐτοῖς ἐκχυθέντων.

ἐπισκοπεῖσθαι δὲ χρὴ μάλιστα μέλλοντας φλέβα
τέμνειν, πότερον κατὰ θώρακα καὶ νῶτα καὶ ὀσφῦν
ἐρείδουσιν αἱ τάσεις τε καὶ αἱ νυγματώδεις ὀδύναι ἢ
κατὰ κεφαλὴν καὶ τράχηλον μᾶλλον. οὕτω μὲν γὰρ
οὖν τὴν ὠμιαίαν διαιρήσεις, καὶ μᾶλλον εἰ πλήρους
αἰσθάνονται καὶ θερμῆς τῆς κεφαλῆς, ἐκείνων δὲ τὴν
ἔνδον· εἰ δ᾽ ὅλον ὁμαλῶς ὑπὸ τοῦ κόπου τὸ σῶμα
κατέχοιτο, τὴν μέσην ἀμφοῖν. εἰ μὲν οὖν ἐπὶ τῇ φλεβοτομίᾳ πυρέσσειν ἄρξαιντο, τῆς θεραπευτικῆς ἔργον ἤδη μεθόδου προνοήσασθαι τούτων· εἰ δ᾽ ἀπύρε-
τοι διαμένοιεν, ἐν μὲν τῇ πρώτῃ τῶν ἡμερῶν ἐπὶ τῇ
φλεβοτομίᾳ πτισάνης χυλὸν ἢ ἐκ χόνδρου ῥόφημα
298K διδόναι μόνον, ἐν δὲ τῇ δευτέρᾳ καὶ λούειν ἂν ἤδη
δύναιο σὺν ἐλαίῳ δαψιλεῖ, διαιτᾶν δέ, κἂν λούσῃς,
μετριώτατα θριδακίνης, εἰ βούλοιτο, διδόντα καὶ κο-
λοκύνθης, εἰ παρείη, καὶ πτισάνης. ἀγαθὸς δὲ καὶ ὁ
χόνδρος, εἴθ᾽ ὡς πτισάνην τις αὐτὸν ἡδύνας ὄξει
σκευάσειεν εἴτε χωρὶς ὄξους. εἰ δὲ μὴ παρείη κολο-
κύνθη, μαλάχῃ καὶ τεύτλῳ καὶ λαπάθῳ καὶ ἀνδρα-
φάξϊ χρῆσθαι. εἰ δὲ καὶ σαρκῶν γεύσασθαι βού-

to swoon, whereas on the second occasion, it is fitting not
to be afraid of swooning. For if the person is brought down
by the first evacuation, he would not endure the second,
whereas if he suffers some such thing in this, recovery
occurs. However, those in such a state who are not phle-
botomized need good fortune to be saved; and they are
not saved unless in some other way, either from nose
bleeds or fierce sweats pouring forth from them.

It is particularly necessary for those who are going to
cut veins to consider whether the tensions and pricking
pains are fixed in the chest, back or loins or more in the
head and neck. For in the latter you will cut the humero-
cephalic vein, especially if they feel fullness and heat in
the head. In the former you will cut the internal (basilic)
vein, whereas if the whole body is equally afflicted by the
fatigue, cut the vein in between both. If, then, they begin
to be febrile after phlebotomy, the task is now to give
forethought to the method of their treatment. If, however,
they remain afebrile, on the first day after the phlebotomy
give only the juice of ptisane or thick gruel. On the sec-   298K
ond day, you are already able to bathe with abundant oil,
and even if you bathe, feed them very moderately, giving
some lettuce, if you wish, or colocynth, if it is available,
and ptisane. Best is gruel, either making it pleasant like
ptisane with vinegar or without vinegar. If colocynth is not
available, use mallow, beet, monk's rhubarb and orach. If
the person wishes to taste flesh, give that of rock fish or

λοιτο, τῶν πετραίων ἰχθύων ἢ ὀνίσκων ἐν λευκῷ
ζωμῷ καλῶς ἑψήσαντα διδόναι. προσαγορεύω δὲ λευκὸν ζωμόν, ὅταν ἄνευ γάρου τε καὶ πολὺ δὴ μᾶλλον
ἔτι τῆς ἄλλης καρυκείας σκευασθῇ, ἀνήθου τε καὶ
ἁλῶν ἐμβληθέντων εἰς ὕδωρ σὺν ἐλαίῳ καὶ πράσῳ
βραχεῖ.

κάλλιον δέ, εἰ καὶ τῇ δευτέρᾳ τῶν ἡμερῶν οἴνου
φείσαιτο. τῇ τρίτῃ δέ, εἰ μὲν εὐπέπτως φέρει τὴν
ὑδροποσίαν, εἴργειν οἴνου καὶ τότε· μὴ φέροντος δέ,
μάλιστα μὲν ἀπόμελι δοτέον· καὶ γὰρ ἐμψύχει πως
ἠρέμα τοῦτο τὸ ποτόν, οὗ καὶ αὐτοῦ χρῄζουσιν οἱ
φλεγμονώδεις κόποι· μὴ παρόντος δὲ τούτου, λευκὸν
299K καὶ λεπτὸν οἶνον διδόναι καὶ τἆλλα κατὰ λόγον ἀνα-
κομίζειν εὐχύμῳ τε καὶ μηδαμῶς θερμαινούσῃ διαίτῃ
χρώμενον. εὐλαβεῖσθαι δὲ μάλιστα πάντων ἀθρόως
ἀνατρέφειν. ὅσοι γὰρ ἐπὶ τοιαύταις κενώσεσιν εἰς τὴν
ἐξ ἀρχῆς δίαιταν εὐθέως ἐπανῆλθον, ἐμπίπλαται τούτοις ἡ ἕξις ἀπέπτων χυμῶν, οὓς ἀναρπάζει, πρὶν πεφθῆναι καλῶς ἔν τε τῇ γαστρὶ καὶ κατὰ τὰς φλέβας,
ὁ τοῦ σώματος ὄγκος. αὕτη μὲν ἡ ἀρίστη πρόνοια τοῦ
φλεγμονώδους κόπου.

διὰ τί δὲ ἐπαφαιρεῖν κελεύομεν αἵματος ἐπ' αὐτῷ
καὶ μὴ πληροῦν ἀθρόως, ἀρκεῖ μὲν δήπου καὶ τὴν
ἐμπειρικὴν αἰτίαν εἰπεῖν, ὅτι καὶ μᾶλλον ὀνίνανται
κενωθέντες οὕτω καὶ ὑγιαίνουσιν εἰς μακρόν, ὡς εἴρηται, διαιτώμενοι.[52] προσθεῖναι δ' οὐδὲν χεῖρον ἂν εἴη
καὶ τὴν ἀπὸ τῆς φύσεως τῶν πραγμάτων ἔνδειξιν.

birds well boiled in white juice. I call it white juice whenever it is prepared without *garos*[28] and much more the
other rich sauce, dill and salt having been put into the
water with oil and a little leek.

It is better, however, if also on the second day the person refrains from drinking wine. On the third day, if he
tolerates drinking water with good digestion, he should
also avoid wine at that time. If, however, he doesn't tolerate it, you must give him *apomel* especially, for this bever-
age somehow cools gently, which is what the inflammation-like fatigues have need of. If this is not available, give
a thin white wine and restore the other things according   299K
to principle, using a diet that is *euchymous* and not at all
heating. Be particularly careful about feeding everything
all at once. For in those who, after such evacuations, return immediately to their original regimen, the system is
filled with unconcocted humors which the mass of the
body snatches up before they have been concocted properly in the stomach and veins. This is the best care for the
inflammation-like fatigue.

On why I recommend removal of blood for this and not
filling with food all at once, it is sufficient to mention the
empirical reason, which is that those evacuated in this way
are helped more and remain healthy for a long time, as I
said. It would be no bad thing to add also the indication
from the nature of the matters. Moreover, since in the

28 Γάρος is described in LSJ as "a kind of sauce or paste made
of brine and small fish." It is described in detail by Pliny, 31.7.43
no. 93.

---

52 διαιτώμενοι Ko; διαγόμενοι Ku

435

ἐπεὶ τοίνυν ἐν τῷ φλεγμονώδει κόπῳ πλῆθος αἵματος
ἠθροισμένον θερμοῦ κατὰ τὸν ὄγκον ἐστίν, ἡ φλεβο-
τομία δὲ τοὺς ἐν τοῖς ἀγγείοις ἐκκενοῖ χυμούς, ἄμεινον ἐπὶ τῇ προτέρᾳ κενώσει τοσοῦτον χρόνον δια-

300K λιπεῖν, ὡς μεταληφθῆναί τι καὶ εἰς τὰς φλέβας ἐκ τῶν
κατὰ τὸ σῶμα. τοῦτο δ' οὐκ ἐπιτρεπτέον ἐν αὐταῖς
ὑπομένειν, ἡμιμόχθηρον ὑπάρχον, ἀλλ' ἐκκενωτέον
αὐτοῦ τὸ πλεῖστον. ταῦτά τοι καὶ κατὰ τὴν δευτέραν
ἡμέραν ἐπαφαιροῦμεν αἵματος, ἐνίοτε δὲ καὶ κατὰ τὴν
τρίτην, ἐπειδὰν ἀντισπᾶν τε καὶ μετάγειν ἐξ ἑτέρων
εἰς ἕτερα συμφέρειν δόξῃ. λέγεται δὲ καὶ περὶ τῶν
τοιούτων ἁπάντων διορισμῶν ἀκριβέστερον ἐν τοῖς
Περὶ φλεβοτομίας, ὧν οὐκ ἐγχωρεῖ μεμνῆσθαι τὰ νῦν
διὰ τὸ τῆς θεραπευτικῆς πραγματείας ἰδιωτέρους
ὑπάρχειν αὐτοὺς καὶ μέλλειν που καὶ αὖθις ἴσως
ἡμᾶς ἐν τῇδε τῇ πραγματείᾳ περὶ φλεβοτομίας ἐπι-
μελέστερον διέρχεσθαι.

11. Λοιπὸν οὖν ὅτι ταχέως οὐ χρὴ πληροῦν ἐπὶ
φλεβοτομίαις εἰπόντες, ἐνταῦθα καταπαύσομεν τὸν
λόγον. ἕξει δὲ καὶ οὗτος εἰς πίστιν ὑπόθεσίν τινα τῶν
ἤδη προαποδεδειγμένων ἐν τῇ Περὶ τῶν φυσικῶν δυνάμεων πραγματείᾳ· δέδεικται γὰρ ἐν ἐκείνῳ τῷ λόγῳ
πᾶσι τοῖς ὑπὸ φύσεως διοικουμένοις ὑπάρχουσα δύ-

301K ναμις ἔμφυτος ἑλκτικὴ τῶν οἰκείων χυμῶν, ὑφ' ὧν
τρέφεσθαι μέλλει, δέδεικται δὲ καὶ ὡς, ἐπειδὰν ἀπορῇ

inflammation-like fatigue an excess of blood that is hot is collected in the mass [of the body], while phlebotomy evacuates the humors in the vessels, it is better to leave an interval after the first evacuation, such that these humors 300K are transferred from the body into the veins. But this must not be allowed to remain in them, being partially bad; most of it must be evacuated. It is for these reasons, surely, that we remove blood during the second day, and sometimes also during the third day, whenever it might seem advantageous to revulse and transfer the humors from some places to others. I have also spoken more precisely about all such distinctions in the work, *On Phlebotomy*.[29] It is not possible to make mention of these now because they more properly belong to the work on therapeutics,[30] and I am perhaps going to go over phlebotomy again more carefully in this work.

11. What remains then, after saying that you must not fill people quickly with food after phlebotomies, is that I shall bring the discussion to a close here. And this will be a reliable foundation for those things I have already demonstrated in the work *On the Natural Faculties*.[31] For in that work it was shown for all those [organisms] controlled by Nature, there is an innate power (capacity) attractive of the suitable humors by which they are going to be nour- 301K

---

[29] Of the three short works on phlebotomy included in Kühn, *Cur. Rat. Ven. Sect.* (XI.250–316K) is presumably the one referred to here.   [30] This is presumably Galen's *magnum opus* on treatment—*MM*, X.1–1021K (English trans., Johnston and Horsley, *Galen: Method of Medicine;* French trans., J. Boulogne [2009]).   [31] *Nat. Fac* , 1.1–204K (English trans., Brock, *On the Natural Faculties*)

μὲν οἰκείας τε ἅμα καὶ χρηστῆς τροφῆς, ἐπείγεται
καὶ τῶν οὐ χρηστῶν τι συναρπάξαι. τοιοῦτον δ' ἐστί
που καὶ τὸ μήπω πεφθὲν ἐν κοιλίᾳ τε καὶ φλεψίν.

ἀναγκαῖον οὖν, ἐπειδὰν πλείω λαμβάνῃ σιτία κατὰ
τὸν καιρὸν τοῦτον ὁ ἄνθρωπος, ἀναρπάζεσθαι πλεῖστον ὠμὸν χυμὸν εἰς τὸν ὄγκον τοῦ ζῴου διὰ πολλὰς
αἰτίας· ὅτι τε φαυλότερον ἐν τῇ γαστρὶ καὶ ταῖς φλεψὶ
πέττεται,[53] ὅτι τε πλέον εἰς τὸν ὄγκον ἀνέλκεται[54] διὰ
τὸ πλέον ὑπάρχειν, ὅτι τε πρωϊαίτερον ἢ χρὴ διὰ τὸ
συγχωρεῖν μὲν τὴν γαστέρα ταῖς φλεψί, τὰς φλέβας
δὲ ἅπασι τοῖς ἄλλοις τοῦ ζῴου μορίοις ἐπισπᾶσθαι
τὸ μήπω κατειργασμένον, ὅπερ οὐκ ἂν συνεχώρησαν
ὁμοίως, εἴπερ ὀλίγον ἦν. ἐπιδέδεικται γάρ, ὡς αὐτὰ
πρότερον ἀπολαύει τὰ μόρια τῆς οἰκείας τροφῆς, εἶθ'
οὕτως ἑτέροις ἐπιπέμπει. τὸ δὲ δὴ τελευταῖον καὶ μέγιστον αἴτιον τῆς βλάβης τοῖς οὕτω διακειμένοις
ἐστὶ τὸ πολλὴν καὶ ἡμίπεπτον ἐπισπασάμενα τροφὴν
302K τὰ καθ' ὅλον τὸν ὄγκον τοῦ ζῴου μόρια πλῆθος οὐκ
ὀλίγον ἐξ αὐτῆς ἀπογεννᾶν περιττωμάτων. οὐδὲ γὰρ
πέττειν αὐτὴν ἅπασαν ὁμοίως τῇ χρηστῇ δυνατὸν
αὐτοῖς οὔτε προσφύειν οὔθ' ὁμοιοῦν, ἀλλ' ἀτυχεῖν ἐν
ἑκάστῳ τῶν οἰκείων ἔργων ἐπὶ τῇ μοχθηρᾷ τε ἅμα καὶ
πολλῇ τροφῇ. ὅμοιον γάρ τι συμβαίνειν αὐτοῖς ἀναγκαῖον, οἷόν τι καὶ αὐτῇ τῇ γαστρὶ προσενεγκαμένῃ
σιτία πολλὰ κακῶς παρεσκευασμένα.

53 post πέττεται: τὸ πλέον (Ku) om.
54 ἀνέλκεται Ko; ἀνέρχεται Ku

ished. It was also shown that, whenever there is a lack of nutriment that is at once suitable and useful, there is also an urge to seize upon some of those that are not useful. This is sometimes the sort of thing that has not yet been concocted in the stomach and veins.

It is inevitable, then, that when the person takes too much food during this time, most unconcocted humor will be carried away to the mass of the organism for many reasons. This is because the greater part[32] is more poorly concocted in the stomach and veins, and also, because there is an excess, more is drawn back to the mass [of the body]. It is also because it is necessary for the stomach to release [material] earlier to the veins than it should, and the veins permit all the other parts of the organism to draw what has not yet been worked up for use. They would not similarly permit this, if it were less. For it has been shown that the parts themselves first enjoy the benefits of their proper nutriment, and then in this way send it to other parts. Certainly, the ultimate and greatest cause of harm to those in this state is the great amount of semiconcocted nutriment drawn to the parts in the whole mass of the 302K organism, from which a significant amount of superfluities is generated. For it is not possible for the parts to concoct all this nutriment like they do that which is useful, nor to retain or assimilate it. Rather, they fail in each of these proper actions due to the bad and excessive nutriment. Of necessity, the same thing happens to them as also happens to the stomach itself, if it is presented with a large amount of food badly prepared.

[32] The translation here follows the Kühn text.

λέγω δὲ παρεσκευάσθαι κακῶς, ὅσα δεόμενά τινος
ἑψήσεως ἢ ὀπτήσεως οὐκ ἀπήλαυσε τελέως αὐτῆς.
ἄρτον οὖν ἐλλιπῶς ὠπτημένον ἢ κρέας ἢ ὄσπριον
ἀτελῶς ἡψημένον ἀδύνατόν ἐστι πεφθῆναι χρηστῶς
ἐν τῇ γαστρί. τὸν αὐτὸν δὲ λόγον ἔχει τὰ κατὰ τὴν
γαστέρα μοχθηρῶς κατεργασθέντα πρὸς τὴν δευτέραν τὴν ἐν φλεψὶ πέψιν, ὃν ἐξ ἀρχῆς⁵⁵ τὰ φαῦλα
παρασκευασθέντα σιτία πρὸς τὴν ἐν τῇ γαστρί. καὶ
μέντοι τὰ κατὰ τὰς φλέβας οὐκ ὀρθῶς πεφθέντα τὸν
αὐτὸν ἔχει λόγον ὡς πρὸς τὴν ἐν τῇ σαρκὶ πέψιν, ὃν
εἶχεν τὰ μὲν σιτία πρὸς τὴν ἐν τῇ γαστρί, τὰ δ᾽ ἐκ
303K ταύτης ἀναδιδόμενα πρὸς τὴν ἐν ταῖς φλεψίν. οὔτ᾽ οὖν
ἡ γαστὴρ ἀκριβῶς πέττει τὰ ἔξωθεν οὔθ᾽ αἱ φλέβες
τὰ ἐκ τῆς γαστρὸς οὔθ᾽ αἱ σάρκες τὰ ἐκ τῶν φλεβῶν,
ὅταν μὴ καλῶς ᾖ προκατειργασμένα· κἂν τούτῳ πλῆθος ἀναγκαῖόν ἐστιν ἐν τῷ σώματι γεννᾶσθαι περιττωμάτων.

ἐμοὶ μὲν οὖν εἴρηται τὸ σύμπαν, ὡς ἐν βραχυτάτῳ
διελθεῖν· εἰ δέ τις ἑκάστου τῶν εἰρημένων εἰς τὴν ἀπόδειξιν ἐπιστήμην ἀκριβῆ λαβεῖν βούλεται, τούτῳ τὰ
Περὶ τῶν φυσικῶν δυνάμεων ἀναγνωστέον ἐστίν, ἐν
οἷς ἀποδέδεικται πρῶτον μέν, ὡς ἡ γαστὴρ ἑαυτῆς
ἕνεκα λαμβάνει τὴν τροφήν, ἵνα ἀπολαύσῃ τε καὶ
ἀναπληρώσῃ τὸ ἐλλεῖπον ἑαυτῇ, καὶ διὰ τοῦτο περι-
πτύσσεται πανταχόθεν αὐτῇ καὶ κατέχει σύμπασαν,
ἄχριπερ ἂν ἱκανῶς κορεσθῇ· δεύτερον δὲ ὡς, ἐπειδὰν

⁵⁵ ἐξ ἀρχῆς add. Ko

440

I call badly prepared that which requires some boiling or roasting but does not enjoy the benefit of this completely. Thus, bread deficiently baked, or meat or vegetables incompletely boiled cannot be properly concocted in the stomach. The same argument holds for those things badly worked up in the stomach for the second concoction in the veins, as with foods inadequately prepared from the beginning for concoction in the stomach. Furthermore, the same argument holds regarding those things not properly concocted in the veins as regards concoction in the flesh as that which holds for the foods regarding concoction in the stomach and their distribution from this toward the concoction in the veins. Thus, if the stomach does not completely concoct those things from without, the veins do not completely concoct those things from the stomach, and the flesh does not concoct those things from the veins whenever they are not well worked up beforehand. And due to this, an excess is inevitably a generator of superfluities in the body.   303K

Therefore, I have said everything, going through it in the briefest possible way. If someone wishes to take each of the things said to the exactitude of scientific demonstration, he must read the book *On the Natural Faculties*[33] in which it has been shown first, that the stomach takes the nutriment for its own sake, so that it enjoys the benefit of it and fills up what is deficient in itself, and because of this enfolds it on all sides and holds it all until it is sufficiently satisfied. Second it is shown that, when it no longer needs

---

[33] See note 31 above.

μηκέτι δέηται τοῦ τρέφεσθαι, τότ' ἀνοίγνυσι μὲν τὸν
πυλωρόν, ἐκθλίβει δὲ καὶ ὠθεῖ κάτω τὰ περιττὰ τῶν
σιτίων, οἷον ἄχθος ἀλλότριον· εἶθ' ὡς ἐν τῇ διὰ τῶν
ἐντέρων ὁδῷ, καὶ μάλιστα τῶν λεπτῶν, ἀναρπάζουσιν
αἱ καθήκουσαι φλέβες τὴν τροφήν, ἐκ τῆς πρὸς τὴν
κοιλίαν ὁμιλίας ἠλλοιωμένην τε καὶ συμφυλοτέραν
304K τῷ ζῴῳ γενομένην· εἶτα καὶ κατὰ τὰς φλέβας ἐξ ἄλλης εἰς ἄλλην διαδίδοται τὸν ὅμοιον τρόπον, ὃν ἐκ
τῆς γαστρὸς εἰς τὰς φλέβας· ἐντεῦθεν δ' ἤδη κατειργασμένη τέλεον εἰς ἕκαστον ἕλκεται τῶν τοῦ ζῴου
μορίων, ἵνα πέττεται τὴν τρίτην πέψιν ὁμοιοῦταί τε
τῷ τρεφομένῳ. ταῦτ' οὖν ὅστις ἐξ ἐκείνων ἀναλέξεται
τῶν γραμμάτων, οὐκέτ' ἀπορήσει τὴν αἰτίαν, δι' ἣν
πολλοὶ κενωθέντες τὴν ἕξιν, εἰ μὴ μετρίως ἀνατρέ-
φοιντο, πολλὰ περιττώματα καθ' ὅλην αὐτὴν ἀθροί-
ζουσι καὶ νοσοῦσιν ἐξ αὐτῶν οὐκ εἰς μακράν.

to be nourished, at that time the stomach opens the pylorus, and compresses and thrusts downward the residues of the foods like some alien load. Then, in the passage through the intestines, and particularly the small intestines, the appropriate veins carry off the nutriment, which has changed from the association in the stomach and become more suitable for the organism. And then, in the veins, it is distributed from one to the other in the same 304K way as it was from the stomach to the veins. From there, already completely worked upon, it is drawn to each of the parts of the organism, so that it is concocted in respect of the third concoction and assimilated by what is being nourished. Therefore, whoever has picked up these matters from those writings, will no longer be at a loss about the cause, due to which many, if they are not moderately renourished when the system is evacuated, collect many superfluities in the whole system, and from these become diseased before long.